VOLUME 1

INDUSTRIAL FLUID POWER

Third Edition

BASIC TEXTBOOK ON HYDRAULICS, PNEUMATICS, AND VACUUM

Study Text, Circuit Diagrams, Design Data Tables, Troubleshooting.
Arranged for the Beginning Student

Prepared by Charles S. Hedges
Assisted by the Technical Staff of Womack Machine Supply Company

Sponsored by Robert C. Womack
Member Fluid Power Society and Fluid Power Educational Foundation
President and Board Chairman Womack Machine Supply Company
and AAA Products International

Published by

Womack Educational Publications

Department of Womack Machine Supply Company
Mail address: P. O. Box 35027, Dallas TX 75235 • Phone: 214-357-3871 • Fax: 214-350-9322
Write, phone, or fax for latest price list or for ordering educational materials.

Publisher's Information
INDUSTRIAL FLUID POWER – VOLUME 1

This is the Sixth Printing of the Third Edition – August 1990
Library of Congress Catalog Card No. 66-28254
ISBN No. 0-9605644-5-4
Printed in the U.S.A.

Mr. Charles S. Hedges holds an engineering degree from the University of Kansas. He has worked in industry in electrical and mechanical design, and now since 1953 in his association with Womack Machine Supply Company has written the Womack textbooks on fluid power, has worked with customers on a variety of fluid power applications, has prepared and published many design data sheets and technical bulletins, has taught fluid power classes in the larger cities of Texas, Oklahoma, and Louisiana, and has helped train new salesmen in fluid power applications and circuitry.

AUTHOR'S PREFACE TO THE THIRD EDITION

The first edition of this book was published in 1966 for use as a textbook in Womack schools conducted for customers in Texas and Oklahoma. The book was revised, enlarged, and completely re-written in 1972 into the Second Edition. Since then many thousands of books have been purchased by fluid power distributors and manufacturers in the U.S.A., Canada, and in many foreign countries for use in private and public schools and for training employees. Many vocational schools use this book because it is the only one they can find which really deals with fluid power like it is used out in industry.

I have been using this book since 1966 for conducting dozens of fluid power classes in all the larger cities of Texas and Oklahoma. In this Third Edition I have incorporated new material which, from my personal contact with students, I have found they want to know about fluid power.

The book has been completely re-written for this Third Edition. Many illustrations from the Second Edition have been retained. Some have been improved, but many new illustrations and photographs have been added. Text material in all sections of the book has been updated. This book is a broad overview of fluid power, covering principles, major components, and basic circuitry. Volumes 2 and 3 of this series deal with more advanced components and circuitry. I have tried to communicate as simply and clearly as I know how. I welcome your comments, good or bad, and will certainly appreciate your constructive criticism along with your suggestions for improvements which can be made in future printings or editions. Write to me in care of the publisher. Thank you for your interest.

Charles S. Hedges

FLUID POWER...

What is it?

How is it Used?

The art of generating, controlling, and applying smooth, effective power of pumped or compressed fluids (such as oil or air) when used to push, pull, rotate, regulate or drive the mechanisms of modern life.

Illustrations in this section are by permission of the National Fluid Power Association (NFPA)

emergence of fluid power technology

Although the principle of fluid power can be traced back to Pascal and the piston, it is only in the past few decades that the makers of fluid power equipment have become, truly, an industry. It was the great era of engineering development between the two world wars which begot the bulging demand for the pumps, cylinders, valves, and motors of fluid power systems. And thus our industry emerged. It has come a long way since then. It is destined to go even further.

Today, the machines and equipment and appliances which rely upon fluid power systems for their functioning are valued in many billions of dollars. And it is difficult to find a manufactured product which hasn't been formed or treated or handled by fluid power at some stage of its production or distribution.

All along the line, the world's need for faster production, better quality, less waste, lower costs, and more machine muscle has called for a growing use of fluid power. Whether it is the building of a vital transportation artery, the mining of a critical ore, or the tooling up for a new home appliance, the fluid power industry has some part of the job to solve.

Yes! Fluid power has become quite an industry.

why its use is expanding

The success of today's fluid power applications makes it certain that fluid power will carry a bigger and bigger load in tomorrow's world.

One reason is *automation* – remote control, automatic control, more machinery for manipulating and moving parts and products between processes. This is fluid power's bailiwick. A fluid can go anywhere a pipe or tube can go. Valve levers and pushbuttons are all that are required to start, stop, and control a fluid power system. And the air or hydraulic cylinder, and the fluid motor, are among the most versatile "manipulators" the engineer has at his command.

Another is *design simplification* – fewer parts, less complicated mechanisms, putting output of power closer to points of use. Here again, more fluid power! A pump, a fluid, controls, piping, hose, or tubing, and a cylinder or fluid motor, are all there are to a fluid power system. Smooth, effective motion is its prime characteristic.

The engineer recognizes other attributes of fluid power which recommend it: easy speed control, absence of vibration, space saving, low cost, versatility, and simplicity. These, too, indicate even greater usage.

Fluid power is, indeed, an expanding industry.

For further information about fluid power education, careers in fluid power, and the fluid power industry, write to:

The National Fluid Power Association
3333 N. Mayfair Road, Milwaukee, Wis. 53222

panorama of fluid power at work

aviation

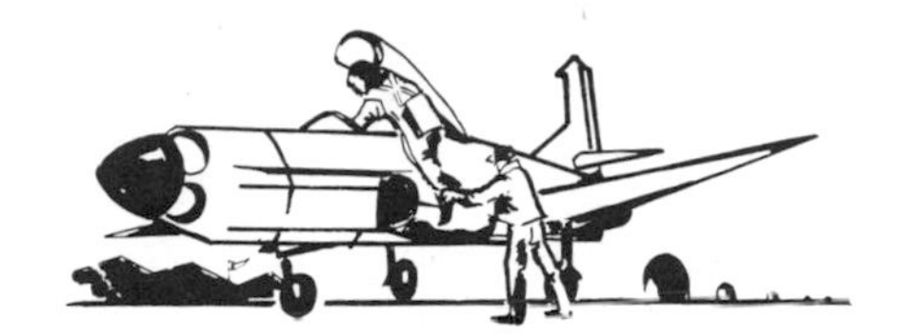

agriculture

chemicals

petroleum

food processing

marine

metalworking

and machine tools

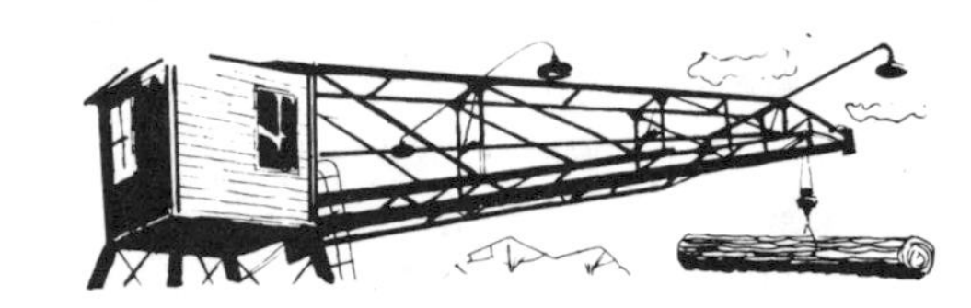

lumber

material handling

construction

mining

steel

Outline Of Book Contents

See complete index on Pages 286 and 287. Abbreviations used in the text appear on Page 257.

CHAPTER 1 – FLUID POWER PRINCIPLES

This is Lesson 1 of the short course in fluid power and orients a student with the principles of transmitting power through fluids as compared to other means of transmission • Laws and principles of fluid power transmission, including units of pressure and flow • Plumbing materials and sizes • Petroleum base hydraulic oil and other hydraulic fluids • Pressure losses through piping • Introduction to the use of vacuum and vacuum applications.

CHAPTER 2 – FLUID POWER CYLINDERS

In Lesson 2 air and hydraulic cylinders are described including standard mounting styles, piston and rod seals • How to calculate cylinder force and speed • Minimum piston rod diameter • Tips on designing with cylinders.

CHAPTER 3 – CONTROL VALVES, 3-WAY AND 4-WAY

Lesson 3 describes the simpler kinds of 2-way and 3-way valves for controlling direction and speed of air and hydraulic cylinders • How to draw schematic symbols for these valves • Simple, illustrative circuits for these valves • Pressure relief valves and how they should be used • Other 2-way types such as check, quick exhaust, and shuttle valves.

CHAPTER 4 – CONTROL VALVES, 4-WAY AND 5-WAY

The study of valves continues in this chapter with 4-way and 5-way directional valves • Circuit symbols • How to draw and read diagrams • Common types of actuators for control valves • Circuits for direction, pressure, and speed control of cylinders • Automatic 1-cycle or continuous reciprocation of cylinders • Bank valves • Power beyond.

CHAPTER 5 – FLUID POWER PUMPS

Air compressors • Various kinds of positive displacement hydraulic pumps • Unloading a hydraulic pump • Pressure compensated pumps • How to calculate power input to a pump • Mounting, alignment, and coupling to a pump • Inlet conditions to a pump • Cavitation • Side drive of a pump • Direction of rotation • Proper operation of a pump.

CHAPTER 6 – OTHER FLUID POWER COMPONENTS

This final lesson of the short course explains filters, regulators, and lubricators for air lines • Air dryers • Hydraulic power unit construction • Filtering in a hydraulic system and where to install filters • Filter beta ratings • Hydraulic accumulators, when and how to use them • Heat exchangers for hydraulic systems and how to size them.

APPENDIX A – DESIGN CALCULATIONS

This section contains calculations of a more advanced nature on how to select the best plumbing size, how to estimate the power input to a hydraulic system, how to select the size of an accumulator, and how to select the capacity for a heat exchanger.

APPENDIX B – TROUBLESHOOTING PROCEDURES

Material in this section shows how to check out a hydraulic system in event of failure, troubleshooting procedures on hydraulic pumps, leakage testing of valves and cylinders, and specific problems with solenoid valves.

APPENDIX C – FLUID POWER DESIGN DATA

Many pages of charts and text data including graphic symbols, abbreviations, fluid power formulae, force charts for cylinders, speed charts for cylinders, plumbing materials, electric motors, vacuum pump-down time and much more – 23 pages in all.

APPENDIX D – METRIC DESIGN DATA

The U.S. system of units is used throughout the book. This section explains the SI international system of units and how to convert units between the systems. Charts on force and speed of metric cylinders and how to make calculations.

Our Position On Metric Units

In the past there have been several different metric systems of units used in various parts of the world. The new international standard (SI) metric system has been developed to replace all of these systems as well as the U.S. (or English) system with a set of units for world-wide use. Most countries have made the conversion to these new units but the U.S.A. has made only a partial conversion and it now appears that it may be many years before the old units are completely abandoned.

We apologize to our Canadian (and other) students for keeping the U.S. units in this book. But at the present time the largest usage of the Womack fluid power books is in the U.S.A, so it is necessary to retain the old units. But we have provided a section, in Appendix D, for easy conversion of U.S. units to the SI system and vice versa. The conversion tables on Pages 280 and 281 will convert all common units used in fluid power. The text material starting on Page 278 explains the SI metric system, how to understand and use it. Subsequent pages explain metric calculations for force and speed of cylinders, and Page 285 lists port threads for metric cylinders, pumps, and valves.

Some students have been confused about the difference between weight and mass. Weight is the force produced by gravity on an object and varies in different locations because of altitude, and becomes zero in far outer space. Force, as from a cylinder, is expressed in the same units as force of gravity but does not change with altitude. Mass is the "substance" of an object and expresses its resistance to being accelerated or decelerated, that is, to any change in its velocity. Mass of an object is always the same even in far outer space.

Force or weight is used for calculations of fluid power force. Mass is used in calculations of braking or accelerating force for vehicles, etc.

In the older metric systems, the kilogram was used to express both weight and mass. In the new SI system the kilogram is now used only for mass. It has been replaced by the Newton to express force. Since the "gravitational constant" in the metric system is 9.81 (instead of 32.2 of the U.S. system), 1 kilogram of mass is equivalent to 9.81 Newtons of weight. If you weighed 80 kg in the older metric system you now weigh 784.8 Newtons. Although the technically correct unit for gauge pressure is the Pascal (Newtons per sq. meter), it is too small for convenient use, and the "bar" is the acceptable unit. See table on Page 281.

DEAR STUDENT:

Welcome to the technology of fluid power. This is the first of a 3-volume series for teaching hydraulics, air, and vacuum, not as a scientific study, but the way it is actually used out in the field. In this book, we have covered basic principles and components along with simple circuitry. You will find many charts of design data at the back of the book. When you have completed this study, and if you are a serious student, you will want to continue your study with the more advanced material in Volumes 2 and 3. See description of these and other Womack books on Page 288. Any Womack book may be purchased directly from the publisher. Send for descriptive brochure and current price list.

Liability Disclaimer. Please remember while studying this book that all information is for educational use only. It has been carefully compiled and checked and we believe it to be accurate. But since any segment of information may apply on one application and not on a similar one because of many factors such as differences in environment, personnel, operating conditions, materials and components, and since errors can occur in circuits, tables, and text, we do not assume any liability for the safe and/or satisfactory performance of any machine or installation designed from information in this book. Thank you.

CHAPTER 1

Fluid Power Principles

The purpose of a fluid power system is to transmit power from one location to another. Actually, there are only three methods used commercially for transmitting power. One is by mechanical transmission through shafts, gears, chains, belts, etc. Another is by electricity flowing through wires. Fluid power is the third. Power is carried by the flow of a fluid (gas or liquid) under pressure.

Some applications are best suited to a particular one of these methods, while other applications may be better suited to another method. For example, fluid power may be better than mechanical transmission for transmitting power to moderate distances or to inaccessible or out of the way places, and may be better than either mechanical or electrical transmission where a fine degree of control, including reversibility and infinite speed variation are important requirements.

How Fluid Power Works. It was nearly three centuries ago that the great French scientist, Pascal, demonstrated that pressure exerted anywhere on a confined fluid (gas or liquid) is transmitted undiminished in all directions and with equal force exerted on all equal areas. Modern fluid power engineering makes use of this principle, in a practical way, to transmit power from one location to another with fluids flowing through pipes.

The input source of power for a fluid system is always electrical (from an electric motor) or mechanical (from an engine). The power is converted to an equivalent amount of fluid power by means of an air compressor or hydraulic pump. After conversion to fluid power it can be transmitted through moderate distances and then be converted back to mechanical power by means of a linear moving piston or a rotating fluid motor. During its transition suitable valving is added to control the direction of flow (and output movement), to regulate the rate of flow, and to limit the maximum pressure in the piston and transmitting pipes. The whole arrangement is called a fluid power circuit or system.

Advantages of Fluid Power. Of course fluid power is not the best answer on every job, but tomorrow's industry will depend more and more on ***automation*** to increase productivity – automation in direct and remote control of production operations, in machinery to speed parts and products through manufacturing processes and distribution channels. Fluid power ***is*** the heart and muscle of automation, and here are some reasons why:

Ease and Accuracy of Control. With simple levers or pushbuttons the operator of a fluid power system can easily start, stop, speed up or slow down, and can position forces which may total hundreds of horsepower – and often to tolerances of a fraction of an inch.

Multiplication of Force. Instead of using cumbersome gears, pulleys, and levers, a fluid power system multiplies force simply and efficiently, from a few ounces to literally thousands of pounds of output. Examples range from power steering or power brakes on your automobile to large hydraulic and pneumatic presses.

Constant Force or Torque. Only fluid power systems provide constant force or torque regardless of speed changes – whether the work output moves a few inches an hour or thousands of revolutions per minute.

Simplicity, Safety, Economy. A fluid power system uses fewer moving parts than comparable electrical or mechanical systems. Hence, they are simpler to maintain, provide maximum safety, are more economical to operate, and more reliable. And fluid power systems have the highest possible ratios of horsepower-to-weight.

Other Advantages. Design engineers will recognize other advantages such as: absence of vibration; space savings; relatively quiet operation; instantly reversible motion; and practically no danger from overloading. Because fluid power can go anywhere a hose or pipe can go, the design of the system is not influenced by the geometry of the machine on which it is used.

Fluid power's versatility assures its increased usage in years ahead in automation, aerospace, construction machines, marine applications – wherever efficiency is the key to improved productivity in the man/machine relationship.

TRANSMITTING POWER THROUGH FLUIDS

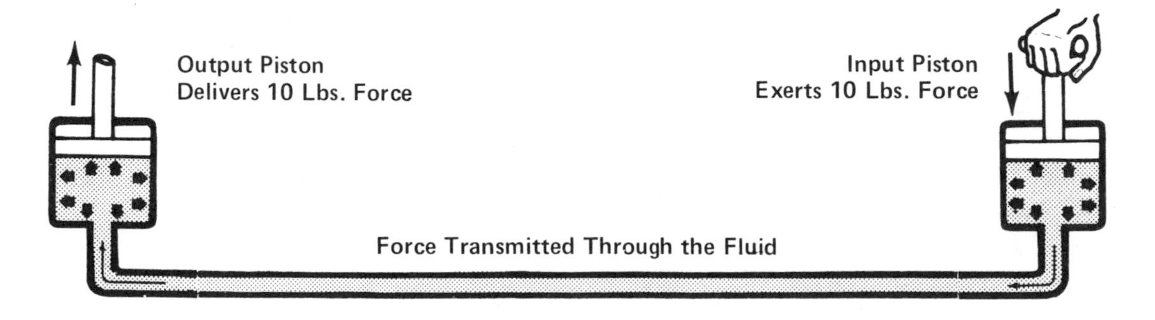

FIGURE 1-1. Input piston and output piston have equal areas exposed to the fluid.

Figure 1-1. The basic purpose of a fluid power system is to transmit power from an input location, through pipes, to an output location. Power input to a fluid power system is nearly always mechanical. Mechanical power is converted into fluid power, in this case, by a hand-operated piston. At the output location it is converted back to mechanical power with the same force and speed as the input power, (except for a small power loss from mechanical and flow friction), assuming input and output pistons are of equal diameter.

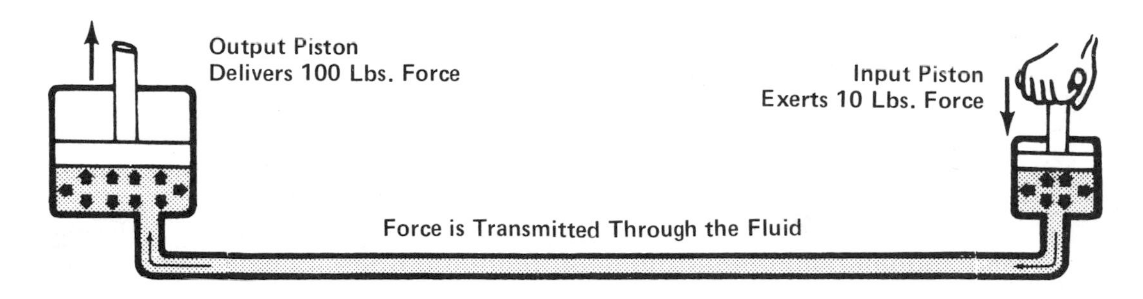

FIGURE 1-2. Input piston has only 1/10th the area of the output piston.

Figure 1-2. But if, for example, the *area* of the output piston is ten times the *area* of the input piston, the force delivered at the output piston will be ten times greater than the force exerted by the input piston, but it will travel only 1/10th as far and with only 1/10th the speed. However, the power delivered by the output piston will be the same as the power applied to the input piston (except for small friction and flow losses) although at a different ratio of force and speed.

Figure 1-3. This sketch illustrates a typical system in its simplest form. The pilot of a small aircraft can raise and lower a landing wheel hydraulically. On the beefier side, a similar system with hydraulic pump, oil reservoir, control valves, and reversible piston would permit the operator of a powerful forming press to shape the entire end of a box car with exceptional precision.

There are many types of pumps, pistons, motors, and valves, and almost as many ingenious ways of hooking them up as there are engineers putting fluid power to work. And what has been accomplished in recent years is remarkable indeed.

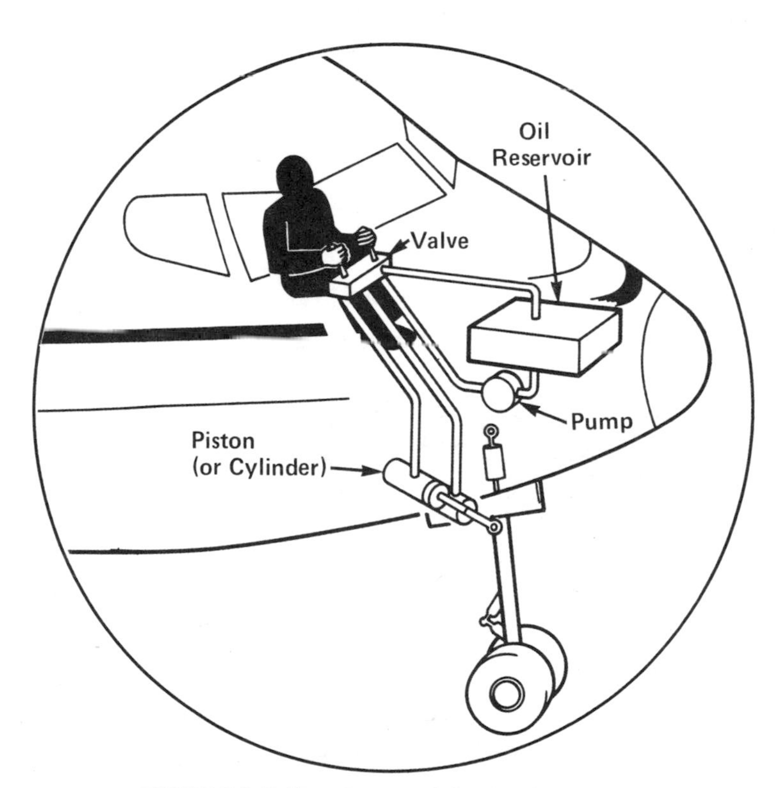

FIGURE 1-3. Simple aircraft hydraulic system.

In Figure 1-3, when the pilot of an airplane moves a small control valve in one direction, fluid under pressure flows to one end of a piston to lower the landing wheel. By moving the same lever in the other direction, fluid under pressure flows into the opposite end of the piston and retracts the landing wheel.

FLUID POWER SYSTEMS IN AN INDUSTRIAL PLANT

FIGURE 1-4 – Compressed Air System

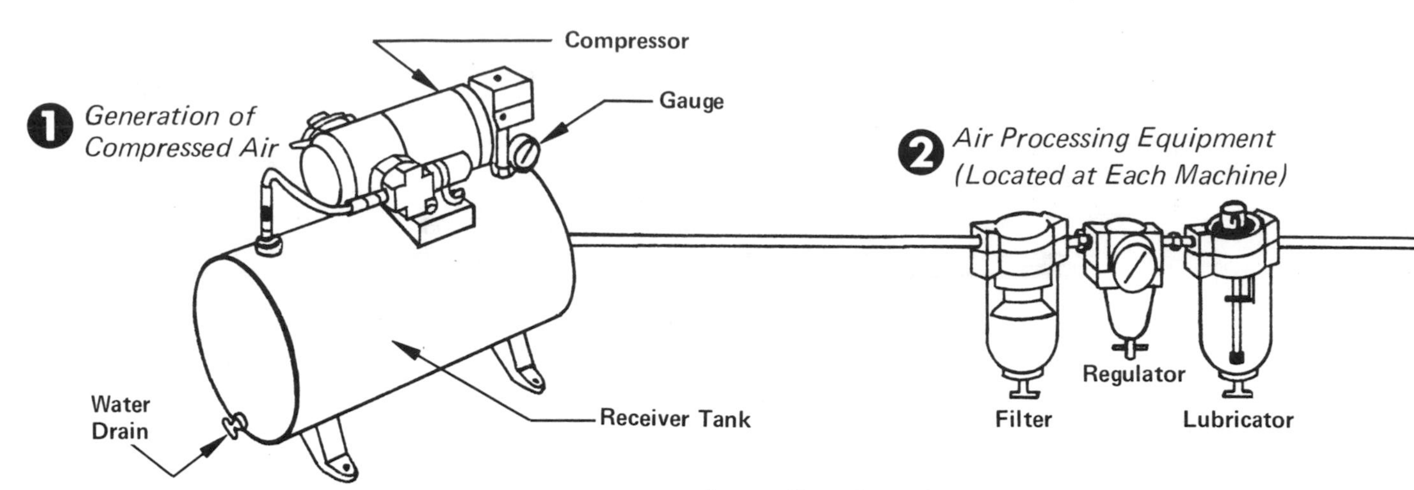

1. Air is pumped by a compressor into a storage tank called a "receiver" and accumulated for future use. Storage pressure is usually about 100 to 150 pounds on the gauge. From the receiver tank a system of piping distributes the compressed air to various points of use in the plant, still at about the same pressure as in the receiver tank. At each machine using the air, pressure is reduced to about 80 to 100 pounds gauge pressure.

2. On each machine the air is processed before being used. It is first filtered, then its pressure is reduced to the desired level and a fine oil mist is added for lubrication of air tools, valves, and pistons downstream.

FIGURE 1-5 – Hydraulic Fluid Power System

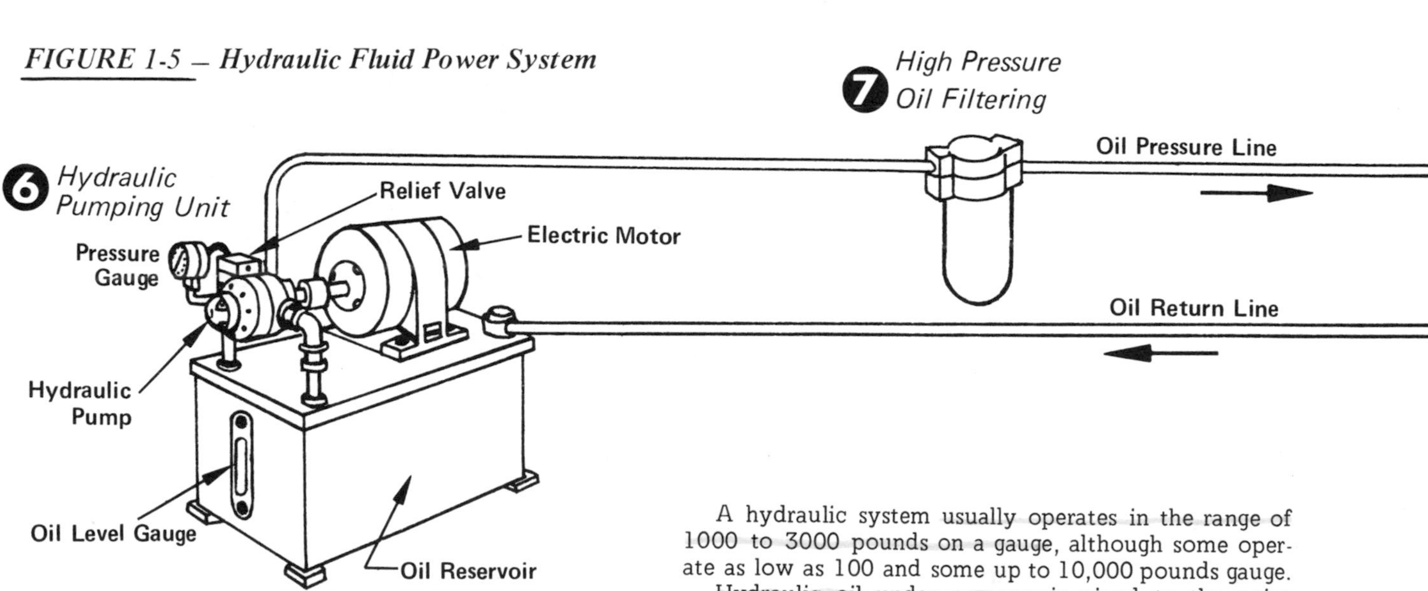

6. The hydraulic pump is driven by an electric motor or an engine. In most systems the pump delivers an oil flow at the rate needed for a desired piston speed, and does not store it like an air system stores compressed air.

A hydraulic system usually operates in the range of 1000 to 3000 pounds on a gauge, although some operate as low as 100 and some up to 10,000 pounds gauge.

Hydraulic oil under pressure is piped to the point of use. After its energy is expended in operation of a hydraulic piston or hydraulic motor it flows back to the reservoir through a return line.

In small to medium power systems the electric motor, hydraulic pump, relief valve, and other valves may be mounted on top of the oil reservoir. This assembly is called a "hydraulic pumping unit" or "hydraulic power unit".

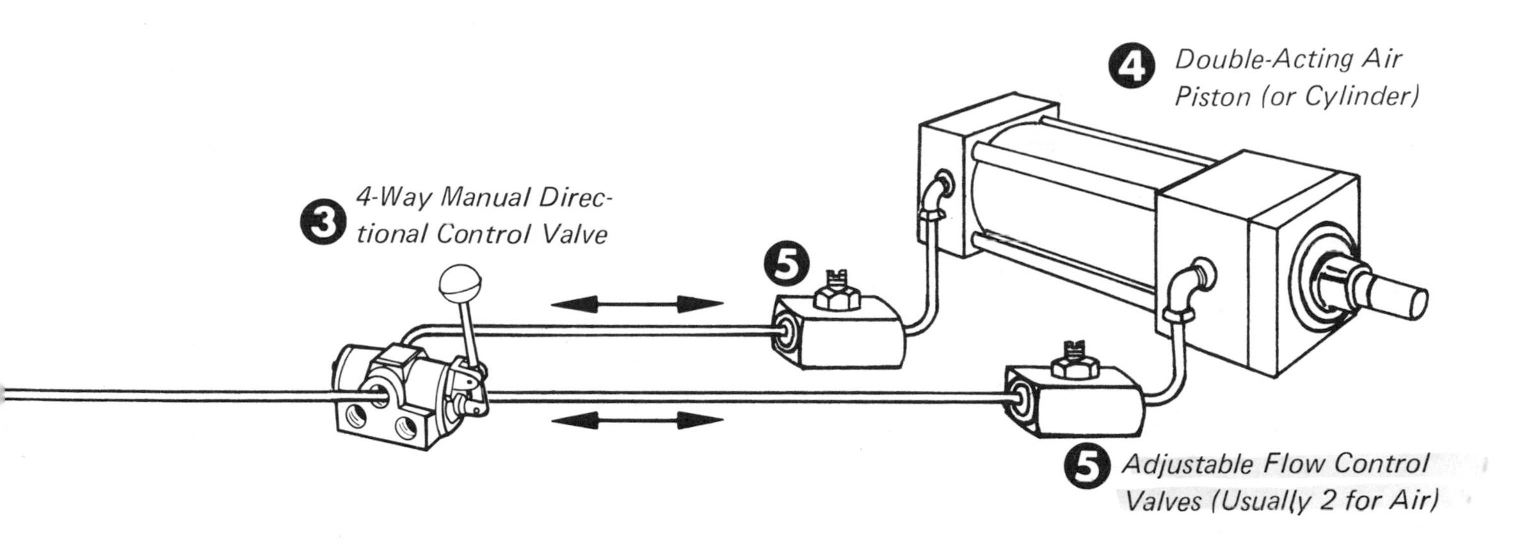

3. A directional control valve, either manually or electrically operated, enables the operator to move the piston forward, to reverse or stop it.

4. Air pressure and flow are converted into straight-line mechanical output by a piston or into rotary output by an air motor or rotary actuator.

5. Flow control valves are used to limit cylinder speed. Since the piston is operating from an unlimited air supply, its speed may have to be limited in both directions of travel, and two flow control valves are normally used, one for each direction of piston travel.

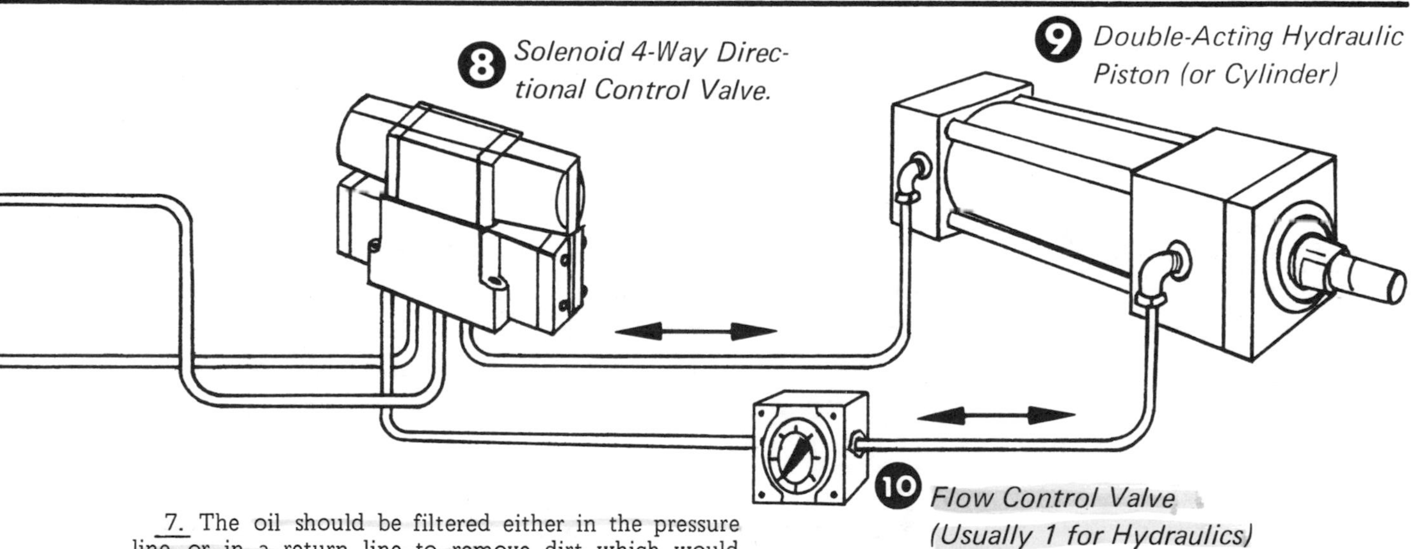

7. The oil should be filtered either in the pressure line or in a return line to remove dirt which would eventually damage system components or cause a breakdown in the flow of hydraulic power.

8. Direction of piston travel is controlled with a 4-way directional control valve, usually a solenoid or a manually operated type.

9. The hydraulic piston (or hydraulic motor) converts fluid power into an equivalent amount of mechanical power output. The piston rod is coupled to the load.

10. In some circuits a flow control valve is used to reduce piston speed during a part of its stroke in one direction, for slow feeding of a cutting tool or to reduce the speed of a press. In most hydraulic circuits the piston is allowed to return at unrestricted speed. Therefore, only one flow control valve is normally used.

To control return speed, a second flow control valve can be added in the other piston line.

COMPARISON OF AN AIR SYSTEM TO A HYDRAULIC SYSTEM

On many jobs it will quickly be apparent which medium, air or hydraulics, is more suitable. But some jobs could be done with either medium, or with vacuum, and a consideration of the following points will help to make the best decision.

1. Power Level. Branch air circuits usually operate in the 1/4 to 1½ HP range while most hydraulic systems operate from 1½ HP and up. This is a general rule, as there are higher powered air systems and lower powered hydraulic systems. But a large air compressor, 100 HP for example, in a large plant is usually feeding numerous low power branch circuits operating independently of one another. But a 100 HP hydraulic system is usually operating only one machine, although this machine may consist of several branch circuits dependent on one another.

2. Noise Level. If the air exhaust hiss is properly muffled, an air system usually operates with less noise than a hydraulic system of the same power. And in between cycles an air system is completely silent while in a hydraulic system the pump remains running but in an unloaded condition. If a hydraulic system must be used, and if noise must be reduced to a minimum, a type of pump with lowest noise level should be used (see Chapter 5), and the system designed to run the pump at a relatively low speed and to operate it at a pressure less than its maximum rating by using pistons of larger bore which will produce the necessary force at a lower pressure.

3. Cleanliness. Normally an air system is very clean provided the air line oiler is not feeding excessive oil. If it is, the excess will be blown out into the atmosphere from the valve exhaust port. A hydraulic system which is carefully designed and constructed can be clean, but there may come a time when a piston rod seal or pump shaft seal may start to leak, or when a line is disconnected to replace a component, or a filter element must be replaced, or make-up oil added to the reservoir. At these times there is a possibility of oil being spilled around the machine.

4. Speed. Light-weight mechanisms can usually be operated faster with compressed air because a large volume of air (an almost unlimited supply) can be drawn from the receiver (storage tank), for short periods, to give a very fast piston speed. A hydraulic system, to match the speed of an equivalent air system, would have to have a very large pump, large size valving, and large size piping, since hydraulic power is usually generated by the pump at no more than the rate of use, with no reserve supply. An exception to this is an accumulator system as described in Chapter 6.

5. Operating Cost. A hydraulic system is usually more efficient than an air system of the same HP output, and should cost less to operate. As air is compressed it becomes hot. This heat of compression radiates from the walls of the compressor, the receiver tank, and the plumbing system, or is removed by an aftercooler. This represents a loss of energy from the system which can never be recovered. There are heat losses, too, in a hydraulic system which can be minimized with good design, but the heat losses in an air system are inherent and cannot be avoided or minimized.

6. First Cost. If the cost of an air compressor is considered, an air system is more costly to build than a hydraulic system. But sometimes a low power air circuit can be added to an existing compressor which has reserve capacity, and this is less costly than a hydraulic system.

7. Rigidity. Some applications must have a high degree of rigidity in the fluid stream. In this case, hydraulics must be used. Examples are lifts and elevators to be stopped at an intermediate point in the piston stroke for adding or removing a part of the load; the feeding of a cutting tool at a slow rate of speed; slow movement of a machine slide or other mechanism having a large area of sliding friction; press applications where two or more pistons are attached to an unguided or inadequately guided platen, particularly if the load is not evenly distributed over all the pistons; any system where close control of piston speed or position must be maintained, or where the moving piston must be accurately stopped at a precise position in mid stroke. These and similar applications, whether high or low power, can only be done satisfactorily with hydraulics.

PHYSICAL LAWS RELATING TO FLUID POWER

Pascal's Law

Blaise Pascal, although living only to the age of 39, was one of the great French scientists and mathematicians of the 17th century. In the last 5 years or so of his life he abandoned science and devoted himself as intensely to a peculiar religious sect as he previously had to science.

Although he was responsible for many important discoveries, we are primarily interested in two of them in relation to fluid power. One was in proving that vacuum power is due to weight of the atmosphere and not to "nature's abhorence of a vacuum". (See Page 44). The other was his principle of transmittal of force through a liquid. He demonstrated that in a fluid at rest, pressure is transmitted equally in all directions. In modern language his principle of hydrostatics can be stated as follows:

"Pressure set up in a confined body of fluid acts equally in all directions, and always at right angles to the containing surfaces".

Blaise Pascal (1623 – 1662). His discoveries are important to the technology of modern fluid power transmission.

Figure 1-6. It is because air and hydraulic fluids act according to Pascal's Law that force and motion can be transmitted through them. If a force is applied to the fluid at Point A, an equal force instantly appears at Point B and at all other points.

The action of Pascal's Law makes a flattened fire hose assume a cylindrical shape under internal pressure.

Pascal's Law applies only to fluid confined in a closed vessel. There is a difference of pressure at points along a pipe while fluid is flowing because of flow resistance, but at any sectional point pressure around the circumference of the pipe is equal.

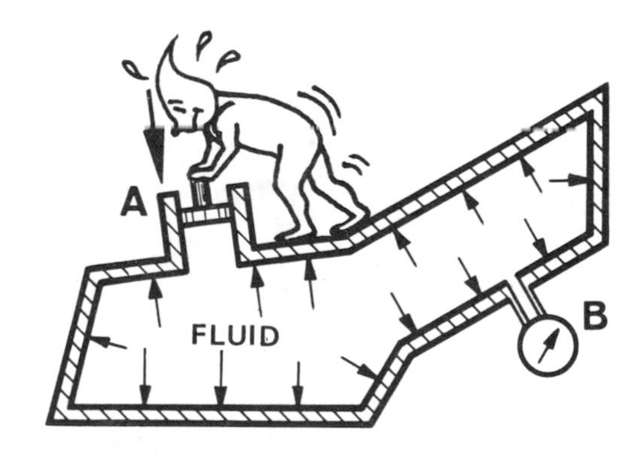

FIGURE 1-6. Pascal's Law states that pressure added to a confined fluid at any point instantly appears equally at all other points.

Figure 1-7. Pascal's Law is demonstrated by transmitting force and power through a column of fluid (any gas or liquid) in the same way it is transmitted, for example, through a steel rod. But since the fluid has no shape of its own, and takes the shape of its container, it must be confined in a pipe, hose, or tube. The same fluid which works against the face of the pistons also works against the inside walls of the pipe or tube, according to Pascal. Therefore, the tube walls must have sufficient strength to contain the pressure.

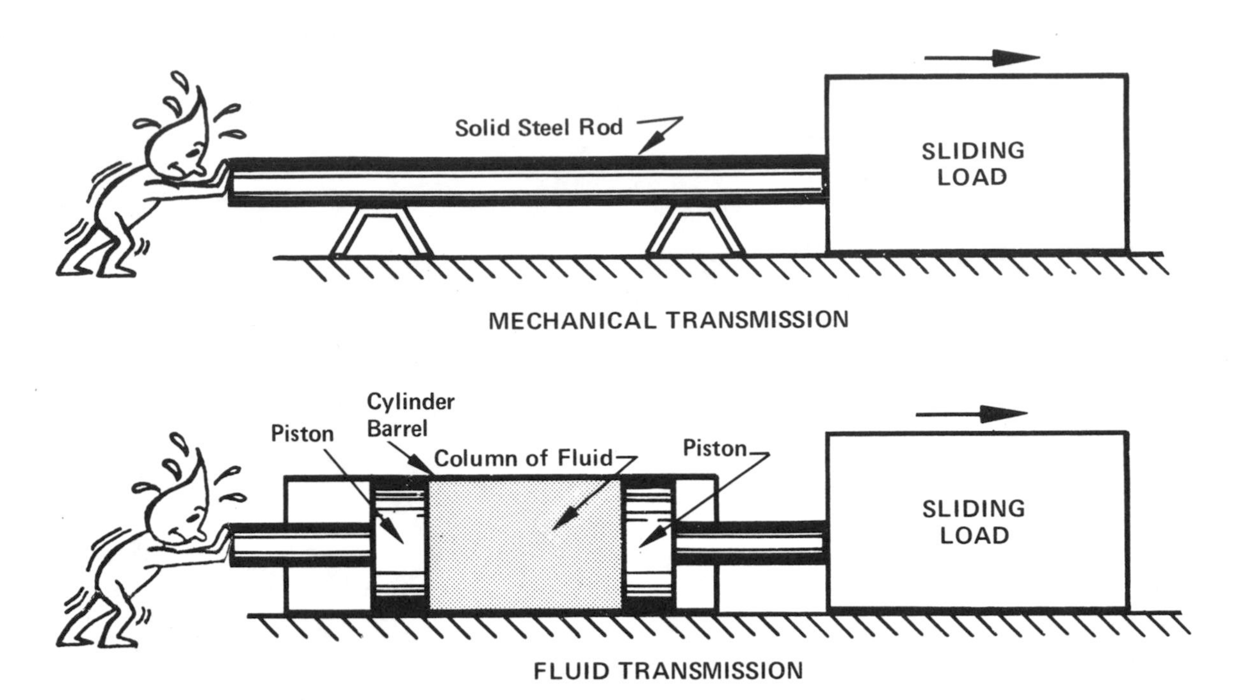

FIGURE 1-7. Power Transmission. Force, motion, and power can be transmitted through a column of gas or liquid just like it can be transmitted through a steel rod, but the fluid must be confined so it will retain its shape.

Compressibility of Fluids

Hydraulic oil is slightly compressible but is treated as being non-compressible for most fluid power systems. But for special systems such as precision machine control, servo systems, or where plumbing runs are unusually long, oil compressibility must be considered. It is often the limiting factor on distance to which fluid power can be transmitted successfully. Compressibility of oil is less at high pressures but is approximately propoprtional to pressure up to 2000 PSI, its reduction in volume being about 1/2% per 1000 PSI. At 3000 PSI its volume is reduced by about 1.2%. These values are approximate. Exact compressibility must be determined mathematically. Water is slightly less compressible than oil, a rule-of-thumb being a reduction in volume of about 1/3% per 1000 PSI at low pressures.

The steel rod of Figure 1-7 will compress about 0.00036 inches for every foot of its length under a mechanical force (strain) of 1000 PSI. The oil column in the lower part of the same figure will compress about 0.06 inches per foot of length under the same strain of 1000 PSI.

Boyle's Law – Compressibility of Air

We already know, from our experience with air compressors, that when the volume of air is reduced by compression, its pressure increases. Boyle's Law clearly defines the relationship between the volume of a gas and its pressure:

"The absolute pressure of a confined body of gas varies inversely as its volume, provided its temperature remains constant".

This means that if the volume of a confined gas is reduced to 1/4th by compression, as in Figure 1-8 on the next page, its absolute pressure after it has been allowed to cool to its original temperature, will be 4 times its original pressure (compressing air raises its temperature according to Charles' Law).

Boyle's Law is not one of the invariable laws of nature like the law of gravity. It describes the behavior of a "perfect gas". Compressed air very nearly follows the law at pressures less than 1000 PSI, and the calculation error using Boyle's Law is usually unimportant. However, other gases, particularly hydrocarbon gases like methane and propane, do not follow Boyle's Law as closely.

Units of Pressure in Fluid Power Systems

In the English system of units which is in predominant use in the United States at this time, pressure in both air and hydraulic systems is measured in units of PSI (pounds per square inch). Gauge pressure readings show the pressure level above the reference base of the surrounding atmosphere, and for clarity they may be stated as PSIG to distinguish from PSIA (absolute) pressure values.

Fluid pressure used for air and hydraulic systems for producing force, torque, and power must always be stated and calculated in gauge pressure readings, not in absolute pressure values. An unconnected gauge always reads 0 regardless of atmospheric pressure, elevation, or barometric reading.

Absolute pressure (PSIA) is a measure of the pressure above the absolute base of a perfect vacuum, the pressure existing in far outer space. It cannot be read on an ordinary gauge but can be measured with barometric-type instruments. Under standard sea level conditions, atmospheric pressure is 14.7 PSI higher than a perfect vacuum. This is due to weight of the atmosphere. Therefore, under standard conditions, gauge readings are always 14.7 PSI less than absolute pressure values. At higher elevations the difference between absolute and gauge pressure is less than 14.7 PSI. Absolute pressure cannot be used for calculating fluid power systems. It is used primarily in scientific work which includes calculations involving Boyle's Law which must be referenced to an absolute pressure base.

Most other countries are adopting the international standard (S. I.) measuring units. These are based on the metric system. Fluid pressure is measured in "bars". 1 bar = 14.5 PSI. A cross reference between all pressure units, U. S. A. and S.I., is on Page 281.

In regard to international metric units of pressure, at this time ANSI (American National Standards Institute) states that the "Pascal" will become the standard for pressure, but that the "bar" may be used as an alternate unit. The problem with the Pascal in fluid power is that it is too small a unit to be easily handled. One atmosphere (14.7 PSI) is equal to more than 100,000 Pascals.

Vacuum can also be measured in PSI but is more commonly measured in "Hg (inches of mercury). 30 "Hg is equal to approximately 14.7 PSI.

Some hydraulic presses have a gauge calibrated in "tons". This is usually the U.S. ton or 2000 lbs. This calibration is based on the cross sectional area of the ram piston having (so many) square inches of exposure to the pressure. Such a gauge would read incorrectly if used on another press having a ram with different area of exposure.

Pressure gauge readings are a measure of intensity of the force or torque which can be produced by the fluid system when acting through a cylinder or fluid motor. In this book we will use the prevailing U.S.A. unit of pressure, PSI. Refer to Page 281 for conversion to other units of pressure.

Relationship Between Bars, Gauge Pressure, and Absolute Pressure

Bars (Gauge Pressure)	0	1	2	3	4	5	6	7	8
Bars (Absolute Pressure)	1	2	3	4	5	6	7	8	9
PSIG (Gauge Pressure)	0	14.5	29.0	43.5	58.0	72.5	87.0	101.5	116.0
PSIA (Absolute Pressure)	14.5	29.0	43.5	58.0	72.5	87.0	101.5	116.0	130.5

Sample Calculation Illustrating Boyle's Law

Boyle's Law must, of course, use absolute pressure as a reference base. In order to apply the law, any gauge readings involved must be converted to absolute values before any calculations can be made. Remember that absolute pressure values are higher than pressure gauge readings by the amount of barometric pressure which prevails at the working location. For average work we normally use 14.7 PSI (1 atmosphere) as the barometric pressure although at elevations higher than sea level the barometric pressure will be slightly less.

In working Boyle's Law problems, then, convert gauge readings to absolute pressure by adding 14.7. To convert absolute pressure to gauge pressure, subtract 14.7.

Figure 1-8. The vessel on the left has a volume of 4 cubic feet under the piston. Air in this space is under a pressure of 30 PSIG as shown by the gauge. If the piston is forced downward to compress the air into a volume of 1/4th the original, the absolute pressure will increase by 4 times according to Boyle's Law. In this example, solve for the new gauge reading at the reduced volume.

Solution: First, convert 30 PSIG to its equivalent absolute pressure: 30 PSIG + 14.7 = 44.7 PSIA. Next, use Boyle's Law: Since the volume was reduced to 1/4th, the absolute pressure will increase 4 times: 44.7 x 4 = 178.8 PSIA. Finally, convert back to gauge pressure: 178.8 - 14.7 = 164.1. This will be the new gauge pressure.

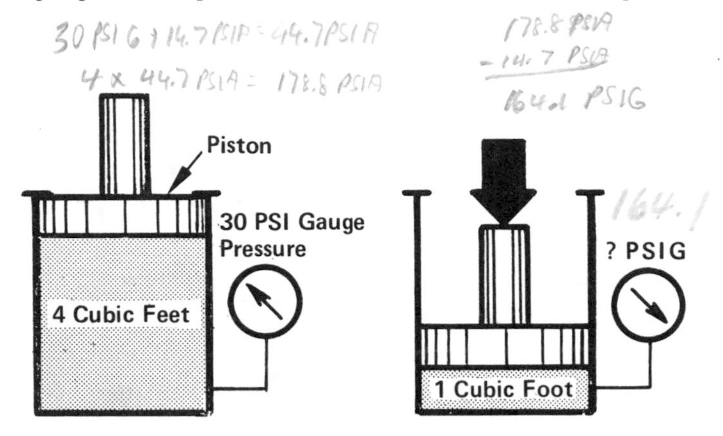

FIGURE 1-8. According to Boyle's Law the absolute pressure will increase 4 times when the volume is reduced to 1/4th. But what will this amount to on a gauge?

Charles' Law

This law defines the realtionship between the temperature of a gas and its volume or pressure or both. While it is a very important law of physics, it is not often significant in the design of air components and systems which work at normal ambient temperatures.

Refer to Page 257 for Charles' Law formula. Charles' Law states that if the temperature of a gas is raised, its volume also increases by the same ratio, providing its pressure remains constant. Or, with an increase in temperature, its pressure increases by the same ratio providing its volume remains the same.

Charles' Law has little effect on air systems because it is referenced to a zero base of -460°F. Any change of a few degrees at +85°F has very little effect on either volume or pressure. While absolute pressure is always higher than gauge pressure by the amount of the barometer reading, usually 14.7 PSI, absolute temperature is always 460° higher than the reading of a Fahrenheit thermometer.

Bernoulli's Principle

Bernoulli's Principle is an important principle of fluid flow which applies to non-compressible fluids like hydraulic oil. It is based on the law of conservation of energy, which states that energy can neither be created nor destroyed, although it can sometimes be changed from one form into another. The value of Bernoulli's Principle is in helping to understand the ***theory*** of fluid flow. But it

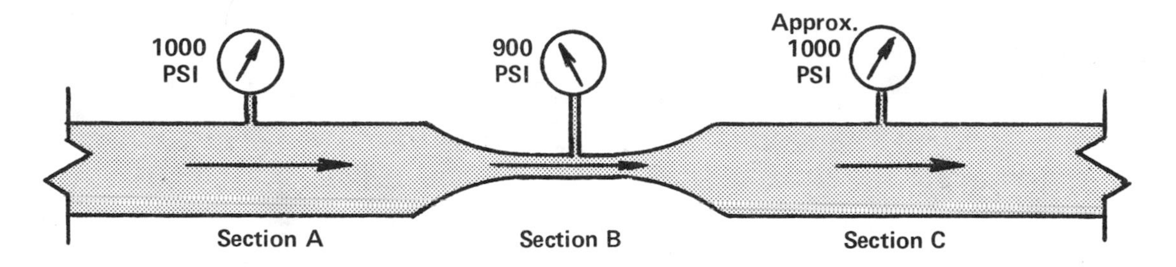

FIGURE 1-9. A venturi orifice illustrates Bernoulli's Principle. Energy flow is constant through all sections. As velocity increases through Section B, static pressure goes down.

cannot readily be applied to solving many of the ***practical*** kinds of problems described in this book.

Figure 1-9. One example of the Bernoulli Principle is a venturi orifice which is sometimes used to produce a lower gauge pressure, without power loss, at a particular point in a fluid stream. The energy level remains the same through Sections A, B, and C, and is represented by a combination of static head pressure (which can be read on a pressure gauge) and velocity head pressure (which ordinarily does not show on a gauge). The volume of flow is the same through all three sections. Therefore, as the velocity increases, as it does through Section B, more of the flowing energy changes from potential energy to kinetic energy, and the static head pressure as read on the gauge, decreases. As the fluid enters Section C, its velocity decreases, and if velocity in this section is the same as in Section A, the gauge pressure returns to its original level, although there may be a slight loss of gauge pressure as a result of flow friction.

Bernoulli's Principle explains other phenomena such as why the spool in a directional control valve may drift out of position, especially when the valve is forced to handle flows higher than its rating.

Units of Flow in Fluid Power Systems

Power flows through a pipe only when the fluid is moving, and only when it is under pressure. Unless both conditions exist, no power is being transferred. If the fluid is under pressure but is in a static (non-flowing) condition, pressure is transferred through the pipe but the power flow is zero.

In the system of units which is in predominant use in the United States at this time, the flow of hydraulic oil is most often stated and measured in units of GPM (gallons per minute). An alternate unit used occasionally is CIM (cubic inches per minute). Pump catalog ratings are usually given in GPM. Hydraulic motor ratings are usually stated in CIR (cubic inches displacement per shaft revolution). The SI (international standard) system of units is based on the metric system. The preferred unit for oil flow is "litres per minute". A litre is equivalent to a "cubic decimetre". An interchange chart between previous metric, S.I. metric, and conventional U.S.A. units is on Pages 280 and 281.

The unit for air (or gas) flow in the U.S. system is SCFM (standard cubic feet per minute). One SCFM is defined as the flow of one cubic foot of air or gas per minute at standard conditions of 14.696 PSIA pressure, 60° F temperature, and 36% relative humidity. These are conditions sometimes referred to as sea level conditions. Air flow ratings must always be based on standard (sea level) conditions rather than on compressed air conditions, because 1 cubic foot of compressed air contains more air molecules, by reason of being compressed, than 1 cubic foot of air at standard conditions.

The preferred SI (international standard) unit for compressed air flow is cubic decimetres per second under the same standard conditions stated above. Refer to Pages 280 and 281 for conversions.

TRANSMISSION OF FORCE, WORK, AND POWER THROUGH FLUIDS

Figure 1-10. Pascal recognized that pressure developed within the fluid must necessarily be equal at all points within the fluid provided the fluid is confined. Any pressure developed within the fluid by mechanical force applied at Piston A will be transmitted through the fluid and will develop force against Piston B, and this will be the same force applied at Piston A if the two pistons have equal surface exposure to the fluid.

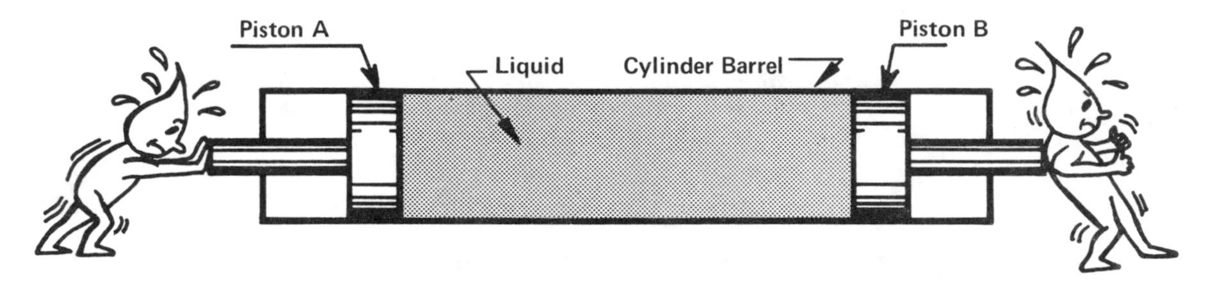

FIGURE 1-10. Pascal's Law is involved here in transmitting force, work, and power from one piston to another through a column of fluid, either air or hydraulic.

Force (and movement) can be transmitted to reasonable distances by making the cylinder barrel as long as necessary to connect the point of transmission (Piston A) to the point of reception (Piston B). If sufficient force is applied on Piston A to cause it to move, Piston B will move the same distance. Thus, the work applied to Piston A (force times distance moved) will be transmitted to Piston B as system output. The amount of power (horsepower) transmitted through the fluid will depend on how fast the pistons move. A fast movement will transfer a greater HP than a slower movement since HP is defined as the rate at which work is performed.

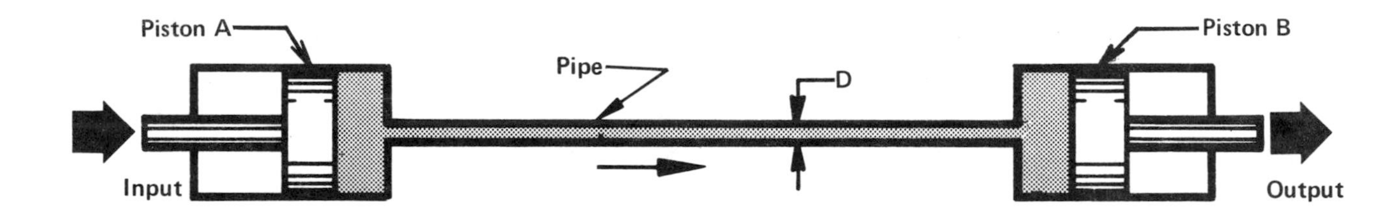

FIGURE 1-11. Pure force (with no flow) can be transmitted through a pipe of small diameter, but if the fluid must move through the pipe, the pipe diameter must be large enough to keep flow loss to a low value.

Figure 1-11. If Pistons A and B of the preceding figure are separated by quite a distance, they can be connected with a small diameter pipe and Pascal's Law will still apply. The system can still transmit the same amount of force and power up to the stroke limit of the pistons. An example is the hydraulic braking system of an automobile where a small tube connects the master cylinder (Piston A) to the wheel cylinders (Pistons B).

How large in diameter must the connecting pipe be? If pistons A and B are used simply to transmit force with no movement of the pistons, they will do so equally well no matter how small the diameter of the connecting pipe. Pascal's Law still applies. But if the fluid must move through the pipe for transmitting work or power, the pipe diameter should be selected according to how fast the fluid must move. If the diameter is too small, flow losses might be too high. If the diameter is unnecessarily large, the plumbing cost might be excessive. A happy medium on pipe size is one which balances flow and power loss against cost of the plumbing. In selecting pipe size for a hydraulic system, the flow velocity of the fluid is used as a guide. This is further explained on Page 239.

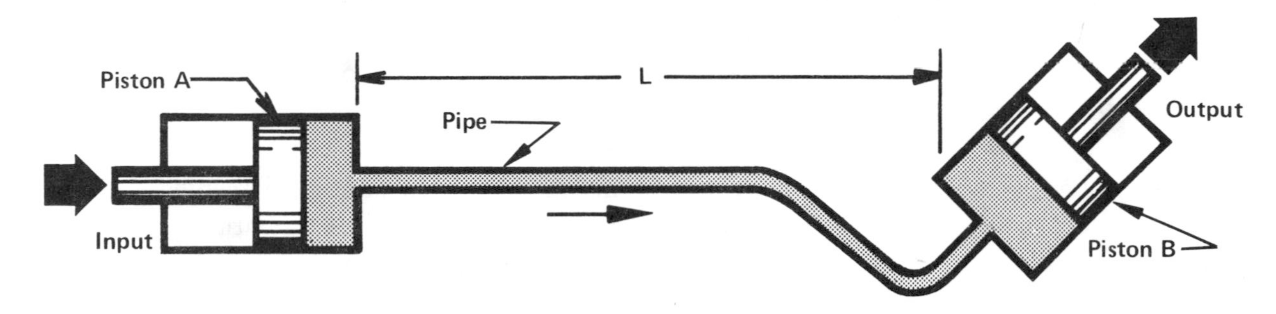

FIGURE 1-12. Fluid power has great flexibility. It can be transmitted up or down, around corners and to greater distances than it can by mechanical transmission means.

Figure 1-12. Fluids are versatile media for transmitting power because the input and output pistons can be widely separated, and the generated power can be transmitted with small losses. Transmitted power can be carried up or down, around corners or obstacles, and through sealed partitions. It can be carried through relatively small pipes with minor losses. It is superior to mechanical power transmission with respect to the ease with which it can be generated, transmitted, controlled, and used.

However, there is a practical limit to the distance over which hydraulic fluid power can be transmitted. Oil compression is a factor which may limit transmission range on some types of applications. An industrial hydraulic system is expected to operate over a separation distance of 20 or 30 feet between pump and cylinder. On applications where oil compressibility is unimportant, the separation distance can be much greater; for example, aircraft hydraulic systems. The practical range for compressed air is limited by separation distance between directional control valve and the cylinder, and the volume of air trapped between them. For practical purposes the maximum separation distance is about the same as for hydraulics. In both air and hydraulic systems, the critical piping is the connecting line between control valve and cylinder. If this line is kept short, the power feed line between pump and cylinder or between air compressor and cylinder can be quite long without seriously affecting system performance provided it is large enough to keep flow losses to a practical minimum.

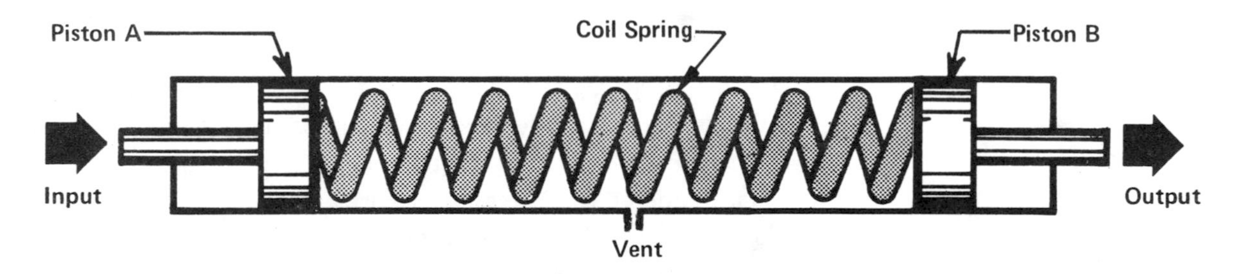

FIGURE 1-13. Transmission of power through a column of fluid is much like transmitting mechanical power through a coil spring. All fluids are compressible to some degree.

Figure 1-13. Transmitting force through a column of fluid is like transmitting force through a spring which is supported by a vented tube. Force applied on Piston A will produce a force output on Piston B although the spring will compress to some extent while transmitting the force. If the spring is constructed of small diameter wire it will compress a great deal while transmitting the force. In this respect it is similar to transmitting force through a column of compressed air. If the spring is constructed of very heavy wire it will transmit the same force while compressing only slightly. In this respect it is similar to transmitting force through hydraulic oil which compresses very little.

While compressed air is an excellent medium for transmitting power, its greater compressibility does limit its use on certain applications where fluid rigidity is required to get a smooth (non-erratic) movement of the output piston. Hydraulic power with its greater rigidity is always preferred on applications like these:

(a). Very slow, and accurate, movement of the output piston, especially for feeding cutting tools into the work. Compressed air is better suited for quick movements but is entirely unsuitable for smooth feed at a slow rate. Hydraulics must be used for these applications.

(b). Moving sliding mechanisms where most of the load consists of friction. For example, moving a mechanism on a machine slide which has a large area of metal-to-metal contact. Friction will cause a highly erratic movement of the output piston when air power is used. Chatter (erratic movement) occurs with compressed air because the sliding load, after starting against high static friction, jumps ahead of the air pressure, then stops until the air pressure builds up again behind the piston.

CONVERSION OF FLUID PRESSURE INTO MECHANICAL FORCE

Since a fluid has no shape of its own, it takes the shape of the vessel which contains it. The only way to measure the intensity of pressure within the fluid is to measure the force which it can exert on a unit area. A pressure gauge is the usual means of measuring the force capability within the fluid. In the English system of units, gauges are calibrated in PSI (pounds per square inch). Other units could be used but the PSI was adopted many years ago and has proved to be a good means of measurment. If, for example, a gauge reads 500 PSI, this means the fluid is capable of exerting a force of 500 pounds on every square inch of surface exposed to it. This force, according to Pascal's Law, is not only against the output piston, but against all internal surfaces in the hydraulic pump, control valves, plumbing, and wall surfaces of the output piston. All components must be designed with sufficient physical strength to withstand this internal pressure which is trying to blow them apart.

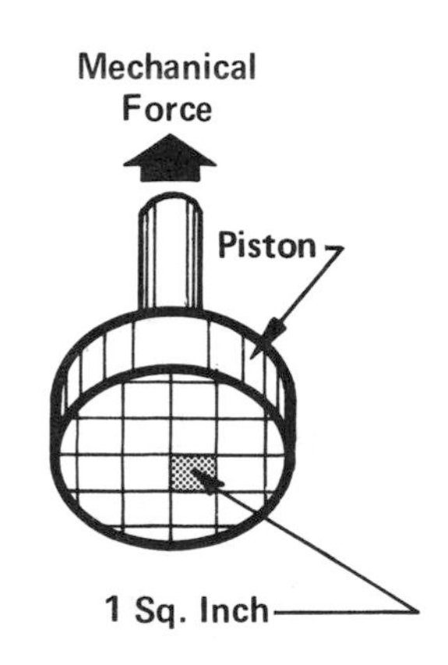

FIGURE 1-14. Gauge pressure works against each square inch of piston surface.

Figure 1-14. The fluid works with gauge pressure intensity against every square inch of surface on the output piston. Therefore, to calculate the overall mechanical force delivered by the output piston, the piston area, in square inches, first must be known or calculated. Then, the piston area can be multiplied times the gauge reading of the fluid. Large diameter pistons working at the same gauge pressure produce more force than smaller diameter pistons by the ratio, not of their diameter, but their surface area.

Square inch surface of a round piston can be found from tables in this book or elsewhere, or it can easily be calculated from the basic formula: $A = \pi \times R^2$ in which A is the area in square inches, and R is the radius in inches. The mathematical symbol π is always 3.14.

Output pistons, or cylinders, are presented in Chapter 2. Tables on Pages 260, 262, 282, and 283 show piston surface areas of standard size cylinders and the force which can be produced at various pressures.

Figure 1-15. One concept of a pump which will be considered in more detail in Chapter 5 is that its primary function is to produce a flow of oil. It produces pressure only as a secondary function and only when a load or restriction is placed against its flow. Then it produces only sufficient pressure to overcome the resistance so it can produce its rated flow.

In this example a load of 9600 pounds is to be lifted. Since the cylinder piston has 12 square inches of surface area, the pump must build up to 800 PSI on every square inch to develop a total of 9600 pounds to equal the load resistance:

9600 ÷ 12 = 800 PSI

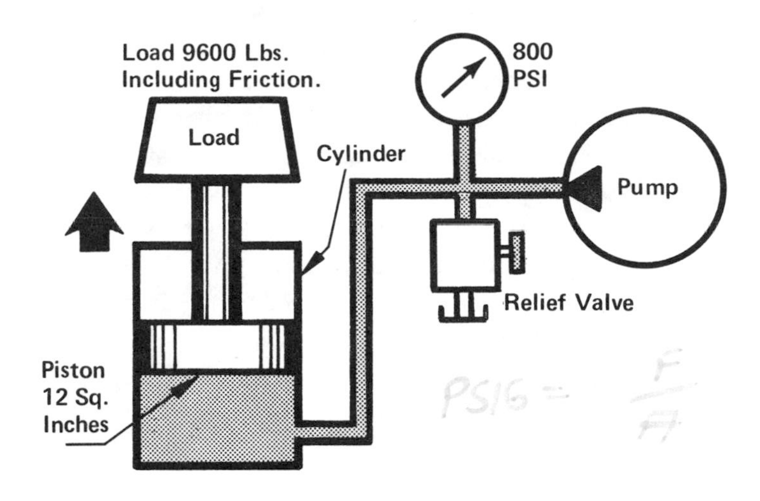

FIGURE 1-15. The load weight causes a fluid pressure to be reflected back to the pump which is sufficient, in combination with the piston area, to support the load.

In a practical situation the pump will also have to build up some additional pressure to force the oil through connecting lines. The amount of additional pressure required depends entirely on the flow resistance of the piping.

CONVERSION OF MECHANICAL FORCE INTO FLUID PRESSURE

Figure 1-16. In any fluid system the fluid force on one side of an output piston is equal to the mechanical opposition force. This was true in the previous figure. It also works in reverse as shown in this figure. The pressure generated in the fluid by a load weight on the output piston is equal to but no more than equivalent to the load weight.

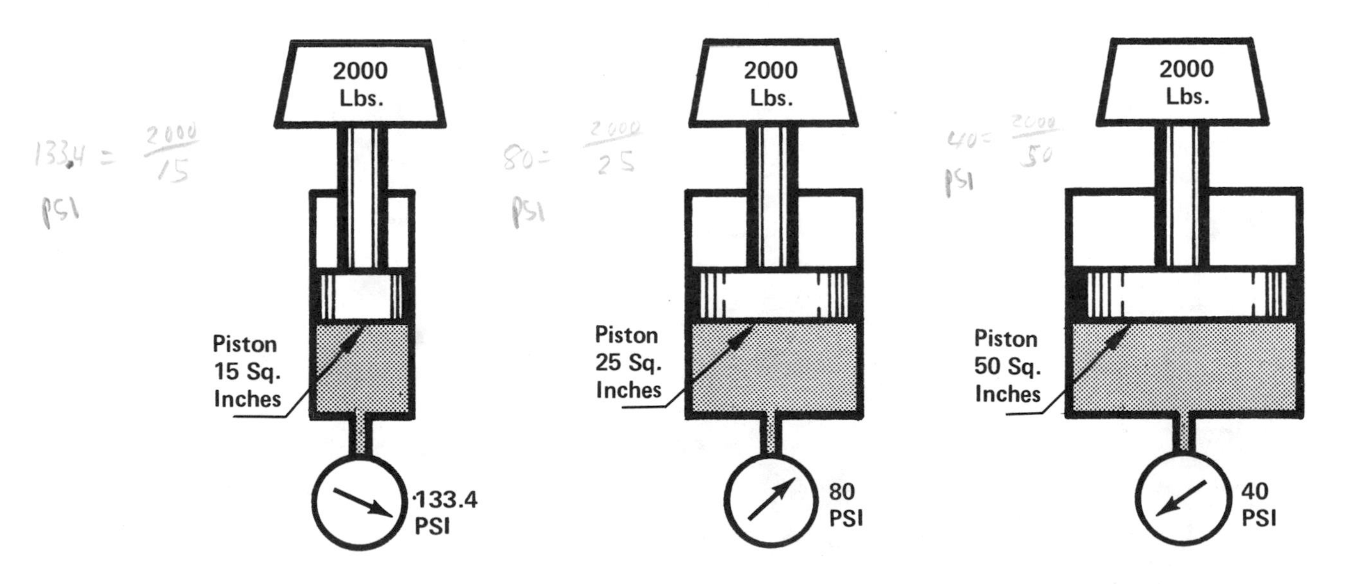

FIGURE 1-16. These figures illustrate the concept of equivalent fluid and mechanical forces. The greater the square inch surface of the fluid, the less internal pressure will be developed.

To illustrate the concept of equivalent mechanical and fluid forces, Figure 1-16 shows three pistons of different diameters supporting the same load weight. These pistons are not connected to a hydraulic pump; they are deadended into a pressure gauge to measure the internal pressure of the fluid.

The smallest piston has an area of 15 square inches. Therefore, the 2000 pounds of weight will be distributed equally on each square inch of piston surface, with each square inch supporting 133.4 pounds (2000 ÷ 15 = 133.4). This can be verified with a pressure gauge. The medium diameter piston has an area of 25 square inches to support the same 2000 pounds. Therefore, each square inch must support only 80 pounds and this also can be verified with a pressure gauge. The largest piston, with an area of 50 square inches, will be subject to a fluid pressure of 40 PSI.

The greater the piston area, the lower the hydraulic pressure needed to balance a given load and the less the pressure that must be produced by a hydraulic pump. This principle is used in designing a hydraulic system – to select the size of the cylinder so the load can be balanced at a selected pressure produced by the pump.

HYDRAULIC LEVERAGE COMPARED TO MECHANICAL LEVERAGE

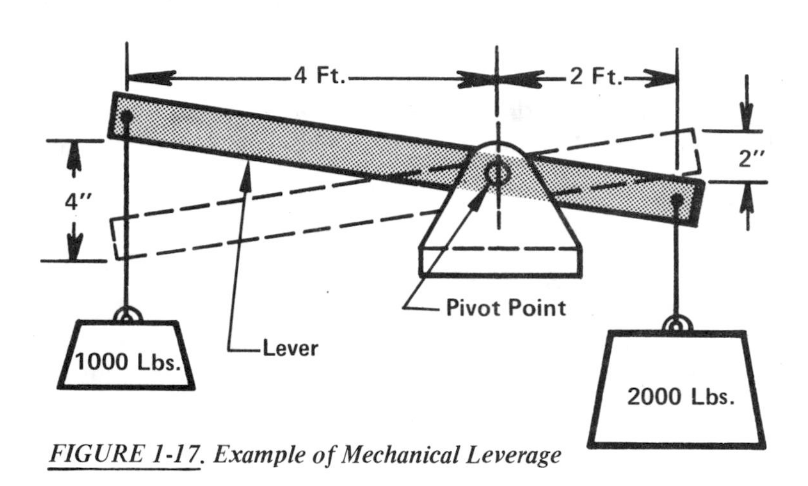

FIGURE 1-17. Example of Mechanical Leverage

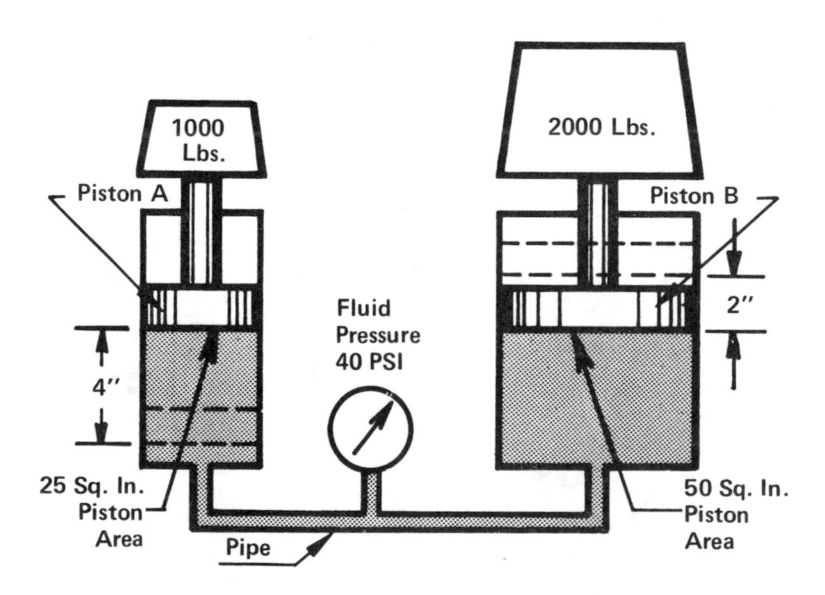

FIGURE 1-18. Example of Hydraulic Leverage

Hydraulic leverage or the multiplication of force by hydraulics can be illustrated by comparing a mechanical system, Figure 1-17, with which everyone is familiar, with an equivalent hydraulic system, Figure 1-18.

Figure 1-17. This mechanical system is in balance when the "moment" is equal around both sides of a pivot point. That is, when the load weight multiplied times the distance from the pivot point is equal for both sides. If a small additional weight should then be placed on either side, the system will overbalance in that direction. In this example, 1000 pounds times 4 feet, or 4000 foot-pounds on the left side is equal to 2000 pounds times 2 feet on the right side. Of course the greater the weights the greater the stress in the lever, and when sufficiently overloaded it will break.

Figure 1-18. This equivalent hydraulic system is also in balance, with a weight of 1000 lbs. on a piston area of 25 square inches producing the same effect as the 2000 lbs. on the right side working on a 50 square inch piston. If a small additional weight were to be added to either side, the piston on that side would move downward.

The pressure in the fluid is calculated as shown for Figure 1-16, and in this case amounts to 40 PSI. The heavier the weights, the greater the fluid pressure, and if it should become more than the pistons or connecting pipe can contain, a rupture would occur.

Speed and Distance Traveled. Multiplication of force in a hydraulic leverage system, Figure 1-18, is achieved through a sacrifice of speed and distance traveled. In Figure 1-18, a movement of 4 inches on the small piston will produce a movement of only 2 inches on the larger piston. Also, since both pistons travel for the same period of time, and since the large one goes only half as far, its rate of travel, or speed, will be only one-half that of the smaller piston. This distance and speed concept is the same for the mechanical lever system of Figure 1-17.

A jacking ram with hand pump built into the same housing is a good illustration of multiplication of force. But the operator may have to stroke the pump many times through a total handle travel of many inches to raise the ram 1 inch. See Pages 188 and 189 for more information on hand pumps.

STATIC HEAD PRESSURE

Pressure is developed in a liquid or gas by the weight of fluid above the point of measurement. For example, an atmospheric pressure of 14.7 PSIA (absolute) at sea level is caused by the weight of the atmosphere above. Likewise, pressure below the surface of a lake or ocean increases with depth in exact proportion to the weight of water above the measuring point. In some industrial hydraulic systems this static head pressure may be important. If the oil must be pumped to a significantly higher elevation, the hydraulic pump must produce, not only the pressure needed for the work, but sufficient extra pressure to raise the oil to the higher elevation. Water, with a specific gravity of 1.000, weighs 62.4 pounds per cubic foot (0.0361 pounds per cubic inch). Therefore, every inch of elevation represents a static head pressure of 0.0361 PSI. Hydraulic oil has a specific gravity of 0.9. Therefore, the rule-of-thumb for oil static head pressure is $0.9 \times 0.0361 = 0.0325$ PSI per inch of elevation, or approximately 0.4 PSI for every foot of elevation. When this oil eventually returns to the lower elevation, the head pressure is recovered. Sometimes it can produce useful work, or in other cases it can be detrimental depending on the particular application.

When pumping oil to a lower elevation, head pressure is produced which adds to the pump pressure. Hydraulic pistons operating at a lower elevation than the pump may become over-pressured.

When pumping oil to a higher or lower elevation, static head pressure is not affected by pipe diameter or actual volume of oil in the pipe. It depends solely on difference in elevation as we will show next.

Figure 1-19. In this figure we will show that tank diameter or volume of liquid in the tank makes no difference in head pressure. Two square tanks, one 24" x 24", the other 1" x 1" are filled with water to a height of 25 feet.The same gauge reading of 10.8 PSI shows at the bottom of both tanks even though one contains a great deal more water than the other. The pressure gauge does not know how large the tank size may be. It is calibrated in PSI and is only interested in measuring the weight of a column of water of 1" x 1" in cross section (1 square inch).

The anticipated gauge reading can be figured as follows: every cubic inch of water weighs 0.0361 pounds. So, a 25-foot head of 1 sq. inch area weighs $0.0361 \times 12 \times 25 = 10.83$ lbs. or 10.83 PSI gauge.

In the larger tank there is a greater volume of water but there is also a greater number of square inches on the tank bottom to support it, so the unit pressure or PSI against each square inch of bottom surface remains the same no matter how large the tank may be.

25 X 12 X .0361 = 10.83 PSI

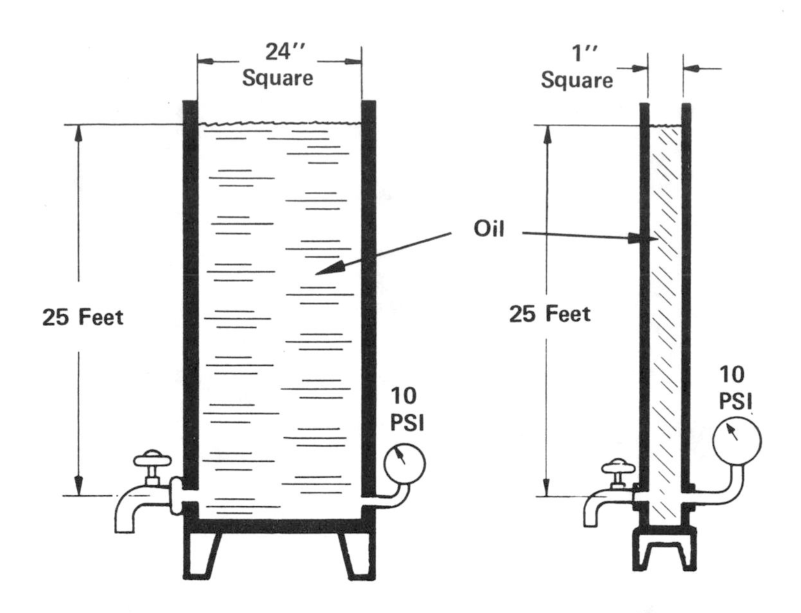

FIGURE 1-19. Head pressure is a function only of liquid height regardless of the size of the tank or the volume it will hold.

Head pressure is also important as related to the inlet or suction port of a hydraulic pump. Although some pumps are capable of pulling a moderate vacuum, most pumps should not be required to operate with more than 2 to 3 PSI (4 to 6 inches of mercury) vacuum. High vacuum on a pump inlet causes excessive wear due to cavitation. This means they should be located at an elevation of not more than 6 feet above the surface of the hydraulic reservoir, and actually this height should be limited to 3 feet because other restrictions in the inlet line will add to the vacuum. Pump suction is considered in greater detail in Chapter 5.

FLUID FLOW THROUGH PIPES, TUBING, AND HOSE

FIGURE 1-20. Examples of smooth laminar flow (above) and turbulent flow (below) caused by excessive flow velocity in the fluid.

Figure 1-20. Fluid Flow. If fluid velocity through piping can be kept moderate by using large pipe with a minimum number of bends and restrictions, the flow will be smooth, in a "laminar" condition. It will flow in a "telescopic" pattern, as a movement of one layer on another, with inner layers traveling faster than outer layers, and with a minimum of cross currents to create turbulence. Pressure and power loss will be less if the fluid flows smoothly rather than in a turbulent condition.

If piping is made too small for the flow rate, turbulence will be set up, especially as the fluid goes around bends. Recommended pipe sizes are given later in this chapter, for both air and hydraulic systems, to minimize power loss due not only to turbulence but also to flow friction.

FIGURE 1-21. Any restrictions or irregularities in the flow path may cause the fluid to temporarily go into a turbulent flow condition.

Figure 1-21. Restrictions. Any restriction placed in a fluid stream which normally is smooth flowing (laminar), may cause turbulence for a short distance, with the fluid then returning to its smooth flowing state. These restrictions are undesirable but sometimes unavoidable, as each one adds to the overall power loss through the system. All unnecessary bends and fittings should be eliminated.

Figure 1-22. Fittings. Every fitting adds its share to the overall power loss in the system. A certain amount of energy is lost every time the fluid changes direction. It must decelerate to zero velocity in the direction it was going, then accelerate back up to speed in the new direction. The kinetic energy lost in changing direction escapes from the system in the form of heat.

While a few bends in the piping are necessary, any unnecessary bends should be eliminated. On a 90° turn, the use of large-radius bends in steel tubing instead of the sharp bends in elbows will help reduce power loss by minimizing turbulence.

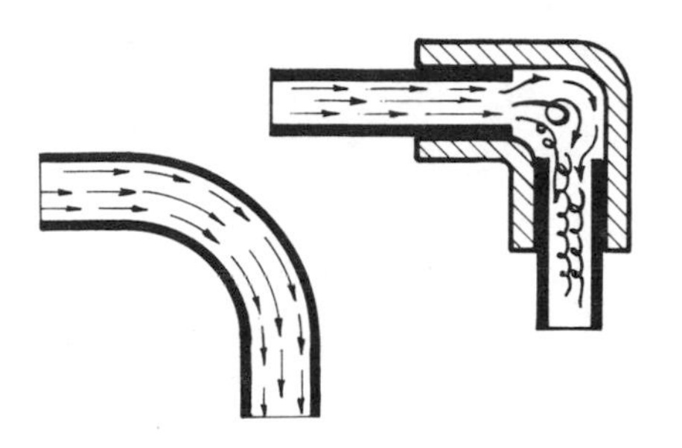

FIGURE 1-22. The fluid has a better chance of remaining in laminar flow if bends are kept smooth and gradual.

Figure 1-23. Hose Fittings. A restriction to flow is introduced wherever a hose fitting is attached to a length of hose. The hose barb must fit inside the hose and this reduces the inside diameter for a short distance. Unlike rubber tubing, fluid power hose does not stretch to any degree to go over a fitting. While the restriction through one hose fitting may be insignificant, if a large number of hoses are used on a machine, the total restriction may be important. In an air system the output piston may move more slowly than it could. While speed is not affected in a hydraulic system, an additional power loss will be present and a small amount of heat will be added in the oil.

One answer to hose fitting restrictions is to use one size larger hose than the size of the porthole in the component, then reduce size, if necessary, at the porthole. For example, when connecting to a 1/4" NPT porthole, use 3/8" I.D. hose (which will usually have a 3/8" pipe thread on the fitting). Then, use a 3/8 to 1/4" reducing bushing at the porthole.

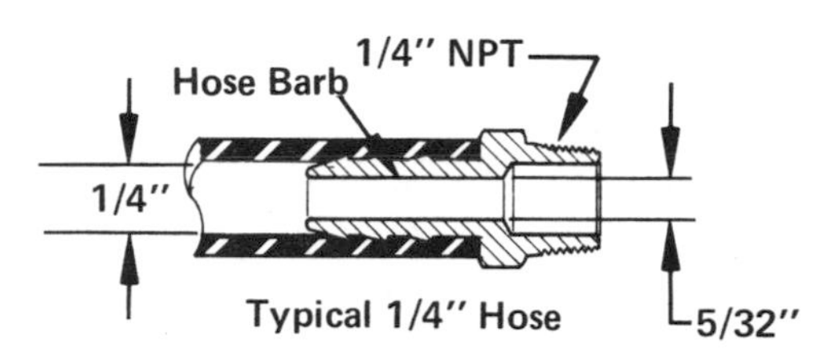

FIGURE 1-23(A). A fitting on the end of a hose always reduces the flow diameter.

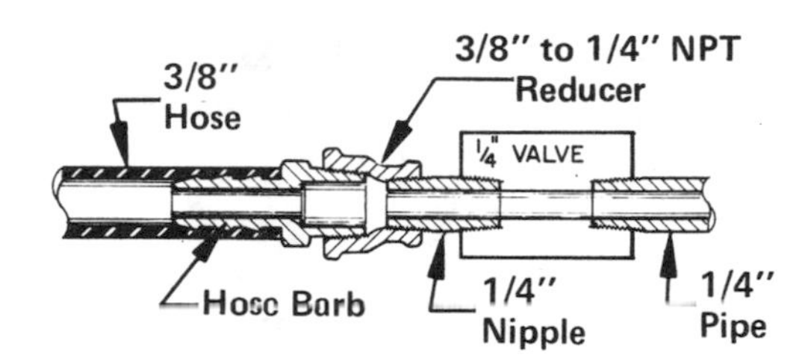

FIGURE 1-23(B). To reduce the restriction at a hose fitting, use a larger size hose and bush down at the porthole.

PRESSURE LOSS THROUGH PIPING

The loss of a part of the compressor or pump pressure as fluid flows through piping and valving represents a power loss of the same percentage. A small loss of pressure is unavoidable, but in a well-designed system a small loss of pressure is permissible to offset the higher cost of larger components and piping to reduce the pressure loss to a lower value. System design is then, a balance between the high cost of large components to reduce the pressure loss, weighed against the higher pressure loss which would be present if smaller valving and piping were used to save money.

A very important point to remember about pressure loss is that it only appears when fluid starts to flow. As long as the fluid is static (non-moving) there is no pressure or power loss even though the fluid pressure may be very high. Pure force, without movement of fluid, can be transmitted through even a long column of fluid with virtually no pressure or power loss. Refer to the statement of Pascal's Law, Page 15. This is analogous to an electrical circuit where there is no loss of voltage until current starts to flow. The three diagrams to follow illustrate the principle that pressure drop in a fluid system is directly related to flow.

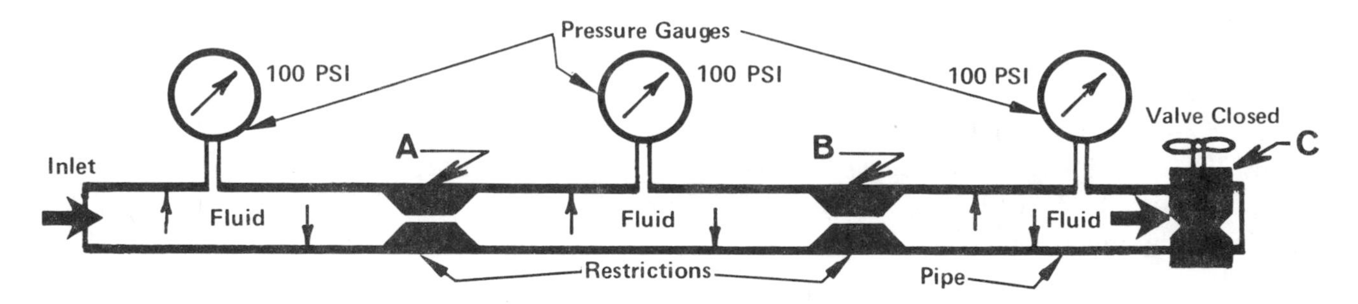

FIGURE 1-24. When fluid is not moving through the pipe, pressure (only) can be transmitted without loss through the pipe no matter how long or restricted it may be. Note that all gauges show the same pressure.

Figure 1-24. A section of a long pipe is shown. The left end of this pipe is exposed to a source of pressure of 100 PSI gauge (air or oil). The far end of the pipe is closed by Valve C. Therefore, the fluid in the pipe is under pressure but is not flowing. This represents a confined body of fluid which obeys Pascal's Law. The pipe may be very long and may be very small. It may have many restrictions such as components with small passageways, or fittings such as elbows and tees. After Valve C is closed, almost instantly in a hydraulic system, or after a brief period if the fluid is air, the pressure will equalize over the entire length of pipe, and will be equal to the inlet pressure of 100 PSI. Notice the equal pressure readings on the three gauges.

Closure of Valve C at the end of the pipe illustrates what happens on an output piston if connected to the end of the pipe in place of Valve C. If the output piston should stall against an immovable load, fluid pressure in the pipe would very quickly rise to the full inlet pressure because the confined fluid in the pipe acts in accordance with Pascal's Law. Full pressure would appear on the output piston against the stalled load regardless of the number and degree of all restrictions in the pipe.

Pressure Loss Only Appears When Fluid Starts To Flow

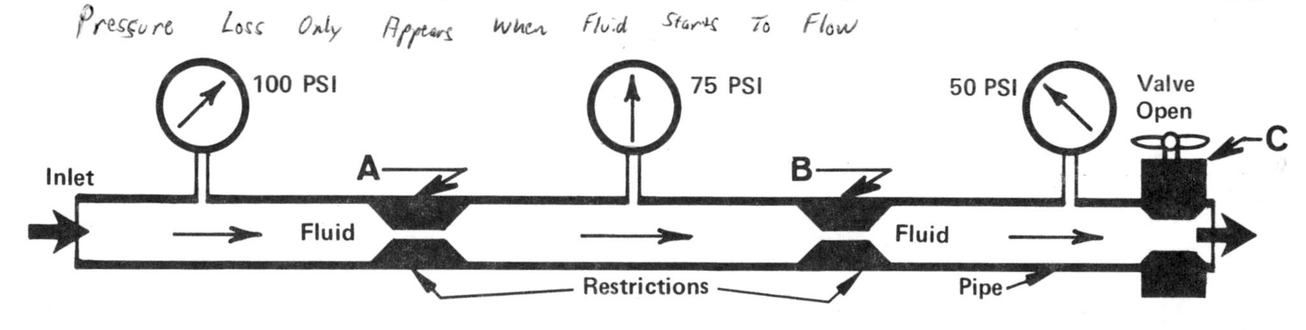

FIGURE 1-25. When Valve C is opened and fluid starts to flow through the pipe, all restrictions in the pipe become significant. Note the progressively diminishing pressure readings down the pipe.

Figure 1-25. When an operator opens Valve C, fluid starts moving through the pipe. The system no longer qualifies as a "confined body" under Pascal's Law. Now, pipe diameter and restrictions in the pipe become very significant. A large part of the inlet 100 PSI may be used up in pushing the flow of fluid through the various restrictions. Full inlet pressure is no longer available on the outlet end of the pipe. Since the restrictions consume a part of the inlet pressure they also consume the same proportion of fluid horsepower which is being transported through the pipe. *Note: The pressure gauge readings are assumed only for illustration. In an actual system these gauge readings would vary according to the amount of pipe restrictions.*

The pressure loss illustrated in this figure shows up in the operation of a piston if it is connected to the outlet end of the pipe. The greater the pressure losses through the pipe, the less the pressure that is available for moving the piston against a load. While the piston is running in free travel (with no load), restrictions in the pipe may not be particularly important, but the force which a piston can produce to move a load is directly reduced by the amount of pressure lost in the pipe.

Now, refer back to Figure 1-24. When Valve C is again closed, full pressure will be restored through the entire length of the pipe as fluid ceases to flow.

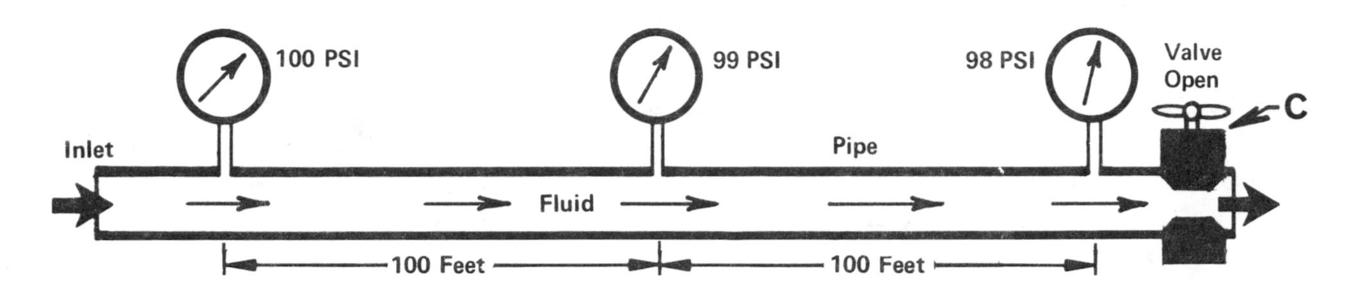

FIGURE 1-26. For transmitting horsepower through a pipe, a pipe size should be selected which will keep pressure losses to an acceptable minimum.

Figure 1-26. When major restrictions are removed and the pipe is made sufficiently large to efficiently carry the flow, a relatively small proportion of inlet pressure will be sacrificed to flow losses. For conducting compressed air, the pipe size is selected to limit flow losses to about 1 PSI for every 100 feet of pipe. A guide for pipe size is on Page 33. For conducting hydraulic oil, pipe size is selected to keep flow velocity within certain limits. More information on Pages 37 and 238.

PRESSURE DROPS IN AUTOMATION CIRCUITS

Pressure losses are usually undesirable while horsepower is being transported through a pipe, but there are some cases where pressure losses can actually be used to advantage. These are cases in which the air piston is free running; during its free running stroke, pressure losses are unimportant. But they can be used to generate a sharp pressure rise when the piston reaches the end of its stroke. One application is shown with the following series of diagrams.

Figure 1-27. This diagram shows a pneumatic vise using an air piston (single-acting cylinder). This piston is normally held in a retracted position by an internal spring. It is powered by an air compressor which can deliver 100 PSI gauge pressure. There is also a control valve (3-way) for the operator by which he can extend or retract the air piston. In the line connecting air compressor to air piston there may be restrictions to air flow, or the connecting line may simply be very small. Pressure gauges installed along the line will show pressures in the line as the air piston goes through its cycle. There is

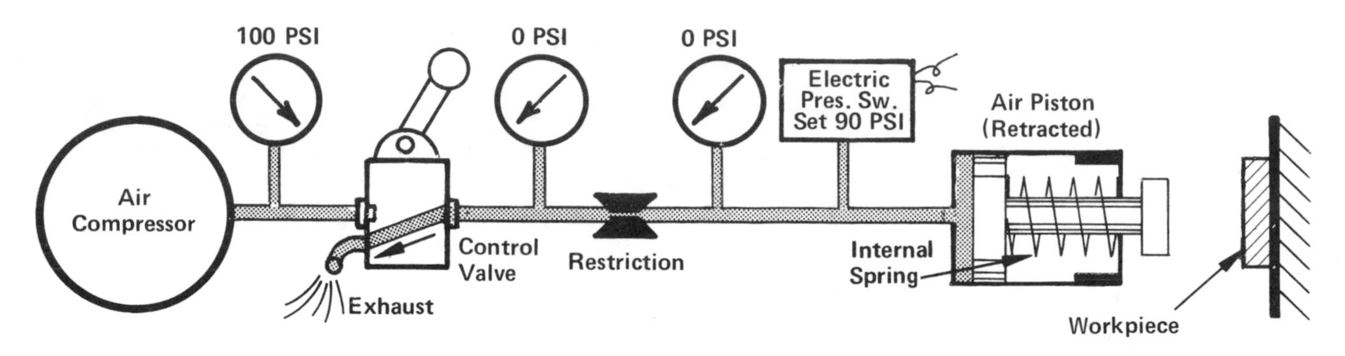

FIGURE 1-27. Starting position. Control valve shifted to retract position. Air piston has been retracted by an internal spring. Pressure switch is set to trip at 90 PSI. Pressure gauges in the line read zero pressure.

an adjustable pressure switch installed in the line directly at the piston port. This switch should be adjusted to trip at about 90% of system inlet pressure, in this case it would be adjusted to trip at 90 PSI. When the switch trips, its electric contacts close. This switching circuit could be used to initiate almost any kind of electrical action when air pressure at the piston port has built up to 90 PSI or higher. It could be used to light a lamp bulb. More often it would be used to start a secondary operation such as starting the electric motor on a drill press, or energizing the solenoid of an air valve.

The important thing to remember about this circuit is its priority action. The air piston is the primary action. The secondary action, started by closure of the pressure switch, cannot start until the primary action has been completed.

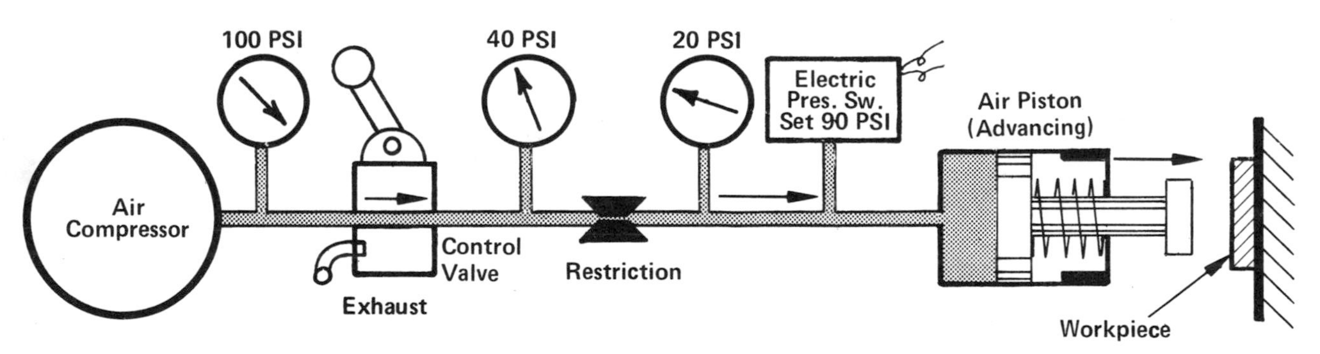

FIGURE 1-28. Vise closing position. Control valve shifted to work position. Air piston advancing toward workpiece. Pressure too low to trip pressure switch. Gauges show progressive loss of pressure down the line.

Figure 1-28. The operator has shifted the control valve to start the air piston forward, and the pressure gauge readings show the pressure at several points in the circuit while the air piston is moving forward and before it reaches the workpiece. There is a progressive loss of pressure down the line caused by various flow resistances in the line. The gauge nearest the piston shows 20 PSI. In this particular case no more than 20 PSI is required to move the piston against its internal friction and to compress its internal spring. The remainder of the pressure is lost in line and valve flow resistance.

A very important point which is illustrated by this diagram is that when a piston (either air or hydraulic) is free to move, pressure behind it can only build up high enough to make up for internal friction, plus enough to compress the spring, plus enough to overcome a load (if any) against the piston. Pressure cannot exceed this level because the piston would simply move faster to keep ahead of the incoming flow.

In Figure 1-28, the 20 PSI which is enough to keep the piston moving, is not high enough to trip the pressure switch which has been set to trip at 90 PSI. *Note: 20 PSI has been used as an example in this case. Other pistons working at other conditions might require more or less than 20 PSI.*

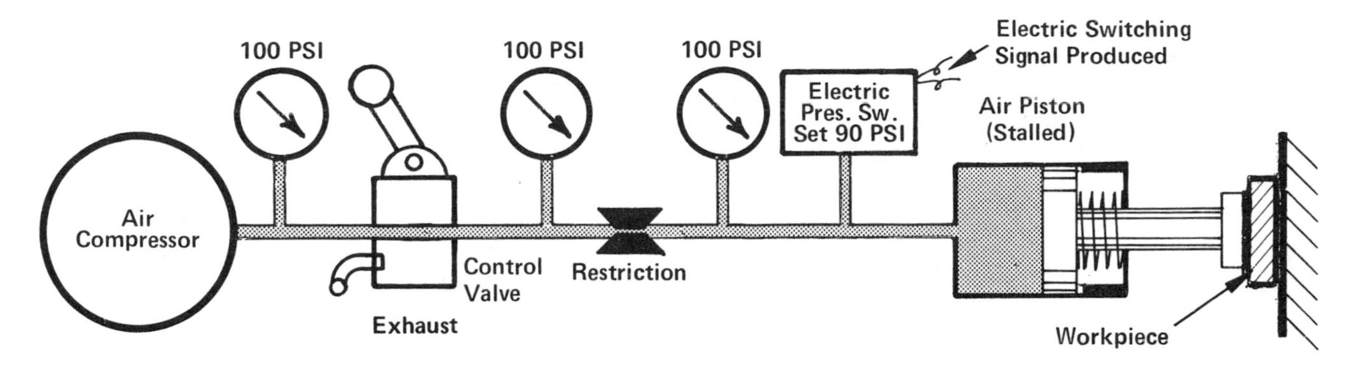

FIGURE 1-29. Air piston has stalled against workpiece. Air ceases to flow in the line and pressure rises to full inlet pressure down the line. Pressure switch has tripped, producing an electrical switching signal.

Figure 1-29. In this diagram the air piston has contacted the workpiece and has stalled against it. When the piston is no longer free to move, air flow which continues to enter the piston will cause pressure to rise behind the piston. After a brief time period the entire inlet pressure of 100 PSI will appear at the piston as air flow ceases (Pascal's Law). During the rise in pressure, the pressure switch will trip when a level of 90 PSI has been reached. The contacts of the pressure switch can be wired to start a secondary action such as lighting a lamp, starting an electric motor, or energizing a solenoid valve.

Remember again that in a circuit of this type, the flow resistances in the line are beneficial in giving the priority action of delaying the start of the secondary action until the primary action has been virtually completed. *Note: If the piston must work against full load resistance during its forward travel, this circuit should not be used. It is appropriate only for pistons that move in free travel (or at less than full load), finally reaching a stall at the end of their stroke.*

PLUMBING COMPRESSED AIR SYSTEMS

Figure 1-30. Water may condense out of any compressed air system, especially at certain times, and is harmful to most processes using compressed air. In a well designed distribution system a great deal of the water can be expelled before it gets into air operated equipment.

The preferred layout of distribution plumbing is to make a complete loop around the plant, returning to and joining the starting point of the loop. If this is not practical, one long header, running the length of the plant, and deadended at the far end of the plant may be used. Then observe these additional rules for good design:

(1). Use a large air storage (receiver) tank with the compressor. Heat of compression is radiated from the walls of the tank and as the air cools, most of the entrained water will condense and collect at the bottom of the tank. It can be expelled daily by manual or automatic draining.

At the far end of a distribution system where pressure at times may be low, additional receiver tanks may be installed to maintain the pressure.

(2). The air distribution piping should be slightly sloped downward in the direction of air flow, about 1°. Gravity plus air flow will carry condensed water along until it reaches a water drop leg.

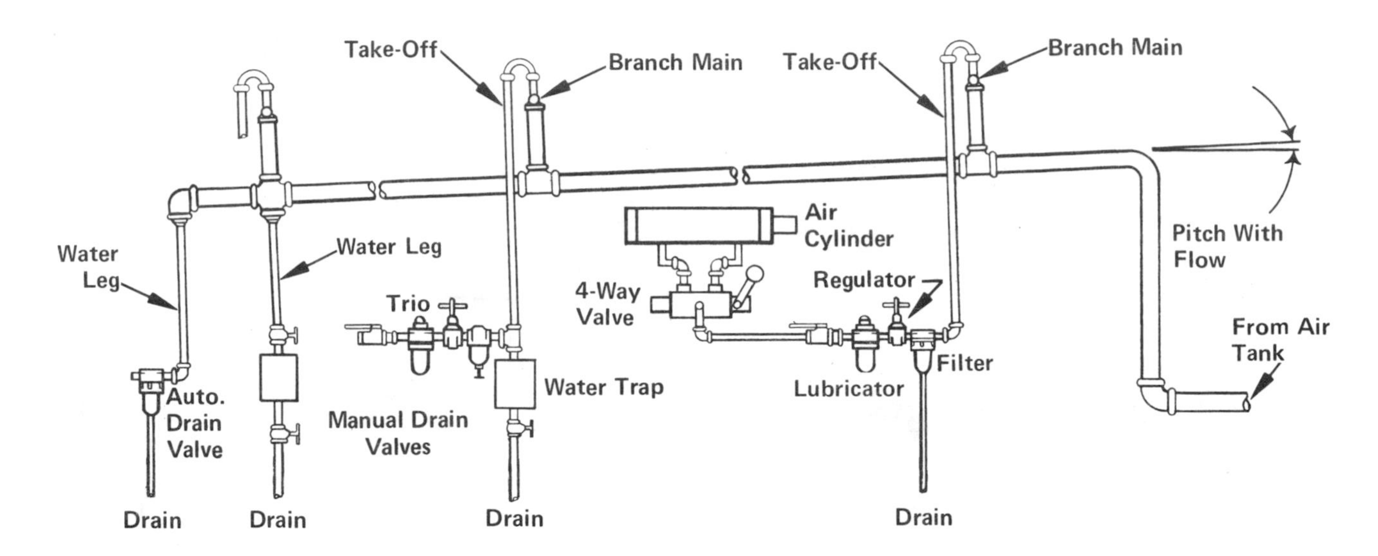

FIGURE 1-30. A great deal of the water which normally condenses in an air distribution system can be expelled by good design of the distribution system. Compressor aftercoolers and air dryers may also be used.

(3). Distribution plumbing should be run overhead, at an elevation higher than any machine which it supplies. If run at floor elevation it will be difficult to expel condensed water.

(4). Each take-off to an air operated machine should be taken off upward from the distribution plumbing before being taken down to the machine. This keeps water which may be flowing in the line from following down into a machine.

(5). A water drop leg should be installed about every 50 to 100 feet along the distribution header. Connection should be made to the under side of the header so water will follow downward and can be expelled by manual or automatic draining.

(6). Every machine using air from the main air supply should be preceded by a "trio" unit consisting of a filter-regulator-lubricator unit. An individual trio should be used for each machine; two or more machines should not use the same trio. Trio units are described in Chapter 6.

(7). Water cooled compressor aftercoolers help to expel water from air lines, and should be used on all systems of 50 HP and larger.

Attention to these design pointers will probably remove enough water for non-critical uses. But on branch circuits where air must be absolutely dry for critical applications like spray painting, food processing, medical or dental use, instrumentation, or fluidics, an air dryer must be used. These are described in Chapter 6.

SELECTING PIPE SIZE FOR COMPRESSED AIR FLOW

Standard weight (Schedule 40) black pipe is economical and quite satisfactory for plumbing the compressed air distribution system in an industrial plant. For specifications see Page 263. Heavier pipe including Schedules 80 and 160 need not be used; its higher pressure rating is not needed on air pressures of 175 PSI and less, and it cannot carry as high a flow because of its smaller inside area. Black pipe is usually preferred to galvanized to avoid the possibility of zinc flakes getting into the air stream. These flakes could be destructive to air and hydraulic components.

Suggested Pipe Size for Compressed Air Flow at 100 PSI

SCFM Air Flow	Length of Run, Feet 25	50	75	100	150	200	300	500	1000	Compr. HP
4	1/2	1/2	1/2	1/2	1/2	1/2	1/2	3/4	3/4	1
12	1/2	1/2	1/2	3/4	3/4	3/4	3/4	1	1	3
20	3/4	3/4	3/4	3/4	1	1	1	1¼	1¼	5
30	3/4	3/4	1	1	1	1	1¼	1¼	1¼	7½
40	3/4	1	1	1	1¼	1¼	1¼	1½	1½	10
60	1	1	1¼	1¼	1¼	1¼	1½	1½	2	15
80	1	1¼	1¼	1¼	1½	1½	1½	2	2	20
100	1¼	1¼	1¼	1½	1½	2	2	2	2½	25
120	1¼	1½	1½	1½	2	2	2	2½	2½	30
160	1¼	1½	1½	2	2	2	2½	2½	3	40
200	1½	2	2	2	2	2½	2½	3	3	50
240	1½	2	2	2	2½	2½	2½	3	3	60
300	2	2	2	2½	2½	3	3	3	3½	75
400	2	2½	2½	2½	3	3	3	3½	4	100
500	2	2½	2½	3	3	3	3½	3½	4	125

On a compressed air distribution system, pressure losses greater than 3% are considered excessive, and a well-designed system having a steady rate of air flow is usually designed for not more than a 1% loss or 1 PSI for a 100 PSI system. The pipe size depends not only on the volume of air flow but how far it must be carried. To hold the distribution loss to 1 PSI, pipes of larger diameter must be used on longer runs to carry the same flow that can be handled by smaller pipes on shorter runs.

Figures in the body of this chart are pipe sizes recommended on a 100 PSI system to carry air with less than 1 PSI loss. When measuring lengths of runs, add 5 feet of length for each pipe fitting. If carrying 120 PSI pressure these sizes will carry slightly ***more*** air than shown, or pressure loss will be slightly less than 1 PSI. If carrying 80 PSI pressure these pipe sizes will carry slightly ***less*** air at 1 PSI pressure loss than shown in the chart.

The left column of the chart shows the volume of air to be carried. It is difficult to estimate the air flow volume to be carried in each leg of the distribution system. This varies with the application. On some applications, like in a large plant with many legs in the distribution system serving dozens of air-operated machines, the air usage may be at a fairly steady rate. Other applications, usually on small systems, may have to carry a high surge of air if several machines happen to be operated at the same time. Then there may be a period with almost no flow.

To make a realistic estimate of air flow volume, the far right column of the chart showing compressor HP may be used. On steady pumping, a compressor will produce a minimum of 4 SCFM air flow for each 1 HP of capacity. This is a conservative figure, as most compressors will produce 5 or 6 SCFM.

For example, a 25 HP compressor will produce at least 100 SCFM of air. This is shown in the far left column on the same line as 25 HP.

PLUMBING COMPRESSED AIR SYSTEMS

Plumbing Materials. For compressed air systems, recommended plumbing materials include iron pipe, copper tubing or pipe, plastic tubing or pipe, and rubber hose. Carbon steel tubing is not recommended because of the possible presence of water in the air lines. Brass and stainless steel tubing are acceptable but are seldom used. All plumbing materials should be pressure rated to the maximum working pressure of the system and with a safety factor of 6. For example, on a 120 PSI system, the plumbing material should have a burst rating of 720 PSI or higher.

Pipe. Black iron pipe is preferred from the standpoint of permanence and cost. Galvanized pipe may be used but there is some danger of getting loose zinc flakes into the system.

Pipe is specified by its ***nominal*** outside diameter, but its actual flow capacity is determined by its inside area. For example, Schedules 40, 80, and 160 in the 1/2" size all have the same ***outside*** diameter and can be threaded by the same pipe die. The difference is in the inside diameter (and area). Schedule 40 pipe is standard and has the thinnest wall with more flow area than the other schedules, but of course has a lower pressure rating. For most compressed air systems, Schedule 40 pipe has a sufficient pressure rating; the larger schedules are seldom used.

Schedule 40 pipe up to 1" nominal size has an inside diameter and flow area larger than would be indicated by its nominal size rating. For example the inside diameter of 1/4" pipe is .364", nearly 3/8". Pipe sizes larger than 1" have about the same inside diameter as their size rating.

U. S. pipe threads are tapered and mate with corresponding tapered threads on many compressed air components. Sizes are specified (for example) as 1/2" NPTF (National Pipe Taper Fuel), also called "dryseal pipe threads", or NPT (National Pipe Taper). These thread forms are physically interchangeable, but NPTF has a slightly modified thread form to close up spiral leakage around the crest of the threads. The seal is achieved by actual distortion of the mating threads, and usually no pipe sealant is needed for compressed air. While NPTF is preferred, NPT threads can be used on the low pressures encountered in a compressed air system by using a pipe sealant. Threads must be cleanly cut.

Some manufacturers are replacing the taper pipe threads on components with straight threads which seal with an O-ring. When pipe is used, its ends are always threaded with a taper thread. To connect these taper threads into a straight thread port an adaptor fitting must be used. See Page 267 for more information on piping materials. See also pipe information in Appendix C.

Tubing. Carbon steel tubing is not recommended for air systems because of the water condensation which takes place in most air systems, causing rust. Copper tubing can be used but has a bad tendency to work harden under machine vibration or shock as the machine goes through its cycle. If used, it should be clamped down to prevent it from breaking off at a fitting after a short period of use. Flaring on copper tubing should be to the SAE flare angle of 45 degrees. For permanence, flare fittings are preferred to the ferrule or flareless type. Plastic tubing and pipe are acceptable but are not considered as permanent as metal tubing. Brass, aluminum, and stainless steel tubing are good but are not as readily available. They are harder to flare than copper tubing.

Hose. Rubber hose is widely used because of its convenience and the time and labor which can be saved, although it is not considered as permanent as metal pipe or tubing. Lengths of hose are usually purchased from a supply house with fittings attached to both ends. Usually the fittings have tapered pipe threads for screwing into component ports. Be sure to order hose with a swivel fitting on at least one end to avoid twisting the hose when connecting it.

Hose material should be synthetic rubber, buna-N being the most common rubber type. Natural rubber is not compatible with the petroleum base lubricating oil used in most air systems, and should not be used.

Hose is used primarily for short connections inside a machine. Iron pipe is nearly always used for long plumbing runs such as distribution systems and connections from the distribution system into each air-operated machine.

Dimensions of Schedule 40 (Standard Weight) Iron Pipe

Nominal Size	1/8	1/4	3/8	1/2	3/4	1	1¼	1½
I.D. (Inches)	*.269*	*.364*	*.493*	*.622*	*.824*	*1.049*	*1.380*	*1.610*
O.D. (Inches)	*.405*	*.540*	*.675*	*.840*	*1.050*	*1.320*	*1.660*	*1.900*
Circumference	*1-9/32*	*1-11/16*	*2-1/8*	*2-5/8*	*3-5/16*	*4-1/8*	*5-7/32*	*5-31/32*

The above table shows dimensions of Schedule 40 (standard) pipe. It may be used to identify the nominal size of an existing line or piece of pipe. Wrap a piece of paper, string, or wire around the pipe to measure its circumference. Measure the length of the string and compare with the bottom line of the chart. See Pages 264 through 271 for more information on pipe, tubing, and hose.

PLUMBING HYDRAULIC SYSTEMS

Plumbing Materials. For oil hydraulic systems, the most often used plumbing materials include steel tubing, steel pipe, or hose. Materials not recommended include copper and copper products, aluminum, and plastic.

We recommend a safety factor of at least 4 on the pressure rating of the plumbing materials, although on shockless systems lower safety factors are sometimes used at the discretion of the design engineer. The safe working pressure of plumbing materials is the rated burst pressure divided by the factor of safety selected for the system. Tables on Pages 264 through 271 give specifications on steel pipe, carbon steel tubing, and hose.

Tubing. Low carbon steel tubing is ordinarily used except on lines where a stronger material is required to handle the pressure. When ordering, "hydraulic grade" should be specified. This is a dead soft steel tubing which has been annealed after drawing to size. It is easy to bend and to flare. A harder grade of carbon steel tubing, or stainless steel tubing, can be used on lines where the pressure is too high for low carbon steel. When using these harder grades, flaring and bending must be done carefully to avoid splitting or cracking the tubing.

Tubing is measured and specified by its wall thickness and outside diameter (O.D.). Standard sizes and pressure ratings are shown on Page 264. A connector is used on each end of every piece of tubing. One piece of tubing is joined to another with "tube-to-tube" connectors. Tubing is connected into the ports of components with "tube-to-pipe" or "tube-to-straight-thread" connectors, whichever is appropriate. Connectors are attached to the ends of tubing either (1), by pre-flaring the end of the tubing and using flare-type connectors, (2), by using connectors which automatically flare the tubing during assembly, (3), by using "flareless" or "ferrule" type connectors which have a sleeve which bites into the outside wall of the tubing during assembly, (4), by brazing or welding the tubing into a "socket-type" connector which can then be bolted to or screwed into the port of a component.

For permanence, leaktightness, and lower flow resistance, Methods (1) and (2) are preferred over Method (3). Method (4) is used primarily on larger tubing sizes which are hard to bend and flare.

For hydraulic work, tubing is flared to the JIC (Joint Industry Council) angle of 37 degrees. The SAE flare angle of 45 degrees is used in automotive and refrigeration work but is not used on hydraulic plumbing. If fittings with different flare angles are mated, a leak is almost sure to result.

Copper and brass tubing are not recommended for oil hydraulic plumbing primarily because the tubing work hardens under machine vibration and hydraulic shock. Unless firmly supported the tubing will break off near a fitting. These materials are subject to possible chemical reactions with certain impurities in the oil. The rule is to keep copper products out of the hydraulic oil as much as possible. One notable exception is the use of bronze tubes in water cooled heat exchangers where maximum heat transfer is more important than possible chemical reactions. But aside from the above limitations, copper products do not have the tensile strength to carry high pressures.

Other materials including plastic and aluminum are seldom used because either they work harden or they do not have sufficient tensile strength for high pressure hydraulic systems.

Pipe. Refer to Page 34 for general description and to Pages 266 through 271 for ratings. Pipe has been used for many years and continues to be popular, although the modern trend is away from taper pipe threads to straight threads which seal with an O-ring. The objection to pipe plumbing is not to the pipe itself but to the taper threads which must be put on the ends of the pipe. These threads often develop leaks which are hard to repair in the field. Components built in foreign countries often have BSP (British Standard Pipe) ports which have a straight thread sealed with some kind of soft packing.

To minimize thread leaks, the NPTF (National Pipe Taper Fuel) thread form (also known as "Dryseal") has been developed to replace the standard NPT (National Pipe Taper) thread form used for so many years. Both of these thread forms are physically interchangeable but the thread shape of the NPTF has been modified to reduce spiral leakage around the crest of the threads. For hydraulic plumbing the NPTF thread form should always be used. Nearly all components which are built with pipe thread ports use the NPTF form. Threads are cut with sharp crests and the seal is achieved by the actual distortion of the mating threads. To further minimize thread leakage, a joint compound may be used. We recommend Teflon Sealant which comes in paste form. It will make a very tight joint, and if some of it should accidentally get inside the pipe it will dissolve in the oil and will be harmless to the system. We do not recommend Teflon tape, especially with NPTF threads. It over-lubricates the threads and may cause the joint to be over-torqued with damage to the threads, or distortion of the component into which the pipe is screwed. And sometimes part of the tape may be squeezed off inside the pipe causing contamination in the system. Good, clean, and sharp threads are important. Pipe nipples of various lengths can be purchased with both ends pre-threaded, and these usually have high quality threads. Threading done on the job should be with a pipe threading machine using the proper lubricant. Hand threading may produce ragged threads which are apt to leak.

When connecting pipe into a component which has a straight thread porthole, an adaptor must be used which has a straight thread on one end to fit the porthole, and an NPTF pipe thread on the other end. Union joints in pipe plumbing are made by screwing or welding a 4-bolt flange on each pipe end to be joined. One flange has an O-ring recessed in a groove, the other flange has a flat surface for the O-ring to seal against. The two flanges are bolted together. On all pipe plumbing, steel fittings should be used rather than the cast iron fittings used for low pressure plumbing.

Flow losses tend to be higher in a system plumbed with pipe compared with one plumbed with steel tubing. For one thing, the inside surface of a pipe may not be as smooth as the inside of tubing, and may offer higher flow resistance. Also, the turbulence (and power loss) at each sharp right angle bend will be higher than in a piece of tubing in which the bend is made without fittings, using a long radius.

Standard (Schedule 40) pipe has a pressure rating high enough for many systems, especially in sizes less than 1 inch. Schedules 80 and 160 are available with heavier wall for higher pressures, but the flow capacity is less than for Schedule 40. On those applications where the pressure and flow specifications of these schedules will not meet job requirements, medium carbon steel pipe can be used. It can be assembled either by threading the ends and using steel fittings, or by socket welding it to fittings terminating in NPTF pipe threads, straight threads, split flange ends, or 4-bolt flanges.

Hose. Although more expensive than pipe or tubing, hose is used extensively in these situations: (1), to simplify connections between components mounted at odd angles to each other, (2), in making plumbing runs in confined compartments where rigid plumbing would be difficult to install, (3), to connect hinge mounted pistons (cylinders) which must swing during their stroke, (4), to minimize hydraulic shock such as may occur when a high pressure piston on a press is suddenly released, (5), to reduce noise level such as may be generated by a hydraulic pump.

Hose is measured and specified by its inside diameter. Its outside diameter will vary according to the number of layers of wire braid and rubber which must be used to obtain the pressure rating. We recommend using hose with a safety factor of 4, although on relatively shockless systems a lower safety factor is sometimes used at the discretion of the design engineer. Working pressure is calculated by using the burst pressure rating and dividing by the safety factor selected for the system.

A metal fitting must be used on each end of a hose to connect it into the system. For industrial hydraulic service the more popular end fittings are, (1), male NPTF pipe thread, with or without swivel action, (2), female NPTF pipe thread, (3), male straight thread with O-ring seal, with or without swivel action, to screw into a straight thread port on a component, (4), male JIC 37° flared nose, without swivel, for mating with steel tube fittings, (5), female JIC 37° flare, swivel type, in straight, 45° or 90° elbow, for mating with steel tubing fittings, (6), split flange either straight, or 22½, 30, 45, 60, 67½, or 90° elbow. They bolt to flat port pads with two brackets and are sealed with an O-ring. All these fittings, and perhaps others, are described in supplier catalogs.

When ordering hose from a supply house the following information should be given: (1), desired working or burst pressure, (2), temperature of fluid, (3), type of fluid, (4), type of hose material, (5), type and size of the fitting on each end, (6), overall length of hose measured from tip to tip on the end fittings. If both end fittings are threaded types, one of them must be a swivel type so the hose can be installed without twisting.

Hose life is good but all rubber slowly deteriorates from contact with various substances such as solvents, water, ozone, and exposure to sunlight and heat. Hoses are not as permanent as metal plumbing and should be replaced every few years.

Hoses are available with pressure ratings from vacuum to 45,000 PSI or higher burst rating. High pressure hoses are very stiff and difficult to install; they will not bend in a small radius. For vacuum service, use only hoses designed for vacuum; they have a spiral wire on the inside to support the hose walls from collapse.

SELECTION OF PLUMBING SIZE FOR HYDRAULIC SYSTEMS

Choice of the optimum size of pipe, tubing, or hose for a hydraulic system is a matter of good balance between cost of plumbing materials, labor cost for its installation, and the acceptable power loss through it. If plumbing lines are too small, material cost will be less but there will be a greater loss of power through the lines because of flow friction. If the lines are overly large, the power loss will be reduced but material and labor costs will be higher.

Probably the best key to a good balance between plumbing cost and power loss is the flow velocity permitted through the lines. Higher velocity produces higher flow losses. Velocity is calculated from the inside area of the pipe in relation to the volume of oil flowing through it.

Flow velocities for hydraulic oil which are generally accepted as producing a good balance between material cost and system efficiency are as follows:

10 feet per second on systems where maximum operating pressure is less than 1000 PSI. At these low pressures, flow loss must be low to remain in an acceptable proportion to the power transmitted.

15 feet per second on systems operating in the range of 1000 to 2000 PSI. At these higher pressures, flow losses can be a little higher and still be in an acceptable proportion to power transmitted.

20 feet per second on systems in the range of 2000 to 3500 PSI. This gives about the same percentage of power loss as 15 feet per second velocity at lower system pressures.

30 feet per second is acceptable on a system operating above 3500 PSI. Plumbing materials large enough to produce a low velocity and still meet pressure requirements may be quite expensive.

Mobile systems are usually designed for velocities of 20 to 30 feet per second even at pressures as low as 1500 PSI. Cost of components on equipment built for the resale market must be kept low. And, small size and light weight, not only of plumbing materials but of all components, are important requirements for a hydraulic system on moving vehicles.

On simple hydraulic systems an alternate method of selecting plumbing size is usually satisfactory. All pressure and return lines can be of one size, the size of the pump outlet port. If the pump has NPTF pipe threads, use the same size pipe throughout the system (except for the pump inlet line). If tubing or hose is used, the internal area should be at least as large as the internal area on Schedule 40 pipe of the same size as the pump outlet port. See tables on Pages 267 and 268.

Plumbing size to the pump inlet should be much larger than in the rest of the system to keep flow velocity to 2 feet per second or not to exceed 4 feet per second. The pump cannot develop enough vacuum to draw in a full flow of oil through a small pipe. The table on Page 268 will show the pipe size for a velocity of 2 or 4 feet per second. An undersize inlet line will cause the pump to run in a state of cavitation which will result in accelerated wear and a shorter pump life.

Some pumps have an inlet port which is larger than the outlet port. These pumps cannot (or should not) be modified in the field to run in the opposite rotation; the outlet port will become the inlet port and will be too small to carry full rated flow. Other pumps may have oversize ports on both inlet and outlet so they can be modified in the field to change direction of rotation. The outlet port may be larger than necessary to carry rated flow. This can be checked with the flow velocity charts on Page 268.

Note that cost of hoses may be higher than for iron pipe or steel tubing, but the reduced labor cost for installing them may more than offset the higher cost of the hose.

On simple hydraulic systems the cost of plumbing may be relatively low in proportion to cost of components, but on large systems the plumbing cost may run into thousands of dollars. The method of precisely calculating plumbing size in various parts of the system for optimum balance between cost and power loss is treated in more detail on Page 238.

PETROLEUM BASE HYDRAULIC OIL

For many years petroleum oil has been the preferred fluid for hydraulic systems, and still is, because of its many desirable characteristics, such as its lubricating quality, low specific gravity, and oil film strength. But due to its increasing cost, flammable nature, and possible shortage, the use of other fluids is increasing. These fluids are described in the next section.

For use in hydraulic fluid power systems, petroleum oil is highly refined to remove chemicals which are undesirable. Certain other chemicals are added to improve performance. These include rust inhibitors, foam suppressants, and viscosity index improvers. Although most brands of the same type of oil will mix without harm, the oil supplier should be consulted before doing so. Sometimes, although the two oils may be compatible, the chemical additives may not.

Ordinary motor oil, although of high quality, is not recommended for general use because it may not have some of the characteristics which are desirable in hydraulic systems. However, it is often used in simple systems because it is readily available in small quantities. Again, it should not be mixed with regular hydraulic oil. Types ATA or ATF transmission fluids are excellent for systems using piston pumps, either hand operated or power driven. It is available world-wide in small or large quantities. However, because it is relatively thin (low viscosity), it may not perform well with gear or vane pumps because these pumps usually have a higher internal slippage than piston pumps.

Caution: Automobile brake fluid should never be used in an industrial hydraulic system unless the system manufacturer specifically calls for its use. It is a non-petroleum oil which is compatible with natural rubber (as used in automobile brake systems) but is incompatible with the synthetic rubber seals used in industrial hydraulic systems. Its use in such systems may cause extensive damage.

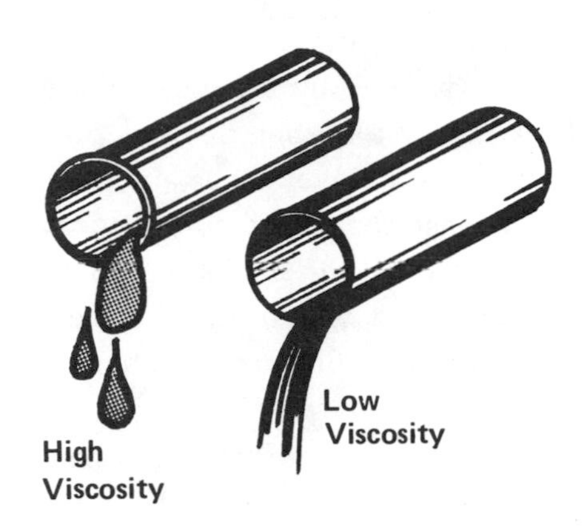

Viscosity of an oil is a measure of its flow resistance.

Viscosity. Viscosity rating is a measure of the oil's "thickness" or resistance to flow, and is probably the most important single characteristic of a hydraulic oil. Since oil viscosity changes with temperature, it is always expressed at a reference temperature, usually 100° F. In the United States the most common unit of viscosity is the Saybolt Second Universal, abbreviated SSU or SUS.

To assign an SSU rating to an oil, a quantity of the oil is brought to a temperature of 100° F, then poured through a standard orifice of 0.0695" diameter. The time, in seconds, for a volume of 60 cubic centimetres to run through the orifice is the rating of the oil. For example, if it takes 150 seconds for 60cc to run through, the oil is rated at 150 SSU at 100° F. To prevent any misunderstanding, this rating may be written as 150/100. This same oil, at a higher temperature would run through faster, and at a lower temperature would require more than 150 seconds to run through.

The hydraulic pump is usually the component most sensitive to viscosity. When choosing a suitable oil, be sure its viscosity rating is according to the pump manufacturers specification. Of course in systems using a hydraulic motor, the motor is equally sensitive to viscosity. Other components in the system are much less sensitive to viscosity and will work with most any viscosity which is compatible with the pump or motor. If the oil is too thin (low viscosity) it is hard to pump because it will excessively slip back through the clearance spaces between gears, vanes, or pistons to the pump inlet. To a lesser degree, slippage will also be higher past cylinder piston seals and valve spools with thin oil. The result is excessive pressure and power loss, poor efficiency, oil heating, and cylinder piston slow-down. But if the oil is too thick (high viscosity), the pump will have difficulty in drawing a full flow of oil into its inlet, and will run in a cavitated state. The result is a shortened pump life, cylinder slow-down, erratic operation of the cylinder pistons, poor efficiency, and there will be excessive flow losses throughout the system. Some systems can operate with a wide range of viscosity, from 45 to 4000 SSU. but in most systems the usual viscosity range is from 100 to 750 SSU.

If a component manufacturer specifies a certain viscosity for his product, find whether this rating is at the standard temperature of 100° F or at the system operating temperature. If specification is at operating temperature it may be necessary to convert to equivalent viscosity at 100° F when selecting a suitable oil.

Viscosity also increases with increased pressure in the oil. However, this is unimportant in most systems operating at less than 5000 PSI. For example, the viscosity of an oil may increase from 150 SSU at atmospheric pressure to about 750 SSU at a pressure of 10,000 PSI, the oil temperature being the same in both cases.

Viscosity Index (V.I.). The viscosity index should not be confused with viscosity; it is a different kind of rating. Viscosity of an oil changes with temperature, becoming less at higher temperatures. The V.I. index number indicates the extent of the viscosity change. Many years ago an arbitrary V.I. rating system was devised on a scale of 0 to 100. An oil was selected which had the greatest viscosity change over a certain temperature, and this oil was assigned a rating of 0. Another oil which had the least change of viscosity over the same temperature range was assigned a rating of 100. It was thought that all other oils would fall within these extremes. Since that time chemical additives have been developed which extend the V.I. rating considerably above 100. For most hydraulic systems the least amount of viscosity change is desirable, and oils with a high V.I. rating are preferred.

Mobile systems which operate out-of-doors are subject to wide variations in ambient temperature. Cold starts in the morning may cause the pump to cavitate until the oil warms up. Then the system may get very hot under continuous running. For these systems the V.I. of the oil is extremely important, and should be as high as possible consistent with other desirable characteristics. But for industrial systems running indoors under fairly uniform temperature conditions, the V.I. of the oil is less important.

V.I. ratings have no units. They are relative terms for comparing variation in viscosity of one oil with another. A good hydraulic oil will usually have a V.I. of 95 or higher.

A high oil temperature may cause many problems in the system.

High Temperature in the Oil. Each day, as a hydraulic system is put into operation the oil gradually increases in temperature until it reaches a "levelling-off" temperature where heat is generated at the same rate it is being removed by radiation to atmosphere or by a heat exchanger. This temperature may be reached in an hour or so of operation. If this "levelling-off" temperature is higher than it should be, it can only be reduced by the addition of a heat exchanger.

Keeping the oil to a moderate temperature level is an important requirement for reliable operation. Machine tool hydraulic systems are usually designed for a "levelling-off" temperature of 130° F. On systems where the duty is not as severe, higher temperatures, perhaps up to 160° F, can usually be tolerated. Still other systems, like those on road machinery and similar mobile applications, are forced to operate at higher temperatures because there is no good way to remove excess heat. But even in these systems the oil should never exceed 200° F. At this temperature or higher, oil life is greatly shortened, viscosity may become so low that power loss becomes high, lubrication of components may be inadequate, and components like rubber seals and filter elements may deteriorate rapidly.

It is said that useful oil life is cut in half for every 20° F rise in temperature. If this is true, oil life at 210° F is only 1/16th of that at 130° F. During the design stage it is difficult to determine just what the levelling-off temperature may be. A heat exchanger can be added later if necessary. During construction it is a very good idea to provide connections in the plumbing for the future addition of a heat exchanger.

Oil life is shortened at high temperatures because many chemical reactions can occur such as oxidation (reaction with oxygen from the air), reaction with minute quantities of acids which were not removed during the refining process or which formed from moisture condensing in the oil. A family of chemical compounds may be produced in the oil. Some of these resemble varnish and are deposited on internal metal surfaces. They may clog filters, strainers, and small orifices. They reduce lubricity of the oil which in turn accelerates wear on metal-to-metal moving parts.

The final result of excessive oil temperature is to cause early failure of the pump and hydraulic motor, excessive valve spool leakage, oil leaks as seals deteriorate, spool sticking in valves, and general malfunctioning of the system.

Low Temperature in the Oil. A hydraulic system can be operated with cold oil or can be located in a cold environment if the oil viscosity after it comes up to operating temperature, is suitable for the pump. What the final operating temperature will be depends largely on the ambient temperature. This is the temperature which surrounds and is in contact with the radiating surfaces of the hydraulic reservoir and component surfaces. As a rule-of-thumb, a well-designed system may operate at a temperature about 60° F above the surrounding atmospheric temperature. Cold weather damage usually comes from trying to run on oil which is so cold that the pump cannot draw it through the feed line to its inlet. However, if the oil is chosen for these cold operating conditions there is no problem. Its viscosity *index* should be as high as possible consistent with other requirements.

Viscosity and viscosity index are critical for low temperature operation.

One solution to cold weather operation is to use an electric heating element mounted in the reservoir below oil level. The heater can be left running overnight, set at a temperature which will permit the oil to be readily pumped in the morning. More of this on Page 206. Note: Be sure to use an element designed for oil heating; an element designed for water heating will operate with too high a surface temperature and may oxidize any oil which stagnates around it.

Another solution to cold weather oil flow is to feed oil to the pump inlet from an elevated tank, 10 feet or more above the pump.

Caution! Warming the pump oil by discharging it across the pump relief valve for an hour or two before starting production may damage the pump if the cold oil is viscous enough to cavitate the pump.

Contaminants in the Oil. Oil never "wears out" but may become so contaminated that it is unfit for further use. Solid contaminants consist of dirt particles which (a) were left in the components when the system was constructed, (b), have been generated within the system through normal wear in moving components or produced by chemical action. or (c), have been ingested; that is, have entered the system from the outside. Solid contamination can be removed with suitable filters. Methods of filtering are presented in Chapter 5.

Oil is also rendered unfit by soluble contaminants which are produced by chemical reactions. Since these contaminants are dissolved in the oil they cannot be removed by filtering, centrifuging, settling, or any method of mechanical separation. Removing them by refining methods is usually impractical; the oil must be replaced.

Soluble contaminants are produced primarily by a high heat level in the oil. The destructive effect of heat on oil is illustrated in the case of an internal combustion engine. An automobile, running at an average speed of 40 miles per hour could only run 100 hours between 4000 mile oil changes.

The importance of filtration to keep solid contaminants removed varies according to the kind of hydraulic pump (or hydraulic motor) used. Systems of many years ago, using loosely fitted gear pumps did not require a highly efficient filtering system, but modern piston pumps of the kind used in hydrostatic transmissions require almost microscopically clean oil. In some systems, valves or other components with very small orifices also require well filtered oil.

Some simple tests on the oil can be made by the user.

Oil Cleanliness. The average user can make only a few simple tests to estimate condition of the oil. After the system has been running a short time, a sample can be drawn from the reservoir in a clean glass jar. Any foam should dissipate in a minute or two. Water which is emulsified with the oil should separate in a short time and settle to the bottom. Likewise large dirt particles. Microscopic sediment suspended in the oil may take several hours, or days, to separate and settle.

When the sample has become reasonably clear, its color can be compared with that of a new sample of the same oil. Actual color of an oil sample may not mean much, but the ***change*** in color is significant. Any oil will darken with use and a slight darkening is not important. But a severe darkening, to the color of dark molasses, may indicate a high level of soluble contamination. The oil can be tested in the laboratory of an oil company, an independent laboratory, or by one of the large manufacturers of hydraulic filters. Consult with your local fluid power distributor for the best place to have your oil tested. It is usually impractical to re-refine small volumes of oil to remove chemical impurities. An oil which has darkened to an opaque condition is usually unfit for further service and should be replaced.

People with sensitive fingertips can detect grit by feeling of an oil sample. But if grit can be detected in this way the oil is highly contaminated. Dirt can be removed by installing micronic-type filters.

If it should become necessary to completely change the oil, clear oil can be siphoned off the top after the oil has been allowed to settle a few hours. This clear oil can be set aside and used for flushing the system after the reservoir has been cleaned. Flushing oil should then be discarded and a completely new oil charge should be added. New oil added to the reservoir should be filtered as it is added even though it comes from a container which has not previously been opened.

A complete change of oil should seldom, if ever, be needed on indoor systems with micronic filters which are serviced regularly, provided the system operating temperature is kept below 150° F.

Anti-Foaming Additives. Oil circulates very rapidly through the plumbing, 10 to 30 feet per second. At any point where it is agitated in the presence of air, bubbles of foam will form. Aeration (or foaming) may take place in the reservoir caused by turbulence as the oil is discharged at high velocity into the reservoir. Aeration may also take place in the plumbing if air is drawn into the system through leaky pump shaft or piston rod seals. One purpose of the hydraulic reservoir is to retain the oil long enough for foam to dissipate before the oil is picked up by the pump and re-circulated. Oils compounded for hydraulic use have a chemical compound added to help the foam dissipate more quickly. If an excess amount of foam appears in a system, enough to cause erratic operation of the piston, this may indicate the use of an improper oil, insufficient reservoir capacity, excessive velocity of the oil through the plumbing, or a continual leak of air into some part of the plumbing.

Oxidation Resistance. This property of an oil determines its resistance to chemical deterioration; that is, its resistance to forming insoluble varnishes and sludges. When these undesirable compounds are formed they can interfere with a systems performance by plugging small orifices or passages, clogging filters, and causing valve spools to stick.

Oxidation of an oil is a reaction with oxygen, usually from air. It progresses slowly at first, then more rapidly toward the end of the oil's useful life. The oxidation process proceeds more rapidly at higher temperatures and in oil where foaming takes place, in the presence of certain metals, notably copper and copper alloys, and in the presence of water and other contaminants.

Oil darkens in color as oxidation progresses, and darkening to the state of opacity may mean that it should be discarded. A chemical laboratory can check the neutralization number of an oil sample to see whether it should be discarded.

Pour Point. This is the lowest temperature at which a given oil will flow by gravity. The pour point should be considered on hydraulic systems designed to be started when the ambient temperature is low. Needless to say the pour point of the oil must be ***well below*** the lowest start-up temperature which will be encountered.

Other Additives. Water may enter the system by condensation in the reservoir. Rust inhibitors are added to retard formation of rust on the sides of the reservoir and on the underside of its lid. Demulsifying agents are also added to resist emulsification of water with the oil. The water can settle quickly in the bottom of the reservoir and be drained away.

OTHER HYDRAULIC FLUIDS

Water was the first hydraulic fluid because, at the time, there was no other liquid available in sufficient quantity and at an affordable price. After the discovery of huge underground reservoirs of petroleum oil, this oil replaced water because it offered several advantages and was relatively cheap at that time. It has a lower specific gravity than any other liquid which has been used in hydraulic systems. For this reason it can be circulated through piping with less power loss. It has excellent lubricity. This is of major importance because the circulating liquid in a hydraulic system must also lubricate the working parts of components. It can be refined to have any desired viscosity. The viscosity of a hydraulic liquid should be high enough so it can be pumped efficiently, yet not so high that the circulation losses due to flow resistance will be excessive. Water has a viscosity of about 30 SSU and is difficult to pump except with piston-type pumps. "Wire drawing" is an effect in which molecules of metal on working surfaces of pumps and valves are actually torn away by molecules of a liquid traveling at high velocity. Oil has a minimum wire drawing effect while water rapidly wears away sharp corners on valve spools, edges of gears, vanes, cam rings, and housings in pumps.

Other hydraulic fluids have been in use for a number of years and many new ones are coming on the market. All-synthetic fluids, phosphate ester type, have been around for a long time and are still popular. They were developed, not for any superior operating advantage, but for their fire resistance. They have been used in aircraft hydraulic systems, die-casting machines, on systems operating near furnaces. In fact, government mining regulations state that in certain underground mines where diesel engines drive hydraulic pumps, the use of fire resistant fluid is compulsory in mobile mining machinery operated by hydraulics. All-synthetic fluids have excellent fire resistance but their cost is considerably higher than petroleum oil.

Many other fluids have been developed, most of which are water base with various amounts (20% up to 95%) of water with oils or synthetic material to give them desirable characteristics. Most of these fluids are much less expensive than petroleum oil but all of them (and the all-synthetics) have certain disadvantages. While they work well in systems operating at low to moderate pressures, none of them works as well as petroleum oil at very high pressures. All of them have a specific gravity higher than oil. This causes higher power losses in the system, more cavitation in the hydraulic pump, and larger filters, plumbing lines, and components may be required. The water base fluids are limited to a temperature range of 40 to 120° F, so they should not be used in either hot or cold locations. Some fluids may require special seal material or special paint or finish on reservoir, plumbing fittings, and components. Some of them may be incompatible with the metal in certain components such as zinc, lead, or copper. They all have a greater tendency to foam than oil. Most of the water base fluids have such a low viscosity that they are hard to pump except with piston pumps. They will leak at joints which are tight on fluids of higher viscosity. All of the water base fluids require monitoring to maintain their proper ratio of water to the additive material. This requires in-plant testing and mixing of fluids to make up fluid which is lost in leaks or by evaporation of the water.

It is not an easy job to convert from petroleum oil to one of these synthetic or water base fluids. In some cases components must be replaced with larger size components; seal materials may have to be replaced; heat exchangers added to reduce operating temperature to no higher than 120° F; the reservoir may have to be elevated or a supercharge pump added to prevent cavitation of the main pump; even the pump may have to be changed depending on the viscosity of the new fluid or the maximum operating pressure of the system; reservoirs should be opened and cleaned; in fact, the entire system should be thoroughly flushed to remove all possible oil.

Petroleum oil is still the best fluid for most systems and will probably continue to be. Synthetic fluids require less monitoring and mixing than water base fluids and should be considered for new systems operating near a source of heat. Water base fluids are much less expensive than oil and should be considered for their fire resistance, first cost, and replacement cost.

The choice of synthetic or water base fluids is so wide and the state of the art development of fluids is progressing so rapidly that when designing a new system, several fluid suppliers should be consulted to see whether all-synthetic or water base fluids would offer any advantages.

INTRODUCTION TO THE USE OF VACUUM

Pascal finally proved in 1647 that vacuum force was caused by atmospheric pressure and not to Nature's "limited horror of a vacuum" as was believed for more than 2000 years before his time. He had induced his brother-in-law, Perier, to carry a barometer to the top of a 4000-foot mountain. The drastic reduction in barometer reading could only be due to a reduction in air pressure, as he could not see why Nature would abhor a vacuum any more at the foot of a mountain than at its top.

Otto von Guericke invented a crude vacuum pump in 1650, and in 1654 he made a spectacular demonstration of air pressure before the parliament at Ratisbon Germany. The "Magdeburg Hemispheres" were two hollow bronze hemispheres carefully fitted together at their edges. When air was pumped out from between them, pressure from the surrounding air held them so tightly together that two teams of eight horses each could not pull them apart. These hemispheres may have been about 2 feet in diameter, and if highly evacuated would adhere with a force of about 2½ tons.

This demonstration by Otto von Guericke before the parliament of Ratisbon, Germany in 1654 proved the power of vacuum was actually due to air pressure on the outside of these hemispheres holding them together.

Vacuum is a form of fluid power, and the maximum force available is dependent on the barometric pressure, about 14.7 PSI under standard sea level conditions. This is a small force as compared to a 100 PSI compressed air system which would cause the Magdeburg Hemispheres to adhere with a force of 90 tons, or to the 500 to 10,000 PSI of a hydraulic system. Nevertheless, there are useful jobs which can be done better, cheaper, or faster with vacuum, and some of them could hardly be done any other way. The following illustrations show some of these jobs.

A vacuum pump is simply an air pump used to remove air pressure from one side of an object so atmospheric pressure can act against the opposite side. Some standard compressed air components will work equally well on vacuum, but hose and receiver tanks which can contain very high internal pressure may collapse under relatively low external pressure. Vacuum hoses have an internal spiral wire to prevent collapse. Receiver tanks to "store" vacuum are specially constructed with closely spaced internal supports. Any air component used for vacuum should be evaluated for suitability.

MEASUREMENT OF VACUUM

Vacuum can be thought of as negative air pressure. Like positive pressure its intensity must be measured by the force which it can exert on a unit area of surface. It could be measured with any set of force and area units, like pounds per square inch, tons per square foot, etc. But in the customary U.S. system of units, vacuum pump force output is measured in "inches of mercury", abbreviated "Hg. This is the height of a column of mercury which can be supported by the vacuum. This method of measurement was probably adopted because the early mercury barometer balanced a column of pure mercury against the weight of atmosphere overhead. One PSI is roughly equivalent to 2" Hg pressure.

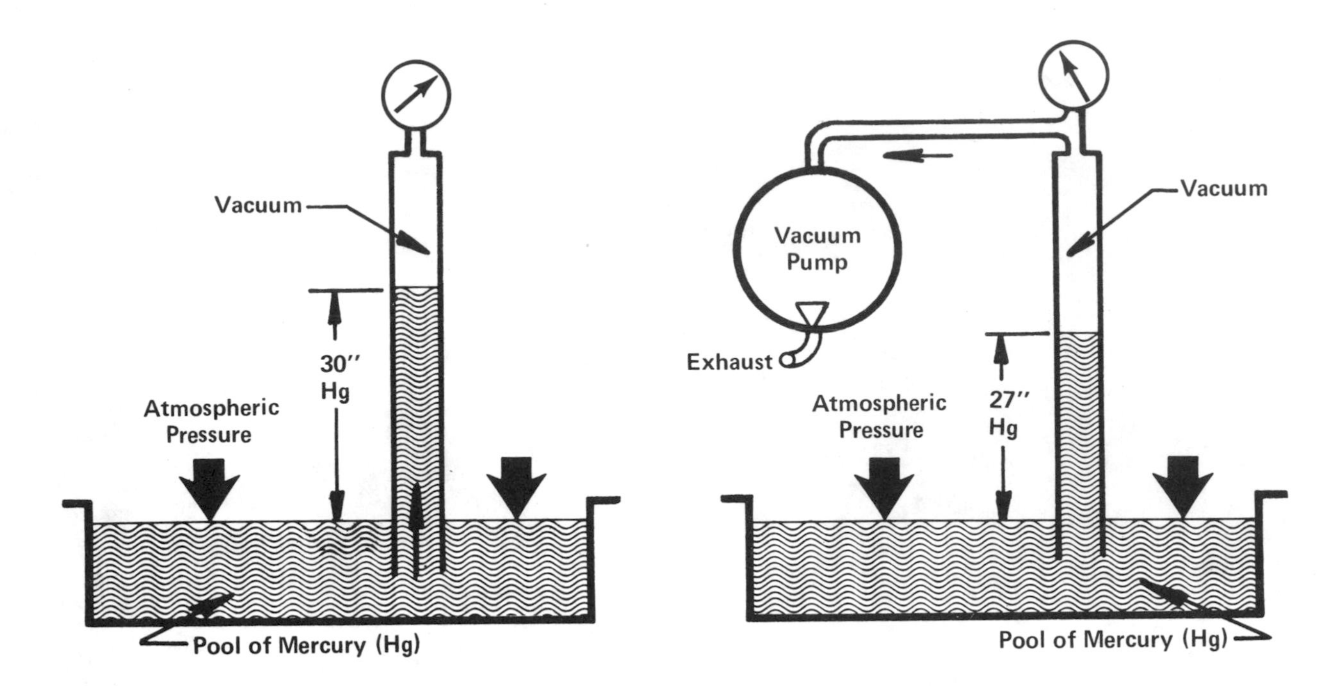

FIGURE 1-31. Mercury barometer. Height of mercury column indicates barometric pressure.

FIGURE 1-32. Test for measuring the performance of a vacuum pump.

Mercury Barometer. Figure 1-31. A mercury barometer can be made with a long tube, closed on the top end, open on the bottom, immersed in a pool of mercury. The tube must first be laid flat in the mercury to expel all air. It should then be raised to a vertical position, while keeping the open lower end immersed. The mercury will drop in the tube (no matter how long the tube) until a balance is reached between the weight of mercury in the column and the atmospheric pressure against the surface of the mercury pool. At sea level elevation the mercury will stand about 30 inches (actually 29.9" or 760 mm) above the surface of the pool, but there would be a little variation from day to day due to weather changes. The space above the mercury is a perfect vacuum except for a small mercury vapor pressure, and a vacuum gauge, at sea level conditions, would show about 30" on its scale.

An interesting experiment would be to take this barometer to the top of Pikes Peak where the air pressure is less than at sea level. The height of the mercury column would be about 17.5 inches to balance the lower air pressure at this altitude, even though a perfect vacuum still exists above the mercury. The pressure gauge would also read 17.5 because it, like the barometer, is referenced to surounding atmospheric pressure.

Vacuum Pump Performance. Figure 1-32. If a vacuum pump is connected to the top of the mercury column and allowed to run until the maximum vacuum of which the pump is capable is obtained, the mercury column would not rise as high as in the barometer of Figure 1-31 because the pump cannot pull a perfect vacuum. Industrial vacuum pumps are usually rated on the degree of vacuum which they can pull based on a normal atmospheric pressure of 30"Hg. A pump rated for 27"Hg could pull the mercury column up to within 3" of a perfect vacuum when the barometer stood at 30"Hg. In other words, it can pull 90% of a perfect vacuum (27 ÷ 30 = .90 or 90%). At other locations the pump could pull 90% of the barometric pressure in vacuum at that location. For example, on Pikes Peak, if the barometer reads 17.5"Hg, the maximum vacuum obtainable with this pump would be 90% x 17.5 = 15.75"Hg, and the vacuum gauge would show this amount.

Vacuum power, then, decreases with an increase in altitude. Some industrial regions of the U. S. are as high as 5000 feet elevation, with a barometric reading of about 25 "Hg. A vacuum pump is not as effective at this elevation as it is at sea level. Therefore, industrial vacuum applications are usually designed to work at no higher than 20" Hg vacuum.

Inexpensive vane or piston-type industrial vacuum pumps are used on the applications described in this section. They will develop about 90% of a perfect vacuum and are usable on 20 "Hg vacuums at 5000 feet elevation.

Units for Measuring Vacuum. Any of the units in the accompanying table may be used for expressing degree of vacuum, although "Hg is the most common in the U.S. On industrial applications vacuum is usually expressed as the level ***below*** normal atmospheric pressure.

Equivalent Vacuum Levels

"Hg	mm Hg	PSI	kPa
30	762	14.73	102
28	711	13.75	94.8
26	660	12.77	88.1
24	610	11.79	81.3
22	559	10.81	74.5
20	508	9.82	67.7
18	457	8.84	61.0
16	406	7.86	54.2
14	356	6.88	47.4
12	305	5.89	40.6
10	254	4.91	33.9
8	203	3.93	27.1

1" Hg (Inches of Mercury) = *3.39 kPa = 0.491 PSI = 25.4 mm Hg*

1 PSI (Pounds per Square Inch) = *51.7 mm Hg = 2.04 "Hg = 6.90 kPa*

1 mm Hg (Millimetres of Mercury) = *0.039 "Hg = 0.133 kPa = 0.019 PSI*

1 kPa (kilopascals) = *0.145 PSI = 7.50 mm Hg = 0.295 "Hg*

VACUUM APPLICATIONS

Suction Cups. Figure 1-33. Suction cups can be used for lifting, transferring, or separating many types of sheet materials including paper, sheet metal, cloth, plastic, leather, glass, and even concrete slabs. Vacuum lifting or holding is extensively used in printing, photography, in wrapping or packaging consumer products where paper, metal foil, cellophane, or cardboard must be handled. Vacuum may not work well in handling porous materials.

Vacuum cups are used only to tightly grip the material to be handled. Actual lifting or movement of the material is done by external mechanical power.

A vacuum release valve must be used to break the vacuum when the material must be released. For extra fast release on high speed operations, or if the material will not release by gravity, a low

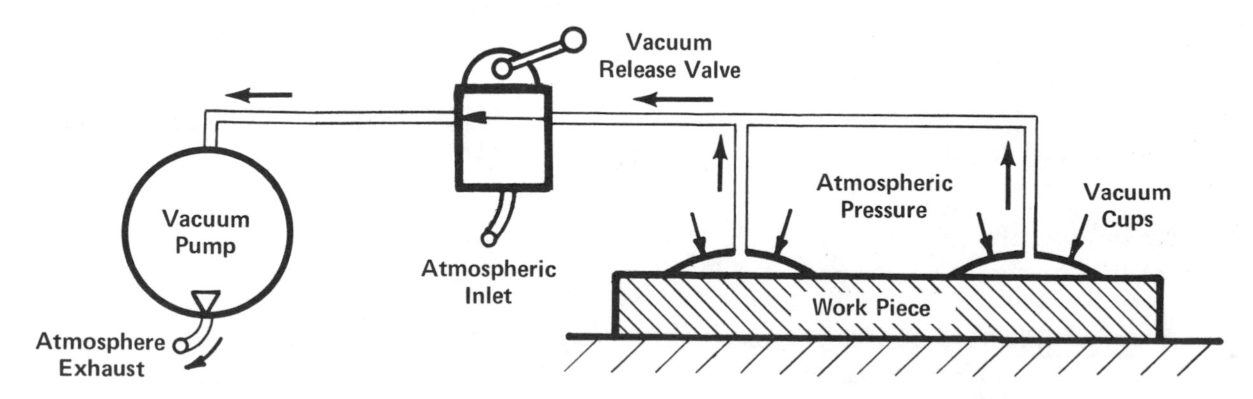

FIGURE 1-33. Sheet or slabs of many kinds can be gripped with suction cups powered with a vacuum pump for lifting, separating, transferring, or feeding.

air pressure is applied to the cups at the moment the vacuum is released. The vacuum release valve is one form of a 3-way directional control valve which is described in Chapter 3.

Vacuum cups can be made by the user for a particular job, or can be purchased ready made at an industrial supply house. Connecting hoses should be those designed for vacuum service, and should be relatively small in diameter to reduce the amount of dead space to be evacuated on each cycle.

Heavy objects with a slick surface have a tendency to slip sidewise off the cups if tilted to a vertical position. If this is a problem, sidewise supports should be provided.

Vacuum cups have a surprising amount of grip. The accompanying chart shows the force with which they will hold. The chart is calculated for circular cups working at a 20 "Hg industrial vacuum. To calculate other cups, take the holding area in square inches and multiply by the degree of vacuum in PSI. Every 2 "Hg is approximately 1 PSI, or use conversion formula on Page 47.

Gripping Force – Circular Cups

Cup Dia. Inches	Cup Area Sq. Inches	Lift, Lbs.
1	785	7.8
2	3.14	31
3	7.07	71
4	12.57	126
5	19.63	196
6	28.27	283
8	50.27	503
10	78.50	785
12	113.1	1131

Calculated at Vacuum of 20 "Hg

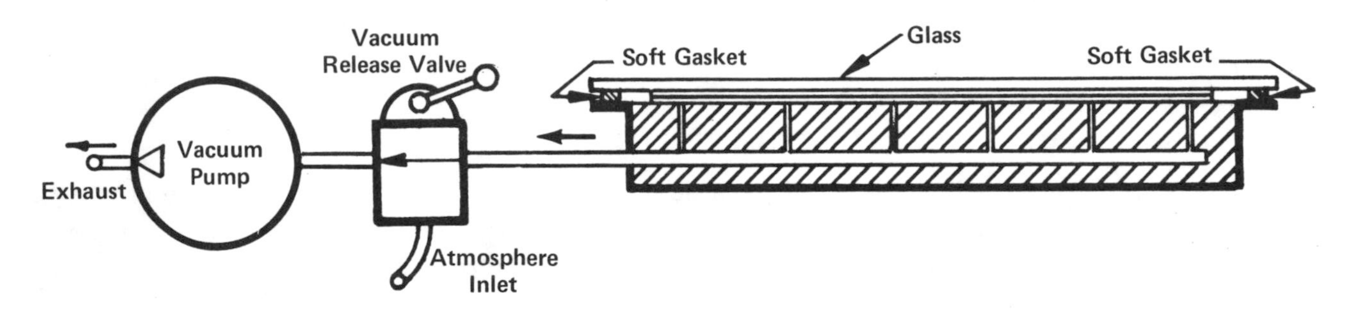

FIGURE 1-34. Vacuum frames are used for laminating, bonding, and in the printing and photographic industries.

Vacuum Frames. Figure 1-34. In manufacturing industries, large sheets of pliable material can be drawn into close contact over their entire surface, using a low degree of vacuum. Typical uses are for bonding, cementing, or laminating two (sometimes more) sheets of material. In the printing industry, vacuum frames are used for making contact prints, for printing large posters, for holding copy on vertical copy boards in front of the camera, and for holding film negatives in tight contact with aluminum or zinc plates.

If glass is used in vacuum frames, it may be supported on compressible gaskets. When vacuum is applied, the glass moves down, compressing the gasket and pressing the sheets of material into tight contact.

A vacuum release valve must be used to relieve the vacuum when the operation has been completed. This is basically a 3-way directional control valve which is described in Chapter 3.

Plastic Forming. Figure 1-35. Sheet plastic can be formed into many shapes by using vacuum to pull hot plastic sheets into a female mold cavity, and holding it in position for a short time until it cools sufficiently to retain the new shape.

Most plastic sheets can be formed after they are heated but for best results a plastic supplier should

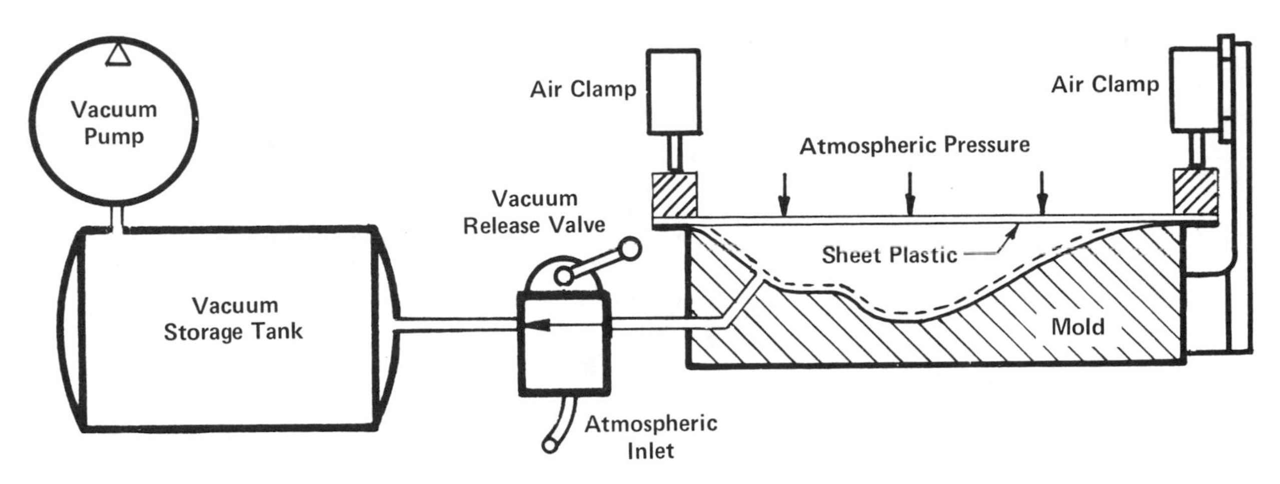

FIGURE 1-35. Molding or shaping hot plastic sheets in a vacuum frame.

be consulted for the best grade to use on a particular application.

The plastic sheets are heated in an oven to about 250° F. They are then placed across the vacuum box containing the mold. After clamping the sheet around its periphery, vacuum is applied between the sheet and the mold.

The operation must be performed quickly while the plastic is hot. For quickly clamping the sheet to the mold, small air pistons may be used around the outside edges of the mold, spaced as closely as necessary. The clamp pistons are connected together and supplied through one air valve.

The vacuum must also be applied quickly while the plastic is hot. On large jobs, vacuum should be stored in large vacuum tanks, similar to the way compressed air is stored in receiver tanks. Vacuum should be applied through a very large manual valve so that the sheet is drawn down quickly when the mold is evacuated. Between operations the vacuum tank can be re-charged with a relatively small vacuum pump which runs continuously.

For experimental work, especially on small jobs, plastic sheets can be heated in boiling water, or a cold sheet can be placed across the mold and heated with an overhead electric heating element.

Vacuum, by itself, has sufficient power to form sweeping contours, but may not have sufficient force to form sharp corners. In production work where sharp corners must be formed, the male counterpart to the female molding die can be mounted on an overhead press, either air or hydraulically powered, and pressed into the mold after vacuum drawing has been applied.

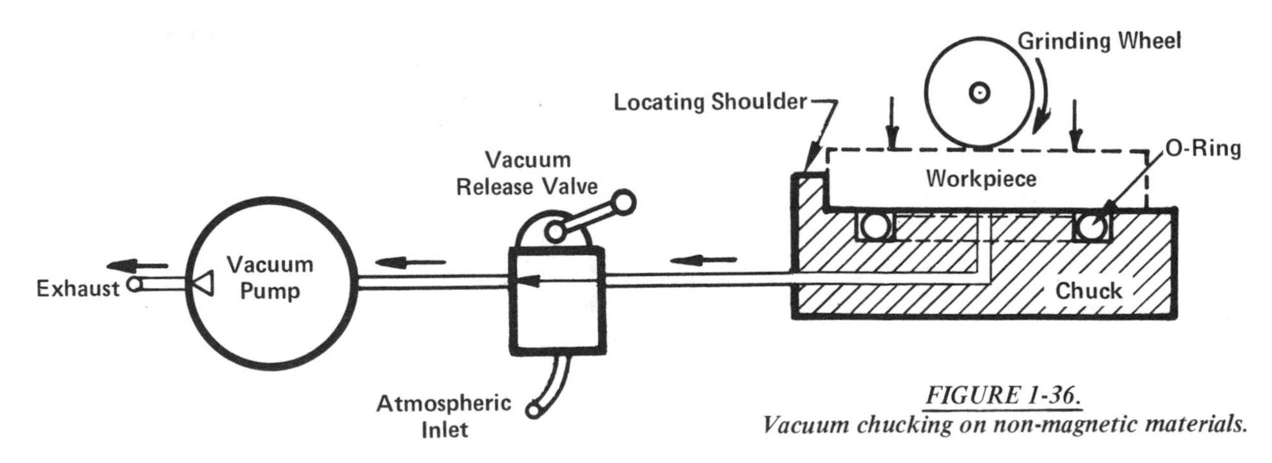

FIGURE 1-36. Vacuum chucking on non-magnetic materials.

Vacuum Chucking. Figure 1-36. A vacuum chuck is a fixture to hold a piece of work while machining is done on it. A vacuum valve between vacuum pump and the chuck applies and releases the vacuum. This is basically a 3-way directional control valve which is described in Chapter 3.

Vacuum chucking is particularly useful for non-magnetic workpieces such as brass, aluminum, copper, zinc, plastic, etc., which cannot be held in a magnetic chuck, but it can also be used to advantage many times with magnetic materials. It is also useful for clamping delicate workpieces which could be damaged by mechanical clamping.

This illustration shows a simple fixture, constructed by the user, which has a circular machined groove into which a standard O-ring is placed as a vacuum seal. If a standard O-ring is not suitable, continuous lengths of O-ring can be purchased and cut to the proper length.

Machining on a workpiece should be toward a locating shoulder to prevent the work from sliding sidewise off the O-ring seal. Usually a very moderate vacuum has ample force to hold down a workpiece for almost any kind of machining.

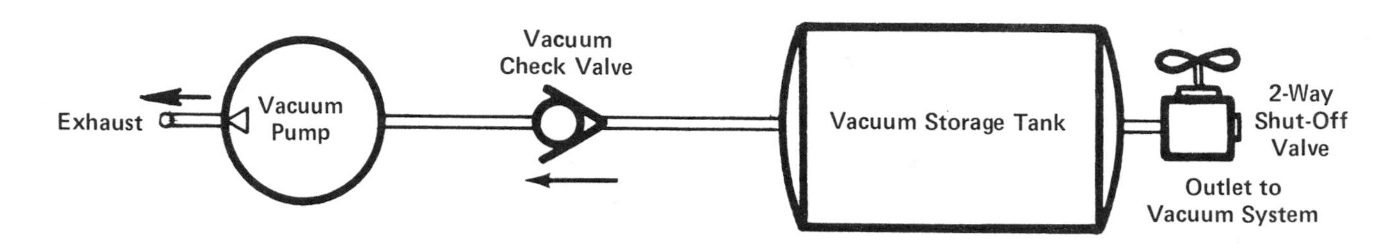

FIGURE 1-37. Storing a large volume of vacuum for future use.

Vacuum Storage System. Figure 1-37. The vacuum forming application of Figure 1-35, and similar ones, may require a quick evacuation of a large space in a very short time. Then there may be a fairly long interval before the start of the next evacuation.

To handle this kind of application a vacuum pump of very high displacement would be required. Or, it could be handled with a much smaller pump if a large storage tank were used to "store" vacuum during the idle periods. On this diagram a relatively small vacuum pump runs continuously. When the tank has reached the degree of vacuum desired, a vacuum operated pressure switch stops the electric motor which drives the vacuum pump, and the pump remains off until a drop in vacuum level starts it again.

A tank designed for storage of compressed air should not be used for vacuum service. Although such a tank may be capable of working at high internal pressure, it would probably collapse on a relatively low external pressure created by a vacuum. Tanks for vacuum service are constructed with internal bracing which is not needed for storage of compressed air.

Hoses used for connecting a vacuum system should be those specifically constructed for vacuum service. Usually they have an internal spiral spring to support the walls from collapse.

Any check valve used in a vacuum line should

Time (in minutes) required to evacuate tanks to 20 "Hg vacuum using a pump with 1 SCFM free running displacement.

Tank Vol., Gallons	Tank Vol., Cu. Feet	Mins. to Evacuate
20	2.7	3.4
30	4.0	5.0
50	6.7	8.4
75	10	12.5
100	13	16.3
150	20	25.1
200	27	33.8
300	40	50.1
500	67	83.9

have a very low cracking pressure – less than 1 PSI. A check valve designed for compressed air may have a cracking pressure of 2 to 5 PSI, and this is undesirable for vacuum because it reduces the available vacuum force by the amount of the cracking pressure. Check valves of the flapper or swin type are preferred.

Evacuation Time. The chart on Page 50 shows the approximate time, in minutes, to evacuate storage tanks, starting at atmospheric pressure, to an industrial vacuum of 20 "Hg (assuming standard 30 "Hg barometric conditions). The times shown are for a pump with 1 SCFM free running displacement. For example, a vacuum pump which has a 5 SCFM free running (physical) displacement will require only 1/5th of the times shown, etc.

A more complete evacuation table will be found on Page 276. Please refer to that table for the following discussion. That table, too, is based on the performance of a vacuum pump with 1 SCFM free running displacement.

The table was calculated from a formula published by Gast Manufacturing Co., and the same formula, shown below, may be used for calculating values which do not appear in the table.

$$T = \frac{V}{D} \times Log_e \left(\frac{A}{A - B} \right)$$

T is pumping time required, in minutes.
V is volume to be evacuated, in cubic feet.
D is free running displacement of pump, SCFM.
A is rated deadhead vacuum of the pump (with inlet blocked).
B is the desired vacuum level in tank, in "Hg.

The table can be used to calculate running time starting from atmospheric pressure up to a desired vacuum level, or can be used to find running time to raise vacuum level from a lower to a higher level:

Example: Estimate evacuation time starting from 12"Hg up to 24"Hg in a 25 cubic foot tank. Tank volume is 25 cubic feet.

Solution: This problem must be solved with two calculations. First find the running time from atmospheric up to a 12"Hg vacuum. Also find the running time from atmospheric up to a 24"Hg vacuum. Subtract these two times:

Atmosphere up to 12"Hg = 14.0 minutes for a 1 SCFM pump.
Atmosphere up to 24"Hg = 48.6 minutes for a 1 SCFM pump.
48.6 - 14.0 = 34.6 minutes for a 1 SCFM capacity.
Adjust for a 6 SCFM pump: 34.6 ÷ 6 = 5.77 minutes.

In determining the deadhead pressure of the pump for use in the formula, the catalog rating may be used, but a better way is to make an actual test by blocking the pump inlet with a vacuum gauge, then running the pump a few minutes until the vacuum gauge rises no higher.

Please remember that calculated running time is approximate and may become inaccurate at higher vacuums because of internal leakage in the vacuum pump. Even pumps of the same brand and model may have different leakage rates. Vacuum pumps should not be operated too closely to their catalog rated deadhead pressure. They operate inefficiently and tend to overheat.

Hard Vacuum. The vacuum applications described in this section are performed with vacuum levels of 20 to 28"Hg. But for scientific or other industrial applications like evacuation of X-ray or TV picture tubes, refrigeration systems, etc., these vacuum levels are useless. In hard vacuum work, the "micron" is a common unit. A micron is one-millionth of a metre or one-thousandth (.001) of a millimetre. A vacuum level described as 50 microns is within 50 microns (.050 mm) of a perfect vacuum based on a 760 mm barometer.

Any 8 questions

STUDY QUESTIONS FOR CHAPTER 1

1. What is the purpose of a fluid power system? – *See Page 9.*

2. Name several possible advantages of using a fluid power system instead of an electrical or mechanical system on certain applications for which it is suited? – *See Page 10.*

3. In your own words, state Pascal's Law. –*See Page 15.*

4. For most applications how does the power level of an air system compare with that of a hydraulic system? – *See Page 14.*

More Pwr Hydraulic

5. Can Boyle's Law be used on hydraulics as well as on air? – *See Page 16.*

6. Does Pascal's Law work for hydraulic fluids as it does for compressed air? – *See Page 15.*

7. What is the approximate atmospheric pressure under standard sea level conditions? – *See Page 17.*

8. What are the common units in the U. S. system for stating pressure in an air system? In a hydraulic system? In a vacuum system? – *See Page 17.*

9. What are the common units in the U. S. system for stating flow in an air system? In a hydraulic system? – *See Page 17.*

10. What is the primary purpose of a pump in a hydraulic system? What is its secondary purpose. – *See Page 23.*

11. How can the force produced on the piston rod of a cylinder be calculated? – *See Page 22.*

12. Horsepower whether in a fluid, electrical, or mechanical system must consist of force multiplied by rate of movement. What are the factors in a fluid system which are equivalent to force and rate of movement? – *See Page 19.*

13. What is meant by "static head pressure?". – *See Page 25.*

14. To achieve force multiplication through a hydraulic system, one small and one large piston is used. Is the small piston used on the input or on the output side? – *See Page 24.*

15. Name at least two ways the flow loss through a fluid power system can be reduced. – *See Pages 26 and 27.*

16. Approximately to what extent is petroleum type hydraulic oil compressible? – *See Page 16.*

17. In Figure 1-24, explain why there is no loss of pressure from one end of the pipe to the other.

18. In Figure 1-28 why do the pressure switch contacts not trip until the piston reaches the end of its stroke?

19. What plumbing materials are recommended for an air system? What materials are not recommended? –*See Pages 33 to 37.*

20. What plumbing materials are recommended for a hydraulic system? Which are not recommended? – *See Pages 33 to 37.*

21. What advantages does petroleum oil have over water as a hydraulic fluid? – *See Page 38.*

22. What is the difference between laminar flow and turbulent flow? How is laminar flow best achieved in a hydraulic system? – *See Page 26.*

23. Why is the piping size to the inlet side of a hydraulic pump of such great importance? – *See Page 38.*

24. Why is the inside diameter of piping so important? What happens if it is too small? If it is too large? – *See Page 37.*

25. How is gauge pressure (PSIG) related to absolute pressure (PSIA)? Is this relationship true at all locations in the world? – *See Page 17.*

26. Using the chart on Page 33, what is the minimum recommended pipe size to carry an air flow of 100 SCFM to a distance of 500 feet? – *Answer: 2" NPT.*

27. When laying out a plant air distribution system, what design specifications should be observed to minimize the water problem? The text on Pages 31 to 33 names at least a half dozen recommendations.

28. Describe what is meant by these terms in relation to hydraulic oil: viscosity; viscosity index. – *See Page 39.*

29. Explain why excessively high oil temperature is destructive to a hydraulic system. – *See Page 40.*

30. Explain the danger to a pump when operating with an oil which is too cold. – *See Page 41.*

31. If the hydraulic oil foams excessively, what are the possible causes? – *See Page 42.*

32. Where does the power of vacuum come from? – *See Page 45.*

33. How many pounds of lifting force can be obtained from a circular suction cup of 6" I.D. when operating at a vacuum of 16 "Hg? – *Answer: 222 lbs.*

34. From the chart on Page 50, estimate the pump-down time for a 500 gallon tank from atmospheric pressure to a vacuum of 20 "Hg, using a pump with 3½ SCFM free running displacement. – *Answer: 24 minutes.*

Metal Piston Rings have the longest life expectancy are
are for pistons only, not piston rod seals
They do not create a tight seal

CHAPTER 2

Fluid Power Cylinders

CYLINDER DESCRIPTION

In Chapter 1, an output "piston" was used to recover fluid power transmitted through a pipe. The preferred terminology is "cylinder" and will be used throughout the remainder of this book.

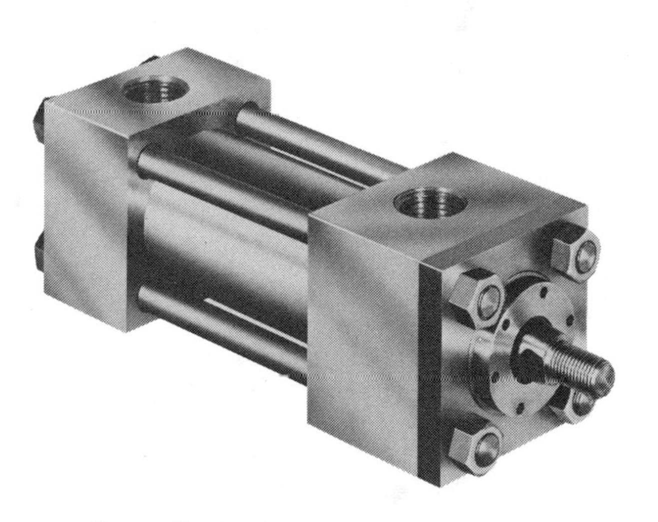

Square-Head Industrial Fluid Power Cylinder

After being transmitted to the point of use, fluid power must be converted to mechanical power by means of an "actuator". Actuators include single and double-acting cylinders, hydraulic or air motors, or rotary oscillators. Cylinders deliver a linear, push-pull, motion. Fluid motors deliver a continuous rotary motion, and oscillators produce a back and forth rotation through a limited arc, usually less than one revolution. This volume of the INDUSTRIAL FLUID POWER series is devoted entirely to cylinder actuators. Fluid motors and oscillators are covered in Volume 3.

Whether to use a cylinder, a motor, or an oscillator depends on the kind of motion which will work best with the application. Cylinders are probably used more often than any other actuator. Although they produce a straight-line motion, this can be converted to limited rotary motion by levers, racks and pinions, or other means.

Either force or power (or both) can be transmitted through a pipe and converted to mechanical force or power by a cylinder. If the fluid is under pressure but is not moving, as a cylinder stalled against a load too great for it to move, only force is being transmitted. When the fluid starts to move, power as well as force is being transmitted. Transmission of power through a pipe from one location to another requires a movement of the fluid as well as pressure in the fluid.

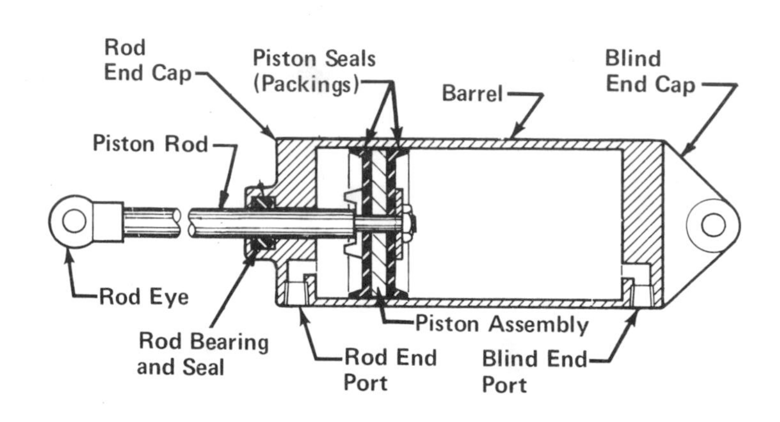

FIGURE 2-1. Standard Double-Acting Cylinder With Hinge Mount.

Cylinder Construction. Fig. 2-1. Although there are many variations in cylinder construction, the essential working parts are shown in this figure. This model is a "double-acting" cylinder because it can produce force and power in either direction of piston movement.

A piston works back and forth in a barrel which is enclosed between two end caps. A piston rod is attached to the piston and extends through one end cap to bring the power out to the load.

The barrel is usually made from a steel tube although other materials such as brass, aluminum, or plastic are occasionally used. The inside of the barrel is polished to a mirror finish to make a tight fluid seal and to minimize wear on the piston seal. On a double-acting cylinder such as this one, the piston should seal fluid from leaking past the piston in either direction. If cup seals are used, two of them are required, facing in opposite directions, or sometimes a single O-ring is used. The lips of cup seals must be prevented from bumping against the end caps, and a piston bumper may be used for this purpose.

On most double-acting cylinders only one piston rod is used as illustrated in the figure, but on special cylinders two rods may be used, both of them attached to the piston, one coming through each end cap. The rod (or rods) must be sealed against leakage where they come through the end caps and for this purpose there is a rod gland consisting of a soft seal, a rod bearing to support the rod against side load, and most cylinders have a wiper or scraper ring to prevent solid particles which may have fallen on the extended rod, from being drawn into the rod seals. The piston rod should have a mirror finish, not only to make a good fluid seal but to minimize wear on the rod seal. Nicks, scratches, or wrench marks on the piston rod will allow pressurized fluid to leak to the outside and will also ruin the rod seal in a short time.

Cylinder Terminology. Please refer again to Figure 2-1 where the parts of a cylinder are named. The end of the cylinder where the rod comes out is the "rod end", and the opposite end is the "blind end". We find these terms much more descriptive and easier to remember than the official designations of "head end" where the rod comes out and "cap end" opposite the piston rod. The terms "rod end" and "blind end" will be used in this book.

The two end caps are the "rod end cap" and the "blind end cap". The fluid ports are "rod end port" and "blind end port".

There are a number of mounting styles which will be described later. The mounting in Figure 2-1 is a "hinge", "clevis", or "tang" mount. When the blind end of a cylinder is pivoted, a hinge is usually required on the end of the piston rod.

Producing Movement From a Cylinder. Figure 2-2. To produce motion of the piston in a double-acting cylinder, a sufficient difference of pressure must be produced between opposite sides of the piston. Usually a fluid pressure is applied to one side while the opposite side is vented to atmospheric pressure with as little flow resistance as possible.

In Part A of this figure, air or hydraulic pressure is connected to the blind end port. The rod end port is vented to a lower pressure or to atmospheric pressure. This produces a force tending to move

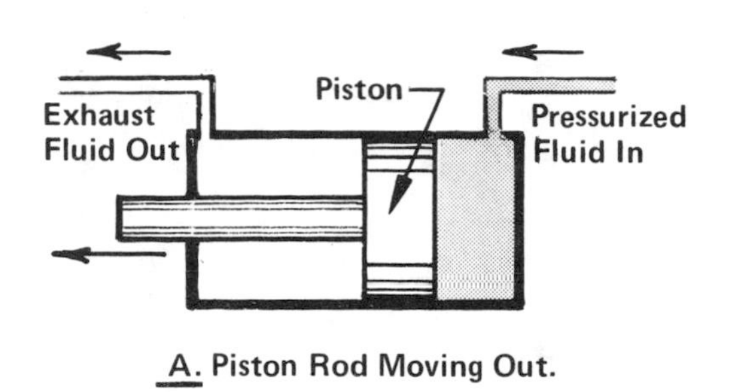

A. Piston Rod Moving Out.

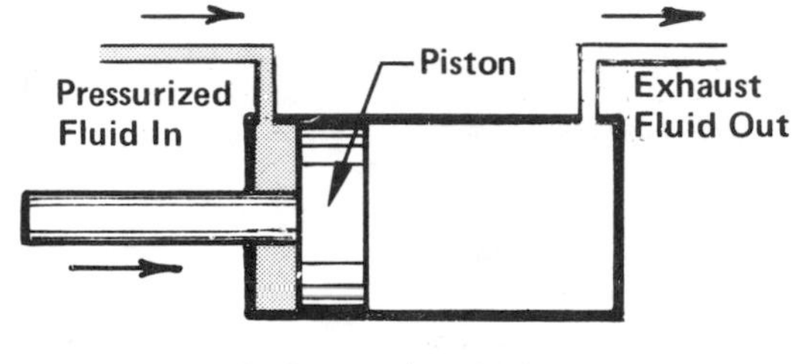

B. Piston Rod Moving In.

FIGURE 2-2. Motion is produced with a double-acting cylinder by pressurizing one side of its piston while venting the opposite side.

the piston to the left. To reverse motion of the piston, the pressure and vent connections must be interchanged to opposite end ports as shown in Part B of the figure. The reversing action of the fluid lines is done with directional control valving to be described later.

Significant Working Areas. Figure 2-3. Inside a cylinder there are three surface areas which are significant. The most important area is the full piston area shown in Part A of the figure. Air or hydraulic pressure working against this area produces force to extend the piston rod.

The second important area is the "net" area, the piston area which surrounds the piston rod, as shown in Part B of the figure. Fluid pressure working against this area produces force to retract the piston rod. Since the full piston area is greater than the "net" area, a cylinder will produce a greater force while the rod is extending than when it is retracting.

However, the cubic volume in the rod end of the cylinder is less than in the blind end. On a hydraulic cylinder this will cause the piston to travel with greater speed (at the same oil flow) while retracting.

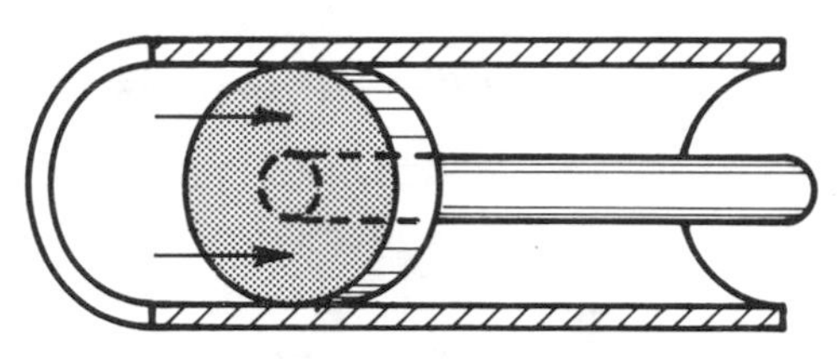

A. Full piston area is effective on the extension stroke.

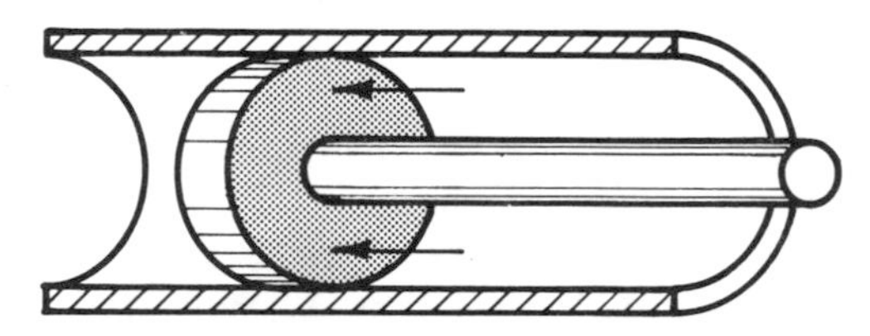

B. Pressure works against the "net" area on retraction stroke.

FIGURE 2-3. The square inch working area is different on opposite sides of the pison of a double-acting cylinder.

The third significant area is the sectional area of the piston rod. Normally the smallest rod listed in a catalog is considered "standard" and is used except where a larger rod must be used for greater mechanical strength. On a hydraulic cylinder, the larger the rod, the smaller the force and the greater the speed of the piston when retracting as compared with force and speed while extending.

With air cylinders, the relative force on extension and retraction is proportional to the full piston and "net" areas just as on a hydraulic cylinder, but piston speed is determined largely by the magnitude of the load being moved.

A cylinder is a "constant horsepower" device. It can produce equal ***horsepower*** either while extending or retracting, although the power may be in a different ratio of ***force*** and ***speed***. For example, it will produce a greater force at less speed while extending and a greater speed with less force while retracting. But the horsepower in each case will be identical.

CALCULATION OF CYLINDER FORCE

Figure 2-4. The magnitude of force, in pounds, produced by a cylinder can be calculated by multiplying gauge pressure (PSIG) times the square inch area of the piston surface against which it is working.

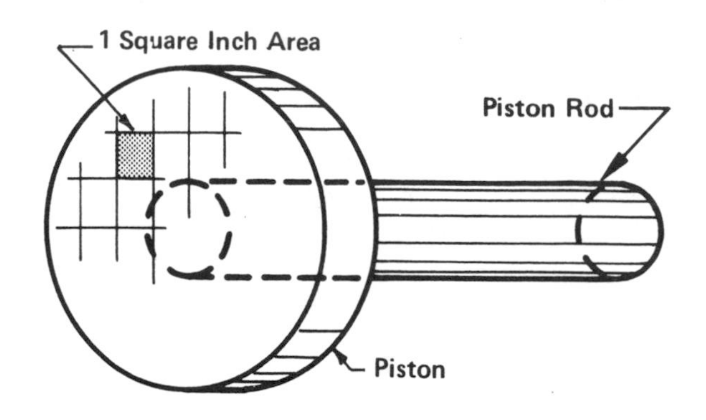

FIGURE 2-4. Total force is calculated by multiplying piston square inch area times pressure difference on opposite sides of the piston.

The square inch areas of circular piston surfaces are listed in machinery and mathematical handbooks, and a chart for all standard piston areas is shown on the following page. Areas may be calculated with this formula:

$$A\ (area,\ sq.\ in.) = \pi D^2 \div 4$$

The diameter, D, must be in inches. The Symbol π is a mathematical concept and always equal to 3.14.

A force equal to the gauge pressure will be produced on every 1 square inch of piston surface.

To find the force produced on the retraction stroke of a piston, multiply gauge pressure times the "net" area. Net area is full piston area minus rod area. Standard air cylinders have relatively small piston rods in which "net" area is about 90% of full piston area, and the rod area is often disregarded in calculations. Hydraulic cylinders may have larger rods, with area up to one-half of full piston area. Rod area on a hydraulic cylinder should not be disregarded in making calculations.

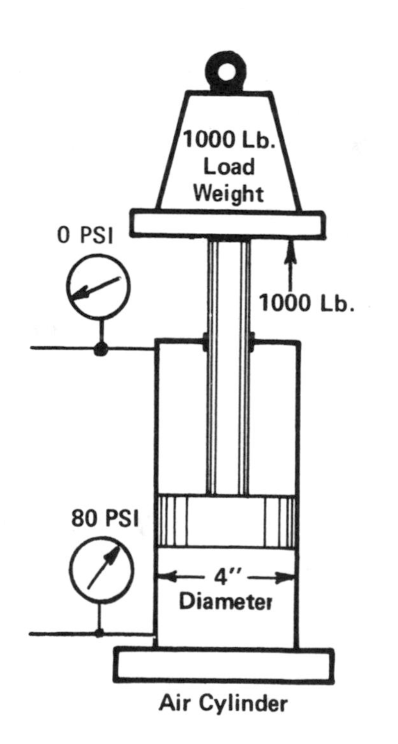

FIGURE 2-5. This cylinder is in a state of balance with the load and will not move up or down.

UNDERSTANDING CYLINDER OPERATION

Figure 2-5. The pressure available to operate any cylinder, whether air or hydraulic, must be greater than the minimum pressure to just equal the load resistance. If not, the cylinder cannot move the load. Additional pressure equal to the flow loss through lines and valving must be available.

Air Cylinder. In Figure 2-5 the cylinder has a bore (piston diameter) of 4 inches, which gives it a piston surface area of 12.57 square inches exposed to pressure for development of force. As shown on the pressure gauges, the pressure difference between bottom and top piston surfaces is 80 PSIG. So the piston has an upward force of 12.57 x 80 = 1000 lbs. This is exactly the force to support a 1000 lb. weight. Therefore, the cylinder is in a state of balance with the load; it cannot move the load upward but will prevent it from dropping.

Additional pressure, above the 80 PSIG balance pressure, must be provided to raise the load. On air operation, any additional pressure, even though small, will produce motion. But the higher the pressure the faster the cylinder will move. The air is supplied from a large reservoir (receiver tank) and the only factor which limits cylinder speed is flow resistance in lines and valving. A higher pressure supplied from the reservoir will force a greater volume of air through the lines to make the cylinder move faster.

The amount of additional pressure above the load balance pressure to give an air cylinder a desired travel speed cannot readily be calculated. A recommended design procedure is to supply a generous amount of overpressure. Then, if the cylinder piston travels faster than desired, its speed can be reduced with flow control valves placed in the lines near the cylinder ports.

A rule-of-thumb states that an additional 25% of pressure above the load balance pressure will usually produce sufficient speed. But for those applications requiring unusually high speed, an allowance of up to 100% over the balance pressure may be necessary.

Hydraulic Cylinders. Unlike air cylinders, hydraulic cylinders are not usually working from a large reservoir of pressurized fluid. They are supplied from a hydraulic pump which is producing a steady flow. The volume of the steady flow is the factor which limits maximum speed. The pressure level of the pump must be sufficient to produce the load balance pressure plus enough additional pressure to make up for all pressure losses in the piping and valving.

SIZING A CYLINDER

Any cylinder, air or hydraulic, should first be sized for bore diameter in combination with a pressure which will produce a force equal to load resistance. This has been called the load balance pressure. However, it must be well below the minimum pressure level available from the air line so that enough additional pressure will be available to force a sufficient volume of air to the cylinder to obtain the desired speed. Pressure losses in the line are not present under load balance conditions but appear as soon as air starts to flow. The tables in Appendices C and D, Pages 260, 262, 282, and 283 may be used to select a combination of bore diameter and load balance pressure.

COMPARISON OF UNIT PRESSURE TO TOTAL PRESSURE

Figure 2-6. In some applications performed through fluid power, like this bonding operation, unit force in the fluid expressed in PSIG is first converted to a summation of total force on the piston rod of a cylinder and is then converted to unit mechanical force exerted on the square inch area of a bonding platen. Bonding consists of gluing two or more pieces together under pressure. Bonding requirements are usually given in pounds per square inch or square foot of mechanical force.

Unit force in the fluid is measured with a pressure gauge and shows the intensity of the fluid force on each square inch of exposed piston surface. Unit fluid force is independent of the size of a cylinder or its length of stroke. On the other hand a weighing scale measures total force without regard for unit size, shape, or square inch exposure area of the object being weighed.

In Figure 2-6, the hydraulic cylinder is supplied with a pressure of 750 PSIG from a hydraulic pump. This unit pressure works against a piston surface area of 50 square inches to produce a total force of 37,500 lbs. on its piston rod. Rod diameter makes no difference provided it has sufficient strength to carry the force without bending.

The piston rod transfers the 37,500 lbs. force to the top platen of a bonding press where it is distributed evenly over the 8'' x 9'' = 72 square inch platen area. If the platen is rigid, each square inch will have a bonding force of 37,500 ÷ 72 = 520 lbs. Or, the bonding force can be expressed as 520 pounds per square inch of physical surface.

Example: A certain material requires 20 pounds force per square inch to make a proper bond. What size cylinder could be used to provide proper bonding force?

Solution: This depends on the size of the sheets to be bonded. Assuming a 2-foot x 3-foot sheet,

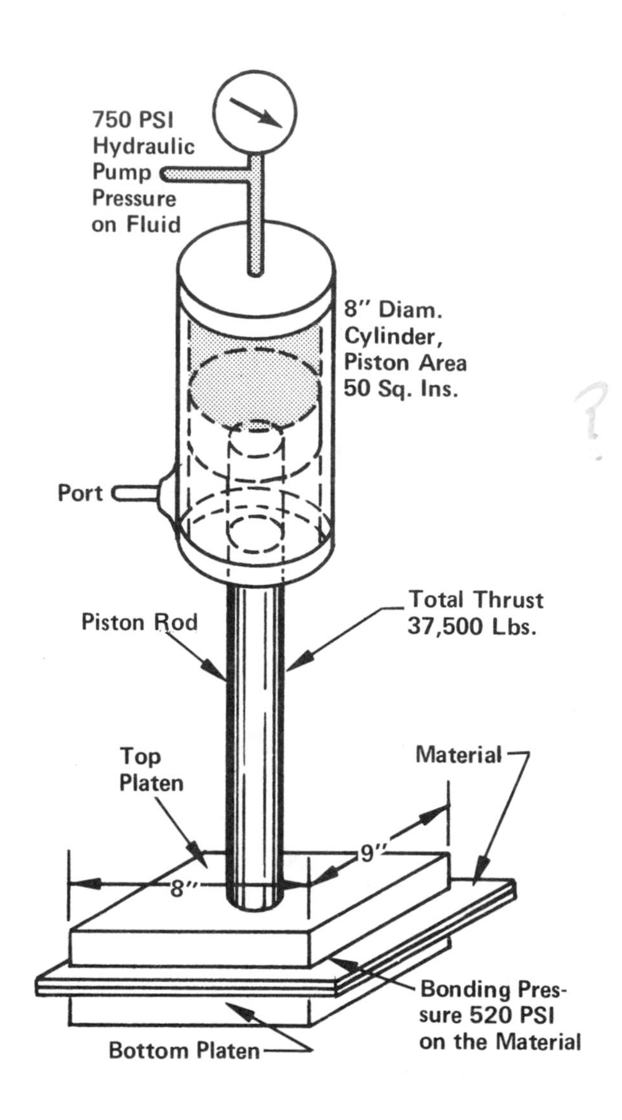

bonding area is 2 x 3 or 6 square feet. Multiply by 144 to convert to square inches: 6 x 144 = 864 square inches. The total force needed from the piston rod of a cylinder is 20 lbs. x 864 square inches = 17,280 lbs. From the cylinder table on Page 260 there is a choice of piston bores in combination with pressure which will produce a total of 18,144 lbs. (17,280 + 5% for cylinder friction). One choice which would produce more than enough force would be a 5-inch bore cylinder working at 1000 PSIG. Another choice would be a 4-inch bore cylinder working at 1500 PSIG. Still another would be a 7-inch bore cylinder working at 500 PSIG, etc.

Note: The pressures in the above paragraph are the actual pressure differences required across the cylinder ports and do not take into account the various pressure losses in the hydraulic system between the pump and the cylinder, nor do they take into account any back pressure from the exhaust port on the cylinder to tank.

FIGURE 2-6. This bonding press illustrates the conversion from unit pressure in the fluid to total force on the cylinder piston rod to unit mechanical force on the bonding platen.

CYLINDER WORKING AT AN ANGLE TO THE LOAD

The full force produced by a cylinder is effective against the load if the cylinder axis is in alignment with the direction of load movement as it was in Figure 2-5. But if the load does not move on the same line as the cylinder piston, only a part of the full cylinder force is effective.

Figure 2-7. In this instance, the cylinder is stroking in a horizontal direction, and is rotating a lever. The end of the lever attached to the cylinder piston rod does not move in the same line as the cylinder motion. At the point in the cylinder stroke where the cylinder axis is at right angles (90°) to the lever, the full cylinder force is effective. But at larger or smaller angles, not all of the cylinder force will be effective for turning the lever.

The full cylinder force, F, is exerted in a horizontal direction. The effective part of the cylinder force is that part which is at right angles to the lever axis. It is shown as Force T on the diagram, and is always less than Force F. The magnitude of Force T in relation to full force F is calculated by the laws of mechanics using trigonometry. However, a simple solution can be worked out with a minimum of mathematics using the table of power factors.

To solve for effective Force T, first solve for total cylinder Force F as shown in preceding pages. Measure or calculate the ***least*** angle between the cylinder and lever axes. This is shown as Angle A on

Angle A Degrees	Power Factor
10	0.174
15	0.259
20	0.342
25	0.423
30	0.500
35	0.573
40	0.643
45	0.707
50	0.766
55	0.819
60	0.867
65	0.906
70	0.940
75	0.966
80	0.985
85	0.996
90	1.000

Power Factors

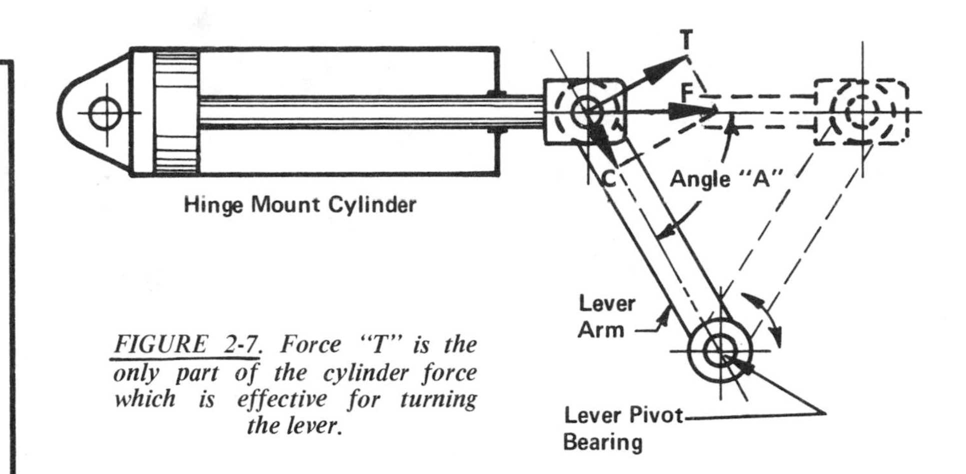

FIGURE 2-7. Force "T" is the only part of the cylinder force which is effective for turning the lever.

the diagram. It could be either on the left side or right side of the lever. Find the factor in the power factor table for this angle. Multiply this factor times the full cylinder Force F.

Although the Force F from the cylinder remains constant throughout the stroke, the magnitude of Force T varies as the angle between cylinder and load axes changes during the cylinder stroke.

Example Calculation. A 4" bore cylinder working at 90 PSIG pressure will produce 1131 lbs. force along its axis. (See force chart on Page 262). From the chart above, the power factor for 65° is 0.906. Therefore, T = 1131 x 0.906 = 1025 lbs. along force line T at right angles to the lever axis.

What happens to the part of the cylinder force which is not effective against the lever? According to the laws of mechanics, cylinder Force F can be replaced by two forces, T and C, which are at right angles to each other, and in combination they produce the same effect as the single Force F. See Figure 2-7. Force T produces torque on the lever; Force C simply adds side loading to the lever pivot bearing without contributing any torque effect on the lever.

Note: Power factors in the chart are sine values of the ***least*** angle between cylinder axis and load axis, on whichever side of the lever it might be. To find power factors for other angles, look up the sine of the desired angle in a table of trigonometric functions in a machine or mathematics handbook.

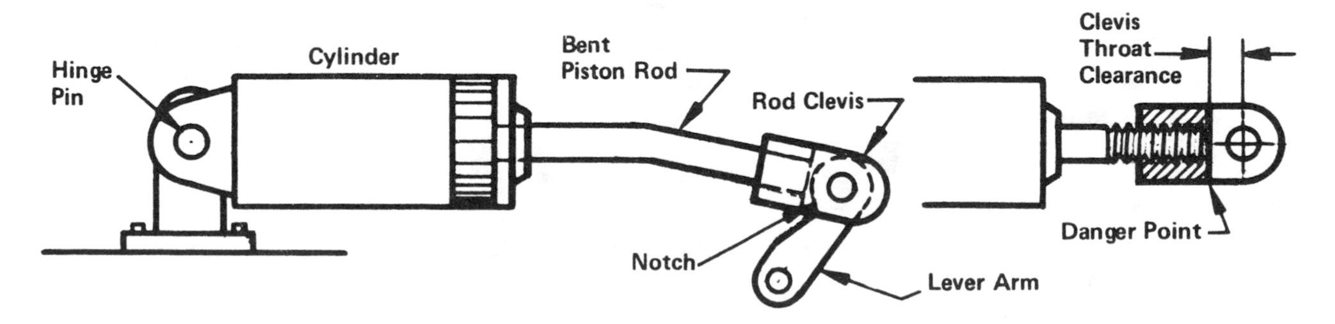

FIGURE 2-8. Levers with a wide swing may bend the cylinder piston rod.

Figure 2-8. Caution! On hinged cylinders with a wide swing, throat clearance on the rod clevis must be sufficient to make sure the lever arm will not hit the bottom of the clevis slot at the end of the cylinder stroke. This could cause the rod to bend or to buckle. It could also cause damage or excessive wear on rod bearings or packings. If necessary, notch the lever or increase the depth of the slot in the clevis.

CYLINDER EXTENSION AND RETRACTION SPEEDS

Speed of an Air Cylinder. Air cylinders operate from a large storage reservoir (receiver tank) which is charged by an air compressor. They have an unlimited supply of air available, and their speed is limited only by the volume of air which can flow through the connecting pipes and valves. The lower the flow resistance in the connecting piping, the more the volume of air which can pass and the faster the cylinder will travel.

It is difficult and usually impractical to calculate the size of plumbing lines and to select valve sizes which will pass sufficient air for a desired travel speed. Since speed depends on pressure supplied to the cylinder, first, sufficient pressure must be available to produce a force from the cylinder which will equal the load resistance. Then, additional pressure must be available to force the air through flow resistances in the feed lines. Rules-of-thumb for sizing are given on Page 57.

A general guide for choosing pipe and valve size to operate an air cylinder is to use the same size as the cylinder port size.

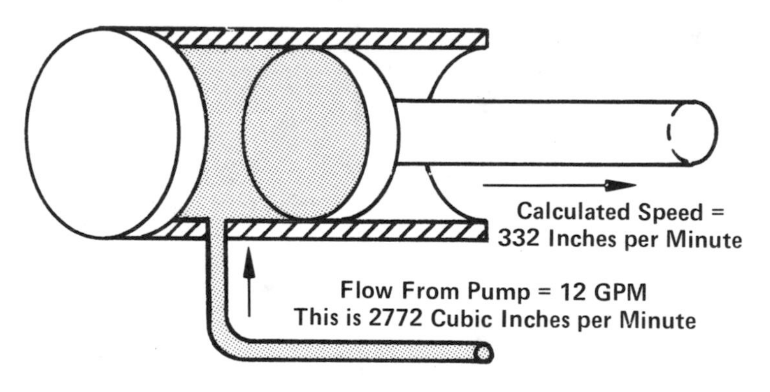

FIGURE 2-9. Example for calculating travel speed of a hydraulic cylinder.

Speed of a Hydraulic Cylinder. We can more accurately predict the speed of a hydraulic cylinder because the fluid is coming at a steady rate from a positive displacement pump. The maximum speed is always determined by the rate of pumping.

Pressure for operation of a hydraulic cylinder is determined in the same way as for an air cylinder: First, sufficient pressure must be available to balance the load resistance. Then, additional pressure must be available to force the hydraulic fluid through the various flow resistances in the system.

Speed Calculation. Figure 2-9. To calculate the travel speed of a hydraulic cylinder, the incoming volume of fluid can be divided by the cross sectional area of the cavity into which it is flowing, being careful to use the proper units:

$$\textit{Speed (inches per minute)} = V \div A$$

A is the square inch area of the cylinder piston; V is the volume of hydraulic fluid entering the cylinder, and must be stated in cubic inches per minute, not GPM. To convert GPM into CIM (cubic inches per minute) multiply by 231.

Example: In Figure 2-9, the cylinder bore is 3¼". Therefore, the piston has an area of 8.30 square inches. Incoming oil is 12 GPM, or 2772 CIM (12 x 231). Speed is 2772 ÷ 8.30 = 332 inches/minute.

This table shows extension speeds for standard bore cylinders. Figures in the body of the table are in inches per minute travel speed. Retraction speeds for the same cylinders will, of course, be faster because of the smaller cavity to be filled with pump oil. Retraction speeds can be calculated with the formula on the preceding page. See Page 261 for a more complete table of speeds for both extension and retraction.

Hydraulic Cylinder Speeds (Inches per Minute)

Piston Dia.	1 GPM	2 GPM	3 GPM	5 GPM	8 GPM	12 GPM	20 GPM	30 GPM	40 GPM
1-1/8	233	466	700						
1-1/2	125	250	375	653					
1-3/4	95	190	286	480	767				
2	73	146	220	368	590				
2-1/4	58	116	174	290	464	686			
2-1/2	47	94	140	235	377	560			
3	32	64	97	163	262	388	653		
3-1/4	28	56	83	139	223	332	556		
3-1/2	24	48	72	120	192	288	480	720	
4	18	36	55	92	147	220	368	550	736
5	12	24	35	59	94	140	235	350	470
6	8	16	24	41	65	96	163	240	326
7	6	12	18	30	48	72	120	180	240
8	5	10	14	23	37	56	92	140	182
10	3	6	9	15	23	36	59	90	118

Figures in this table are piston speeds in inches per minute. Blank area represents unrealistic piston speeds.

DIRECTIONAL CONTROL OF AIR AND HYDRAULIC CYLINDERS

Cylinders are started, stopped and reversed in their direction of travel by one or more directional control valves, 2-way, 3-way, 4-way, or 5-way types. These valves will be described in detail in Chapters 3 and 4.

SPEED CONTROL OF AIR AND HYDRAULIC CYLINDERS

The travel speed of the piston in a cylinder is always controlled by regulating the volume of fluid which flows into the cylinder or which flows out of the cylinder. Control of the flow can be done in several ways, (a), by controlling the RPM of the pump which supplies oil to a hydraulic cylinder, (b), by controlling the displacement (cubic inches per revolution) of the hydraulic pump, or (c), with flow control valves placed either in series or parallel with the cylinder. This last method is more common and the one we will present here. In fact, it is the only method used for air cylinders.

Flow Control Valves. Figure 2-10. Ordinary needle valves can be used to control flow rate into or out of a cylinder, but they are seldom used because they meter equally in both directions of flow. On most applications it is important that the metering valve act in only one direction of flow, and pass fluid in the opposite direction without restriction. Usually, speed is controlled with "flow control" valves rather than with needle valves.

Flow control valves are usually manufactured from bar stock. They contain a metering needle which is by-passed with a check valve for free return flow. Brass valves are preferred for air systems because

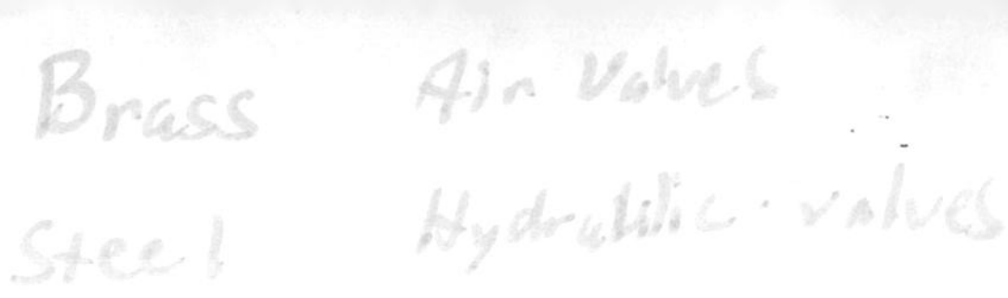

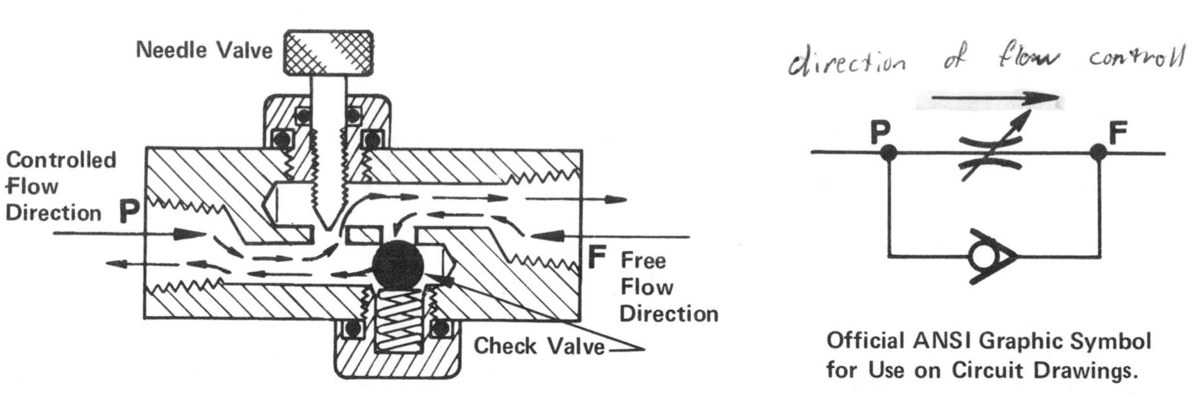

FIGURE 2-10. Simple flow control valve manufactured from brass or steel bar stock.

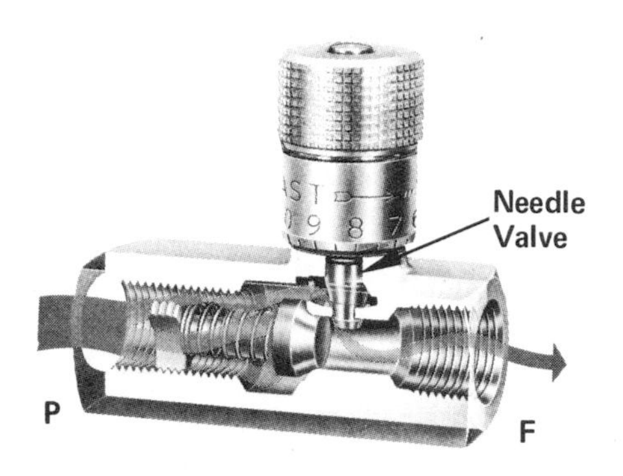

Flow control valve with poppet-type check valve. Flow entering the "P" port on the left is restricted by the adjustment of the needle valve. The check valve is closed.

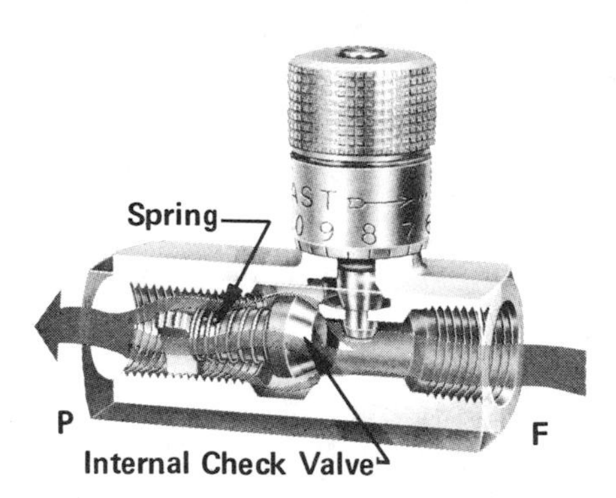

Reverse flow, entering the "F" port from the right can pass unrestricted through the internal check valve.

of the water problem in many systems. Steel valves are preferred for hydaulic systems for their higher pressure ratings and for their better compatibility with hydraulic oil. Flow control valves are offered in sizes starting at 1/8" NPTF (pipe thread) and, depending on the manufacturer, in sizes up to 1" and larger. The metering needle is adjustable, and many of these valves have a micrometer-type calibrated stem to make it possible to return to a previously recorded setting.

Construction of Flow Control Valve. Fluid entering the valve from the left is blocked by the check valve and is forced to pass through the needle opening. Fluid entering from the right can pass unrestricted through the valve with most of it passing through the check valve. Thus, the metering needle can restrict the flow in one direction but not in the other. The standard ANSI graphic symbol for a flow control valve is also shown in this figure.

Note: The direction of controlled flow will be stamped on the body of the flow control valve. In the diagrams to follow, the flow arrow shown adjacent to the valve symbol indicates the direction in which flow can be controlled.

Single-Acting Cylinder. Figure 2-11. A single-acting cylinder is one which delivers power in one direction of piston travel. The piston returns to its starting position by load weight or by spring force. These cylinders are described further starting on Page 70.

The extension speed of a single-acting cylinder can be controlled with flow control Valve V-2 placed in the line between the control valve and the cylinder, and placed so its flow arrow points toward the cylinder. Another flow control valve, V-1, can regulate the retraction speed of the cylinder if it is placed with its flow arrow pointing away from the cylinder. The adjustment of one valve will not interfere with the action of the other because of the check valve built into each valve.

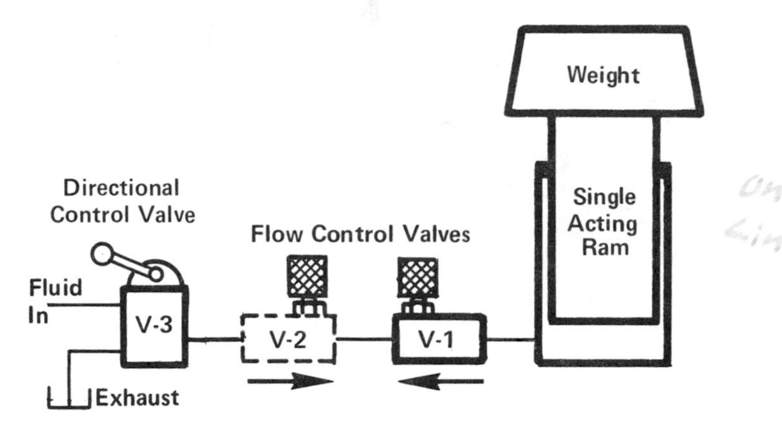

FIGURE 2-11. Single-acting hydraulic cylinder (ram). with flow control valves for control of speed in both directions.

Double-Acting Cylinder. Fig. 2-12. Two flow control valves are usually required for air cylinders, one for each direction of travel. One should be installed in each line connecting the control valve to a cylinder port.

In Figure 2-12, the flow control valves are connected with their flow arrows pointing away from the cylinder. This arrangement is called "meter-out" speed control. On the extension stroke, air is admitted freely to the blind end of the cylinder through the check valve in flow control Valve V-1. Speed is controlled by metering exhaust air through V-2. On the retraction stroke, air enters freely into the rod end through the check valve in V-2 while exhaust air is metered through V-1 to control the speed. On most air systems the "meter-out" speed control arrangement usually gives more stable operation for tool control than the "meter-in" arrangement shown in the next figure because it holds at least partial back pressure on the advancing side of the piston. This reduces the tendency for the piston to "lunge" when a cutting tool driven by the cylinder suddenly breaks through the work. The tendency for an air cylinder to "chatter" when moving a high friction slide at a slow rate of speed is also reduced by using meter-out control, although it is not necessarily eliminated.

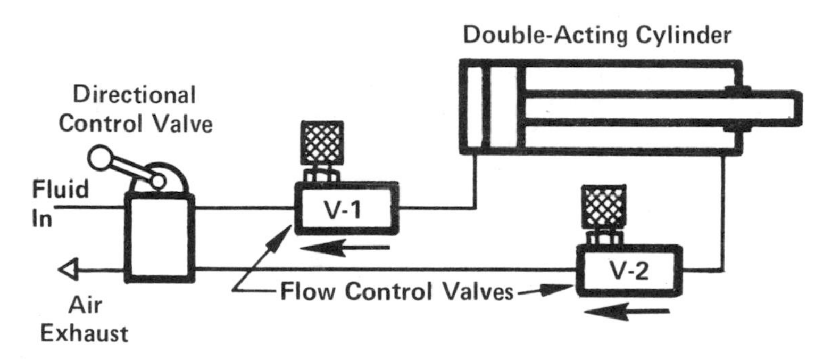

FIGURE 2-12. Speed control of a double-acting air cylinder in both directions. Flow control valves are connected for "meter-out" mode of speed control"

For controlling hydraulic cylinders, only one flow control valve is normally used, either in the rod end or the blind end port, and connected either in a "meter-out" or "meter-in" arrangement. It may be necessary to reduce maximum cylinder speed in the working direction of travel, but it should not be necessary to reduce speed in the return direction. This is limited by the flow volume from the pump. If speed must be reduced in both directions of travel this usually means that the pump has been oversized for the application. To reduce wasted power a smaller pump should be used.

For hydraulic cylinders, meter-out rather than meter-in control is preferred for the same reason of increased stability given above for air cylinders.

Meter-In Speed Control. Figure 2-13. In this diagram, both flow control valves have been turned to face in the opposite direction – with controlled flow arrows pointing toward the cylinder. Extension speed is regulated by metering air into the blind end of the cylinder through Valve V-1. Exhaust air

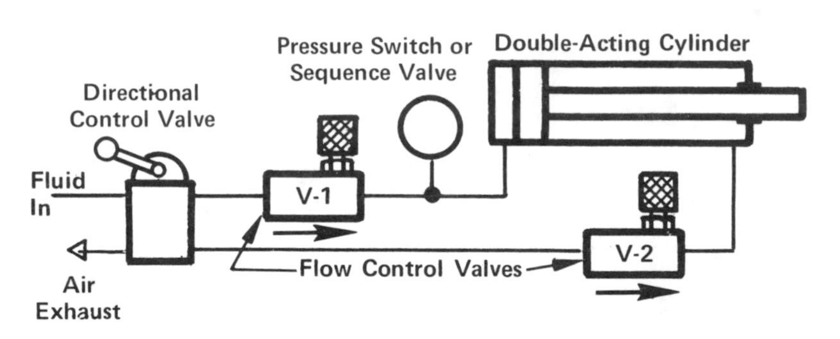

FIGURE 2-13. Speed control of a double-acting air cylinder in both directions. In this diagram the flow control valves are reversed to give "meter-in" speed control.

coming out the rod end of the cylinder can pass unrestricted through the check valve in V-2. Cylinder retraction speed is regulated by metering air into the rod end through V-2. Exhaust air from the blind end can pass unrestricted through the check valve in V-1.

A meter-in speed control arrangement does not offer quite the stability of meter-out control when moving a high friction load at slow speed, and does not offer as great a resistance against "tool lunge" when breaking through the work, but it does have important applications. It usually performs better in circuits where a pressure sensitive device (an electric pressure switch or sequence valve) is teed into the cylinder port behind the advancing piston to give an electric or pressure signal when the piston encounters a heavy load or reaches the end of its stroke. This is an application where two or more cylinders are working together. One cylinder advances and when a signal is received from the pressure switch or sequence valve, a second cylinder is started by the signal. If meter-out speed control were to be used on this application, there would be a high pressure against the piston throughout the entire stroke. The pressure rise when load was encountered would be relatively small and not as reliable for starting the next cylinder.

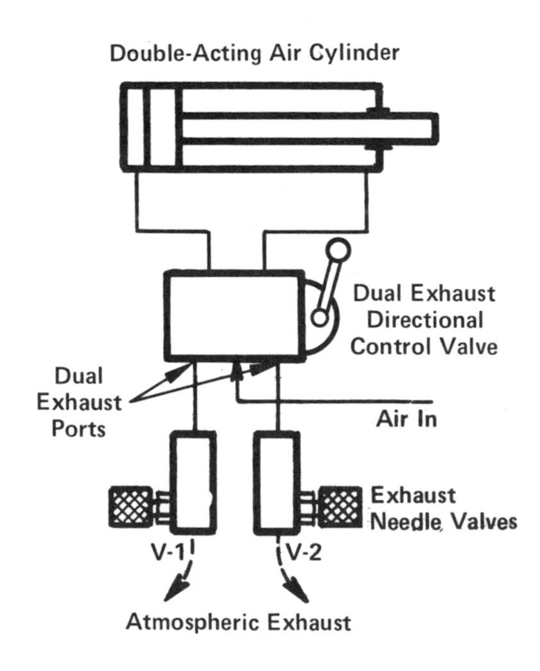

FIGURE 2-14. Speed control of air cylinder with needle valves in dual exhaust valve ports.

Exhaust Speed Controls. Figure 2-14. This is a more economical method of speed control but can only be used for air cylinders which are controlled with a valve which has "dual exhaust" ports. While the cylinder is extending, exhaust air comes out the exhaust port on the right. While it is retracting, exhaust air comes out the port on the left. By separating exhaust air for each direction of travel, a plain needle valve can be screwed into each exhaust port for independent adjustment of speed in each direction.

Dual exhaust needle valves always meter the air coming out of the cylinder. For this reason they cannot be used on applications as described in Figure 2-13 where meter-in control is desired.

Dual exhaust control works best when connecting lines between control valve and cylinder are relatively short, no more than a few feet. On long separations between valve and cylinder a large volume of air is trapped in the lines. Flow control valves placed in the connecting lines near the cylinder ports may work a little better.

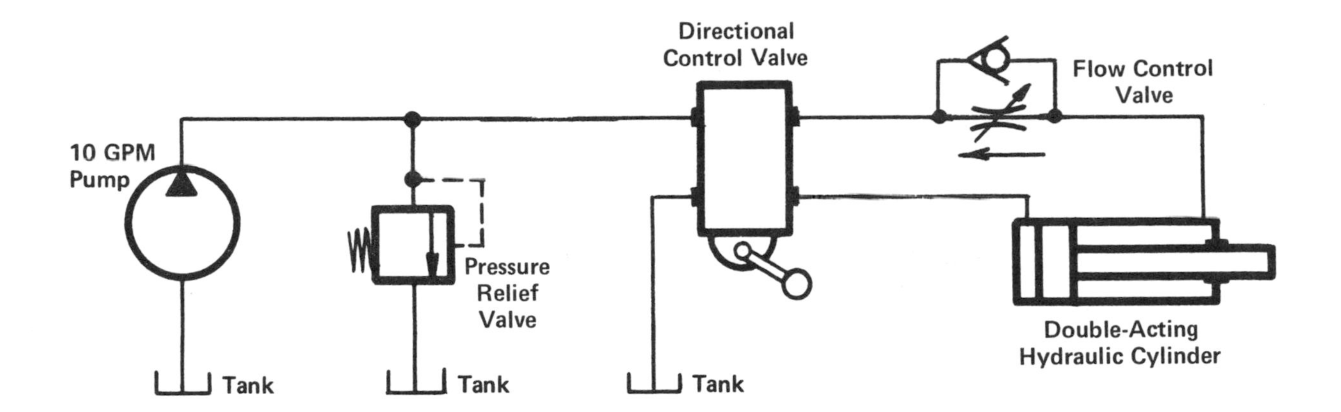

FIGURE 2-15. Series-Connected, Meter-Out Speed Control for a Hydraulic Cylinder.

SPEED CONTROL FOR DOUBLE-ACTING HYDRAULIC CYLINDERS

Most hydraulic circuits are designed to work against a maximum load in one direction of travel, and to return to the starting position as rapidly as possible against little or no load resistance. Usually, a hydraulically powered machine is designed to move the load in the extension direction of cylinder travel because a cylinder will produce more force while extending than when retracting. One notable exception is a broaching machine in which the broach must be pulled through a hole.

Hydraulic cylinder speed is nearly always controlled only in the direction of load movement. Return speed should be as fast as the oil can be supplied from the pump. Cylinder speed is always varied by controlling the rate at which fluid is permitted to flow into or out of it. There are three possible ways of using a flow control valve to meter fluid flow: (a), series connected, meter-out, (b), series connected, meter-in, and (c), by-pass control.

(a). Series Connected, Meter-Out. Figure 2-15. The flow control valve is installed in one of the lines connecting control valve to cylinder. Assuming in this diagram that the working direction is on extension, the flow control valve must be installed in the line from the rod port, and with its controlled flow direction pointing away from the cylinder. On completion of the forward stroke, the piston can retract at maximum speed since pump oil can pass through the flow control valve unrestricted and into the rod end of the cylinder. On some applications, meter-out speed control may help to reduce cylinder "lunge" if a massive load is removed from the cylinder, as by a tool breaking through the work.

With either this series speed control method or the meter-in method to be described, all the oil which is not permitted to pass through the flow control valve will back up and build up sufficent pressure to discharge across the pump relief valve. The energy in this discharged oil converts into heat. Therefore, when the cylinder is operating at less than maximum speed, the surplus energy is heating the oil. When the cylinder is running at full speed, oil heating is at a minimum. Pressure compensated pumps can be used to reduce oil heating at reduced speeds. See Chapter 5.

(b). Series Connected, Meter-In. Figure 2-15. The above circuit can be changed to series meter-in speed control by moving the flow control valve to the blind end port of the cylinder and turning it so its controlled flow direction is toward the cylinder. This is the preferred method of speed control for most hydraulic cylinder applications because it does not produce pressure intensification or high pressure spikes in the rod end of the cylinder, and no more heating than meter-out speed control.

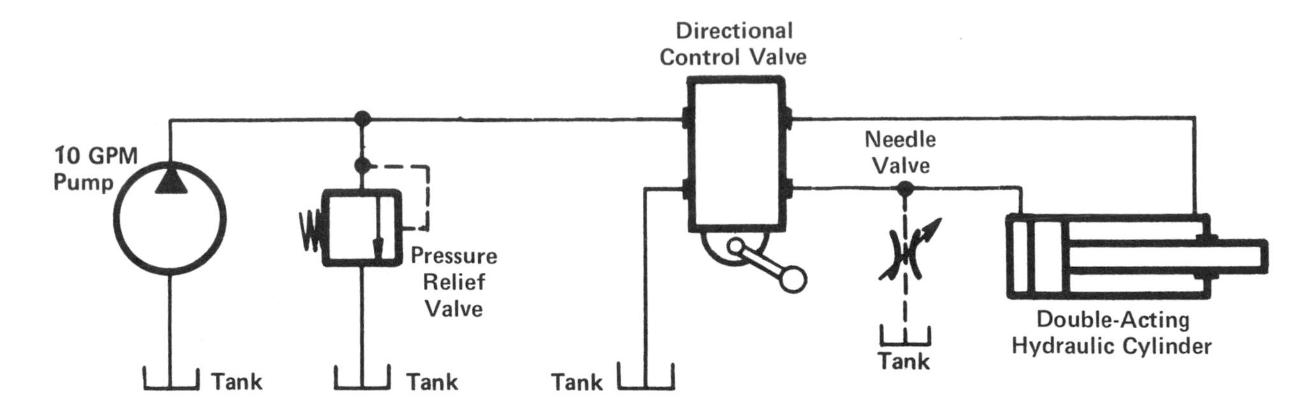

FIGURE 2-16. Parallel, or By-Pass Method of Speed Control for a Hydraulic Cylinder.

(c). By-Pass Speed Control. Figure 2-16. A needle valve connected between the blind end port and tank can vary the cylinder speed by metering off a part of the pump flow to tank. If the valve is completely closed the cylinder will travel at the maximum speed of the pump flow.

Example: If the pump is flowing 10 GPM a certain maximum cylinder speed can be produced when the needle valve is completely closed. If the valve is opened far enough to permit 5 GPM to escape to tank, the cylinder will travel at one-half its maximum speed. The presence of the needle valve or its setting does not affect cylinder retraction speed.

While this is a simple means of controlling cylinder speed, the speed tends to drop more under mid-stroke changes in load resistance than if one of the series speed control methods previously described were used. However, during periods when the cylinder is moving under less than full load resistance, there will be less power waste and less heating in the oil than if a series method were used because the oil discharged to tank through the needle valve can pass to tank at less than relief valve pressure setting. But during periods of cylinder operation under full load resistance, the same amount of heating is produced by any of the speed control methods described.

SYNCHRONIZING TWO OR MORE CYLINDERS

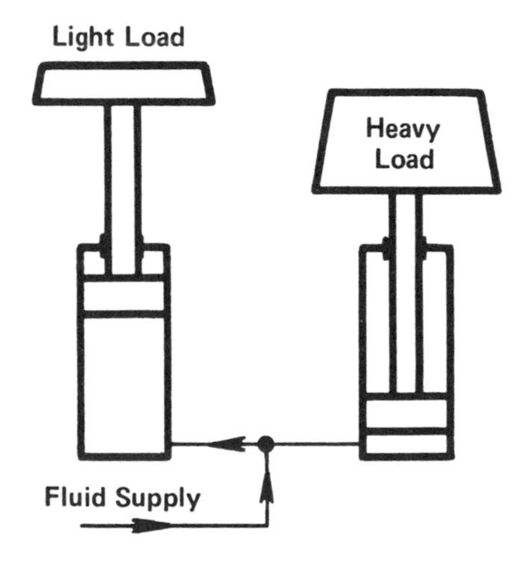

FIGURE 2-17. More of the oil will split off to the more lightly loaded cylinder.

Figure 2-17. If two or more cylinders operate from the same fluid supply, whether air or oil, and receive fluid through the same directional control valve, they will usually travel at different speeds if they are moving separate and unconnected loads. The fluid flows more easily through the path with least resistance, and divides in inverse proportion to the relative loads on the cylinders. The result is that the lightly loaded cylinder will travel faster. In extreme cases if there is a large difference in relative loads, the more heavily loaded cylinder may not move at all until the other one reaches the end of its stroke. Laying out the plumbing so there is a run of equal length to each cylinder helps very little to synchronize their speeds.

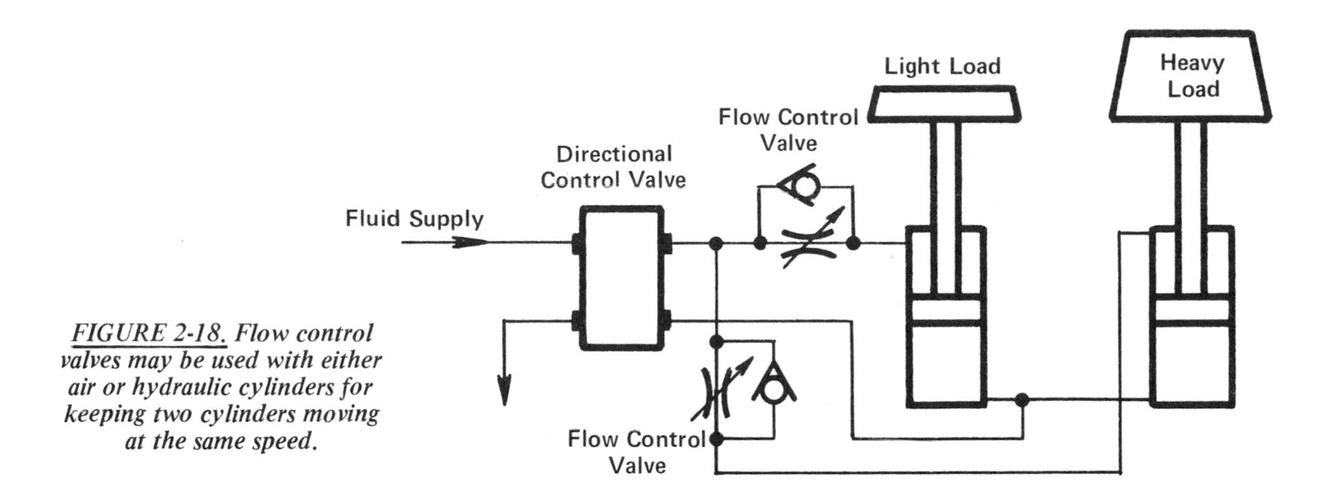

FIGURE 2-18. Flow control valves may be used with either air or hydraulic cylinders for keeping two cylinders moving at the same speed.

Synchronization With Flow Control Valves. Figure 2-18. This method is suitable for air or hydraulic cylinders, and may be used for any number of cylinders operating through one directional valve and which must move at the same speed even though not equally loaded. A flow control valve is installed in the exhaust line of each cylinder. These valves are experimentally adjusted to get the best speed balance. However, if there should be a change in the loading on the cylinders, their speed relation would also change. After the initial adjustment of the flow control valves, if the load on either one or both cylinders should change, or if one cylinder should encounter an increase or decrease of load at some point in its stroke, the speed balance would be affected and the flow control valves would have to be re-adjusted.

For greater accuracy of synchronization on hydraulic cylinders (only), pressure compensated flow control valves can be used. Once adjusted, any change in cylinder loading will not affect cylinder speed. These valves are described in Volume 2 of this textbook series.

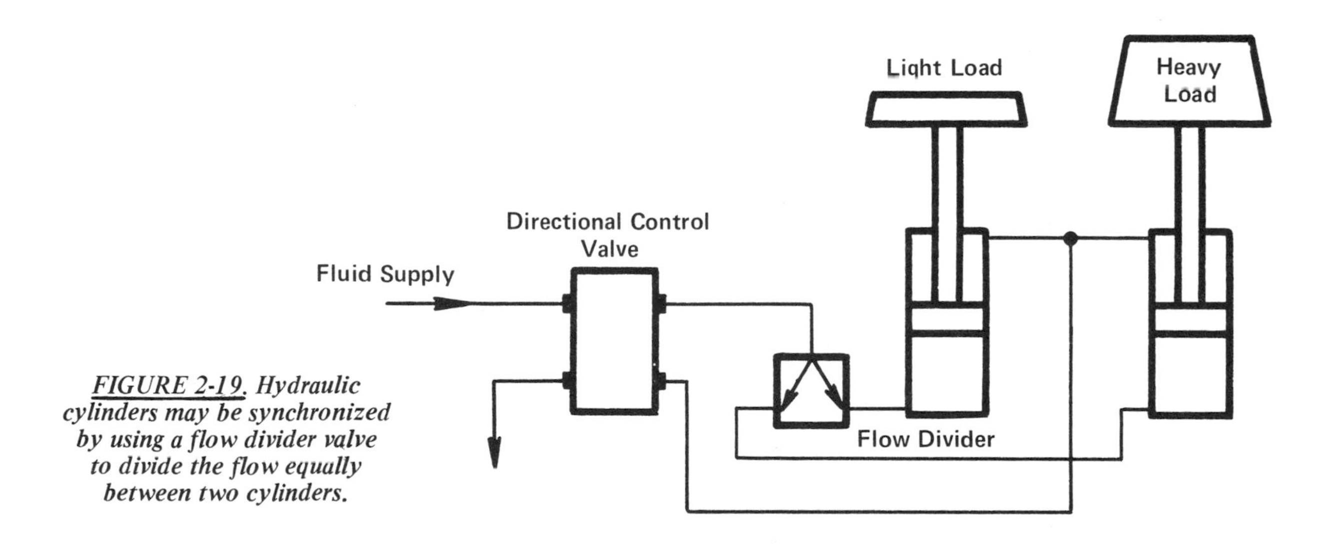

FIGURE 2-19. Hydraulic cylinders may be synchronized by using a flow divider valve to divide the flow equally between two cylinders.

Synchronization With Flow Divider Valves. Figure 2-19. This method is for hydraulic cylinders only. A flow divider valve receives oil flow from the directional control valve and splits it into two equal streams. Speed synchronization between two cylinders is maintained even with changes in load. Flow dividers, spool type and rotary type, are described in Volume 3 of this textbook series.

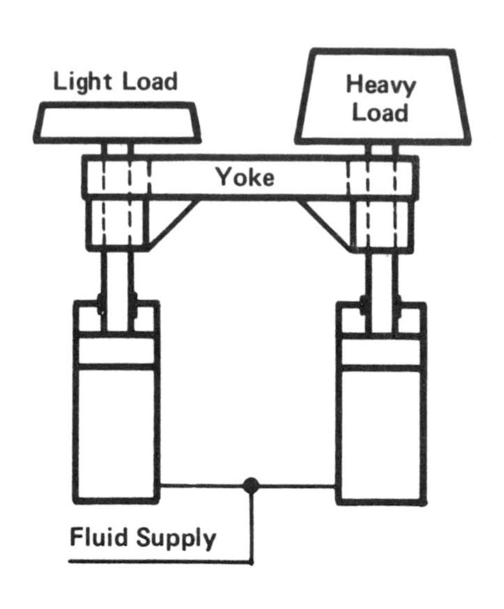

FIGURE 2-20. Mechanical synchronization works better than dividing the flow.

Mechanical Synchronization. Figure 2-20. Where possible, some mechanical means of keeping the two cylinders moving at the same speed is preferred over flow controls or flow dividers, and especially for air cylinders where pressure compensated devices cannot be used. One example of mechanical synchronization is a heavy yoke between the cylinders. This yoke actually transfers some of the load from the more heavily loaded to the more lightly loaded cylinder. In effect, both cylinders are equally sharing the total load. Other methods which are used to transfer load include these:

If each cylinder is working a lever arm, a torsion bar can be added to connect both levers at their hinge points, to force them to rotate through equal arcs.

For rack and pinion synchronization, each cylinder pushes a rack. Pinions of equal diameter mesh with each rack. A torsion bar spans between the pinions, forcing them to rotate in synchronism, and forcing the two cylinders to move at equal speeds.

On mechanisms where the spacing between the cylinders is relatively close, long and sturdy guides at each side of the common platen will keep the cylinders moving together. On all hydraulic presses, the work should be kept as close to the center of the platen as possible. This reduces the strain on any kind of mechanical synchronization method which is used.

On selected applications two cylinders can be connected in series to keep them moving at the same speed. Details of such applications are given in Volume 2 of this textbook series.

Preferred design on a platform lift is to design the structure with sufficient mechanical strength so it can be raised with one large cylinder placed under the center of gravity of the platform. This eliminates the synchronization problem created by using four smaller cylinders, one at each corner.

OTHER CYLINDER TYPES

The standard double-acting cylinder described on Pages 53 and 54 is used in the majority of applications, both air and hydraulic. The cylinders described in this section are variations of the basic model, and may be used on special applications.

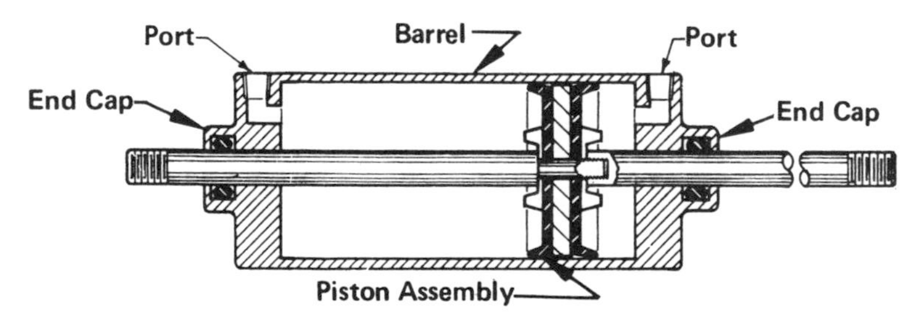

FIGURE 2-21. A double-acting cylinder with double-end-rod construction.

Double-End-Rod Cylinder.

Figure 2-21. This is also a double-acting cylinder but has a piston rod extending from both ends. Both of these rods are the same diameter, and both are attached to the piston. Since both ends of the cylinder are identical, we have to identify them on drawings and in pictures as the "left end" and "right end".

Since both rods are the same diameter, square inch areas are equal on both sides of the piston. At the same pressure, the piston will produce the same force in either direction. In the same circuit, a hydraulic cylinder will travel at the same speed in both directions. An air cylinder may have different speeds in opposite directions depending on magnitude of the load each way.

A double-end-rod cylinder can safely support a much greater side load on the rod than the same size cylinder with a single piston rod. Compare this figure with Figure 2-74 on Page 92.

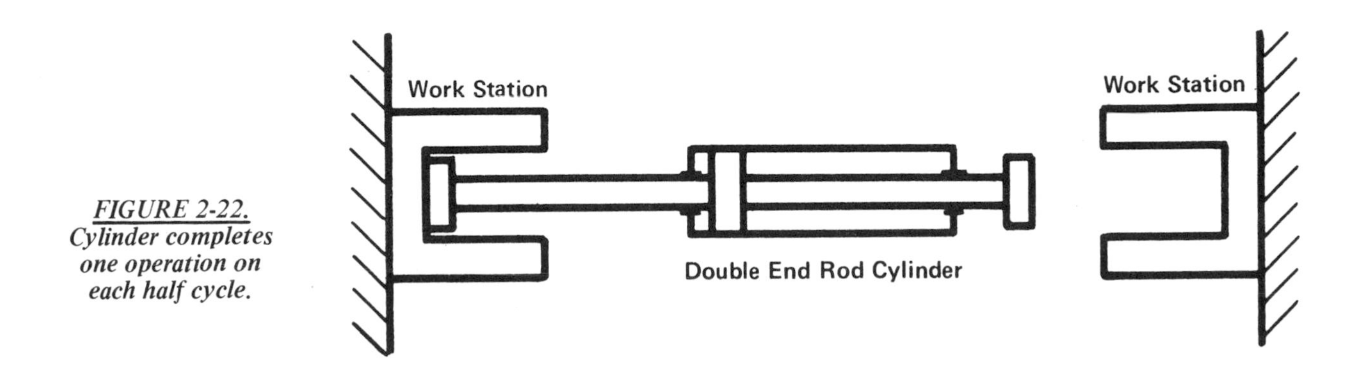

FIGURE 2-22.
Cylinder completes one operation on each half cycle.

Double Work Station. Figure 2-22. Most hydraulic cylinders are working only about half the time because the return stroke takes almost as long as the working stroke. Production rate can be almost doubled on some applications by using a work station at each end of the cylinder stroke. This may be of special advantage with cylinders moving at a slow rate through a long stroke.

The two work stations can have identical tooling or can be set for progressive operations. These cylinders can also be used to clamp one piece (at one end) for machining while a finished piece is being loaded or unloaded at the opposite end.

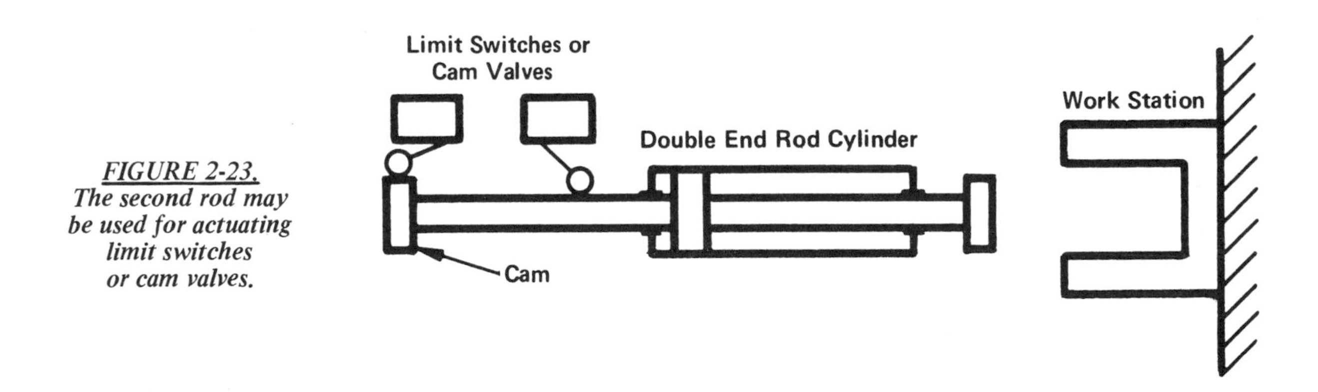

FIGURE 2-23.
The second rod may be used for actuating limit switches or cam valves.

Cam Carrier. Figure 2-23. On some applications the cylinder may be "plugged in" to the machine in such a way that it is difficult or impractical to mount a cam on a moving part of the machine for actuation of a limit switch or cam valve. The extra piston rod can be used to mount an adjustable cam (or cams).

Stroke Limiter. For limiting the forward stroke of the cylinder a sleeve or adjustable split cam can be slipped over the rear piston rod.

Single-Acting Air Cylinders.

Single-acting cylinders produce power in only one direction of piston travel. The piston is returned to its starting ("home") position either by external force reaction from the load, by an external spring, or by a self-contained internal spring. Most single-acting cylinders are available either in small bore and short stroke models for air operation, or in large bore, high pressure models for hydraulics.

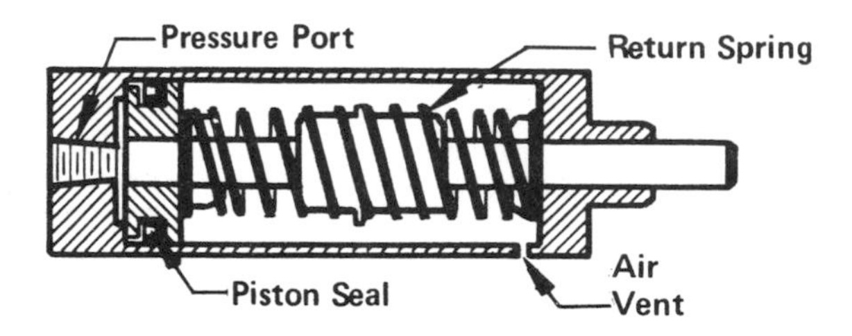

FIGURE 2-24. Single-acting push-type cylinder. Power on extension stroke, return by spring.

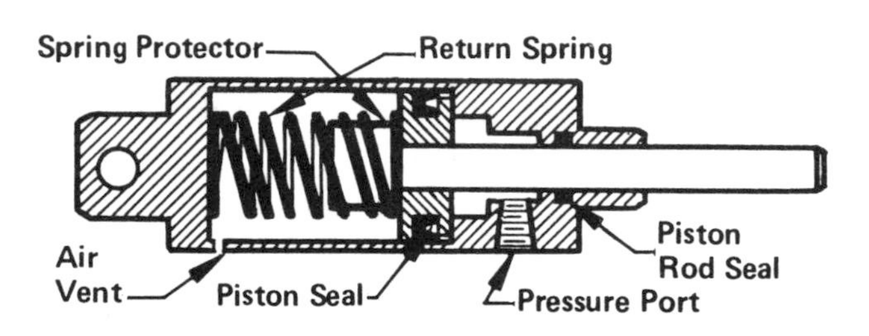

FIGURE 2-25. Single-acting pull-type cylinder. Power on retraction stroke only.

Push-Type. Figure 2-24. Power is on the extension stroke, with spring return to retracted or "home" position. The spring chamber is vented to atmosphere. One rubber cup seal is used on the piston with lips pointing back toward the pressure. No rod seal is needed. Cylinder force is calculated on the full piston area. The spring support keeps spring from sagging and scoring the barrel. It also acts as a forward piston stop and keeps the spring from being over-compressed.

The spring is heavy enough only to retract the piston in free travel (no load). As much as 25% of the cylinder force may be required to compress the spring on the forward stroke, and this is wasted power.

The overall length of the cylinder barrel is greater than that of a comparable double-acting cylinder of the same stroke length; extra length is needed to contain the return spring. The spring chamber should have its vent hole protected with some kind of filter to prevent dirt from being drawn into the barrel.

Control valves for single-acting cylinders must be more than simple shut-off valves. They must have 3-way action. They are described in Chapter 3.

Single-acting cylinders of the kind shown in Figure 2-24 are primarily used for air clamps. They are available only in small bores – from 1/2" to about 2 inches, and in short strokes up to 2½ or 3 inches. They are not usually suited for hydraulic use.

Pull-Type Cylinders. Figure 2-25. Power is on the retraction stroke, with spring return to the "home" position at full extension. One seal is required on the piston, with lips facing the pressure. No seal is required on the rod. Force is calculated on the "net" area – full piston area minus rod area. Spring protector limits cylinder stroke to protect the spring from over-compression.

Non-Rotating Piston Rod. Push-type single-acting cylinders are available with square or oval piston rods to prevent the rod from rotation while it is extending. On these cylinders, a seal is not required on the piston rod.

Displacement-Type Rams.

Figure 2-26. Hydraulic cylinders of high pressure and/or high tonnage can be built as single-acting "rams", but like other single-acting cylinders these rams provide power only while extending, and cannot return to home position under their own power. "Up-acting" rams can sometimes return by their own weight or by weight of the load.

Rams work on the "extrusion" or "displacement" principle. They require a seal only around the rod. The rod itself must be highly polished to preserve the life of the seal, but the inside of the barrel may have only a machine finish. A shoulder is provided on the rod to keep it from accidentally being extruded completely out of the barrel. This shoulder can touch the inside of the barrel to prevent a ram tilt accident. However, the shoulder does not add force to the bottom of the rod because the forces balance on both sides of the shoulder.

Rams, like other single-acting cylinders are controlled with 3-way directional valves which are described in Chapter 3.

Hydraulic Press Ram. Figure 2-27. Hydraulic rams in large sizes are easier to build than double-acting cylinders of equal tonnage capacity. They are also cheaper because the inside of the barrel does not have to have a mirror polish. A further advantage is elimination of piston seal leakage.

When a ram is mounted horizontally or in a "down-acting" position, some external means must be provided to pull it back to its starting position. Two small double-acting cylinders are added for this purpose, and mounted on either side of the ram. They are usually powered from the same hydraulic supply which operates the ram, but air cylinders can be used if they are large enough to provide sufficient force.

Hydraulic rams mounted in an "up-acting" position can sometimes retract by their own weight when their port is vented to tank, but pull-back cylinders can be added if needed.

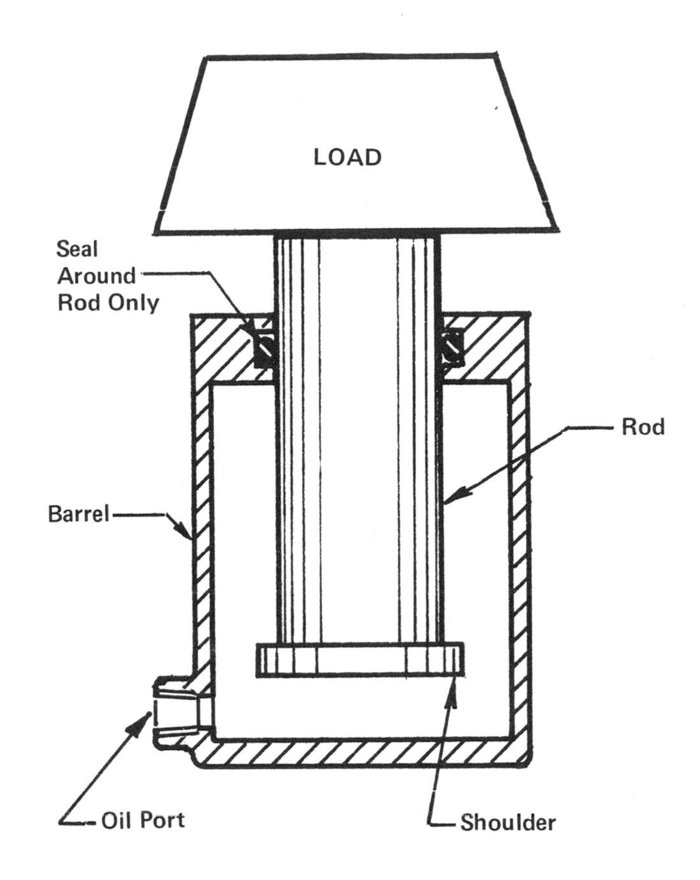

FIGURE 2-26. Displacement-type single-acting cylinder or "ram" for hydraulic service.

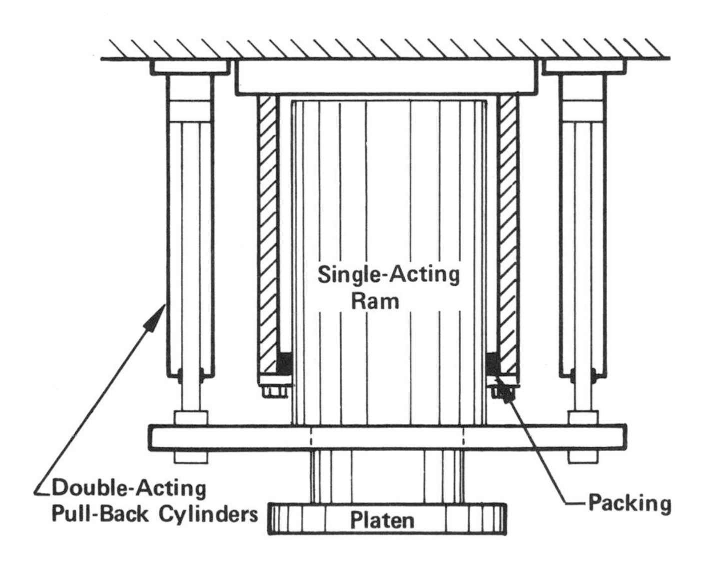

FIGURE 2-27. Hydraulic press with "down acting" ram and small "pull-back" cylinders.

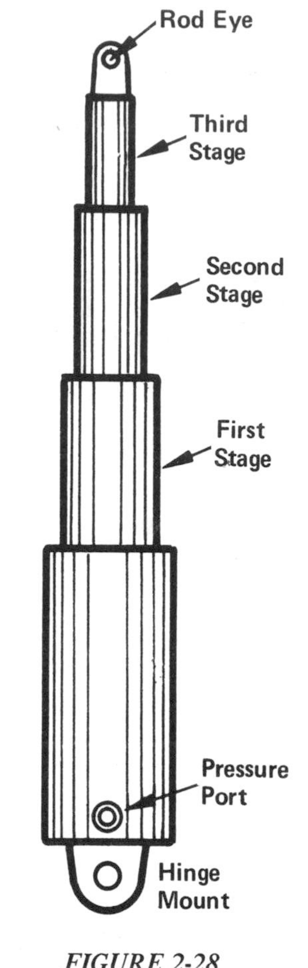

FIGURE 2-28
Three-Stage Telescoping Cylinder

Telescoping Cylinders.

Figure 2-28. Telescoping cylinders are for the purpose of obtaining a long stroke with a relatively short collapsed length. They are normally used only on applications where standard cylinders will not fit, since their cost is several times that of a standard cylinder having comparable lifting force. They are built only for hydraulic service.

Models presently on the market are manufactured with up to five stages (5 sleeves plus barrel) and with maximum strokes from 6 feet to 30 feet, depending on the number of sleeves and the diameter of the largest sleeve. Pressure rating may be from 1000 to 2500 PSI.

When extending, the largest sleeve extends first, then the next largest, etc. Because of the difference in diameter from sleeve to sleeve, cylinder force decreases and speed increases each time the cylinder steps to a smaller sleeve. Lifting force must be calculated separately for each sleeve by using O.D. of the sleeve to calculate working area, then multiplying area times gauge pressure.

Telescoping cylinders are built in both single-acting and double-acting models. Single-acting models are used for pushing. They will not retract under their own power. Usually the weight of the load causes them to collapse when their oil port is vented to tank. The sleeves collapse in reverse order with the smallest sleeve collapsing first.

Double-acting models do their heavy work on extension, but they do have a very small "net" area which can be used to supplement load weight for collapsing them. If they push the load over center for a short distance, they can usually develop enough "pull-back" force to get them back over center so the load weight can collapse them. Pull-back force is different for each sleeve and is calculated from "net" area, which is area of the larger sleeve minus area of the next smaller sleeve. The sleeves may or may not collapse in reverse order, but the sleeve with smallest net area will collapse first, then the sleeve with next smaller net area will collapse next, and so on.

Since speed and load capacity change each time the next sleeve starts to extend, telescoping cylinders must be matched to the load characteristics, and this is illustrated below and on the next page.

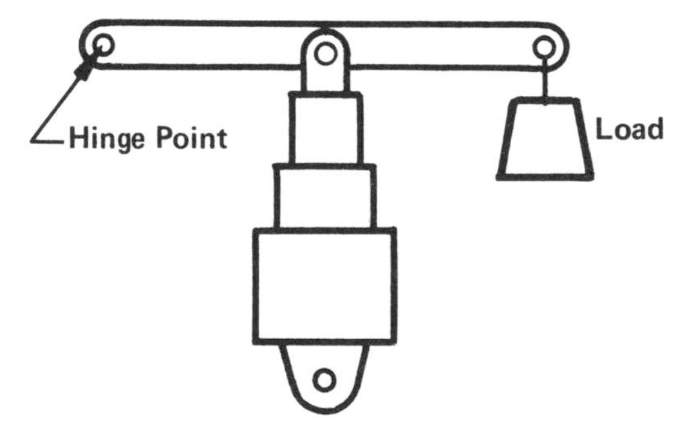

FIGURE 2-29. A Good Application.

Applications for Telescoping Cylinders. A telescoping cylinder starts to extend its first stage with a high force but at a low speed. As each succeeding stage starts to extend, the cylinder force becomes less and its speed becomes greater. For best results the cylinder should be used on applications which have similar load and speed characteristics; that is, maximum force is required to start and move the load during the first part of its travel. Then, the load resistance becomes less the further it moves. For example, in the table below, force at 1000 PSI decreases from 19,000 lbs. on a 5" sleeve to only 12,600 lbs. on a 4" sleeve. An increase in speed on the smaller sleeves, especially on jobs like raising the boom on a crane, might cause the boom to whip.

Good Application. Figure 2-29. Raising a loaded hinged beam from a horizontal to a vertical position is a very good application because the load requirement matches a telescoping cylinder characteristic. The load requires maximum force at the start. Then, the required force diminishes to zero as the beam reaches a vertical position. Speed may become excessive as the beam approaches vertical, so a speed control valve should be available to the operator.

Another Good Application. Figure 2-30. This configuration is used on a dump truck to dump material. Less force is required to raise the load as the cylinder extends because a part of the load is being dumped, and the remaining load requires less force as it moves toward a vertical position.

A Poor Application. Figure 2-31. This is a poor application because very little force is required to start this hinged load from a vertical toward a horizontal position. But the load resistance increases while the cylinder force decreases.

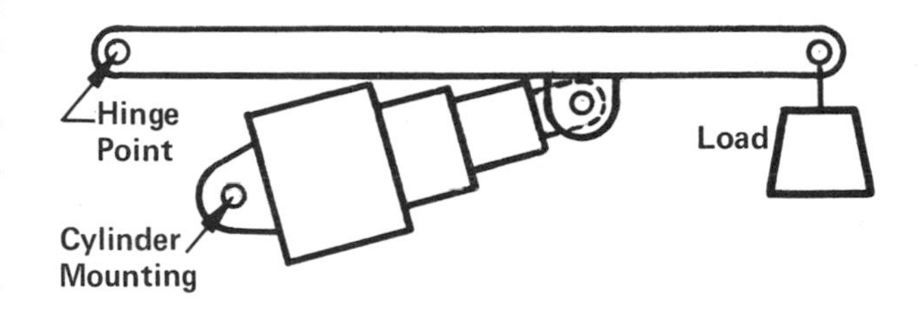

FIGURE 2-30. Another good application for a telescopic cylinder.

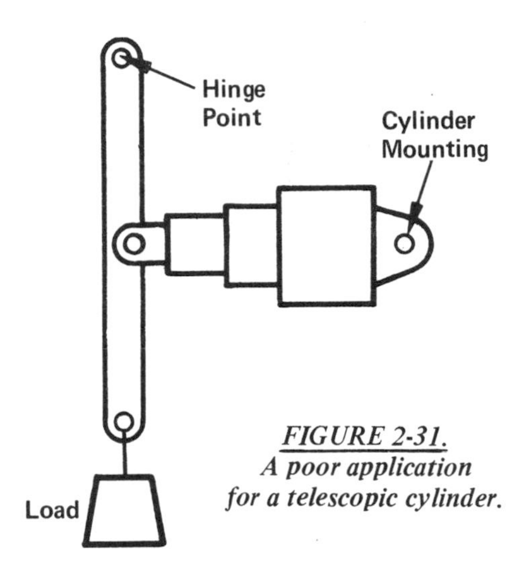

FIGURE 2-31. A poor application for a telescopic cylinder.

Figures in the body of this chart are pounds lifting force. An inspection of the figures in this chart will show that lifting force becomes less on smaller sleeves. Therefore, the number of stages should be kept to the minimum consistent with other requirements of the application. The extension speed becomes progressively greater on the smaller sleeves, in proportion to the decrease in lifting force.

Lifting Force of Telescoping Cylinders

		Fluid Pressure - PSI						
Sleeve Dia.	Sleeve Area Sq.Ins.	500 PSI	750 PSI	1000 PSI	1250 PSI	1500 PSI	1750 PSI	2000 PSI
3	7.07	3500	5300	7000	8800	10,000	12,000	14,000
4	12.6	6300	9400	12,600	15,500	18,800	22,000	25,000
5	19.6	9750	14,700	19,600	24,400	29,400	34,300	39,000
6	28.3	14,000	21,200	28,300	35,300	42,400	49,400	56,400
7	38.5	19,200	28,700	38,500	48,000	57,700	67,300	77,000
8	50.3	25,200	37,700	50,300	62,800	75,300	87,800	100,600

Non-Rotating Piston Rod Cylinders.

Special small air cylinders are available with a piston rod which cannot drift out of a fixed rotational position while it is extending and retracting. The rod has a square, hexagonal, or oval cross section. Because of the difficulty of making a tight fluid seal around the rod, non-rotating rod cylinders are available only in single-acting models, in which no rod seal is required. These cylinders usually have an internal spring for retraction. Such cylinders are often used in tooling applications on jigs, fixtures, and air clamps, where the rod must be secured against rotational drift.

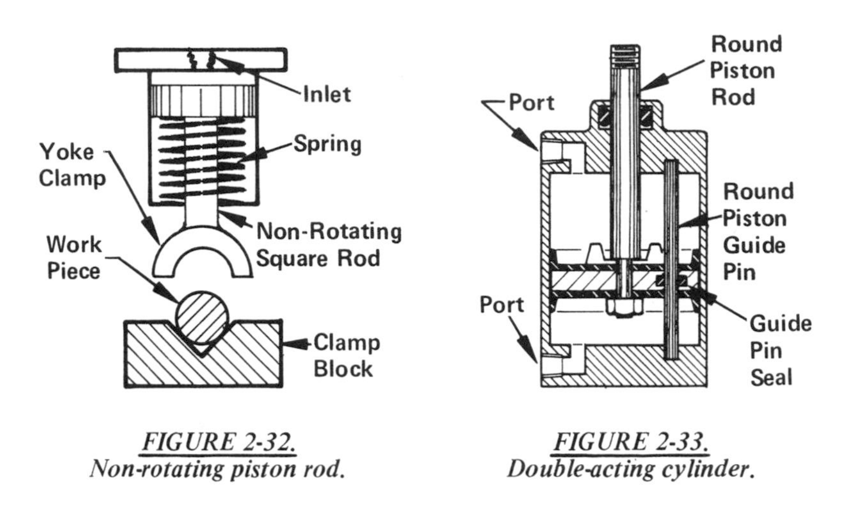

FIGURE 2-32. Non-rotating piston rod.

FIGURE 2-33. Double-acting cylinder.

Figure 2-32. In this tooling application a semi-circular yoke attached to the cylinder rod clamps a workpiece into a circular or V-shaped groove in a clamp block. The cylinder has a square piston rod. No rod seal is required because air pressure is never applied to the rod end. Note: Close tolerance is difficult to hold on rotational accuracy of the piston rod, and these cylinders should not be depended on for precision alignment such as might be required in tool and die work.

Double-Acting Cylinder With Non-Rotating Rod. Figure 2-33. This cylinder is not available as a standard model from any manufacturer. It was constructed by a user who modified a standard double-acting cylinder by running a guide rod between the end caps and through the piston. An O-ring seal was used in a groove in the piston to prevent internal leakage from one end of the cylinder to the other. The cylinder piston must, of course, be locked to the piston rod.

Air Cylinders With a Short Collapsed Length.

Telescoping cylinders are used for hydraulic applications where a short collapsed length is required. But telescoping cylinders are not generally suited to compressed air applications. For air use there are at least three construction methods in use.

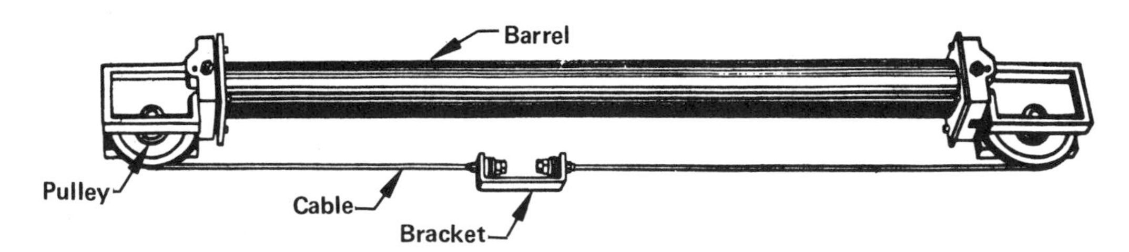

FIGURE 2-34. Cable cylinder for short collapsed length.

Figure 2-34. Cable Cylinder. This cylinder has a piston but no piston rod. Instead, a cable is attached to each side of the piston and goes around pulleys at each end. The load attaches to the bracket. The cable is coated with plastic to insure sealing as it goes through the end caps. This cylinder is double-acting, and while primarily for air, it can be used on low pressure hydraulics.

Another method of constructing a cylinder with short collapsed length is by using a barrel which has a slit running its entire length. Force developed internally by the piston is transmitted through the slit by a power take-off bracket which is attached to the piston. The load is attached to this bracket. A thin steel band covers the entire length of the slit from the inside. This band runs under the power take-off bracket and through the piston. Another steel band covers the slit from the outside and prevents entry of dirt. The two sealing bands are kept in tight contact with the barrel by a series of small permanent magnets which are imbedded in the barrel. This cylinder is double-acting, but is not suitable for hydraulic service.

Still another construction method is the rodless cylinder in which no piston rods are used. The piston travels from one end of the barrel to the other and is magnetically coupled to a yoke which follows the piston movement. The load is coupled to this yoke. The coupling force is from 40 to 265 lbs., depending on the bore size. Stroke length is almost unlimited. This cylinder can also be used on low pressure hydraulics as well as on compressed air.

Special Cylinder Types

Diaphragm Actuator. Figure 2-35. Popular on bus and truck air brakes. Produces a high force on low air pressure because of the large diaphragm area. Usually available only with single action, spring return and only for compressed air use.

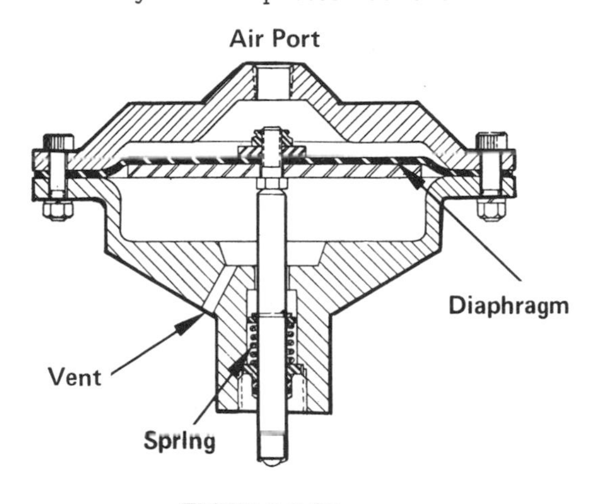

FIGURE 2-35.
Diaphragm brake actuator.

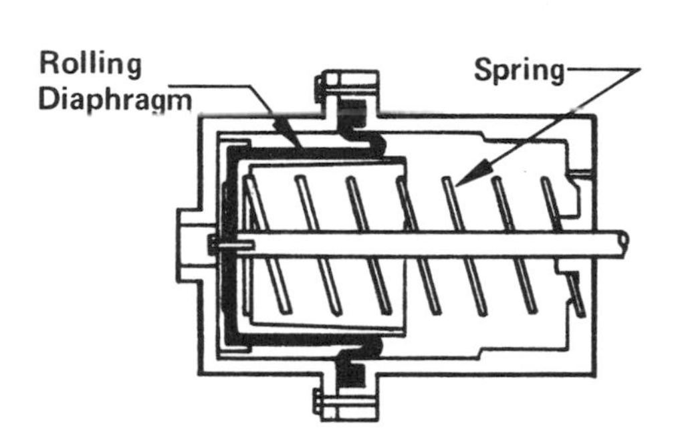

FIGURE 2-36.
Rolling diaphragm actuator.

Rolling Diaphragm Actuator. Figure 2-36. A rolling diaphragm of synthetic rubber instead of a piston gives zero leakage and near-zero breakaway friction. Longer strokes are possible than with diaphragm actuators. These actuators are usually limited to low pressure – less than 500 PSI, but are available in large bores, up to 16 inches, and in strokes up to 12 inches.

Bellows Actuator. Figure 2-37. Synthetic rubber bellows used either for developing force or for shock absorption. Available up to 15,000 lbs. force on 100 PSI air pressure, and with stroke up to9½".

FIGURE 2-37.
Bellows type actuator or air cushion.

CUSHIONED CYLINDERS

The purpose of a cushion on a cylinder is to decelerate the piston as it nears the end of its stroke, to reduce impact and possible damage to the cylinder when the piston contacts the cylinder end cap. A cushion is sometimes needed to decelerate a delicate load to protect it from shock damage as it comes to a stop.

Most industrial cylinders, both air and hydraulic, can be factory ordered with a built-in cushion at either or both ends of its stroke. Cushions cannot be added to a cylinder in the field without extensive modification. Cushions are not available on most mobile and farm machinery cylinders.

A cushion usually increases the overall length of the cylinder and its mounting centers. Cushioning distance varies from 7/8" to 2¼" depending on cylinder bore diameter.

The following diagrams illustrate the operation of a cushion:

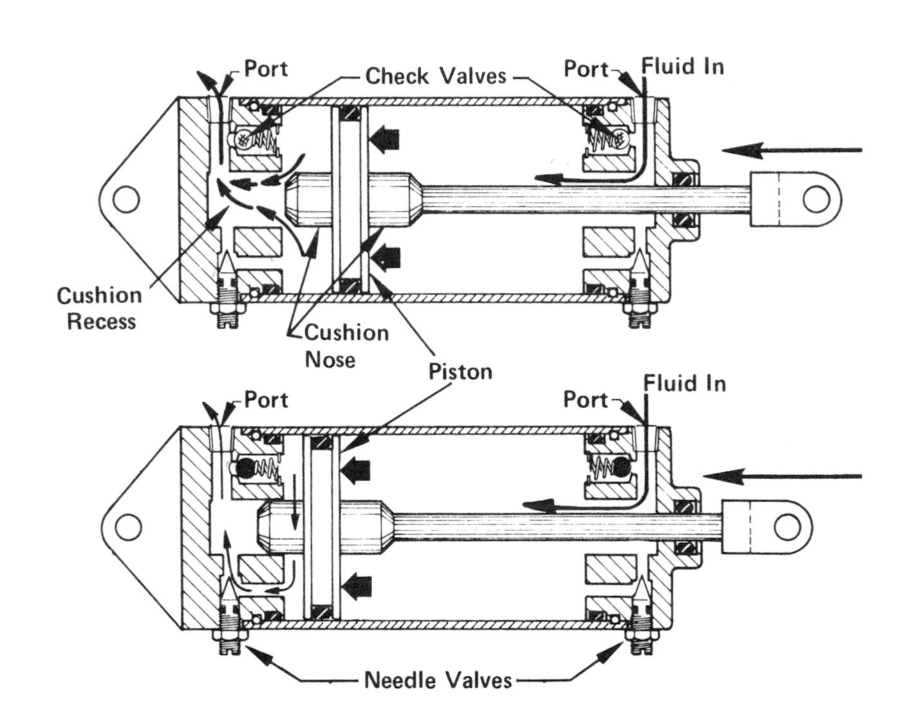

FIGURE 2-38.
The retracting piston is shown just before the cushion nose enters the cushion recess. At this point, fluid is un-restricted in leaving the cylinder.

FIGURE 2-39.
The retracting piston has advanced until the cushion nose has entered the cushion recess. Fluid remaining between piston and end cap must pass through the needle valve to leave the cylinder, and is subject to metering.

Figures 2-38 and 2-39. This cylinder is cushioned on both ends of the stroke. As shown, the piston is traveling to the left (retracting). There is a cushion recess in the blind end cap and a cushion nose attached to the piston. In Figure 2-38 the fluid pushed ahead of the retracting piston can freely leave the cylinder through the normal flow path.

In Figure 2-39, the cushion nose has entered the cushion recess and the normal flow path has been blocked. The remaining fluid is trapped between the piston and the end cap, and to leave the cylinder it must pass through the needle valve. The degree of cushioning can be adjusted with the needle valve. On small bore cylinders, due to lack of space, the cushion may be non-adjustable, especially on the rod end, and on cylinders with larger than standard rod diameters. In these illustrations a cushion is shown on both ends of the cylinder.

Upon starting in the opposite direction, fluid enters the blind end port, passes through the check valve, and works against the full piston area. The purpose of the check valve is to permit the cylinder to start its extension stroke with full force and speed, even before the nose leaves the cushion recess.

On a hydraulic cylinder a cushion is very effective for decelerating a massive load, but should be used with discretion on the rod end of a cylinder because a very high pressure peak may be generated against the rod seal by pressure intensification. See Page 90 for details of pressure intensification.

On most air cylinder applications, cushions are of little value and sometimes useless because of the compressibility of the air trapped between the piston and the end cap. The only applications where cushions are of value are those where the load has little mass and consists primarily of friction. If a massive load must be decelerated, some other method should be used such as cam valves or limit switches placed at a suitable distance from the end of the stroke. Circuits for this type of deceleration are given in Volume 2. An excellent deceleration method for air cylinders is the use of an external hydraulic deceleration device such as a shock absorber. If a cushion is used on an air cylinder, it will be more effective if the flow control valves used for speed control are placed in a meter-out rather than a meter-in mode. This causes the air in the cushion chamber to be pre-compressed, and it will have greater resistance at the moment the cushion nose enters the cushion recess.

MOUNTING STYLES FOR SQUARE HEAD INDUSTRIAL CYLINDERS

Cylinder Standardization. The square head industrial cylinder, with square end caps anchored together with tie rods, is the accepted standard for both air and hydraulic operation of in-plant industrial machinery. It is designed for long service life on indoor equipment, and is seldom used on mobile or farm machinery; other types built for outdoor use are more popular for those applications.

Under the sponsorship of the NFPA (National Fluid Power Association), voluntary standards have been developed which make square head cylinders of one manufacturer physically interchangeable with equivalent models of all other manufacturers who voluntarily comply with the standards. The NFPA recommended standard T3.6.69.7 calls out frame dimensions, mounting centers, overall size, rod size, rod thread, port size, port spacing, etc. But the standards are only for the purpose of interchangeability, and do not specify things which affect the quality of manufacture, such as type or material of seals, materials of construction, fineness of polished surfaces, manufacturing tolerance, piston rod hardness, finish, or plating.

Square head cylinders are grouped into two basic frame sizes: (1), a smaller frame size intended for lighter duty service at lower pressures in the range of 350 PSI in the larger bores to 1500 PSI in the smaller bores. They are offered for both hydraulic and air service, and (2), a larger frame size intended for heavy duty and high pressure hydraulic service. These cylinders usually carry a rating of 3000 PSI for continuous duty under shock conditions, and up to 5000 PSI for non-shock applications. Cylinders within each of these groups are interchangeable between manufacturers, but a low pressure cylinder with Mounting Code MS7, for example, will not interchange with a Code MS7 in the heavy duty series. Standard bore and rod sizes for both of these groups are shown on Page 259.

Outline drawings for the more popular mounting styles are shown on the following three pages. NFPA identification codes are shown on Page 80 for a number of less popular styles.

All major manufacturers can furnish NFPA standard square-head cylinders, and in addition usually have other styles of special cylinders such as mill-type, automotive-type, space-saving type, and others which may not be interchangeable with cylinders by other manufacturers.

Metric Cylinders. The NFPA standards committee has already authorized a set of voluntary standards for bore size, rod size, and mounting dimensions for metric cylinders. These standards are in conformity with ISO (international standards). Metric cylinders are now available from some U.S.A. manufacturers. Eventually they are expected to become world standard sizes, to replace the present inch size cylinders. A list of metric size cylinders is on Pages 282 - 284.

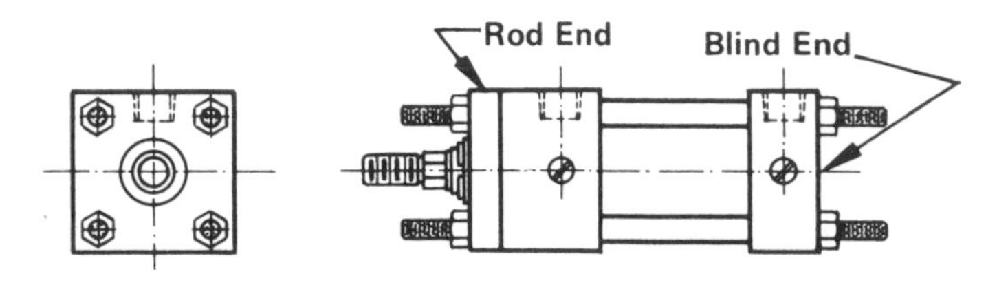

FIGURE 2-40. Extended Tie Rod Mounting. NFPA Mounting Codes MX1, MX2, MX3, and MX4.

Extended Tie Rod Mounting

Figure 2-40. On single-end-rod cylinders, tie rods can be extended on one or both ends as a means of mounting the cylinder. When ordering, use the NFPA mounting code and specify a tie rod extension of (so many) inches on the rod end and/or (so many) inches on the blind end.

Mounting Code MX1. All tie rods are extended on both ends to the length specified by the user.
Mounting Code MX2. All tie rods are extended only on the blind end to the length specified by user.
Mounting Code MX3. All tie rods are extended only on the rod end to the length specified by user.
Mounting Code MX4. Two tie rods are extended on both ends. User must specify which rods are to be extended and must specify the extended length on each end.

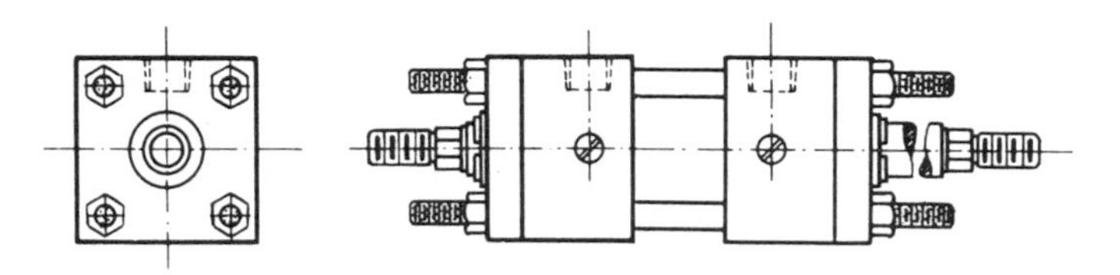

FIGURE 2-41. Double-End Rod Cylinder. Extended Tie Rod Mounting Shown.

Double-End-Rod Cylinders

Figure 2-41. Most cylinders can be furnished with two piston rods of the same diameter. Both are attached to the piston, so as one rod extends, the other retracts. Exceptions are rear flange mounted, fixed clevis, and blind end trunnion mount. The mounting style shown here is extended tie rod mounting.

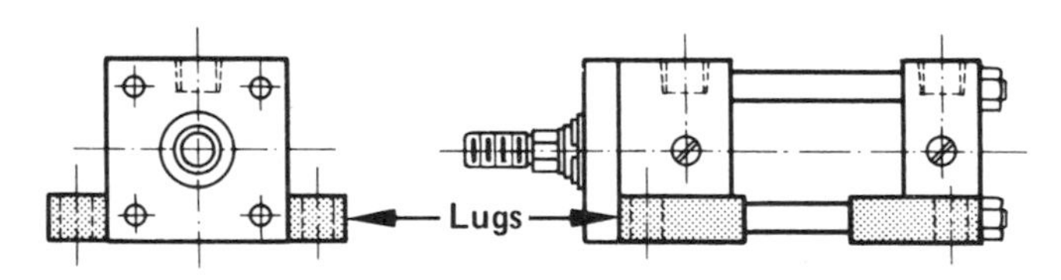

FIGURE 2-42. Side Lug Mounting. NFPA Mounting Code MS2.

Base Mounted Cylinders

Side Lug Mounting. Figure 2-42. General purpose foot mounting for cylinders of small to medium bore operating at no more than 3000 PSI. Mounting lugs are welded to end caps, flush on front and rear, and extended on both sides. All foot mounted cylinders including this and Codes MS3 and MS7 should be pinned or keyed to mounting surface after cylinder alignment so cylinder cannot shift if mounting bolts should loosen.

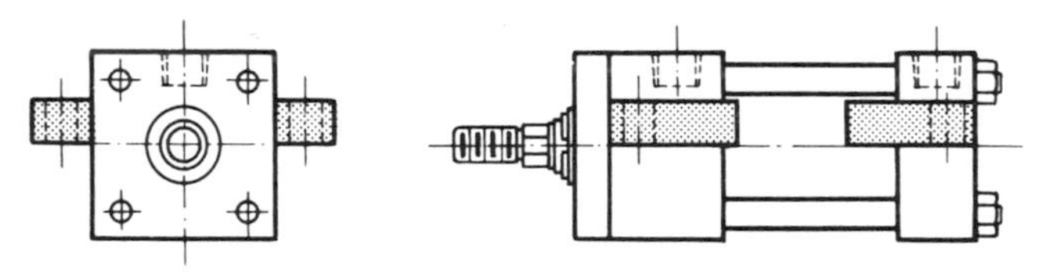

FIGURE 2-43. Centerline Lug Mounting. NFPA Mounting Code MS3.

Centerline Lug Mounting. Figure 2-43. This is a stronger mounting than side lug mounting because the lug mounting surface is aligned with the cylinder axis which eliminates bending moments on cylinder caps, tie rods, and lugs. Recommended for larger bore cylinders, say 5" bore and larger, and for all sizes in which the pressure may peak at 5000 PSI.

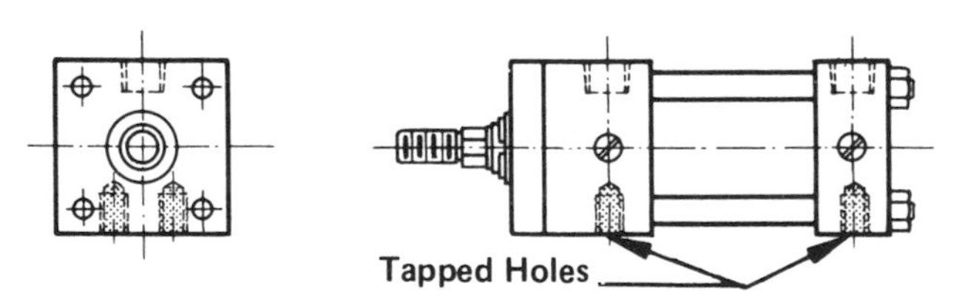

FIGURE 2-44. Tapped Hole Mounting. NFPA Mounting Code MS4.

Tapped Hole Mounting. Figure 2-44. Mounts by two tapped holes in each head. Cylinder should be keyed or pinned to the mounting surface after alignment with the load. Not recommended for larger than 5" bore, nor for more than 3000 PSI.

End Lug Mounting. Figure 2-45. Mounting lugs are flush on the sides and extend front and rear. This mount must be used with care, as the lugs may interfere with clevis on rod (if used), or may make rod cartridge removal difficult.

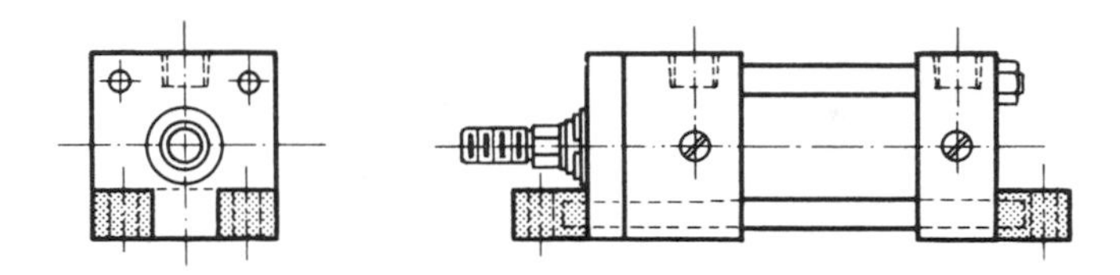

FIGURE 2-45. End Lug Mounting. NFPA Mounting Code MS7.

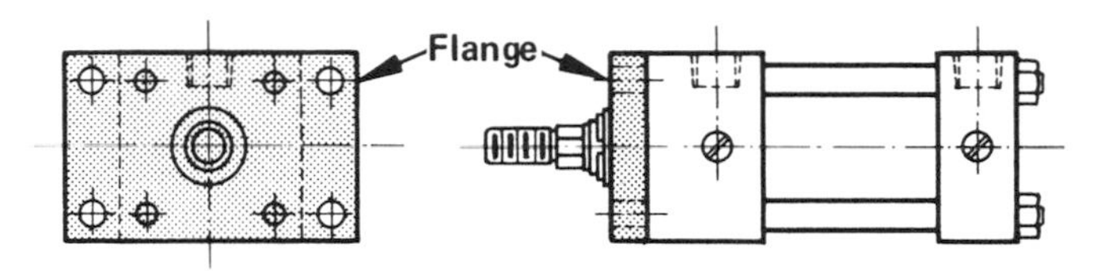

FIGURE 2-46. Rectangular Flange, Rod End. NFPA Mounting Code MF1.

Flange Mounted Cylinders

Rectangular Flange Rod End Mounting. Figure 2-46. This flange bolts to the rod end cap. It may bend if not properly supported on large cylinders operating at high pressure. If space permits, the square flange shown in Figure 2-47 is preferred on heavy duty applications or pressures over 3000 PSI.

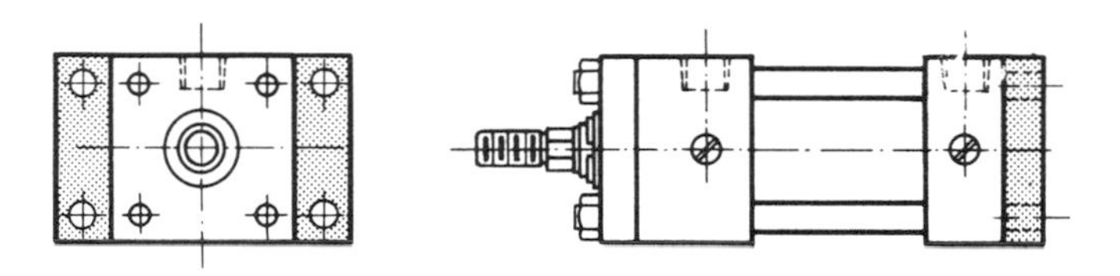

FIGURE 2-47. Rectangular Flange, Blind End. NFPA Mounting Code MF2.

Rectangular Blind Flange Mounting. Figure 2-47. This flange bolts to the blind end cap of the cylinder. It will handle full cylinder force in the extension direction if backed against a structural member of the machine. The larger square flange of Figure 2-49 is preferred for heavy duty service especially at pressures over 3000 PSI.

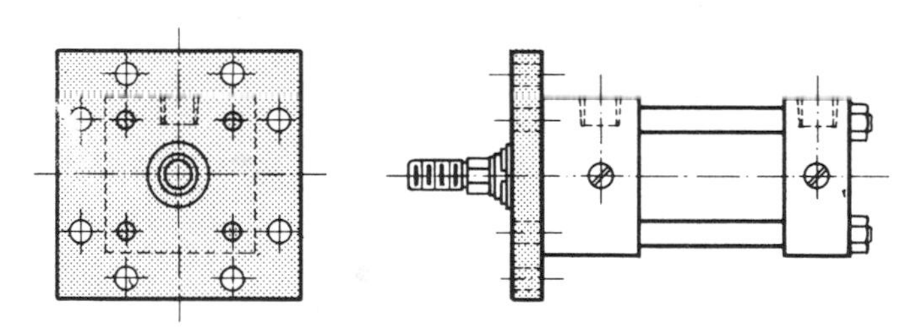

FIGURE 2-48. Square Flange, Rod End. NFPA Mounting Code MF5.

Square Flange Rod End Mounting. Figure 2-48. This large square flange bolts to the rod end cap. It is larger and thicker than a rectangular flange and is recommended if pressures may exceed 3000 PSI or on large bore cylinders, over 5" bore.

Square Flange Blind End Mounting. Figure 2-49. This large square flange bolts to the blind end cap. It is much larger and thicker than a rectangular flange. It is recommended for all cylinders operating over 3000 PSI, and for all cylinders with bores larger than 4" or 5".

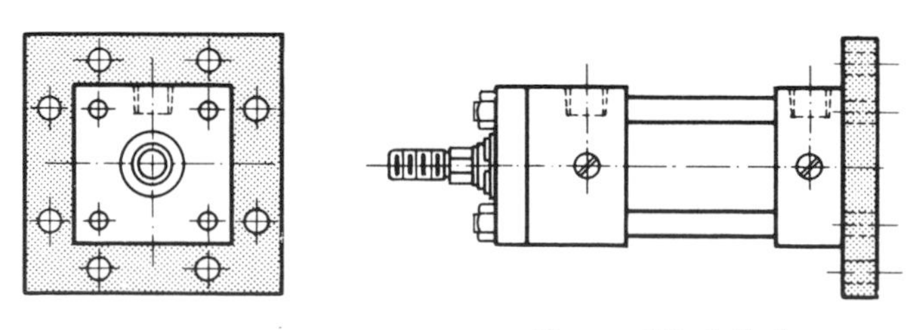

FIGURE 2-49. Square Flange, Blind End. NFPA Mounting Code MF6.

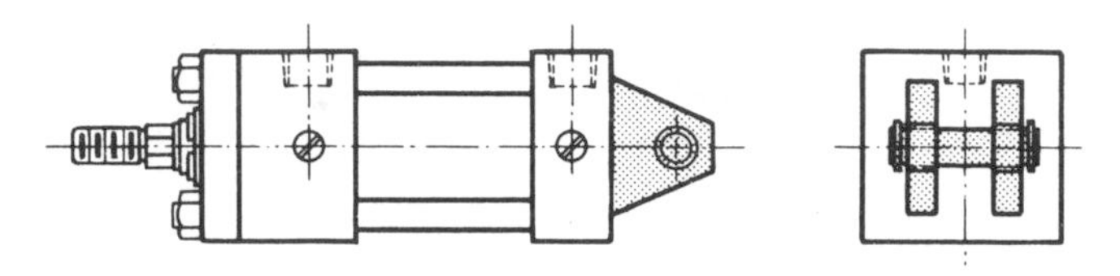

FIGURE 2-50. Fixed Clevis Mounting. NFPA Mounting Code MP1.

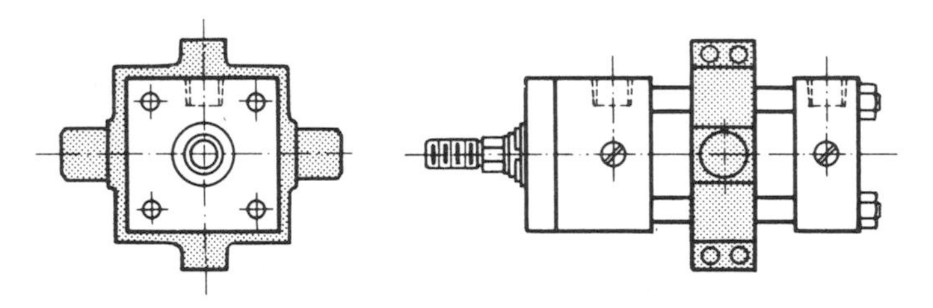

FIGURE 2-51. Intermediate Trunnion Mounting. NFPA Mounting Code MT4.

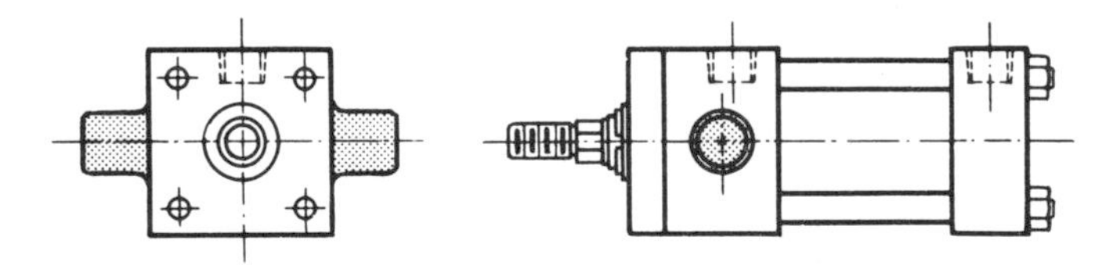

FIGURE 2-52. Trunnion Mounting, Rod End Cap. NFPA Mounting Code MT1.

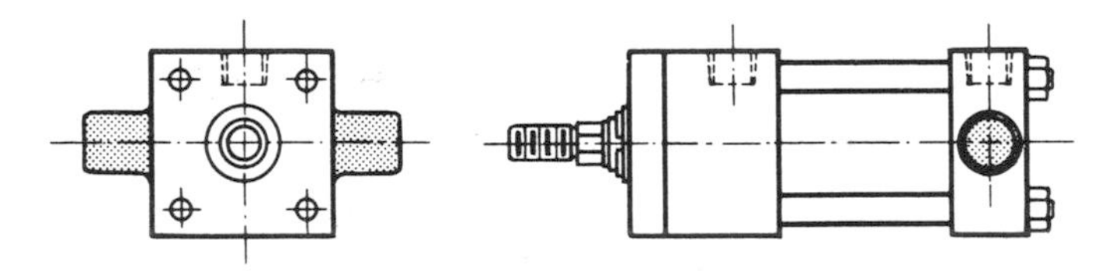

FIGURE 2-53. Trunnion Mounting, Blind End. NFPA Mounting Code MT2.

Hinge Mounted Cylinders

Fixed Clevis Mounting. Figure 2-50. This is the most popular mounting for cylinders which must swing during their stroke. However, large bore cylinders with long stroke may be too heavy for this mounting when they are operating horizontally, and one of the trunnion mounts can be used.

The NFPA mounting code does not include accessories such as rod clevis or eye, nor base or rod pivot mounting brackets. Order separately as needed.

Intermediate Trunnion. Figure 2-51. This is an optional swivel mounting for heavy cylinders with long stroke. By properly locating trunnion position, the cylinder weight can be balanced when rod is extended. Distance of trunnion from front face of cylinder must be specified when ordering. Trunnion position cannot be changed after cylinder is manufactured.

Rod End Trunnion. Fig. 2-52. Trunnion is in a fixed position on the rod end cap.

Blind End Trunnion. Fig. 2-53. Trunnion is in a fixed position on the blind end cap.

Trunnion Note: Pivots are not designed to stand bending loads. They should be supported on pillow blocks placed as close to the cylinder as possible. Heavy cylinders which are mounted horizontally, should not use blind end trunnion mounts.

Other NFPA Mounting Styles

Although the above mounting styles are the more popular, and available from all major cylinder manufacturers, the following styles may also be available from some sources. They are described in NFPA Standards T3.2.60.1, "Dimension Identification Code for Fluid Power Cylinders".

MS1	Side End Angles	**MR1**	Rod End Male Rabbet	**ME4**	Blind End Square Cap
MS5	Through Holes	**MR2**	Rod End Female Rabbet	**MT3**	Intermediate Fixed Trunnion
MS6	End Plates	**ME1**	Rod End Circular Cap	**MP2**	Blind End Detachable Clevis
MF3	Rod End Circular Flange	**ME2**	Blind End Circular Cap	**MP3**	Blind End Fixed Eye
MF4	Blind End Circular Flange	**ME3**	Rod End Square Cap	**MP4**	Blind End Detachable Eye

CYLINDERS FOR MOBILE EQUIPMENT

At this time there are no industry-wide standards for mobile cylinders, so specifications vary between manufacturers. There are a number of catalog standard models available, but where a very large number of cylinders are to be purchased, the cylinders are usually custom designed specifically for the application.

Mobile cylinders are those which are built primarily for use on outdoor moving equipment such as road machinery, trucks and buses, construction machinery, lumbering, material handling, mining, and similar applications. Cylinders on mobile applications usually do not make as many working cycles per day, per month, or per year as compared to cylinders on industrial applications like hydraulic presses. Mobile cylinders may operate in a dirty environment, so they are built to be easily and quickly disassembled for replacement of piston and rod seals or other service. Since they are often used on equipment built for re-sale, they must be built for a very competitive price. For this reason and because they operate at lower pressures than many industrial applications, they are constructed of thinner or lighter materials. They are not recommended for heavy duty in-plant industrial applications because of their shorter service life expectancy.

The majority of mobile cylinders are designed for swivel mounting, with a clevis or eye on the back and with a clevis or eye on the piston rod. They can be obtained for foot or flange mounting but this usually involves a special factory order. They are usually not available with optional features such as a wide choice of piston rod diameters, cushions, etc.

Figure 2-54. A typical mobile cylinder has its blind end cap welded to the barrel, and a clevis or tang welded to the end cap. The rod end cap is keyed or screwed into the barrel for easy removal when replacing the internal seals. The end of the piston rod is fitted with a clevis complete with hinge pin and retainer clip. Oil ports are standard taper pipe threads; straight thread ports have not yet become popular on mobile cylinders. Bore sizes range from 2" to 4", and standard stroke lengths from 6 to 24 inches. Since there are no industry-wide standards, some manufacturers offer bore and stroke sizes outside the ranges given. Usually there is only one rod size available for each bore and stroke. The cylinders are designed with rod diameter large enough to avoid rod buckling from column failure, according to cylinder pressure rating and stroke length.

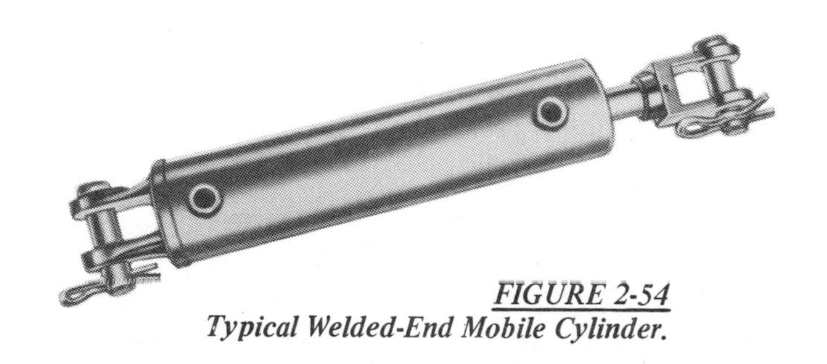

FIGURE 2-54
Typical Welded-End Mobile Cylinder.

FIGURE 2-55
Typical Tie-Rod-Type Mobile Cylinder

Figure 2-55. This is another version of a mobile cylinder. Cast iron end caps are held to the barrel by tie rods. These cylinders are not dimensionally interchangeable with the square head industrial models. This construction appears to be less popular than the tie-rodless type, possibly because they are more costly to manufacture and a little more difficult to service.

FARM MACHINERY CYLINDERS

The ASAE (American Society of Agricultural Engineers) adopted a set of recommended standards in 1949 which applied to hydraulic cylinders used on farm tractors and on trailing-type farm implements. The standards were revised several times, the latest revision being in June 1966. These standards carry the number ASAE S201.4 and are available from the ASAE. Their address at the time of this printing was 420 Main St., St. Joseph, MI 49085.

The ASAE standards apply to dimensions which would affect interchangeability of cylinders from one implement to another or to replacement cylinders to fit existing mountings. The dimensions include closed and open pin-to-pin center distances, clevis throat width, and clevis pin diameter. They do not cover bore or rod diameter, pressure rating, construction, or type of seals.

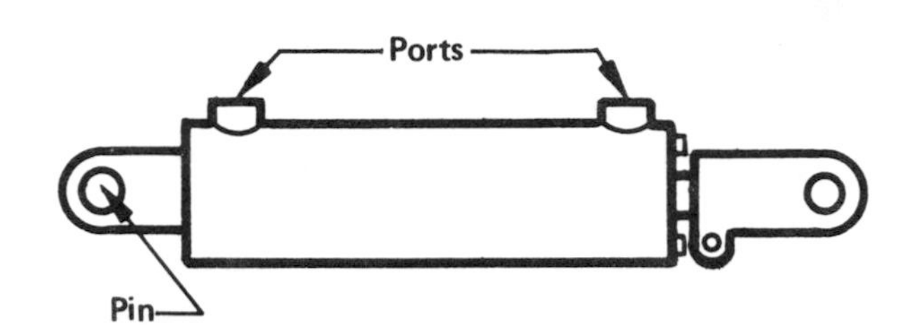

FIGURE 2-56. Farm Machinery Cylinder.

Figure 2-56. Farm cylinders are stripped to the bare essentials, and are standardized in only two stroke lengths, and in a few bore sizes as listed below. Although they are quite adequate for implement use, when used on industrial hydraulic systems they are considered to be very light-duty components. Their construction is simple, with O-ring piston and rod seals, usually with back-up rings. Tie-rodless construction as illustrated in this figure is the most common construction, but some manufacturers offer them in tie rod construction similar to that shown in Figure 2-55. Some of them are available with adjustable stroke feature.

When ordering a farm cylinder for replacement, the system operating pressure should be ascertained. Most tractors have pumps producing from 1500 to 2500 PSI, but some newer systems are designed to operate at higher pressures.

Standard Sizes for Farm Cylinders. Only two stroke lengths have been standardized: 8" and 16". Specifications for each size are:

All 8" Stroke Models: Closed pin-to-pin distance 20¼". Clevis slot 1-1/16" to accommodate a 1" tang. Clevis pin diameter 1". Offered in bore diameters: 2, 2½, 3, 3½, and 4 inches.

All 16" Stroke Models: Closed pin-to-pin distance 31½". Clevis slot 1-1/16 to mate with 1" tang. Clevis pin 1¼" diameter. Offered in bore sizes: 3, 3½, 4, and 5 inches, although some manufacturers may offer additional bore sizes.

MILL-TYPE HYDRAULIC CYLINDERS

Extra heavy duty cylinders are available for use in foundries, steel mills, automotive plants, and other locations where service is severe. No industry standards exist at the present time, although several manufacturers publish catalogs in which models built to their own standards are listed.

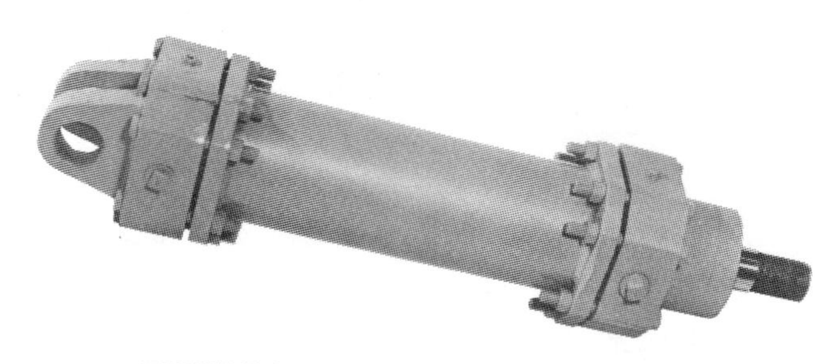

FIGURE 2-57.
Typical Mill-Type Hydraulic Cylinder.

Figure 2-57. In typical mill cylinder construction, steel flanges are welded to each end of the barrel and the end caps are bolted to the flanges. Mill-type cylinders have an extra long rod bearing which will support a certain amount of side load. Typically, piston and rod seals are of two kinds: (1), synthetic rubber cups with an O-ring expander, or (2), chevron rings. Many mill-type cylinders are built to customer specifications.

A swivel mounting is illustrated in Figure 2-57, and other mountings are also available including foot, trunnion, rear flange, and front flange.

CYLINDER PISTON SEALS

Although hydraulic presses may have been used as early as 1800, fluid power equipment in its present dependable form came into being with World War 2 with the development of self-adjusting seals (packings) of synthetic materials to operate at high pressure and/or temperature. The development of techniques for fine metal finishes was another important factor.

In the early days, before 1900, hydraulic machinery was used in several European countries – England, France, Germany, Russia, and perhaps others. For example, water power helped England manufacture fine goods for world trade. Steam-driven water pumps, with crude stuffing box packings, pumped water from the Thames river into a central hydraulic system using piston accumulators with weight-loaded pistons. A heavy cast iron pipeline distributed the pressurized water to nearby factories. Simple valving worked a large press ram and the exhaust water was discharged back into the river. These systems were very temperamental: the packings were leaky and had to be frequently and critically adjusted to prevent leakage and galling of the rams.

Today, cylinder manufacturers carefully select the type and material for piston and rod seals according to design requirements: whether for air or hydraulic use, the kind of fluid, pressure and temperature range, tightness of seal required, duty rating, and other factors. Composition and hardness of rubber seals is often important for a particular application, and the mechanic should select replacement seals exactly like the originals. Since various grades of rubber cannot always be correctly identified, original factory replacement seals should be obtained if possible. If this is not practical, the mechanic can select substitutes if he understands the advantages and limitations of each packing type.

SEAL MATERIALS

Synthetic Rubbers. Each compound listed in the table will work satisfactorily on several fluid media, but the ones which appear to be more popular are buna-N and polyurethane for hydraulic oil; Viton

Suggested Seal Materials for Hydraulic Fluids

Seal Material	Recommended for:	Temp. Range
Leather, wax impregnated	*Air, oil, water at limited temperature*	*-65 to +180° F*
Leather, rubber impreg.	*Air, oil, water at higher temperature*	*-65 to +250° F*
Buna-N, Hycar	*Air, oil, water, water/glycol*	*-40 to +250° F*
Neoprene	*Water, water/glycol*	*-40 to +250° F*
Viton	*Air, oil, water, water/glycol, phosphate ester*	*-40 to +450° F*
Polyurethane	*Oil*	*-20 to +200° F*
Teflon, Kel-F	*Oil, water, phosphate ester, water, water/glycol*	*-320 to +500° F*

for high temperature oil or phosphate ester fluids; buna-N for compressed air; and neoprene for water and water/glycol.

Polyurethane is a relatively recent addition to the synthetic rubber family and is coming into wide usage for petroleum hydraulic oil because it will outwear buna-N if the barrel has a high polish and if the oil is kept well filtered. However, because of its higher friction it is not as good as buna-N for compressed air service or water service.

Natural Rubber. Seals made from natural rubber should not be used in hydraulic or air equipment where they might come into contact with petroleum oil. They are for use with non-petroleum oils such as automotive brake fluid.

Teflon. Due to its hardness it may not be leaktight as a moving seal. Unlike rubber it may not return to original shape after being severely distorted. A spring is sometimes used behind or under it. As a moving seal it has low breakaway and running friction, long life, and the ability to work at very high temperature. It is often used, not as the seal itself, but as a backup ring to protect another seal, and as a thin piston ring with a rubber expander under it to obtain low breakaway friction.

Leather. This is probably the material used in the first self-adjusting packings. It proved to be remarkably good and still has limited use in cylinders, and wide use in devices such as door checks, water pumps, and tire pumps. It will operate reasonably well in barrels which are too rough for satisfactory life of a rubber seal. It has lower breakaway friction than rubber and requires less lubrication. It hones the barrel on every stroke and the second set of "leathers" will often outwear the first. Wax was originally used to fill the leather pores and to stiffen it so the seal would hold its shape. The wax limited the operating temperature to 180° F. Polyurethane rubber is used to impregnate leather for operation to 250° F. The supply of good quality genuine leather today is limited because most cowhides have numerous blowfly holes. This makes the production of large diameter seals impractical. Leather substitutes such as Corfam have been used in the past but have largely been discontinued for use in seals. The most popular use for leather at the present time is for back-up rings for hydraulics, and cup seals for compressed air.

SEAL TYPES

Figure 2-58. Various types of U-seals seem to be favored by the majority of cylinder manufacturers for both piston and rod seals, for mobile as well as industrial cylinders.

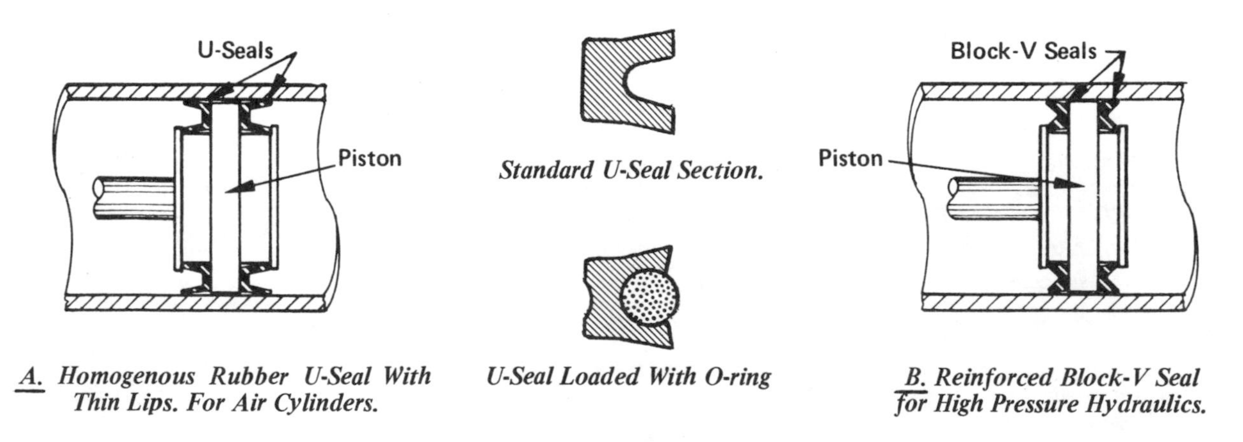

A. Homogenous Rubber U-Seal With Thin Lips. For Air Cylinders.

U-Seal Loaded With O-ring

B. Reinforced Block-V Seal for High Pressure Hydraulics.

FIGURE 2-58. Variations of the U-Seal.

Advantages of U-seals as compared to multiple-ring, adjustable, or spring loaded packings, are economy in first cost and in replacement. Only one sealing ring is required on single-acting, two rings on double-acting cylinders instead of multiple rings with adaptors top and bottom. The seals are easy to install and easy to replace. They are self-loading; they do not require springs, adjustment nuts, or shims. Machining of the packing space is not critical. They are reasonably leaktight, but not considered as leaktight as chevrons or O-rings.

One previous disadvantage of U-seals, when molded from buna-N or Hy-car rubber, was short seal life compared to V-type multiple chevron seals, and the possibility of sudden failure without warning. Since the introduction of polyurethane synthetic rubber, U-seal life has been greatly extended. This new rubber can far outlast buna-N as a piston seal, provided the barrel has a highly polished inside surface, at least 16 to 32 rms, and provided the system is well filtered, to 10μm. Of course heat is an enemy of all synthetic rubbers, and the operation of any rubber seal at elevated temperatures, over 160°F, will substantially shorten life of the rubber.

In Part B of Figure 2-58, a block-V cross section is illustrated. This is a buna-N or polyurethane rubber seal which is molded with cotton duck or other material as a reinforcement. It is used on hydraulic cylinders operating at high pressures, up to 5000 PSI. It is sometimes molded from a harder grade of rubber for additional stiffness. Due to its stiffness it is not suitable for compressed air nor for very low hydraulic pressure. It is too stiff to be successfully used on air-over-oil systems which require a high degree of leaktightness.

Loaded U-Seals. During the last 10 years, several seal manufacturers have developed U-seals which have an O-ring installed between the lips for spring loading them against the sealing surfaces. A cross section of typical O-ring loaded U-seal is shown in the center of Figure 2-58. Advantages claimed for loaded seals are leaktight operation at both low and high pressures, even on vacuum applications. For extremely high pressure, two or more loaded U-seals can be stacked in the same groove to divide the pressure. No top or bottom adaptors are required.

Packing Shelf Life. The U.S. armed services require the cure date on all rubber seals and will not permit their installation after two years from cure date even though they are in a hermetically sealed package, and will not permit the installation of new cylinders which have been in storage for more than two years until all rubber seals have been replaced.

Metal Ring Seals. Figure 2-59. Automotive-type metal piston rings have a longer life expectancy than any other cylinder seal. They are used only for piston seals, never for piston rod seals. They have the ability to withstand high surges and shock loads. On machine tool applications they are the preferred seal where their limitations as described later do not apply. The rings are usually cast iron working in a steel cylinder barrel. Lighter duty cylinders may have only two rings per piston while heavy duty models may have four to six rings.

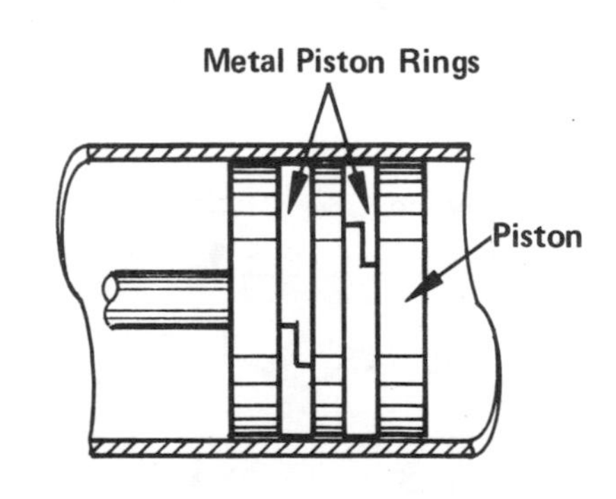

FIGURE 2-59. Step-Cut Cast Iron Rings Running in a Steel Barrel.

Metal ring seals are not leaktight, but can be used on those applications where the cylinder does not have to hold against drift while stopped. They should not be used on these applications: (1), where a cylinder is operating in a vertical position and must stop and hold at some point in mid stroke; (2), where a cylinder is operating horizontally and is subject to a reactionary force against the piston rod while stopped; and (3), where a cylinder is operating from a high pressure manifold which is common to several branch circuits. High pressure on its control valve inlet port may cause the cylinder to drift forward while its control valve is in neutral. This is caused by valve spool leakage into both ends of the cylinder plus leakage across the cylinder piston. Leakage across metal rings may affect the accuracy of feed if the cylinder is pushing a cutting tool at a slow feed rate. Many high speed cylinders successfully use metal ring seals because they are little affected by heat and friction. However, where leaktight operation is more important than long seal life, soft seals rather than metal rings should be used. Metal rings are not suitable for air cylinders because of piston leakage.

O-Ring Seals. O-rings are extensively used as static seals on many industrial, mobile, and farm machinery cylinders, to seal end caps and rod glands to the barrel.

As moving seals, their use is now limited to high pressure service, to short life or inexpensive cylinders, or on applications where virtually leaktight sealing is required. They seal well at high pressure but tend to leak badly when used as moving seals on low pressure, less than 50 PSI. They are inexpensive, easy to install and easy to replace.

O-rings have high breakaway and running friction. They were used extensively at one time on air cylinders but have largely been replaced by cup seals which have far less friction.

They have a relatively short service life as compared to other types of seals. This prohibits their use on industrial cylinders, but is often acceptable on mobile and farm machinery cylinders.

They have been used on aircraft cylinders because of their excellent sealing capability. Their short life is acceptable on these applications because they are routinely replaced at regular intervals.

O-rings are molded from many rubber compounds. When used for moving seals, polyurethane is the preferred compound. As static seals, buna-N rings give excellent service. Viton rings are used for both moving and static seals when operation is on synthetic hydraulic fluids.

O-rings should be used with discretion as moving seals on pistons larger than 4 or 5 inches in diameter. Large diameter moving seal O-rings may twist as the piston strokes, causing high leakage and premature failure.

No more than one O-ring should be used as a piston seal. One ring seals better than two or three. With more than one ring, a pressure block will develop between the rings which will interfere with their normal rolling action, and will pre-load the rings to their "cracking pressure", causing shorter seal life and higher leakage. Two rings are used on some applications if the space between the rings can be vented to tank.

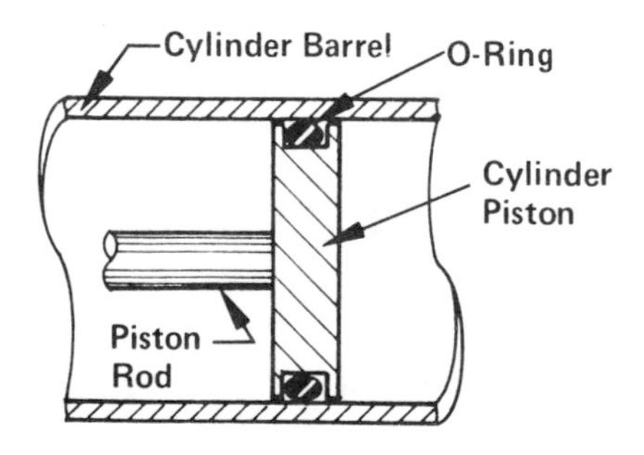

FIGURE 2-60. O-Ring Used Without Back-Up Washers for Pressures Not Over 1500 PSI.

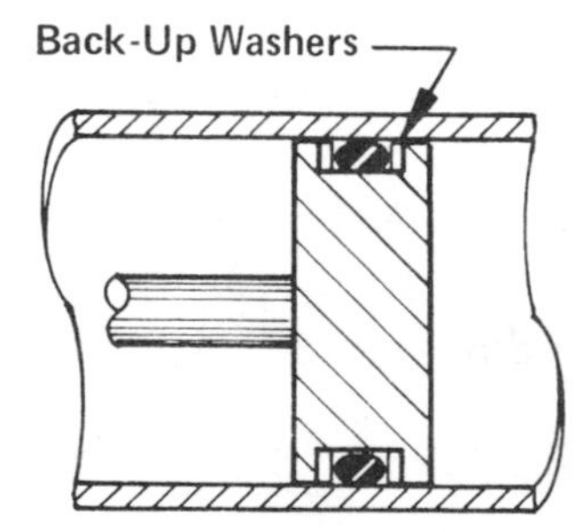

FIGURE 2-61. For Pressures Over 1500 PSI, a Back-Up Washer Should be Used on Each Side of the O-Ring.

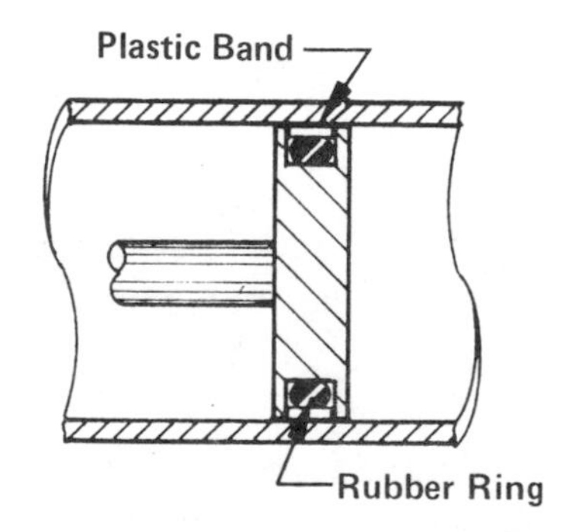

FIGURE 2-62. O-Ring Used as Expander for Thin Plastic Band.

Moving Seal O-Ring Applications at Low Pressure. Figure 2-60. For low pressure applications, under 1500 PSI, an O-ring can be used without protective back-up washers provided the piston is a close fit inside the barrel with a diametral clearance not to exceed 0.020" at any point. The groove width should be about 1½ times the O-ring diameter to give the ring a chance to roll from side to side as the piston reverses its direction of travel. This offers a larger area of O-ring surface to absorb wear, and O-ring life is thereby increased.

High Pressure Applications. Figure 2-61. Back-up washers are recommended for all applications of O-rings as moving seals at pressures over 1500 PSI, and for applications at lower pressures if the piston is not a close fit inside the barrel. For double-acting cylinders a back-up washer should be used on both sides of the O-ring. The washers prevent extrusion of the O-ring into the clearance space between piston and barrel as illustrated in the box diagram below. Only one back-up washer is required for rod seals and for single-acting cylinders. The washer is always installed on the side of the O-ring which is away from the pressure. Groove width should be equal to back-up washer width plus 1½ times the O-ring diameter.

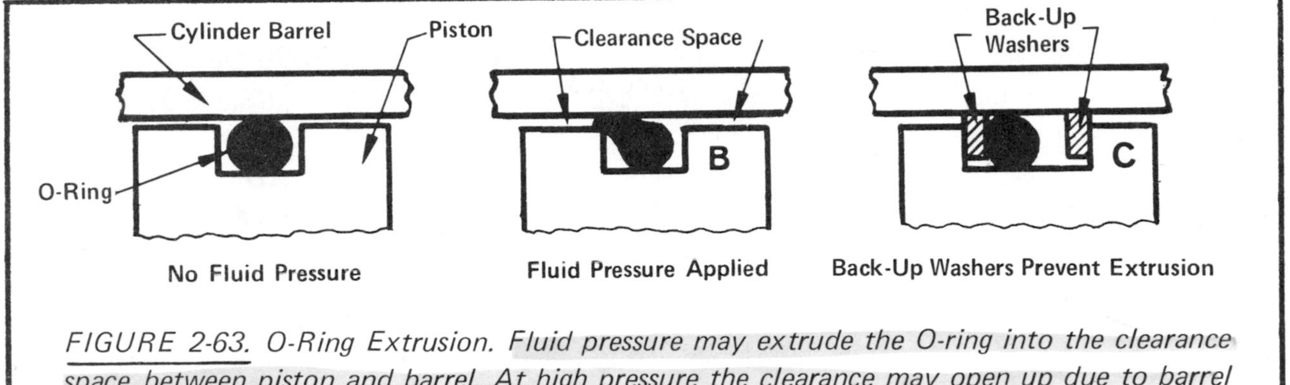

FIGURE 2-63. O-Ring Extrusion. Fluid pressure may extrude the O-ring into the clearance space between piston and barrel. At high pressure the clearance may open up due to barrel expansion. When pressure is released, the metal barrel may bite off pieces of the O-ring as it contracts faster than the rubber can retract out of the clearance space. At pressures over 1500 PSI, leather or Teflon back-up washers expand with the barrel to keep clearance space closed.

Back-up washers are commonly of leather or Teflon. Leather washers can be installed by stretching them over the piston. Teflon rings cannot be stretched. They are 2-turn split spiral construction and can be threaded into the groove without stretching.

O-Ring Expander. Figure 2-62. An O-ring can be used as an expander or "spring" underneath a thin plastic (usually Nylon) band. The plastic band gives low breakaway and running friction and needs little or no lubrication. The O-ring keeps the plastic band in tight contact with the barrel.

Multiple-V (Chevron) Seals. Figure 2-64. Chevron seal rings are used in sets of 2 up to 6 or 8 rings, depending on the pressure to be sealed. One ring is used for every 500 to 700 PSI, with a minimum of two rings. When used as moving seals on the piston of a double-acting cylinder, two sets must be used, one set facing in each direction, with lips facing the pressure. They must be pre-loaded to extend the lips into light contact with the sealing surfaces. Since each ring is too thin to be self-loading, the entire stack must be preloaded in one of 3 ways: (1), the width of the packing gland can be made slightly smaller than the height of the ring stack, and the stack can be compressed at installation; (2), the gland width can be made slightly larger than the stack height and a wave washer installed

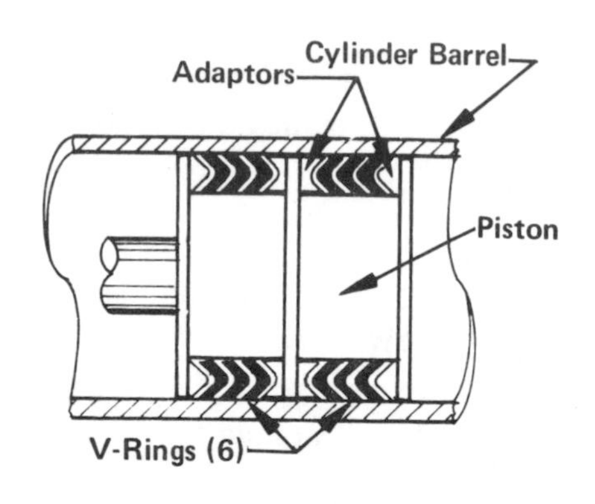

FIGURE 2-64. A Set of V-Type (Chevron) Rings With Top and Bottom Adaptors.

in the gland; and (3), a gland nut can be used and the amount of pre-loading adjusted at assembly. This adjustment should not have to be changed during the life of the seals. The U. S. Navy requires an adjustment nut on all installations, for both piston and rod seals, so the pre-loading can be correctly adjusted.

When repacking with new V-rings, the new set may not be exactly the same height as the old stack because of manufacturing variations. If shorter than the old stack, shims may be added. If longer than the old stack, one ring can be removed and shims added. If pre-loading is too light the new seal will leak. If pre-loading is too heavy the rings will wear out prematurely. The set of rings should always be supported both top and bottom by metal, plastic, or rubber adaptors having a contour to properly expand and support them. If re-packing is not done carefully, the new seals may leak worse than the ones they replaced.

Multiple-V rings, when carefully installed and properly adjusted, are the finest soft seals available. They are valued for their leaktightness and dependability. Seldom do they fail suddenly. They give sufficient warning, by leaking slightly, long before actual failure. This gives time for replacements to be procurred. But in spite of their excellent performance, cylinder manufacturers are discontinuing their use in favor of U-seals which are considerably cheaper, less critical to install, need no pre-loading adjustment, and are less likely to leak when the cylinder is re-packed. V-rings are still used by some manufacturers especially for rod seals where leaktightness and long life are important.

V-rings are molded from various rubber compounds, either homogenous construction or with reinforcing material for use at very high pressures, 10,000 PSI and even higher. Homogenous rings work very well on air but most air cylinders perform satisfactorily on less expensive U-seals or cup seals.

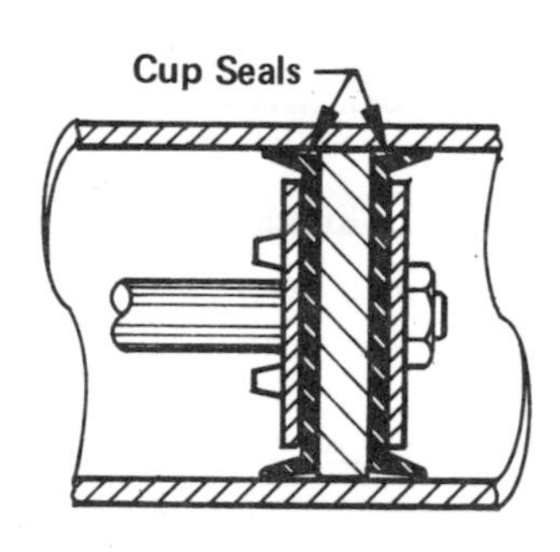

FIGURE 2-65. Form of Cup Seal for Leather or Synthetic Rubber.

Cup Seals. Figure 2-65. Cup seals were probably the first packing shape used for piston seals. They were originally manufactured from leather, and were impregnated with wax to seal the pores against leakage and to help the seal hold its shape. While not used as extensively any more on cylinders, they are preferred on devices like door checks and bicycle pumps because they seal better than rubber on slightly rough surfaces. They are still used on special air cylinders where low breakaway and running friction is important and where they cannot be continuously lubricated.

Cup seals are molded from a variety of rubber compounds for service in various fluids at various temperatures. Homogenous seals (of solid rubber) with thin lips and a lower hardness than used for hydraulics are preferred for air cylinders and even for low pressure hydraulics. Reinforced seals have layers of cotton duck or Nylon mesh molded in with the rubber to give stiffness to the seal to prevent extrusion of the rubber into the clearance space between piston and barrel when the seal is to be used at medium to high pressures. Back-up rings are not generally used with cup seals.

Block V Seals are the most air tight.

PISTON ROD SEALS FOR CYLINDERS

Leaktight operation is an important requirement for rod seals (called rod packings). Any external leakage, especially of hydraulic oil, is undesirable. Leaktight rod seals may be even more important than leaktight piston seals. If the seal is to be leaktight around the rod, the rod must have a fine polish. This will also extend seal life.

U-Seals. Figure 2-66. This is probably the most popular rod seal for both air and hydraulic cylinders. It is self-loading, does not require an expander spring or adjustable gland. It seals reasonably leaktight and is easy to replace. A popular form of this seal is polyurethane rubber with an O-ring expander as illustrated in Figure 2-58.

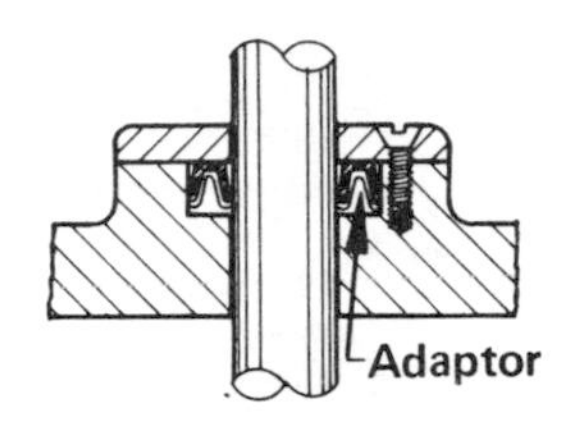

FIGURE 2-66.
U-Seal Rod Packing.

Multiple-V Seals. Figure 2-67. Used in sets of 2 rings for air cylinders and 4 to 6 rings for hydraulic cylinders, plus top and bottom adaptor rings. Lips must be facing toward the pressure.

A V-ring set must be end loaded to force the lips into light contact with the sealing surfaces. In a non-adjustable gland a wave spring may be used behind the rings to keep them loaded. Top and bottom adaptor rings are shaped so they cause the lips to spread when end force is applied. On machine tool cylinders an adjustment nut similar to that shown in Figure 2-69 may be used to maintain a "dry rod".

Homogenous rings molded from solid rubber of low durometer are preferred for air cylinder or low pressure hydraulic seals. Fabric reinforced rings molded from a harder rubber are better for higher pressure hydraulics.

Since multiple-V rings seal over a greater length on the piston rod, they continue to seal more tightly on rods which have become scratched or dented. While single-lip packings may fail suddenly and shut the system down, seldom do all rings in a multiple-V set fail at the same time. They give warning by leaking slightly.

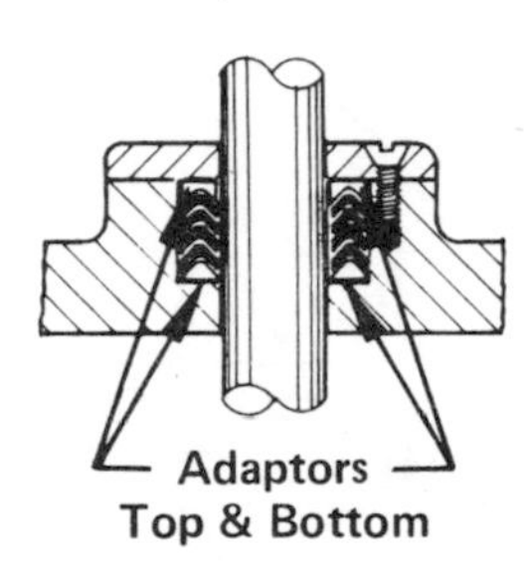

FIGURE 2-67.
Multiple-V Rod Seals.

O-Rings and Quad Rings. Figure 2-68. An inexpensive seal used mostly on farm machinery cylinders. An O-ring seals very tightly but has a relatively short life. It also has a high friction drag. Quad rings are a patented variation of an O-ring. They seal tightly like an O-ring but have much less frictional drag.

For operation above 1500 PSI, a back-up washer should be installed to protect the O-ring from extrusion.

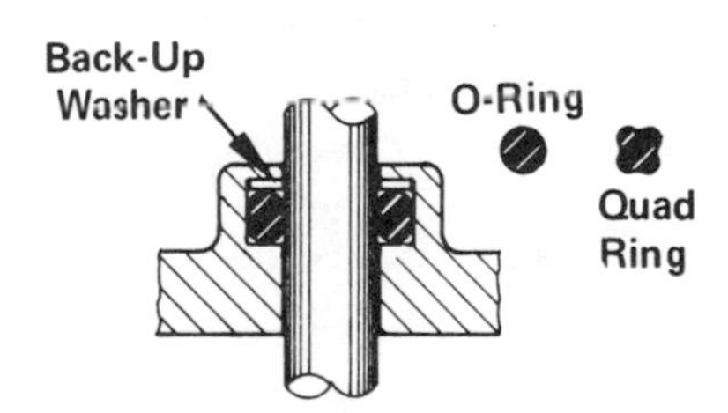

FIGURE 2-68. O-Ring and Quad Ring Seals With Back-Up Washer.

Adjustable Rod Seal. Figure 2-69. Various sealing materials can be used, including solid or split rings, or continuous rope packing impregnated with synthetic rubber. Adjustable seals are used more often on heavy-duty machine tool cylinders to maintain a "dry" rod. They are seldom seen on industrial cylinders, and never on mobile or farm machinery cylinders.

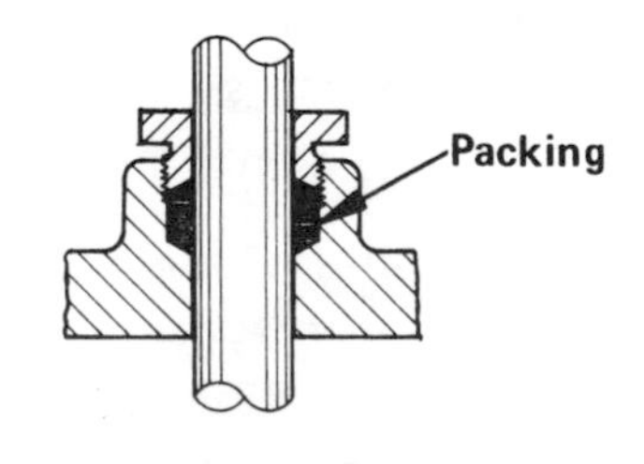

FIGURE 2-69.
Adjustable Packing Gland.

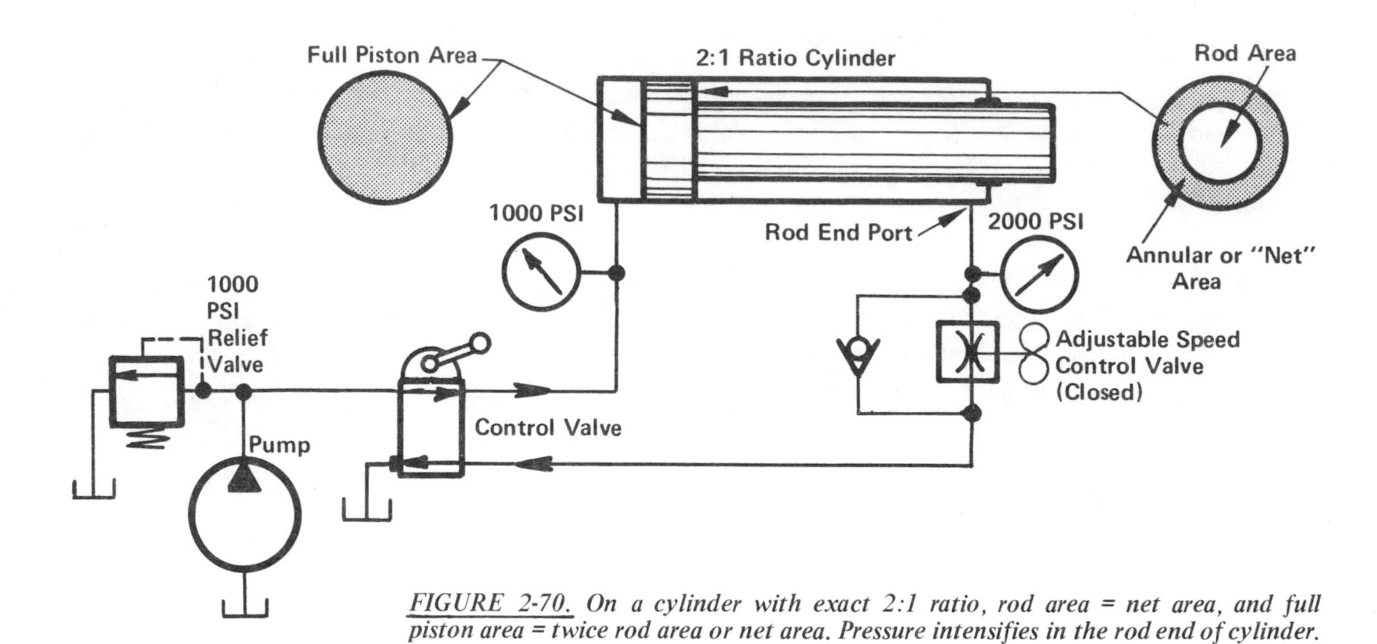

FIGURE 2-70. On a cylinder with exact 2:1 ratio, rod area = net area, and full piston area = twice rod area or net area. Pressure intensifies in the rod end of cylinder.

CHOICE OF PISTON ROD DIAMETER

Referring to the chart on Page 259, there is a choice of rod diameter in each bore size. Standard rod diameter, shown in the second column will be large enough for all "pull" applications provided the cylinder pressure rating is not exceeded, and will be large enough for most "push" applications except where the stroke is unusually long. See "Column Strength", Page 93.

The largest rod available in each bore size is shown in the last column. The area of the piston rod is ***approximately*** one-half the full piston area, and these rods are called "2:1 ratio piston rods". Intermediate size rods are shown in the center section of these tables. Rods larger than standard are used primarily where greater resistance to column failure is required, or where a rapid retraction speed is desirable, or in regenerative circuits (See Volume 2).

Pressure Intensification. Figure 2-70. Pressure intensification may be produced in the rod end of a cylinder if the discharge oil is restricted or blocked. This problem becomes greater the larger the piston rod area, and in a 2:1 ratio cylinder the pressure in the rod end can be intensified up to twice the pressure applied to the blind end, if flow is blocked. Pressure intensification is produced as follows:

Inlet fluid pressure (1000 PSI in this case) is applied to the full round piston area and produces a certain mechanical force which must be supported by the fluid trapped in the rod end on just one-half the surface area. Therefore, the intensity in the fluid on the rod side is twice (2000 PSI in this case) the intensity in the fluid on the blind side. Intensification can always be present in any cylinder which has unequal areas on opposite sides of the piston but is greatest in 2:1 ratio cylinders. All components used in the rod end circuit must be rated for the intensified pressure.

Flow Multiplication. Cylinders with unequal areas on opposite sides of the piston will retract faster than they extend because there is less volume to fill in the rod end for retraction. This effect is greatest on 2:1 ratio cylinders. While the cylinder is retracting the volume of oil discharged will be twice the volume supplied by the pump. All components in the blind end circuit should have a flow rating to handle this multiplied flow.

TIPS ON DESIGNING WITH CYLINDERS

Finding Cylinder Stroke by Chord Factor Method. Figure 2-71. If the cylinder is rotating a lever to an equal angle each side of the perpendicular as in this figure, the length of stroke can easily be determined by multiplying lever length times the chord factor taken from the table below.

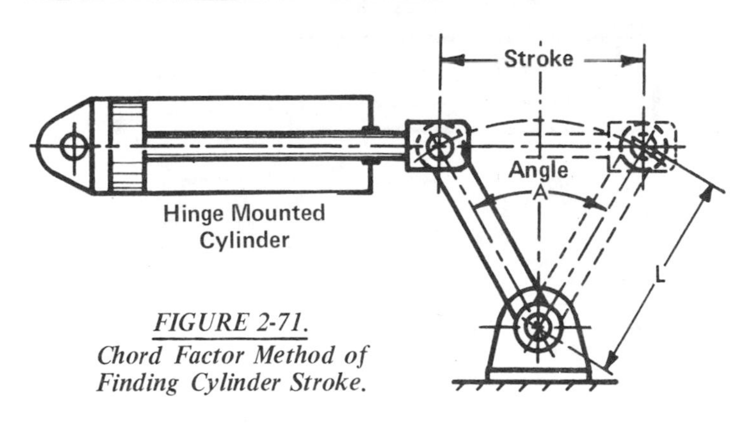

FIGURE 2-71.
Chord Factor Method of Finding Cylinder Stroke.

Example: To find the cylinder stroke needed to swing a 14-inch lever through an arc of 105° (Angle A) when mounted as in the figure, multiply the 14-inch lever length by the chord factor of 1.587" taken from the table: 14 x 1.587 = 22.218" stroke.

Note: Sometimes a stock cylinder with standardized stroke can be used by lengthening or shortening the lever arm for the desired angular travel.

Angle A Degrees	Chord Factor	Angle A Degrees	Chord Factor
30	0.518	95	1.475
35	0.601	100	1.532
40	0.684	105	1.587
45	0.765	110	1.638
50	0.845	115	1.687
55	0.923	120	1.732
60	1.000	125	1.774
65	1.075	130	1.813
70	1.147	135	1.848
75	1.217	140	1.879
80	1.286	145	1.907
85	1.351	150	1.932
90	1.414	155	1.953

Scale Layout Method. Figure 2-72. If the cylinder is not rotating the lever to an equal angle each side of the perpendicular, the stroke must be calculated by mathematics, or may be determined by an accurate scale layout.

For the scale layout method, a sketch must first be made showing length and angular travel of the lever, and the mounting position of the cylinder. All parts must be laid out to exact scale, either full size or reduced size. Pin-to-pin centers on the proposed cylinder can be obtained from the manufacturers drawings.

A ruler, tape, or scale is used to measure the distance from the cylinder rear hinge to the starting and ending points of lever travel. These will be the retracted and extended cylinder lengths. The stroke will be the difference between these measurements.

It may be necessary to experiment with different hinge point locations until the best mounting position can be determined. Pieces of cardboard may be cut to represent parts of the mechanism

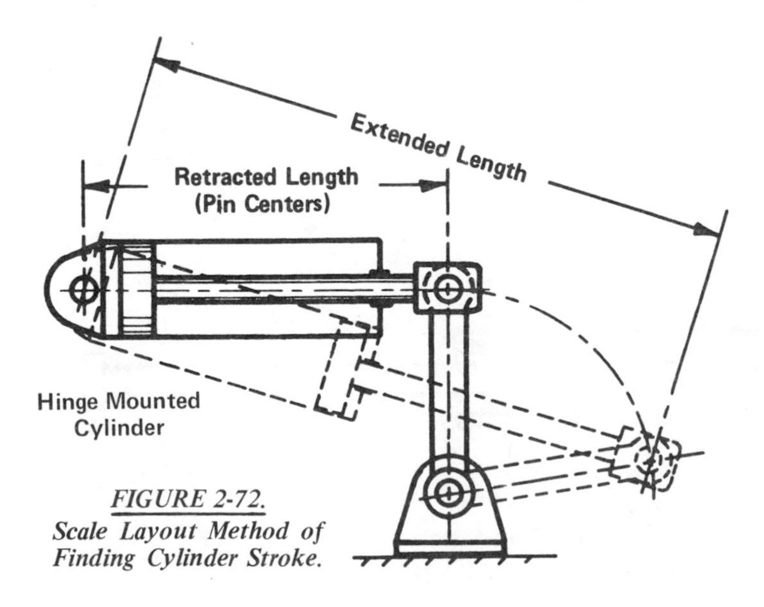

FIGURE 2-72.
Scale Layout Method of Finding Cylinder Stroke.

which move, with thumb tacks for clevis or tang hinge points. Note: Effective cylinder force against the lever varies with the angle between them. See Page 59.

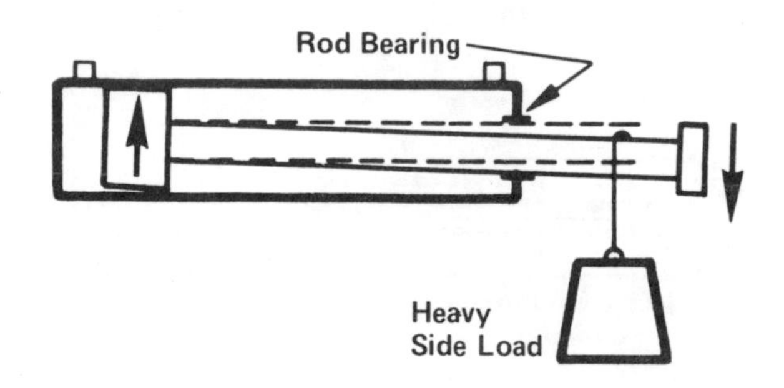

FIGURE 2-73. Avoid Side Load on the Piston Rod.

Side Load. Figure 2-73. Standard catalog cylinders are designed only for thrust loads; the rod bearing is too small to handle heavy side loads, and this is not required on most applications. For those applications where a heavy side load must be supported, this can be handled in one of several ways: (1), if space permits, a double-end-rod cylinder can be used. It has a rod bearing on each end of the cylinder and can handle a moderate amount of side loading; (2), an outboard bearing can be installed on a standard cylinder, but this must be done by the user; and (3), a custom cylinder may be necessary and can be ordered from a number of small companies who make a specialty of building cylinders to the customer's specifications.

To avoid side loading, the cylinder should be carefully and accurately mounted in alignment with the load, and the alignment should be checked in two planes at right angles and at both closed and fully extended positions. During alignment, a hydraulic cylinder, if unloaded, can be extended by compressed air to check the alignment, or can be extended by a hydraulic hand pump if it is connected to a heavy load. Binding at any point in its stroke means that a side load is being transmitted to the rod bearing and the barrel. If the cylinder cannot be perfectly aligned with the load throughout its stroke, it may be necessary to use a hinge mounted cylinder with a linear alignment coupler attached to the end of its piston rod.

Side load may produce several undesirable results. The piston rod can be permanently bent. Usually a bent rod cannot be satisfactorily straightened. It should be replaced with a new rod ordered from the cylinder manufacturer. The rod bearing can become worn to an oval shape. The side clearance thus opened up can cause the rod seals to be distorted or deformed, to be extruded, or to be blown out. Fluid leakage around the rod seal can become excessive. The entire rod bearing or cartridge should be replaced; new sealing rings installed with a worn bearing will not last. Finally, a heavy side load can cause the barrel to be scored, resulting in a loss of power and speed, and short piston seal life.

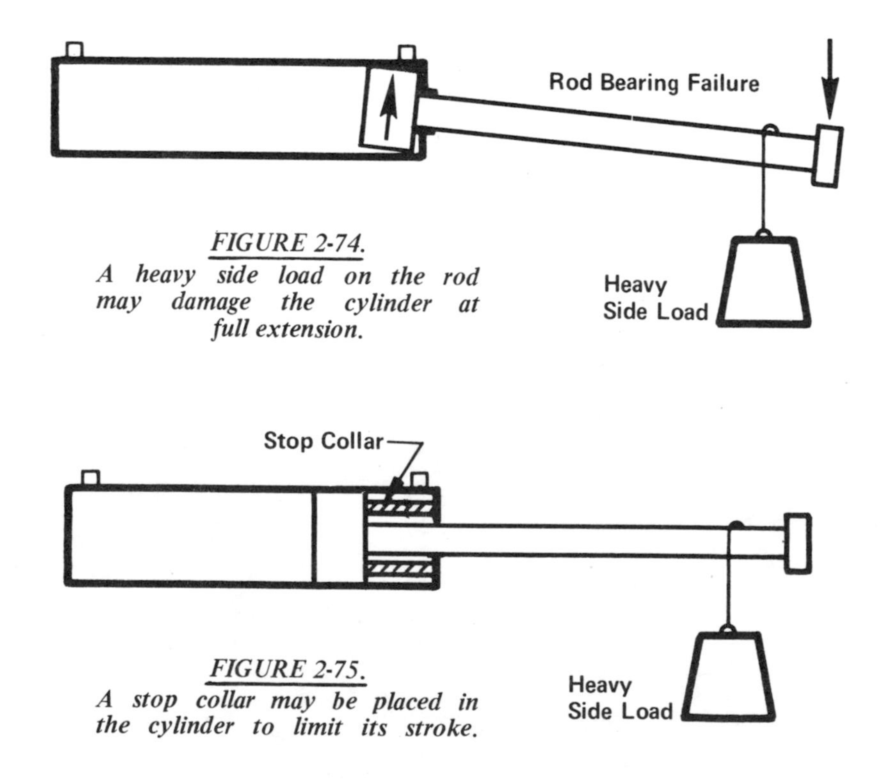

FIGURE 2-74. A heavy side load on the rod may damage the cylinder at full extension.

FIGURE 2-75. A stop collar may be placed in the cylinder to limit its stroke.

Rod Bearing Failure. Figure 2-74. Side loading against a piston rod can be particularly damaging to a cylinder as it approaches full extension because of the increasing leverage between the side load and the rod bearing. A leverage force can be produced which may cause the piston to score the barrel, and can produce accelerated wear of the rod bearing and seals, even to the point of fracture of the bearing and destruction of the seals. The machine on which the cylinder is installed should be designed so the side load can be carried on slide bearings which are external to the cylinder.

Stop Collar. Figure 2-75. On those applications where a certain amount of side loading is unavoidable, the cylinder can be partially protected with an internal stop collar to prevent the piston from approaching too closely to the end cap. When a stop collar is used the cylinder cannot be cushioned on the rod end.

The stop collar should preferably be installed by the manufacturer at original assembly. The user must specify the length of collar desired, and the appropriate length can be calculated with tables in the cylinder manufacturers catalog. The length of the collar reduces the effective stroke by the same amount. The cylinder must be ordered with extra stroke. For example, if an effective 24" stroke is needed, and if a 6" collar is specified, the cylinder must be ordered with a 30" stroke, and the mounting dimensions for a 30" stroke will apply.

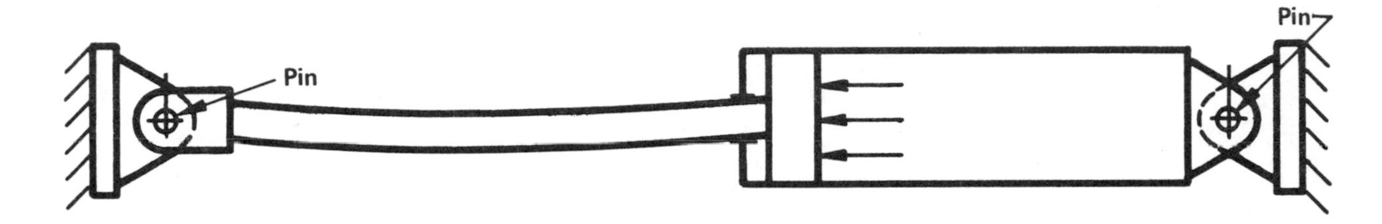

FIGURE 2-76. An example of column failure or "rod buckling".

Column Strength. Figure 2-76. Column failure or rod buckling may occur if the piston rod is too slim for the length of stroke and for the load. The "column strength" of a piston rod cannot be increased by using steel with a higher tensile strength or by heat treating. For this reason it is sometimes necessary to use an oversize piston rod strictly for the purpose of achieving the necessary "column strength".

Nearly all major cylinder manufacturers catalogs carry diagrams and charts to enable the customer to estimate the minimum diameter piston rod which should be used for the cylinder mounting style, stroke, and load. Exact loading that will cause column failure is difficult to calculate, so all these charts are conservative. Our own recommendations for minimum piston rod diameter are in Volume 2 of this textbook series.

Referring to the standard cylinder piston and rod sizes on Page 259, the smallest diameter rod is considered "standard" and should be specified unless one of the oversize rods is needed to protect against column failure.

All cylinders are designed with a piston rod large enough to carry all the "pull" load which the cylinder is capable of producing at its rated maximum pressure, and failure will never occur in this direction. The smallest diameter rod available for each bore size should be used.

REVIEW QUESTIONS AND PRACTICE PROBLEMS FOR CHAPTER 2

– QUESTIONS –

1. To transmit ***power*** through a pipe to a cylinder, what two conditions must be generated in the fluid? *– See Page 53.*

2. Explain why a cylinder can produce the same ***power*** in both directions of movement regardless of the difference in area on opposite sides of its piston. *– See Page 55.*

3. How is cylinder force calculated? *– See Page 56.*

4. Give the rule-of-thumb for oversizing the pressure on an air cylinder to produce a moderate load speed. *– See Page 57.*

5. How is the extension speed of a hydraulic cylinder calculated? The retraction speed? *– See Page 60.*

6. How is a "flow control valve" different from a "needle valve"? *– See Pages 61 and 62.*

7. What is the difference between "meter-out" and "meter-in" speed control? *– See Pages 65 and 66.*

8. Explain the difference between a "double-acting" and a "double-end-rod" cylinder. *– See Page 68.*

9. What is the difference between a "double-acting" and a "single-acting" cylinder? *– See Pages 54 and 70.*

10. What is the "net" area in a cylinder? *– See Page 55.*

11. What factor or factors set the maximum travel speed of an air cylinder? Of a hydraulic cylinder? *– See Page 60.*

12. On a hydraulic cylinder, in which direction will it travel faster on a given oil flow? *– See Pages 60 & 61.*

13. What causes "column failure" in a cylinder? How can it be prevented? *– See Page 93.*

14. What kind of damage can be caused to a cylinder by a heavy side load on its piston rod? *– See Page 92.*

15. What causes "pressure intensification" in a cylinder, and on which side of its piston is it produced? *– See Page 90.*

16. On an air cylinder, why does the available air pressure have to be from 10% to 100% more than the level needed to just equal the load resistance? *– See Page 56.*

17. In comparing "series connected" with "by-pass type" speed control for hydraulic cylinders, which of these methods usually gives more stable control? Which method tends to produce less heat in the system? *– See Pages 65 and 66.*

18. For causing two cylinders, controlled through the same 4-way valve, to move at the same speed, three circuit methods are mentioned in the text. Briefly describe each method. *– See Pages 67 and 68.*

19. Telescoping cylinders are considerably more expensive than standard cylinders. But on what kind of applications do they offer a worthwhile advantage? *– See Pages 72 and 73.*

20. What is a "cushion" in a cylinder and what is its purpose? *– See Pages 76 and 77.*

21. What are the three most common synthetic materials for cylinder piston and rod seals? *– See Page 83.*

22. Although O-rings seal very tightly at high pressures, what are the reasons they are seldom used as cylinder piston and rod seals? *– See Page 86.*

23. What is the purpose of a check valve inside a flow control valve? *– See Page 62.*

–PROBLEMS –

1. Pick from the chart on Page 262 a minimum bore for an air cylinder to raise a load of 2000 lbs. at a moderate speed. Assume a maximum pressure of 90 PSIG is available for operation of the circuit. *–Ans. = 6" bore.*

2(a). Calculate the ***theoretical*** force from a hydraulic cylinder having a 4" bore when operated with a difference of 2000 PSI across its ports. (b). What is the estimated ***actual*** force? *– Ans. (a) = 25,140 lbs. (b) = 23,883 lbs.*

3(a). Find extension speed of a 6" bore hydraulic cylinder operating from a 25 GPM pump. (b). If it has a 4" diameter piston rod, how fast will it retract? *– Ans. (a) = 204"/min. (b) = 368"/min.*

4. Using the power factor chart on Page 59, how much ***effective*** force can be exerted against the end of a lever when Angle A is 30° and the force output from the cylinder is 19,000 lbs? *– Ans. = 9500 lbs.*

5. If a press bonding operation requires 10 lbs. per square inch, and if the surface area to be bonded is 1000 square inches, choose a cylinder bore size to produce the necessary force without exceeding 1200 PSI gauge pressure across the cylinder ports. *–Ans. 4" bore.*

6. Using the chord factor chart on Page 91, find the cylinder stroke necessary to rotate a lever through an angle of 90°. Lever length is 24 inches. *– Ans. = 33.9 inches.*

CHAPTER 3

Control Valves– 2-Way and 3-Way

FLUID POWER VALVES CLASSIFIED

Valves used to control the fluid flow between air compressor or pump and cylinder can be classified into three groups – directional control valves, pressure control valves, and flow control valves. In this chapter we will describe pressure control, flow control and the simpler kinds of directional control valves. In the next chapter we will continue the study of more sophisticated directional control valves, 4-way and 5-way types.

Directional Control Valves

In this group are all valves which primarily handle routing and diversion of the fluid stream – including starting and stopping it – without affecting the pressure level or the flow rate.

Valves for controlling direction of movement of a cylinder, or rotation direction of an air or hydraulic motor have 3-way, 4-way, or sometimes 5-way action. Directional valves for starting and stopping fluid flow have 2-way action. They include shut-off, check, shuttle, and quick exhaust types.

The terminology "3-way", etc., is a little misleading because it does not truly describe the valve action. It refers to the number of active porting connections. Thus, a "2-way" valve has two main portholes, a "3-way" valve has 3, a "4-way" valve has 4, and so on.

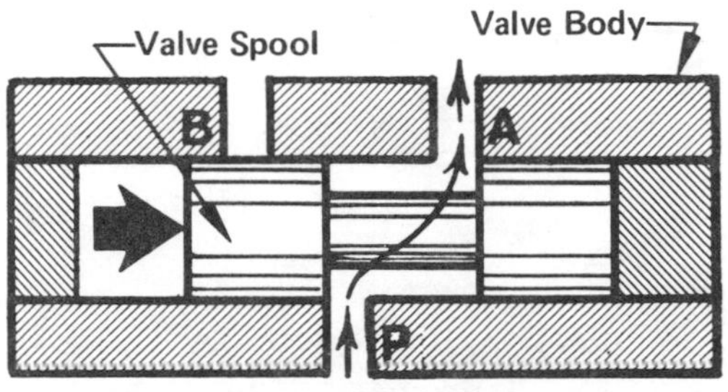

Spool shifted to the right. Fluid entering at Port P is directed to outlet Port A

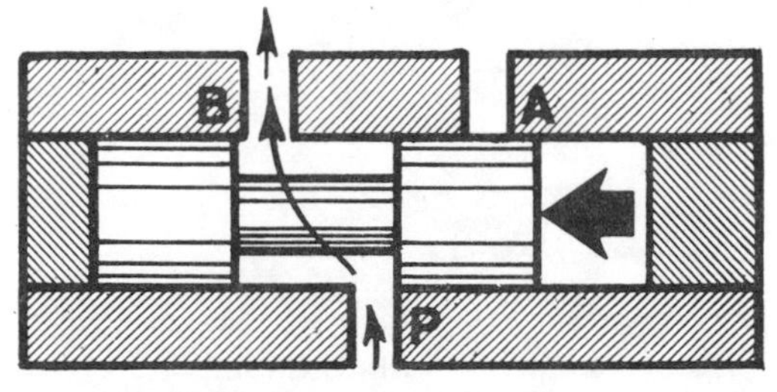

Spool shifted to the left. Fluid entering at Port P is directed to outlet Port B.

FIGURE 3-1.
Spool-Type Directional Control Valve.

Sliding Spool Valves. Figure 3-1. For air operation and oil hydraulic operation the sliding spool principle of valving is often preferred (over poppet-type valving) because the spool is pressure balanced, at least under static conditions, and will not drift in either direction when pressure is applied to its inlet port. The spool moves back and forth in a body which is enclosed with end caps.

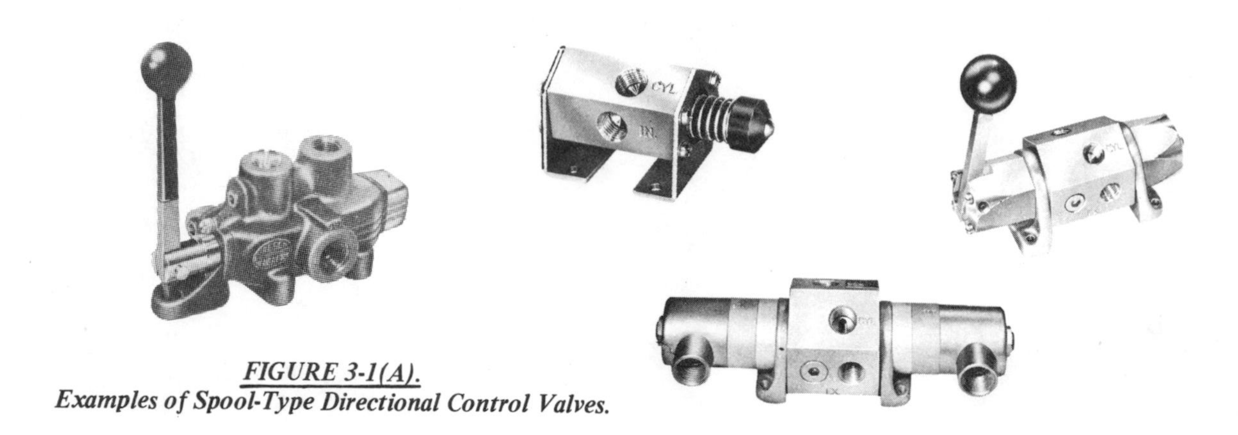

FIGURE 3-1(A).
Examples of Spool-Type Directional Control Valves.

Valves designed to handle only a single stream of fluid, as in Figure 3-1, have only one groove on the spool. As the spool is shifted to one position or the other, the spool groove connects a different set of ports together to start, stop, or divert the flow of the fluid. In the upper view of Figure 3-1, the spool is shifted to the right side of the bore. The groove connects the valve inlet Port P to outlet Port A. With the spool shifted to the left, as in the lower view, the fluid is diverted from Port P to Port B.

Valves designed to handle two separate streams of fluid at the same time have two grooves on their spool. These are the 4-way and 5-way valves described in the next chapter. Special valves can be designed to handle several streams at the same time by providing a groove for each fluid stream and by providing inlet and outlet ports at the right locations. However, standard valves are designed for handling a maximum of two streams at the same time.

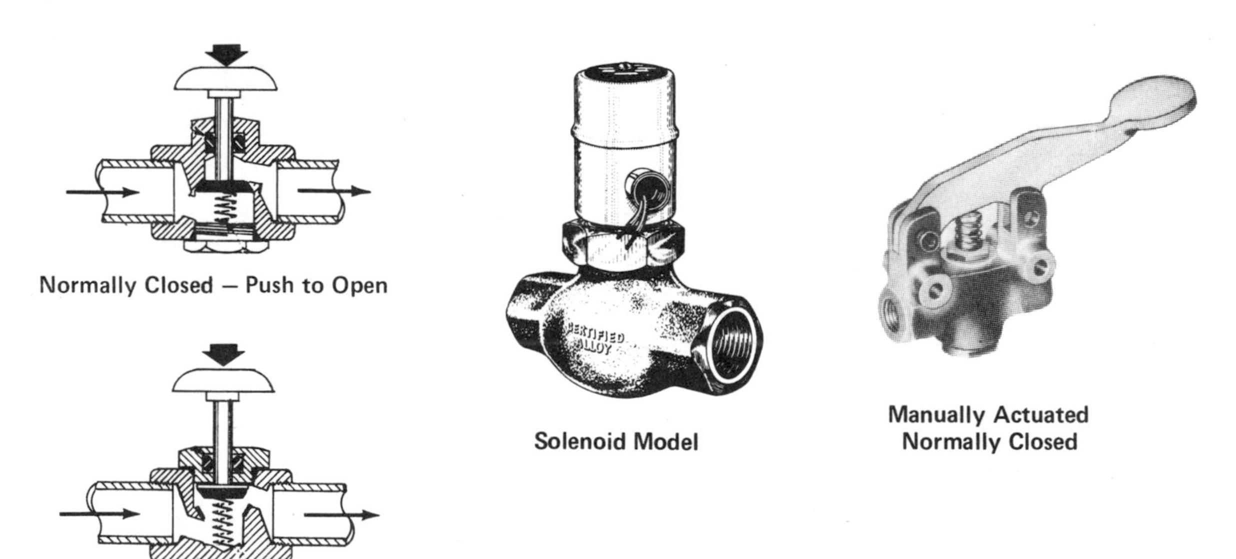

FIGURE 3-2. Examples of Poppet-Type Directional Control Valves.

Poppet-Type Valves. Figure 3-2. For operation on water or other non-lubricating fluids, or on fluids with higher specific gravity than oil, poppet-type action is preferred to sliding spool action. One problem with spool-type valves is that high specific gravity fluids flowing at high velocity through the grooves on the spool wear away the sharp corners at the edges of the groove. This is called "wire drawing" and is the result of steel molecules being torn away by the high velocity fluid. In a relatively short time this will cause excessive leakage across the spool. Poppet-type valves have a greater resistance to wire drawing, or at least the wire drawing which does occur will not as quickly affect valve performance.

Poppet action is simple 2-way action – open and close, or 3-way action – open one orifice and close another with one movement of the valve stem. To get fluid diversion or for handling more than one flow at the same time with a common lever, several poppets may be used, connected to one operating lever.

An important disadvantage to poppet-type valve action is that pressure on poppet surfaces is always unbalanced, both under static and flow conditions. In Figure 3-2, fluid pressure on the inlet port, if high enough, will cause the poppet to open without the control stem having been actuated. A valve of this type would usually be held in a closed position by a spring which is heavy enough to hold the poppet closed up to the maximum pressure rating of the valve. To open this valve, an operator would have to exert sufficient force, over and above the fluid pressure, to compress the spring. For this reason it is difficult to build poppet-type valves for handling high flow at high pressure.

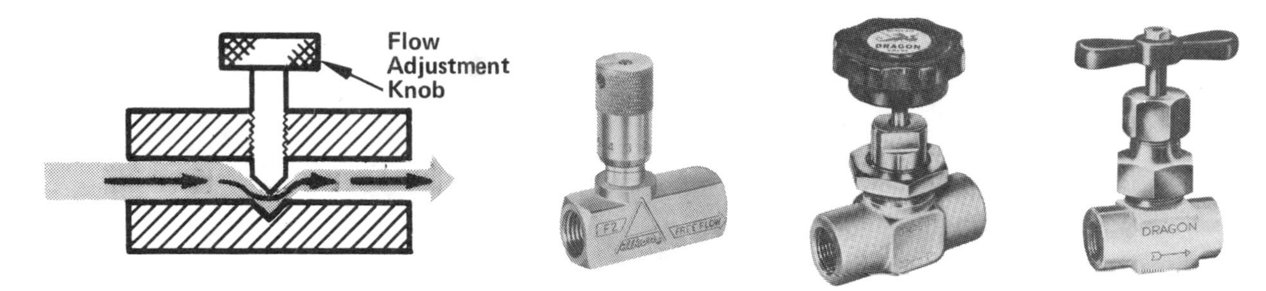

FIGURE 3-3. Examples of Needle-Type Valves.

Flow Control Valves

Figure 3-3. The needle valve is one of a small family of valves which regulate the rate of fluid flow in air and hydraulic circuits. The use of a needle-type flow control valve for controlling cylinder speed was shown, starting on Page 61.

Other flow regulating valves for hydraulic circuits, not covered in this book, include flow dividers and combiners, and pressure compensated flow control valves. These are described in Volumes 2 and 3 of this textbook series.

Pressure Control Valves

Hydraulic Pressure Control. The function of pressure control valves is to control the maximum pressure level either in the pump line or in a branch line. In this book, Volume 1, the hydraulic pressure relief valve will be described later in this chapter. Other valves in the pressure control group – pressure reducing, by-pass, sequence, and counterbalance types – are described in Volume 2 of this textbook series.

Figure 3-4. In hydraulics, a pressure relief valve limits the maximum level to which pressure is permitted to rise. It remains closed during operating periods when pressure is less than the maximum allowed for the circuit, but opens to provide an escape path for oil to discharge back into the oil reservoir if pressure rises too high due to an overload being placed on the system. In this respect it is like a safety valve. All hydraulic circuits using a positive displacement pump should include at least one relief valve, on the pump pressure line. Relief valves may also be needed in other parts of the circuit.

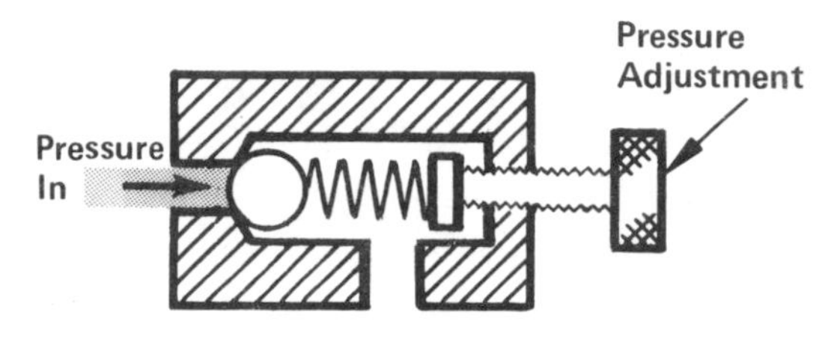

A. Relief valve remains closed and holds oil in the system when fluid pressure is less than the relief valve setting.

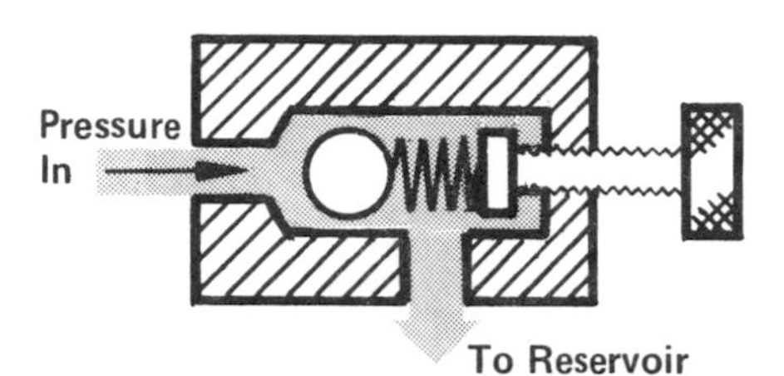

B. Relief valve opens to release part or all of the pressurized oil to reservoir if fluid pressure rises to a dangerous level.

FIGURE 3-4. Adjustable Relief Valve for Pressure Limiting in Hydraulic Systems.

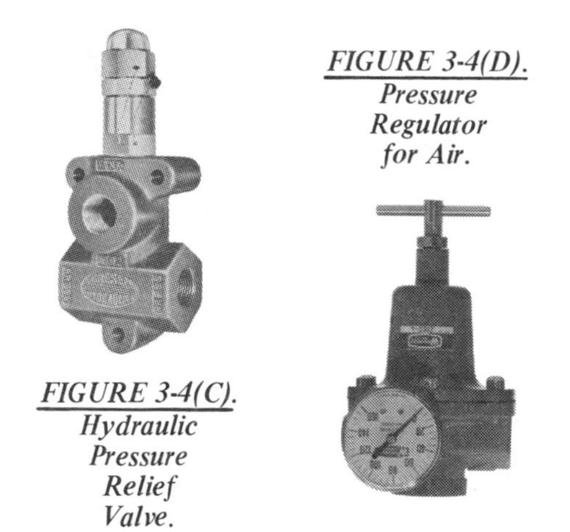

FIGURE 3-4(C). Hydraulic Pressure Relief Valve.

FIGURE 3-4(D). Pressure Regulator for Air.

In Part A of Figure 3-4, a hydraulic pressure relief valve is shown in its normally closed position. Its inlet should be teed into the pump pressure line (or other line to be controlled) and its outlet connected to the system reservoir. Spring tension should be adjusted high enough to hold the poppet closed up to the maximum allowable system pressure. During normal operation, while the system is operating at or below maximum allowable pressure, the valve will remain closed to retain all of the pump flow to the cylinder. But if an overload is placed on the system, as by a machine operator trying to lift more load than the system is designed for, hydraulic pressure will open the poppet, as shown in Part B of the same figure, and a part of the flow (or all of it if necessary) can escape to reservoir to prevent a further rise of pressure in the pump line.

Pressure Control in an Air Line. Relief valves are seldom used on compressed air. Since an air-operated machine usually receives its air supply from a storage tank of pre-compressed air, an overload on the system cannot cause pressure to rise higher than the pressure level in the storage tank.

Part D of Figure 3-4 shows the only pressure control valve normally used on compressed air. It is called a "pressure regulator" and is equivalent to a hydraulic pressure reducing valve. It does not permit any air to escape to atmosphere; it simply reduces the incoming pressure to a lower level which is more suitable for each air-operated machine. It is described in detail in Chapter 6.

MANUAL SHUT-OFF VALVES

Shut-off valves are a group of valves for simply opening or closing a flow passage. Several of these valves can be used in combination to control travel direction of a cylinder or direction of rotation of air and hydraulic motors. In this sense they can qualify as directional control valves. But they are used much more often as simple closures to fluid flow. On occasion one of these shut-off valves may be used to meter the flow as well as to shut it off. When used in that way the valve becomes a flow control rather than a directional control valve.

Shut-off valves have two connections, and are classed as "2-way" valves. They are manufactured in a variety of types, sizes, and pressure ratings. Most are operated manually, but some types have been fitted with solenoid operators. The ones shown in this section are those more often seen in fluid systems.

FIGURE 3-5. Ball-type Plug Valve.

Ball Valve. Figure 3-5. Consists of a ball with a hole through the center. When the hole is turned at right angles to the line, the passageway is blocked. The ball rotates in a synthetic rubber seal.

The ball valve is probably used more often in fluid power than any other type. It has many advantages which include: moderately priced; quick opening and closing with a 1/4 turn of the handle; virtually unrestricted flow when open; leaktight sealing when closed; available in a wide range of sizes and materials, and with moderate to high pressure ratings; and its 1/4 turn action adapts it to power actuation by a fluid cylinder or to padlocking in an open or closed position.

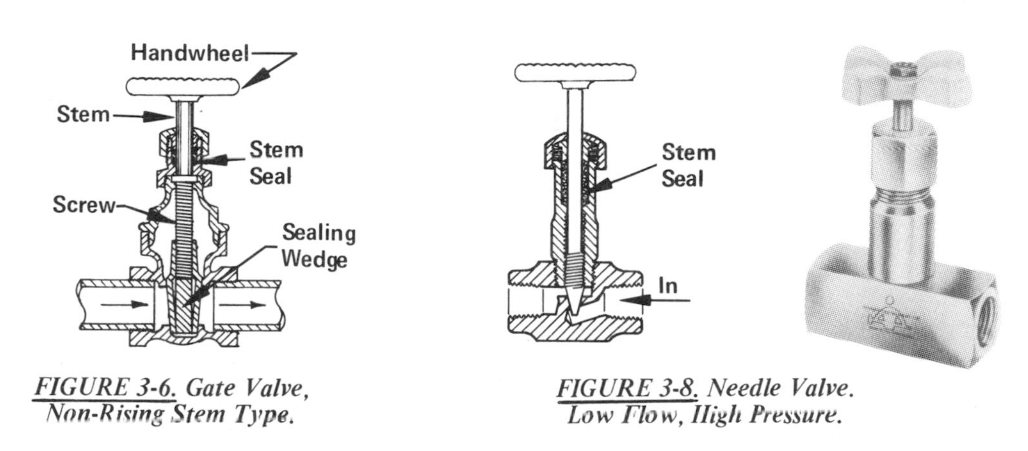

FIGURE 3-6. Gate Valve, Non-Rising Stem Type.

FIGURE 3-8. Needle Valve. Low Flow, High Pressure.

Gate Valve. Figure 3-6. Although more costly than ball valves, gate valves are available in larger sizes, up to 10" or more in diameter and with ratings in the 2000 to 5000 PSI range. A slightly tapered wedge makes a tight metal-to-metal seal when closed, and pulls up completely out of the fluid stream when open. The flow passage, or port, is full size all the way through and without any changes in flow direction, so there is virtually no pressure loss through the valve and little tendency for turbulence to develop. They will usually handle flow in either direction.

The two general types of gate valves are the rising stem type in which the stem and handwheel rise as the sealing wedge rises, and the non-rising stem type shown here in which the handwheel does not rise as the valve is opened.

Needle Valve. Figure 3-8. Can be used as a shut-off valve but the needle may soon become scored if the handle is over-torqued. It is primarily intended as a throttling valve and various models are available with pipe thread connections for pressures up to 15,000 PSI, and with special high pressure connections in ratings to more than 100,000 PSI. The internal passage (port) is very small compared to its connection size, so it is best suited for handling high pressure at very low flow.

The preferred direction of flow is shown in the figure, and this keeps pressure off the stem seal when the valve is closed. The most popular size range is from 1/8 to 3/4" NPT.

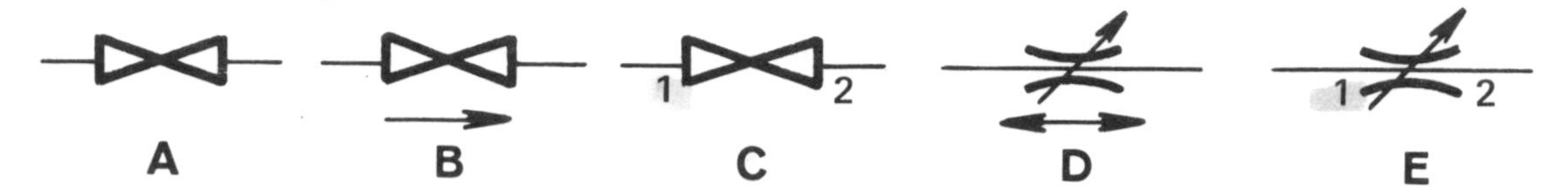

FIGURE 3-9. Graphic Symbols for Shut-Off and Needle Valves. (See Preceding Page).

Graphic Symbols. Figure 3-9. Part A is a general purpose symbol for manually operated 2-way shut-off valves of the type shown on the preceding page. If flow through the valves must be in a specified direction, the symbols of B or C may be used, with "1" indicating the inlet. Two-way valves used for throttling should be drawn as at D or E, with "1" being the inlet. A slash arrow drawn across a symbol, as in D or E, shows that the valve is adjustable and intended for throttling.

APPLICATIONS FOR SHUT-OFF VALVES

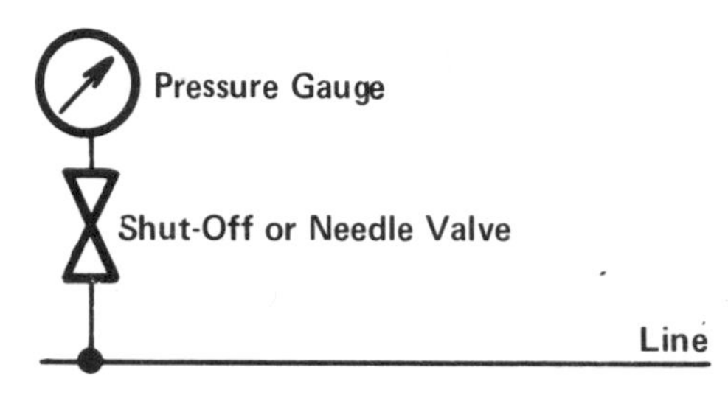

FIGURE 3-10. Gauge Shut-Off Application.

Shut-off valves are selected according to the type of fluid to be handled, system pressure, and flow rate. Brass or bronze valves are preferred for compressed air or water because of the corrosive effect of water on steel. Steel or iron valves are preferred for petroleum oil because copper products may chemically contaminate hydraulic oil. Stainless steel or plastic valves may be used for handling corrosive fluids.

Gauge Shut-Off. Figure 3-10. A shut-off valve should be used with every pressure gauge, and the valve should remain closed except when taking a pressure reading. The life of a gauge will be shortened if left unprotected against severe pulsations and accidental pressure surges. Although it does not make a good permanent gauge snubber (pulsation dampener), a needle valve is often used as a gauge shut-off valve because when only slightly opened it can control severe pulsations while a reading is being taken.

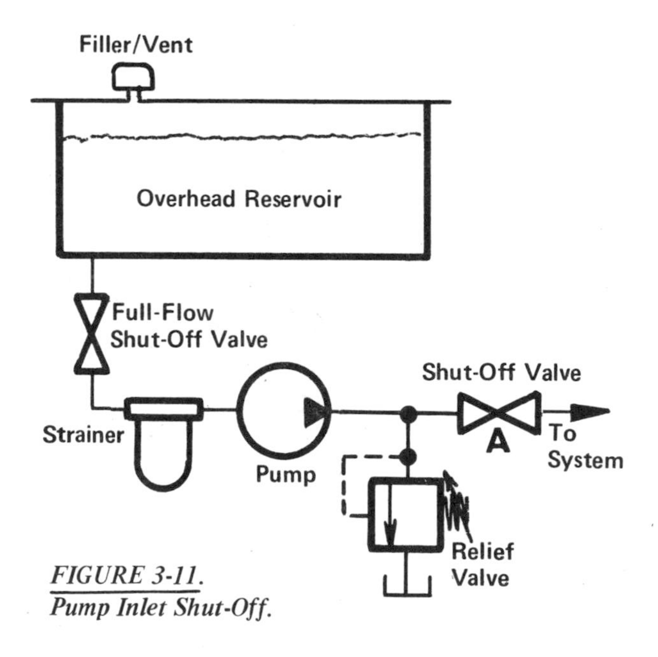

FIGURE 3-11.
Pump Inlet Shut-Off.

Pump Inlet Shut-Off. Figure 3-11. If the oil reservoir is at a higher elevation than the hydraulic pump, a shut-off valve should be installed close to the reservoir. This valve can then be closed while repairs are being made on the system, or while the strainer element is being cleaned or replaced. Extreme care should be used to be sure the valve is open before starting the pump. It should be padlocked in the open position so no one can close it while the pump is running.

If a shut-off valve is to be installed in the pressure line of a pump, it should be placed ***downstream*** of the relief valve, Position A in this figure, and not between relief valve and pump pressure port.

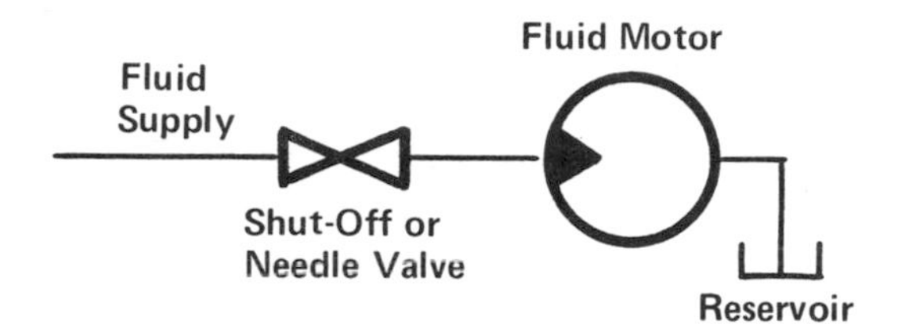

FIGURE 3-12. Fluid Motor Control.

Fluid Motor Control. Figure 3-12. Shut-off valves are used to start and stop devices such as air motors, air-operated vibrators, pressure intensifiers, and sometimes even hydraulic motors. If a needle valve is used for this purpose it could also be used as a speed control. If so used, it should be large enough so it will not restrict the fluid when wide open. The internal orifice is smaller than its connection size, so a needle valve one or two sizes larger than the line size is suggested.

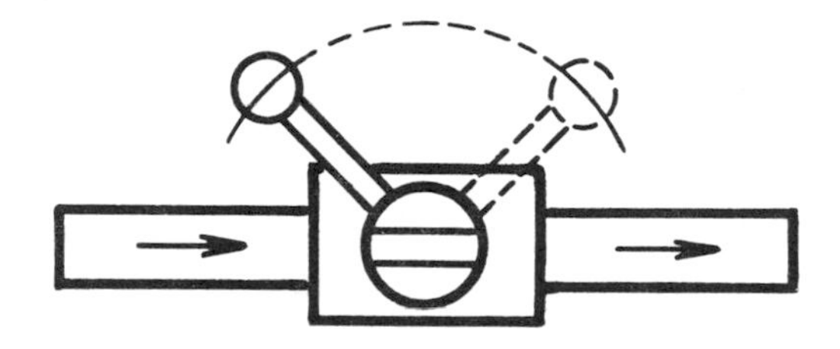

FIGURE 3-13. Pipeline Shut-off.

Shut-Off Service. Figure 3-13. The type of valve is selected according to the nature of the fluid and its pressure. For compressed air, globe valves or 1/4 turn plug valves are popular. For high pressure hydraulics, gate valves or high pressure ball valves are used for shut-off, needle valves for metering the flow. For water service, bronze globe or plug valves are used for shut-off, stainless steel needle valves for metering the flow. Large size pipelines can use ball valves or wedge-type gate valves. For handling food products such as cooking oil or beverages, stainless steel or plastic valves should be selected.

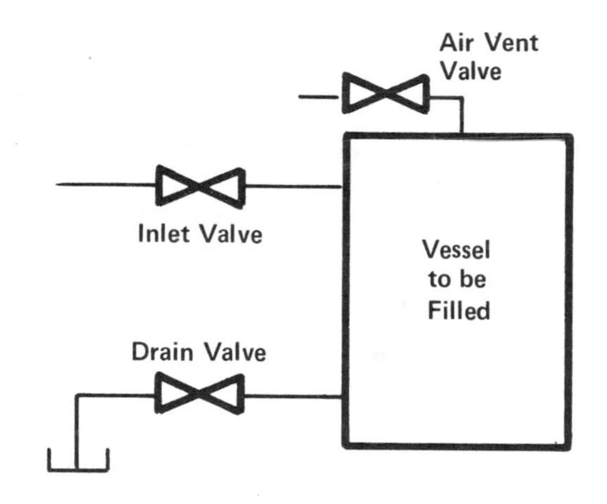

FIGURE 3-14. Filling and Draining.

Filling and Draining Vessels. Figure 3-14. Two 2-way shut-off valves can be used for filling and draining open vessels or closed (and vented) vessels. On sealed vessels an air venting, leaktight, valve may also be required. Air venting can be made automatic by using a "velocity fuse" which readily allows air to escape but closes when liquid starts to flow. This system is adapted to measuring precise amounts of a liquid to be used in a process. Valves can be manually or electrically operated.

A 3-way valve can replace both 2-way valves. See Page 113.

Release for Single-Acting Cylinder. Figure 3-15. Every hydraulic hand pump which has a built-in reservoir uses a shut-off or "release" valve to permit the ram to return when the valve is opened, since the oil cannot return through the check valve. This valve can also be used to control the rate of descent by metering the flow of oil back to the reservoir.

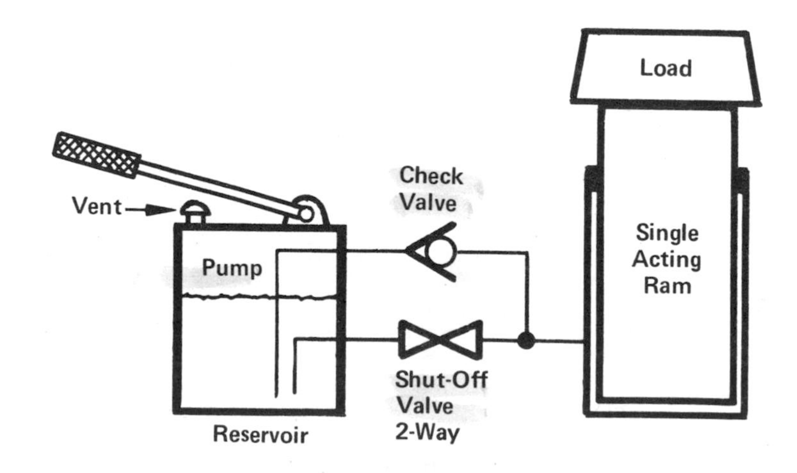

FIGURE 3-15. Release for Single-Acting Ram.

DIRECTIONAL 2-WAY VALVES

While the manual shut-off valves previously described can be used, in combinations of 2 to 4 valves, for directional control of cylinders or direction of rotation of air and hydraulic motors, the solenoid or mechanically actuated 2-way valves to be described in this section are more appropriate for these functions. For control of compressed air valves there is quite an assortment of actuators available including single and double solenoid, manual lever, pedal, treadle, pilot operated, button bleeder, cam roller, and knob. These actuators will be described later in this chapter and in Chapter 4.

For directional control of cylinders, 2-way valves have limited application because a combination of two 2-way valves is required to control single-acting, ram-type cylinders and a combination of four 2-way valves is required for double-acting cylinders. Spool-type and poppet-type valves are sometimes used to meter the flow as well as to start and stop it. But the metering action is not as precise as obtained with needle valves because of the quite abrupt opening and closing of a large porting area as the valve is shifted.

The terms "normally closed" (N. C.) and "normally open" (N. O.) will be used frequently in our descriptions of valves. These terms describe the condition of the inlet port (whether open to flow or closed to flow) when the valve is in its non-actuated or de-energized state. In valves which have built-in return or centering springs, these springs return the spool or poppet to its "normal" state when it is manually released or de-energized. Valves which do not have internal springs do not have a "normal" position. When shifted, they remain in this state until shifted to another position.

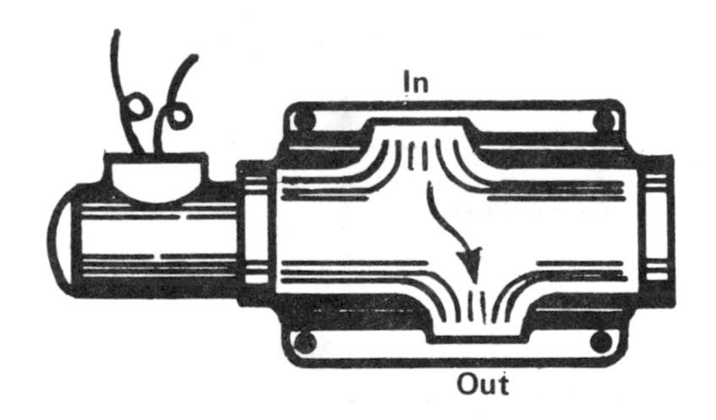

FIGURE 3-16
Solenoid operated sliding spool 2-way valve. Other actuators may be fitted to this body. See Pages 119, 137, and 138.

FIGURE 3-17.
Solenoid 2-way valve with poppet action in which a poppet covers an orifice to stop flow. See Page 96.
Solenoid poppet valves may be directly actuated as described in this section, or may be pilot-operated as described on Page 106.

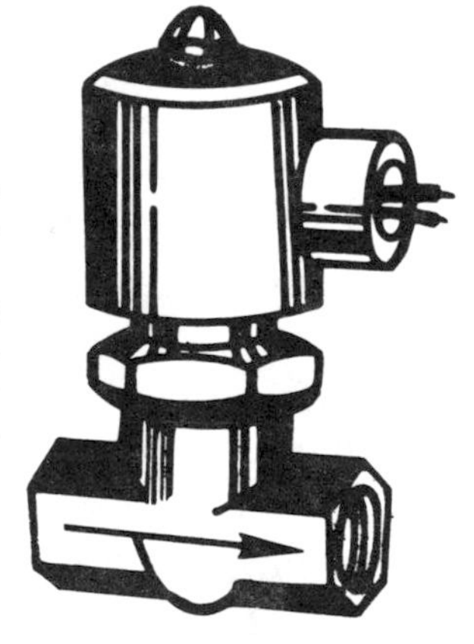

Valve Configurations. Figures 3-16 and 3-17. Two-way valves are offered in as many sizes and shapes as there are valve manufacturers, so these illustrations can only give an idea of appearance of 2-way valves offered by some manufacturers.

SLIDING SPOOL 2-WAY VALVES

Normally Closed 2-Way Valve. Figure 3-18. The diagram shows derivation of the official ANSI (American National Standards Institute) graphic symbol for this kind of valve, which is to be used on schematic circuit diagrams. The diagram starts at the left side, at Positions ❶ and ❸ by showing the same spool-type 2-way valve in both of its working positions.

In constructing the graphic symbol, a square block is used to represent the internal flow passages for each of all working positions, which in this case is two; this valve does not have a neutral.

At ❷ the square block shows that when the valve is actuated, inlet Port P is opened to outlet Port A. This is shown by the directional arrow inside the block. At ❹ the square block shows that when the valve is allowed to spring back into its normal position, the internal passage from inlet to outlet is blocked.

At ❺ all square blocks are put together into one symbol, and to complete the symbol, appropriate actuators are added to one or both ends of the symbol blocks. In this case we are showing a manual lever actuator with spring return. Port markings shown, P for inlet, A for outlet, are according to ANSI standards. A list of other standard port markings is on Page 258.

Actuators can be attached to either end of the block assembly, as desired. On a spring return valve, the circuit connections should be drawn to the block next to the spring. On Page 136 the rules for drawing symbols are summarized.

To summarize the construction of a graphic symbol in Figure 3-18 below, the completed symbol on the right side, if seen on a circuit diagram, would tell the reader he is looking at a 2-way, normally closed, 2-position manual lever actuated valve with spring return.

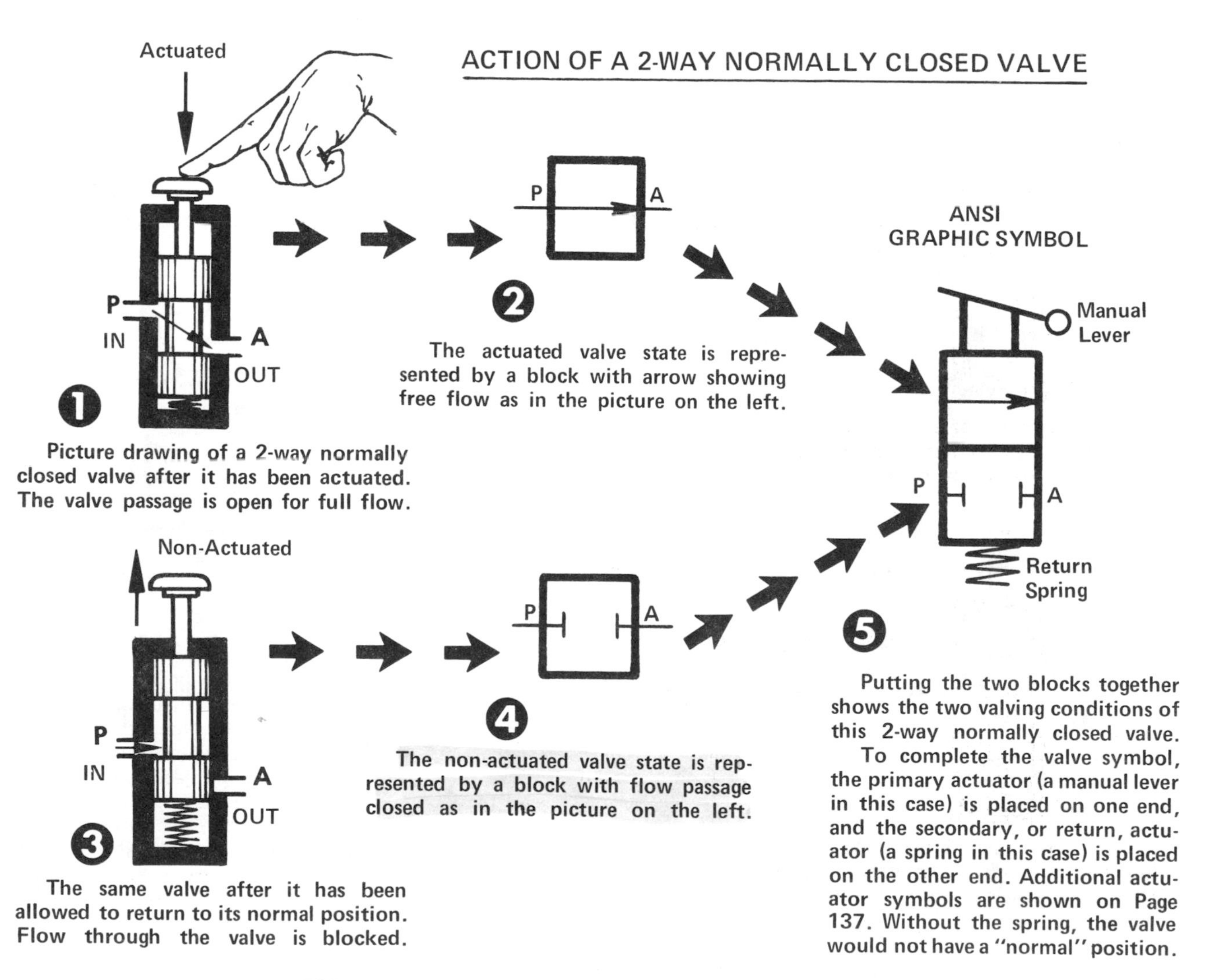

FIGURE 3-18. Method of constructing the graphic symbol for a 2-way, normally closed valve. This is the symbol that should be used on schematic circuit diagrams.

Normally Open 2-Way Valve. Figure 3-19. The action of this valve is opposite to that of the valve shown on the preceding page. When non-actuated or de-energized the flow passage is wide open. Actuating the valve closes the flow passage.

Starting at the left side of the diagram, the pictures at ❶ and ❸ show the same valve in its two valving positions. In constructing a graphic symbol to be used on circuit diagrams, a square block is used to depict the state of flow through the valve in each of its working positions.

At ❷ the square block shows that when the valve is actuated, inlet Port P is internally blocked. When the valve is allowed to spring back to its normal position, the block at ❹ shows the internal flow passage to be open for full flow.

To show the complete valving capability of this valve, both blocks are joined into one symbol, as shown at ❺, and to complete the symbol we are showing a manual lever actuator with spring return. Port markings, P for inlet, A for outlet, are according to ANSI standards for port markings on valves. Other standard port markings are shown on Page 258. Please also see Page 136 for additional rules for the construction and use of graphic symbols. *(Description continued on the next page).*

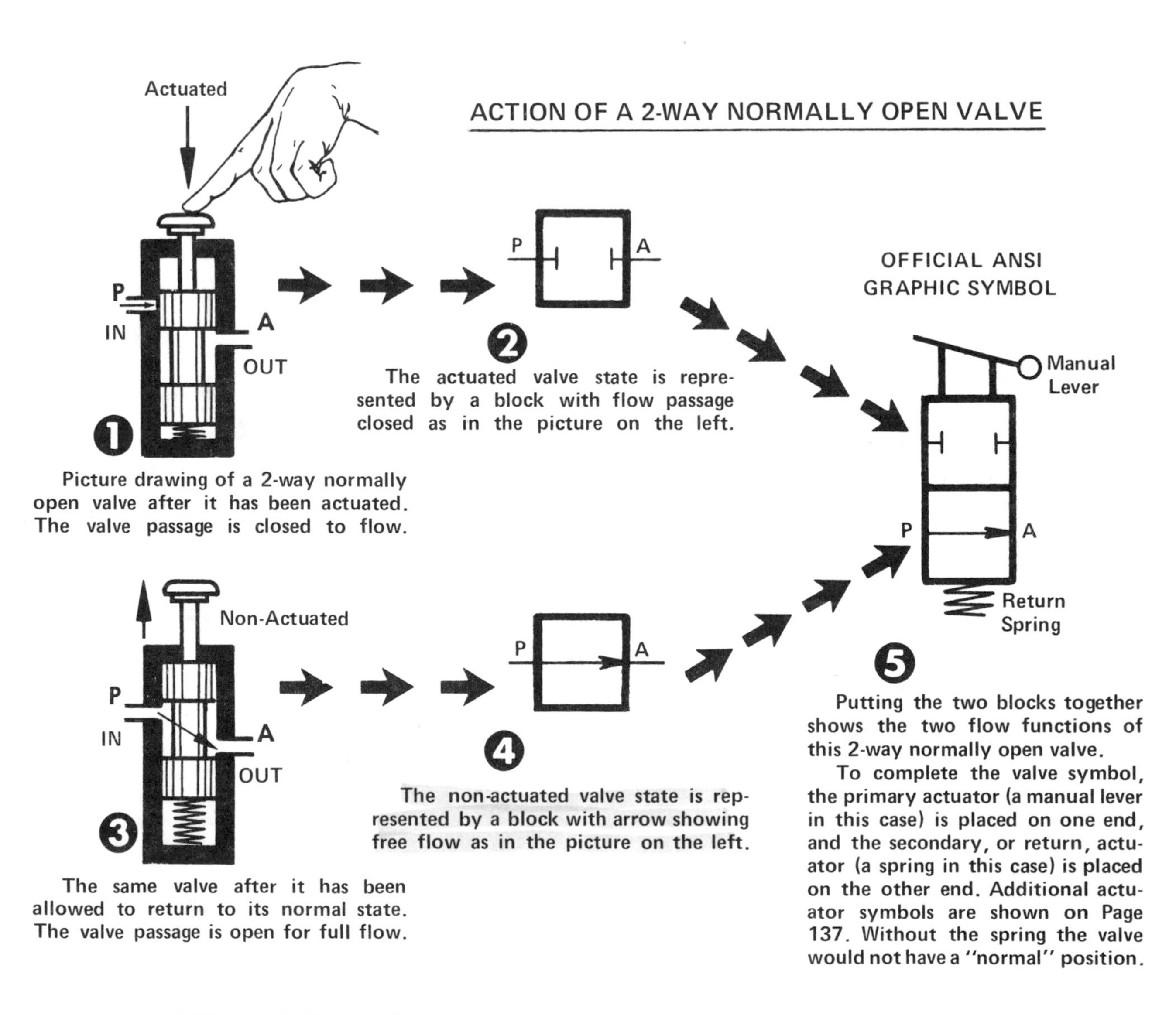

FIGURE 3-19. Method of constructing the graphic symbol for a 2-way normally open valve. This is the symbol that should be used on schematic circuit diagrams.

Actuators can be placed on either or both ends, as required. When using this symbol in a circuit drawing, circuit lines should be drawn to the block nearest the spring.

To summarize the construction of the graphic symbol developed in Figure 3-19, the completed symbol at ❺, if seen on a circuit diagram, would tell the reader that he is looking at a 2-way, normally open, 2-position manual lever actuated valve with spring return.

SOLENOID 2-WAY VALVES – DIRECT-ACTING TYPE

Solenoid valves are of two kinds: direct-acting and pilot-operated. In a direct-acting valve the mechanical force to shift the spool or lift the poppet is supplied by direct push or pull from a solenoid. It is limited to low flow and/or low pressure operation because of the limited force available from a low power solenoid. On the other hand, a pilot-operated solenoid valve uses fluid pressure taken from the line for spool shifting and does not depend on solenoid force. This valve is described on the next page. It is capable of handling very high pressures and very high flows with a low power solenoid.

Direct-acting 2-way solenoid valves can be built with sliding spool or with poppet-type construction, although poppet construction is more common. Graphic symbols for both types are generated by the method shown on the preceding page. Some brands of 4-way valves (4 main ports and 2 independent flow passages) can be used for 2-way service simply by plugging the ports which are not used.

Direct-Acting Solenoid Valve. Figure 3-20. Electrical solenoids which will operate at low amperage to preserve the life of standard limit switches will produce only about 15 to 20 lbs. force for starting. This is insufficient to shift valves larger than 1/4 to 1/2" size. One manufacturer, for example, catalogs direct-acting solenoid valves for pressures to 800 PSI but with only a 1/16" diameter flow orifice. Another model is rated for higher flows, up to 2" capacity, but at only 1 PSI pressure. Some spool-type valves may use larger solenoids but even these are limited to about 3/8" flow capacity at 1000 to 3000 PSI.

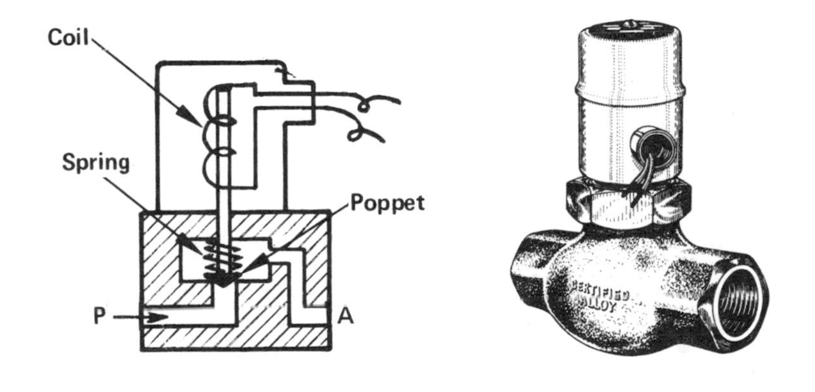

FIGURE 3-20. Direct-Acting, 2-Way Solenoid Valve.

Of course, valves could be built with much larger solenoids, sufficiently powerful to handle both high pressure and high flow, but oversize limit switches and contactors would be required to handle the high inrush current at the moment a solenoid was energized. They would require larger wiring and a larger electrical supply. And, the closing impact of large solenoids is not only noisy but destructive to the valve and to themselves. Pilot-operated solenoid valves, described on the next page, solve most of the problems of handling both high pressure and high flow at the same time.

In Figure 3-20, a normally closed, 2-way, poppet-type, direct-acting solenoid valve is shown. This has the same valving action described on Page 96. The graphic symbol is similar to the valve on Page 103 except the manual lever actuator would be replaced with the symbol for a solenoid. A valve of this kind can also be constructed with normally open (N. O.) action.

SOLENOID 2-WAY VALVES – PILOT-OPERATED TYPE

A "pilot-operated" solenoid valve is one which uses fluid power, instead of the direct push of a solenoid, to open and close its main poppet. The fluid power for valve operation is usually tapped internally from the valve inlet, although it could be taken from an external source of pressure. The solenoid coil itself does not work directly on the main poppet; it merely directs shifting pressure against the poppet.

Pilot operation, as opposed to direct solenoid action, in addition to its ability to handle almost unlimited levels of pressure and flow, has other advantages: the solenoid is relatively small, requiring low amperage; it can be actuated from standard limit switches because of its low current requirement. Opening and closing of the main poppet is relatively quiet; pressure surges in the line can be reduced by tailoring the valve control orifice to open or close the main poppet slowly.

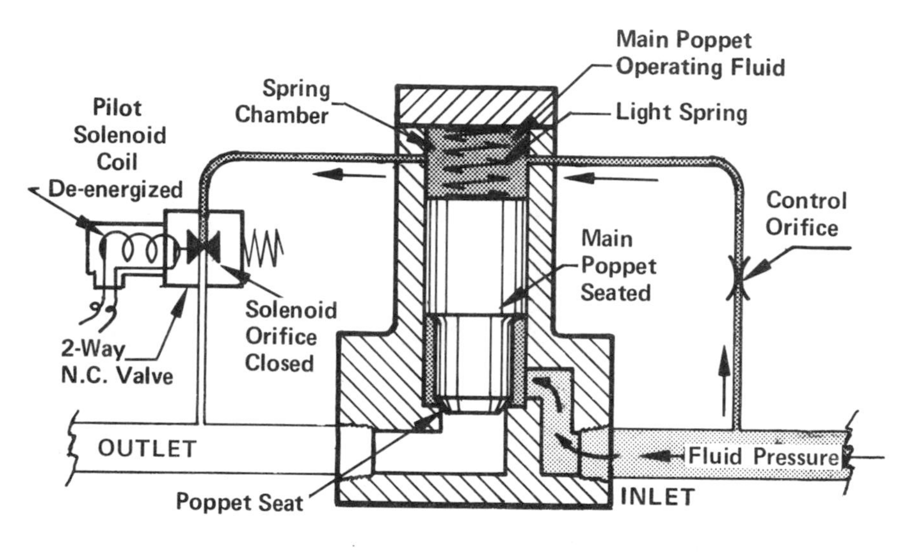

A. Solenoid de-energized; main poppet closed.

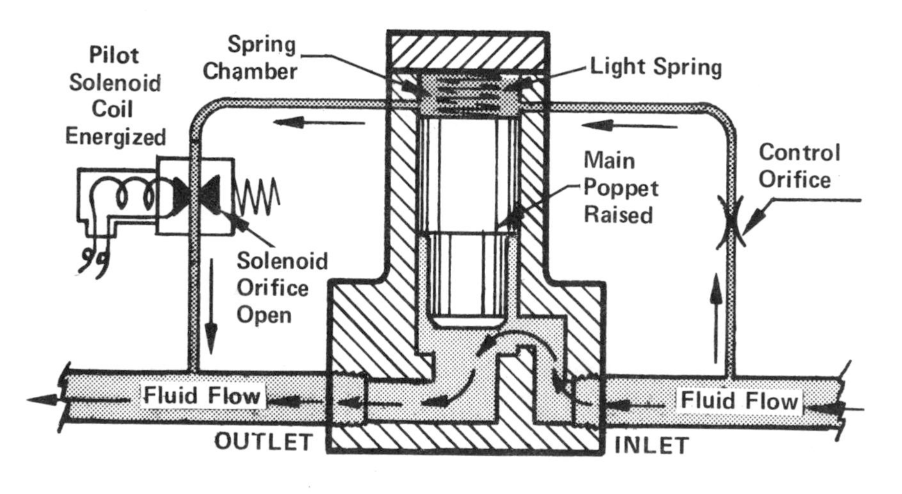

B. Solenoid energized; main poppet open.

FIGURE 3-21. Picture layout of 2-way normally closed, pilot-operated solenoid valve in de-energized state (top view) and energized state (bottom view).

The pilot operating principle, controlled by solenoid action is used extensively on both air and hydraulics over a range of pressures starting at 1 to 5 PSI minimum up to 1000, 3000, 5000, and even 10,000 PSI, and in flow capacities from 1/4 to 6" or greater. The same principle of pilot operation which is described below for 2-way solenoid valves is also used on many other kinds of valves including air and hydraulic 4-way valves, pressure relief, pressure regulator, pressure reducing, by-pass, sequence, unloading, and others.

Operating Principle. Figure 3-21. All parts shown in this schematic diagram are usually contained in one housing. The solenoid coil, when energized, opens a very small orifice, for example 3/32" diameter, which drains the main spring area to the downstream line. This light spring is sometimes omitted in valves designed for vertical mounting, the weight of the poppet taking its place.

In the upper view, with the solenoid de-energized, the main poppet is held closed by the pressure of fluid admitted to the spring chamber from the valve inlet through the control orifice. Exposed area on top of the poppet is larger than exposed annular area around the smaller section of the poppet near the bottom. Therefore, the poppet will remain tightly seated, held on seat primarily by fluid pressure.

To open the poppet the solenoid must be energized. Refer to the lower view in Figure 3-21. This releases fluid pressure on top of the poppet and allows pressure from the valve inlet to unseat the poppet against the force of the light spring. As long as the solenoid remains emergized, the spring chamber of the poppet will remain vented and the valve will remain open for flow except for a small pressure drop of 1 to 5 PSI from inlet to outlet which the valve automatically produces to keep the light spring compressed. The control orifice limits the rate of flow into the spring chamber to a value which can be freely vented by the solenoid.

To close the main poppet the solenoid must be de-energized. This closes the solenoid orifice, blocking the vent path. Fluid continues to flow through the control orifice into the spring chamber and drives the poppet downward. The poppet is able to close in a fraction of a second after the solenoid has been de-energized.

If the control orifice is made too small, the valve will be fast in opening but may be a little sluggish in closing. If the orifice is too large, the valve may be sluggish in opening but will close fast. If a valve with orifice designed for compressed air is used on water or oil, it will be too slow in opening, or may not open at all. The best orifice size is determined according to the fluid viscosity.

Minimum Operating Pressure. Figure 3-22. One limitation to all pilot-operated solenoid valves is that they consume a small amount of pressure from the fluid line for their operation, from 1 to 2 PSI for the 2-way normally closed valve just described and about 5 PSI for the same valve in a normally open version. This amount of pressure is required by the valve to compress the light poppet spring or to support poppet weight, or both, and is not available downstream of the valve to other devices. This means the valve cannot be used on pressures which are less than the values needed to compress the spring. When energized, a normally closed valve will not remain open if the inlet pressure falls too low and a normally open valve will not remain closed.

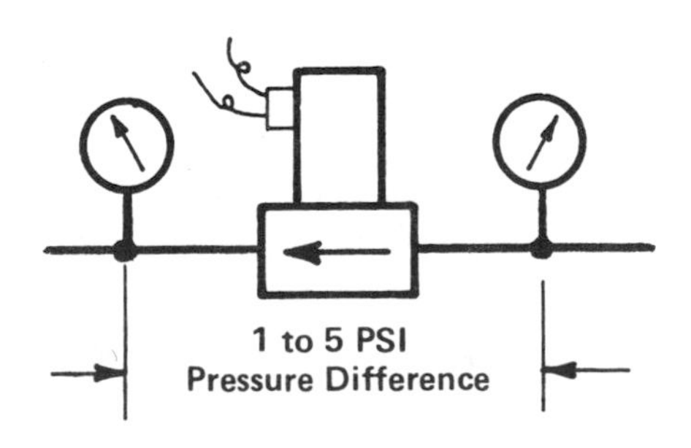

FIGURE 3-22. A pilot-operated solenoid valve will not operate on inlet pressures less than 1 to 5 PSI.

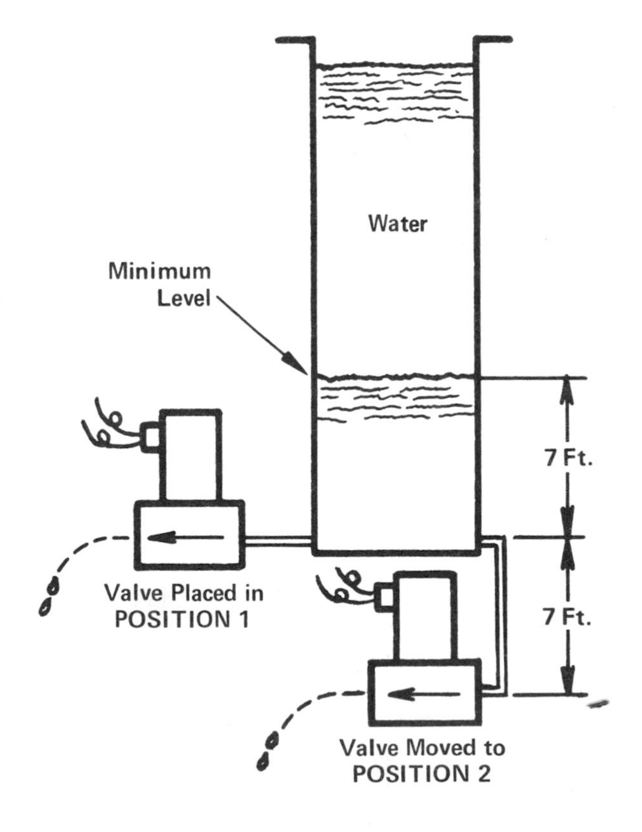

FIGURE 3-23. The action of a pilot-operated solenoid valve in draining a tank to atmospheric pressure.

Draining Open Vessels. Figure 3-23. Pilot-operated solenoid valves should not be used on such applications as draining open vessels to atmospheric pressure. Vacuum applications, too, are not suitable because the valve consumes too great a proportion of the available pressure for its operation.

In this figure we are using a pilot-operated solenoid valve which requires a minimum of 3 PSI for its operation. If the vessel were to be drained from its full level to atmosphere through this valve placed at Position 1, the rate of discharge would diminish as the water reached a minimum level because the available head pressure for keeping the valve open would be insufficient. Its poppet would completely close when the water level had dropped to 7 feet above the valve inlet. At this point, since water weighs 0.43 PSI per foot of head, the valve inlet pressure would be less than required to keep the poppet open. If the solenoid were kept energized, the remainder of the water would drain very slowly through the control orifice and the solenoid orifice in the valve.

A direct-acting solenoid valve used for this application would remain wide open until the water had completely drained. It does not depend on fluid pressure to keep its poppet open.

It is interesting to note that if the same valve were placed at Position 2 which is at an elevation of 7 feet or more below the bottom of the vessel and connected with a pipe to the port on the bottom of the vessel, the valve would remain wide open until the water in the main vessel had completely drained. A 7-foot head of water, even in a small pipe would produce enough pressure to keep the valve open.

3-WAY FLUID POWER VALVES

A 3-way valve has three main connections, all different. It can handle certain valving functions beyond the capability of a shut-off valve or a 2-way valve, or functions that otherwise would require two 2-way valves. Usually, all three connections are full size, although a model is shown in Figure 3-28 which has a small, unthreaded third port. A 3-way spool-type valve has only one working groove on its spool and can handle only a single flow.

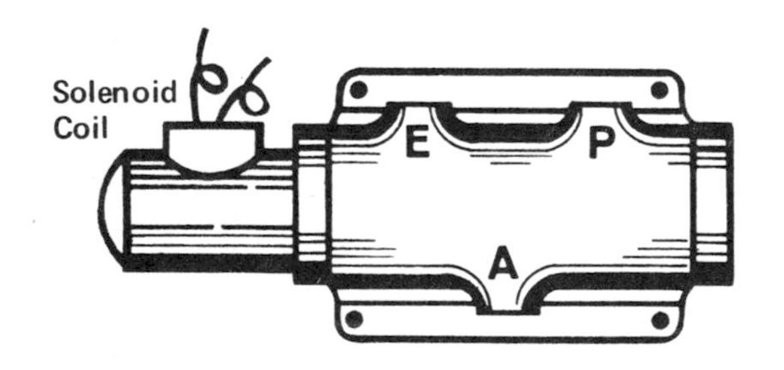

FIGURE 3-24. Spool-Type 3-Way Valve. Most 3-way valves are spool rather than poppet type since this construction lends itself to the use of various actuators. A solenoid actuator is shown on this valve. Other actuators, as shown on Pages 137 and 138 may be fitted to this body.

FIGURE 3-25. Poppet-Type 3-Way Valve. This construction is used mainly for small size solenoid valves. Poppet construction is popular for water hydraulic valves because it will resist "wire drawing" type of corrosion.

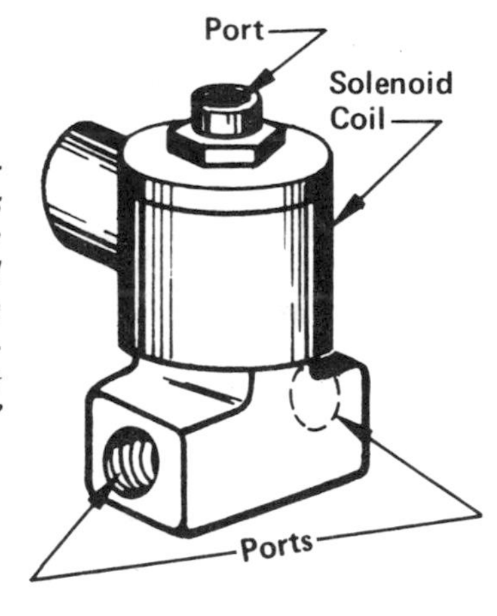

There are several kinds of applications which require 3-way valve action. Some require a valve which can operate with full pressure on any or all ports. Other applications like directional control may use a valve having one port vented to exhaust, and with high pressure on inlet and outlet ports.

On miniature size 3-way valves, duplex construction is occasionally used in which two individual 2-way valves are mounted in the same frame. One valve is normally closed, the other normally open. Each has its own solenoid coil, and these solenoids can either be energized individually or both can be energized at the same time. This construction is rare or non-existent in larger valves.

THREE-WAY VALVES FOR DIRECTIONAL CONTROL

The most important applications in fluid power for 3-way valves are for directional control which includes control of single-acting cylinders, filling and draining vessels, application and release of vacuum, and control of air and hydraulic non-reversible motors.

Figure 3-26 below illustrates a sliding spool valve. A 3-way poppet-type valve has the same action. Those 3-way valves having internal springs to return the spool to a "normal" position when non-actuated or de-energized are classed either as "normally closed" (N. C.) or "normally open". (N. O.) depending on the condition of the pressure inlet port (closed or open to flow) when the valve is not actuated. Those valves which do not have internal springs do not have a "normal" position. The choice between a N. C. or a N. O. valve depends on the circuit action desired.

N. C. 3-Way Valve for Directional Control. Figure 3-26. This diagram shows how the official ANSI symbol is derived for use on schematic diagrams. On the left side of the diagram, in Views ❶ and ❸, the same valve is shown in both of its valving positions, non-actuated in the lower view and actuated in the upper view. A square block is drawn for each of the working positions with flow arrows

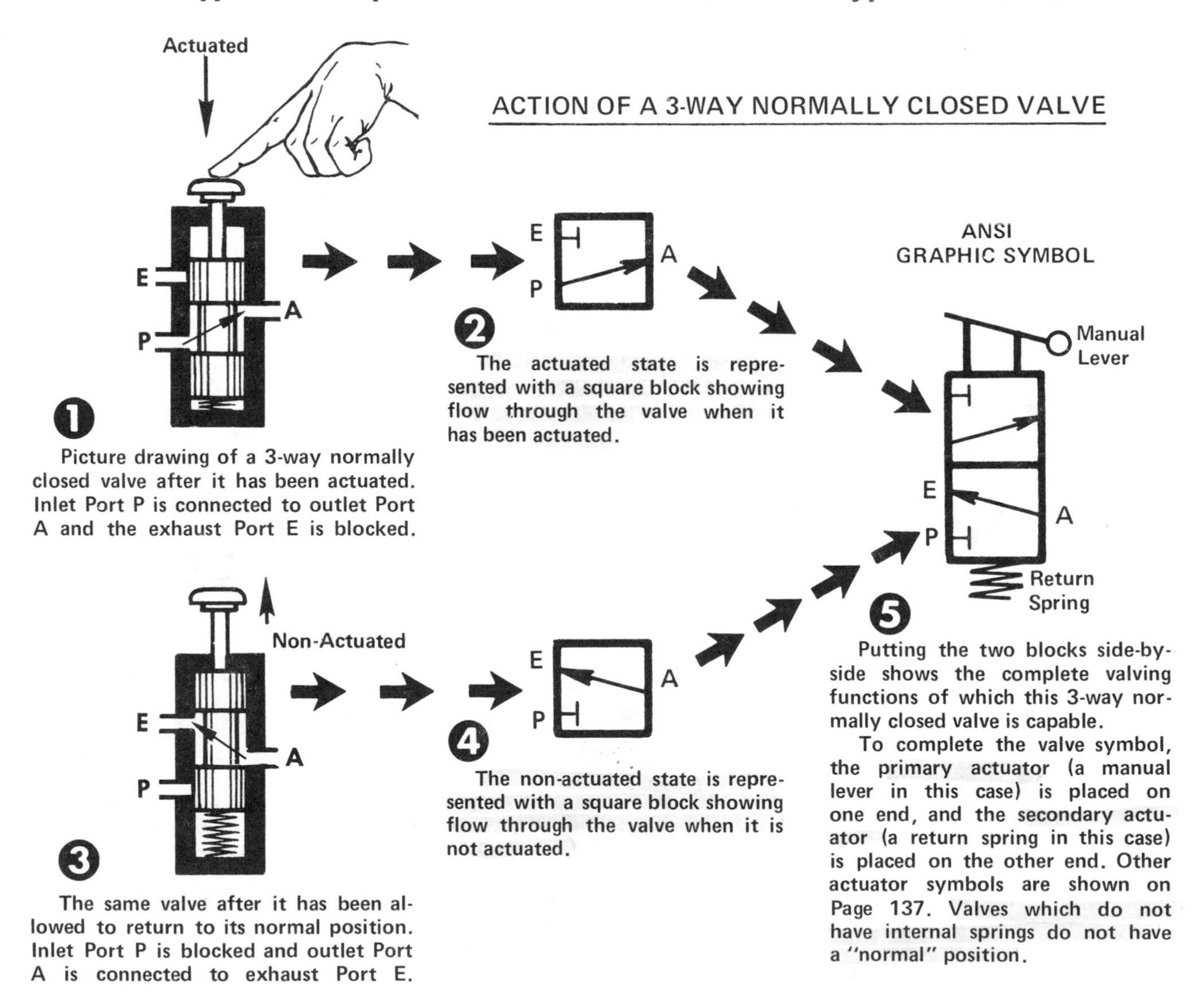

FIGURE 3-26. Development of ANSI graphic symbol from a picture drawing of a 3-way N. C. valve. See text for rules governing construction and proper use of symbols on schematic diagrams.

showing whether internal passages are blocked or open when the valve is shifted to that position. Arrowheads are important because they show the direction of flow for which the valve is designed. Routing flow in a direction against the flow arrows may not be acceptable on some 3-way valves. The valve illustrated does not have a third, or neutral, position.

At ❷ the square block shows that when the valve is actuated, inlet Port P is opened to outlet Port A, while the exhaust Port E is blocked. At ❹ the square block shows that when the valve is allowed to spring back to its "normal" position, pressure Port P is blocked, while outlet Port A is vented to exhaust Port E.

At ❺ all square blocks are joined into one symbol, and to complete the symbol, appropriate actuators are added to either or both ends of the symbol blocks. In this case we are showing a manual lever actuator with spring return. Port markings are according to ANSI standards. Markings are the same for either an air valve or a hydraulic valve with the exception of the exhaust port, in which E is used for exhaust on an air valve and T is used for tank return on a hydraulic valve. Other port markings are shown on Page 258.

When drawing a completed symbol, actuators can be attached to either end of the symbol blocks, at

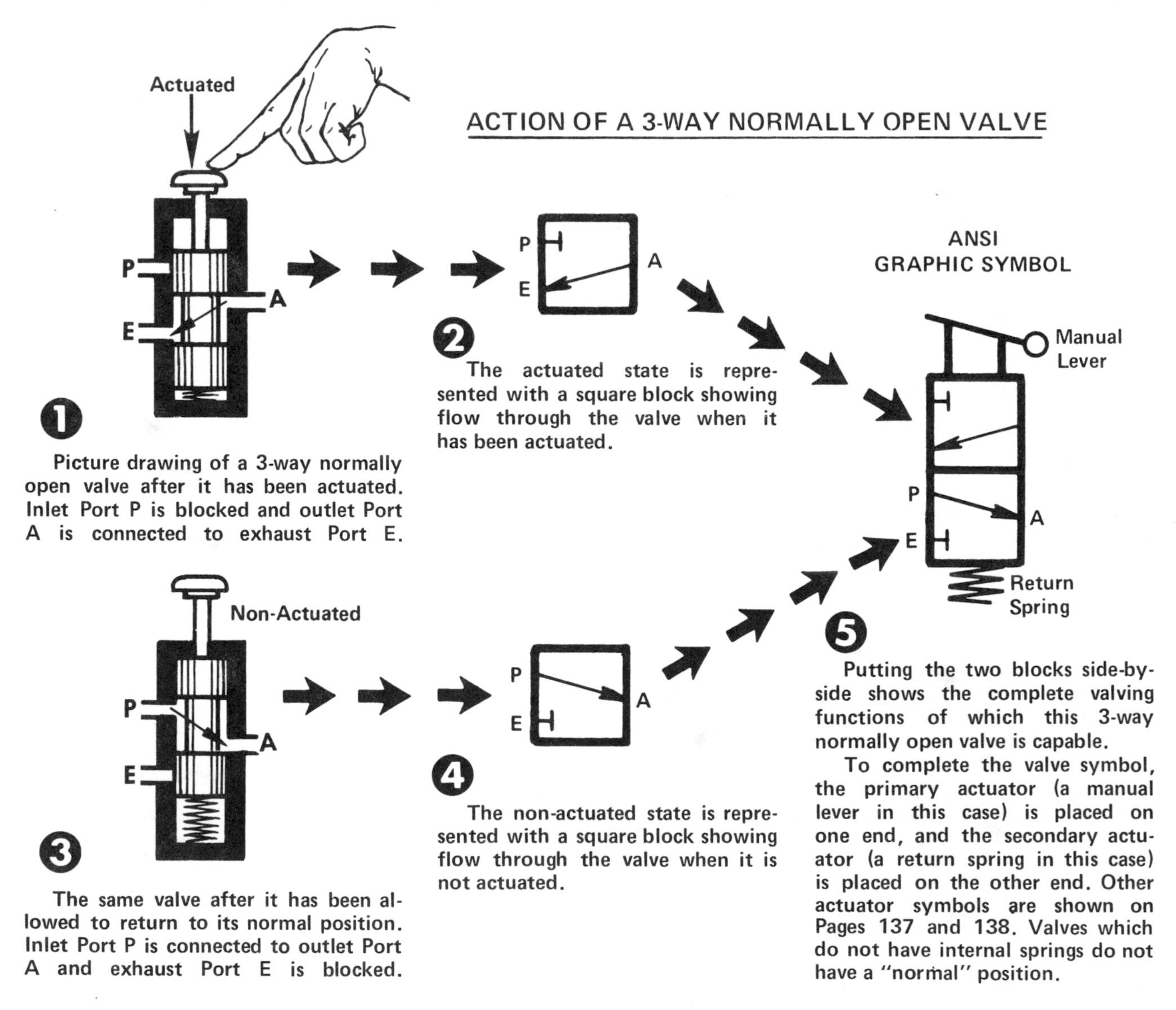

FIGURE 3-27. Development of ANSI graphic symbol from a picture drawing of a 3-way N. O. valve. See text for rules governing construction and proper use of symbols on schematic diagrams.

your convenience. Connecting lines to the rest of the circuit, as a rule, should be drawn to the block farthest from the primary actuator, the manual lever in this case. See Page 136 for complete rules.

To summarize the construction of the graphic symbol in Figure 3-26, the completed symbol at ❺ would tell the reader this is a 3-way, normally closed, 2-position, manual lever valve, spring return.

Normally Open 3-Way Valve for Directional Control. Figure 3-27. The action of this valve is opposite to that of the N. C. valve in the preceding figure. When non-actuated or de-energized, the internal flow passage from Port P to Port A is open. Actuating the valve closes inlet Port P.

Procedure in constructing a graphic symbol is the same as for a N. C. 3-way valve. It can be briefly summarized as follows: The two picture views at the left of the diagram show a valve in its two working positions, actuated in the upper view, released in the lower view. The two blocks at ❷ and ❹ show the porting pattern through the valve in these two working positions. The completed symbol is shown at ❺. Connecting lines to the rest of the circuit are usually drawn from the block which is farthest from the primary actuator, a manual lever in this case. On Page 136 there is a summary for rules to be followed in constructing valve symbols.

DIRECTIONAL CONTROL APPLICATIONS FOR 3-WAY VALVES

Stop and Waste Cock. Figure 3-28. This is a directional control valve application in its simplest form, and is used for draining trapped water from outdoor hydrants when the main valve is shut off. A valve is used which has two main threaded connections and a small unthreaded vent hole.

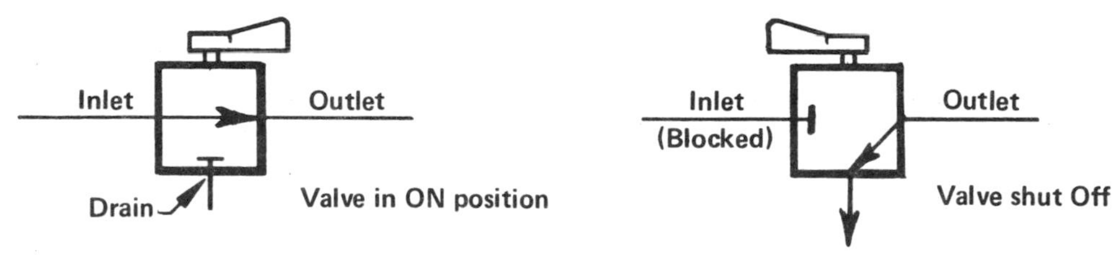

FIGURE 3-28. Stop and waste 3-way valve.

Service Bleed. Figure 3-29. In a branch circuit of a compressed air system, the air must be shut off when it is necessary to service the air line filter, regulator, lubricator, or other component. At the same time, the compressed air trapped in the branch circuit must be vented off to prevent an accident as the plumbing is opened. This should be done with a 3-way valve (instead of a 2-way shut-off valve) installed at the entrance to the branch. A manual lever valve with detent action as shown in Figure 3-43C can be used. Compact, in-line sliding sleeve valves can also be used. They vent trapped air to atmosphere through a small unthreaded hole when inlet air is shut off. Rules for graphic symbols are on Page 136.

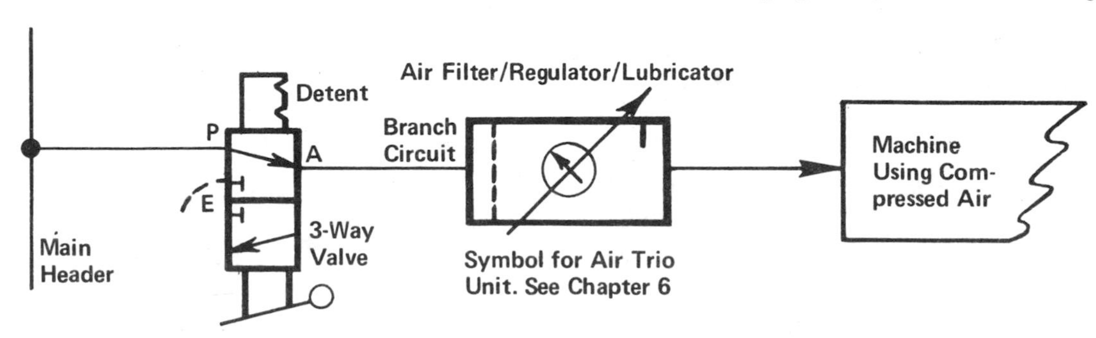

FIGURE 3-29. Shut-off and vent 3-way valve for servicing a compressed air branch circuit.

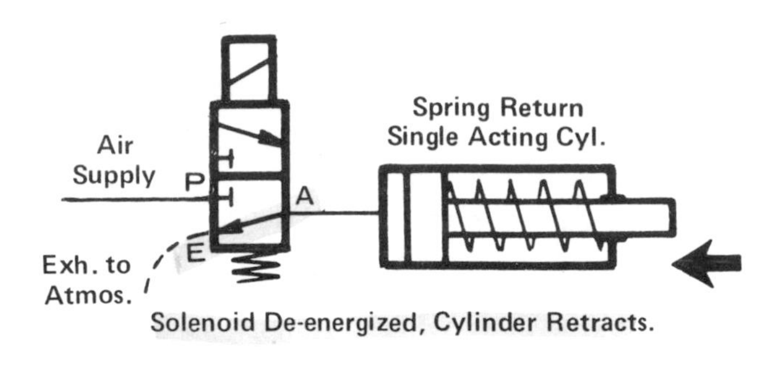

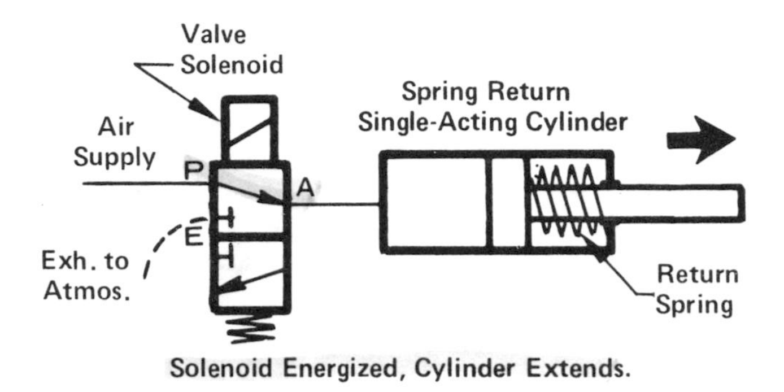

FIGURE 3-30. The action of a 3-way directional valve in controlling a single-acting air cylinder.

Control of Single-Acting Air Cylinder. Figure 3-30. The 3-way directional control valve used here has a solenoid actuator with spring return. The cylinder extends under air power but retracts by its built-in spring.

With the solenoid de-energized (top view), inlet air is blocked and the cylinder has moved to its retracted position by venting trapped air to atmosphere through the valve, as the internal spring pushes the piston back.

When the solenoid is energized, (bottom view), the valve spool moves to this new position, directing inlet air into the cylinder to extend the piston, compressing the internal spring. In this valve position the exhaust is blocked. When the piston completes its forward stroke it remains stalled until the solenoid is de-energized.

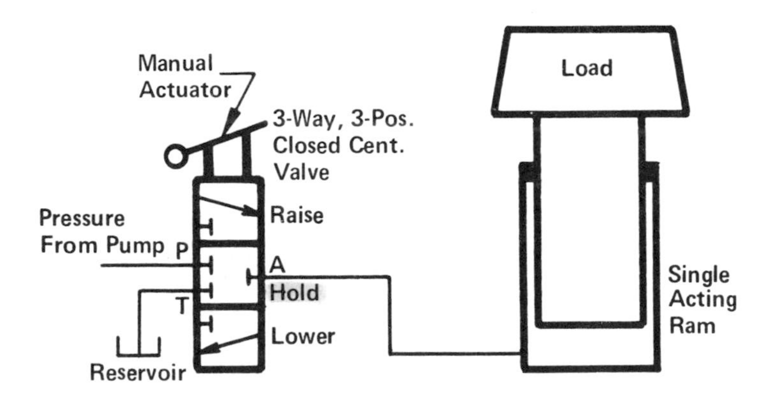

FIGURE 3-31. For controlling a hydraulic ram cylinder, a 3-way valve with center neutral is used.

Hydraulic Ram. Figure 3-31. A single-acting ram which must be raised and held in a partially raised position may be controlled with a 3-way valve having a center neutral position with all ports blocked (closed center neutral). This would normally be a manual valve operating from a hand pump.

The valve has three working positions – raise (upper block), hold (center block), and lower (bottom block). Retraction of the cylinder is by weight of the load. If the load weight will not bring it down, an external spring or other mechanical device must be used.

This circuit is not often used with hydraulic rotary power pumps because it will not "unload" the pump while the ram is holding.

Air Motor Start-Stop Control. Figure 3-32. A 3-way directional control valve may be used instead of a simple shut-off valve to start and stop a non-reversing air motor. The 3-way valve has the advantage that when the motor is stopped, both motor ports are vented to atmosphere, and the motor is in a free wheeling state. To control a reversible air motor, a 4-way valve as described in Chapter 4 would be needed.

In Figure 3-32 a solenoid type 3-way directional control valve is shown. However, other types of actuators such as manual, pedal, or cam could be used if appropriate for the application.

When the solenoid has been de-energized (top view), inlet air is cut off, and the motor stops with both its ports vented to atmosphere, in a free wheeling state.

When the solenoid has been energized (bottom view), the air supply is connected to the motor and the exhaust port is blocked.

Hydraulic motors are not usually controlled with 3-way valves. Additional means should be provided to "unload" the pump when the motor stops. This can be done with a 4-way directional valve as described in Chapter 4. Pump unloading will be covered in due time, in Chapter 5.

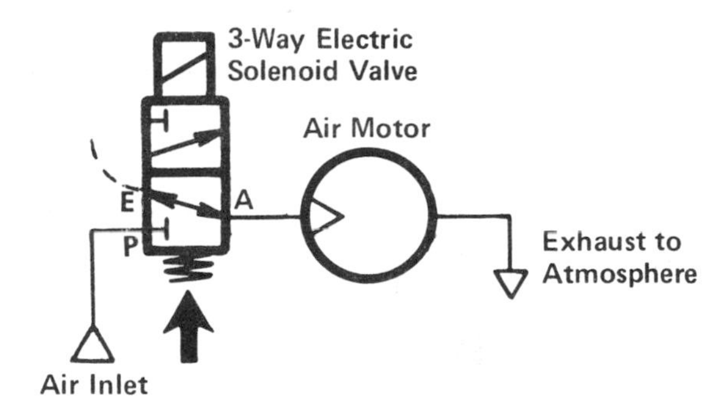

Electric Solenoid De-energized; Motor Stopped.

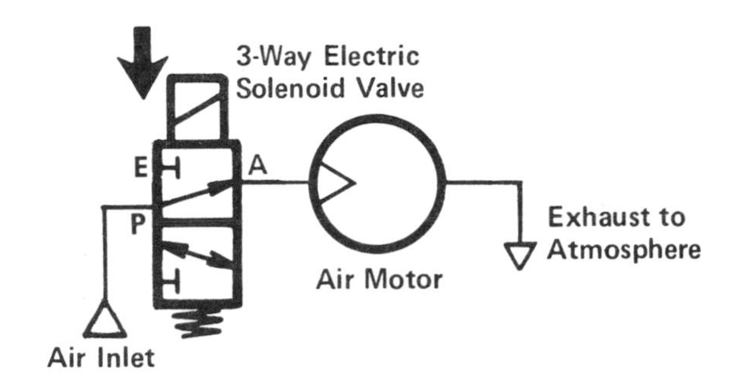

Electric Solenoid Energized; Motor Running.

FIGURE 3-32. A non-reversible air motor may be started and stopped with a 3-way directional valve.

Filling and Draining a Vessel. Figure 3-33. One 3-way, 2-position spool valve can be used for both filling and draining provided the liquid is compatible with materials and seals in the valve. For handling corrosive liquids, two shut-off valves may have to be used, as they are available in a wide choice of body materials.

In this illustration, when the valve spool is shifted upward, the inlet is closed and the vessel is draining. When the spool is shifted downward, the liquid supply is connected into the vessel and the drain is blocked.

A 3-position valve with closed center neutral may be preferred for this application.

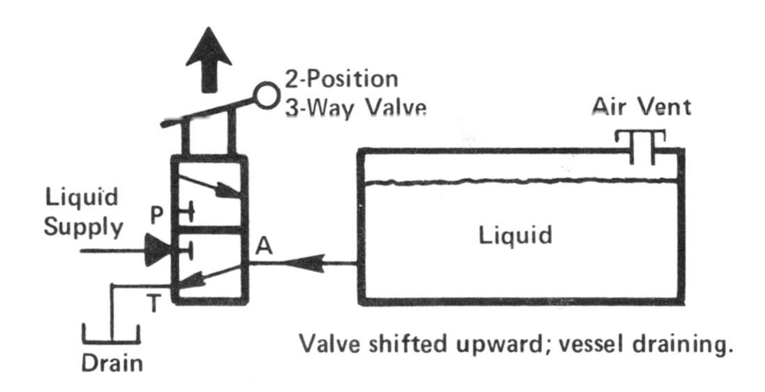

Valve shifted upward; vessel draining.

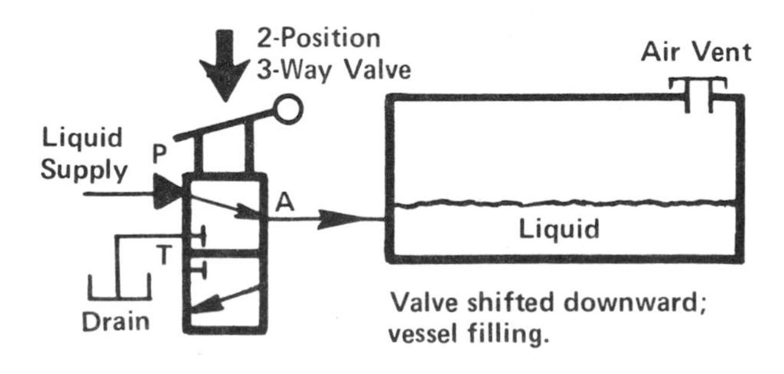

Valve shifted downward; vessel filling.

FIGURE 3-33. A 3-way valve may replace two shut-off valves in filling and draining a vessel.

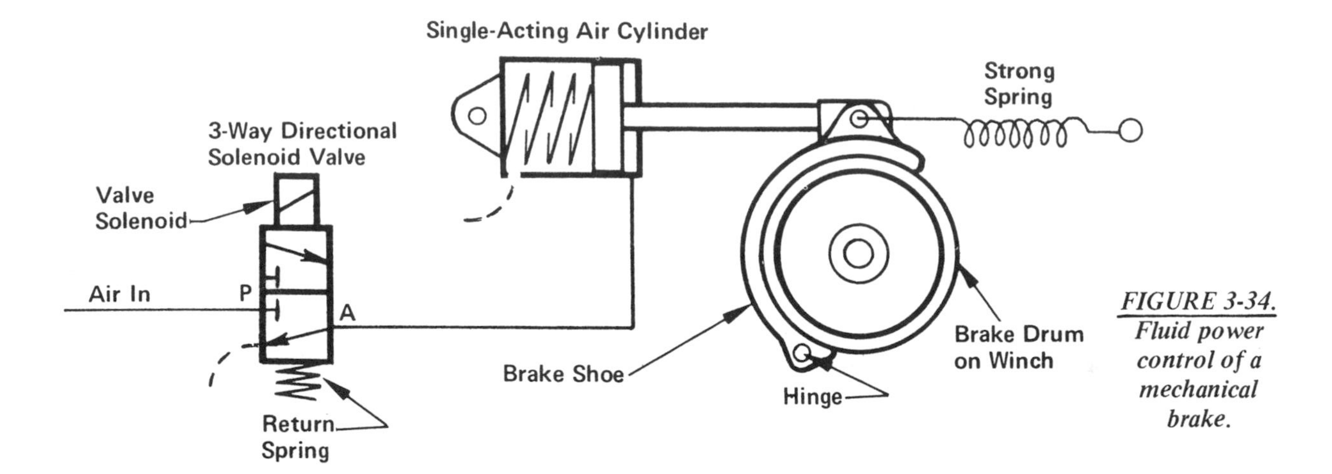

FIGURE 3-34. Fluid power control of a mechanical brake.

Mechanical Brake. Figure 3-34. Such devices as wire lines and winches may have a mechanical brake to hold the drum from unwinding or creeping when the system power is shut off. Brake shoes are spring loaded to the closed position. When power is applied to the winch drum, the 3-way solenoid brake valve is also energized. A cylinder, controlled by the 3-way valve, pulls the brake shoe away from the drum. In this example, a 3-way, normally closed valve is shown. When de-energized, the air cylinder is vented to atmosphere, allowing the brake to be "set" by a strong spring.

Brake circuits should be designed to be "fail safe". In this example, if either fluid pressure or electrical control current should fail, the brake would be automatically set.

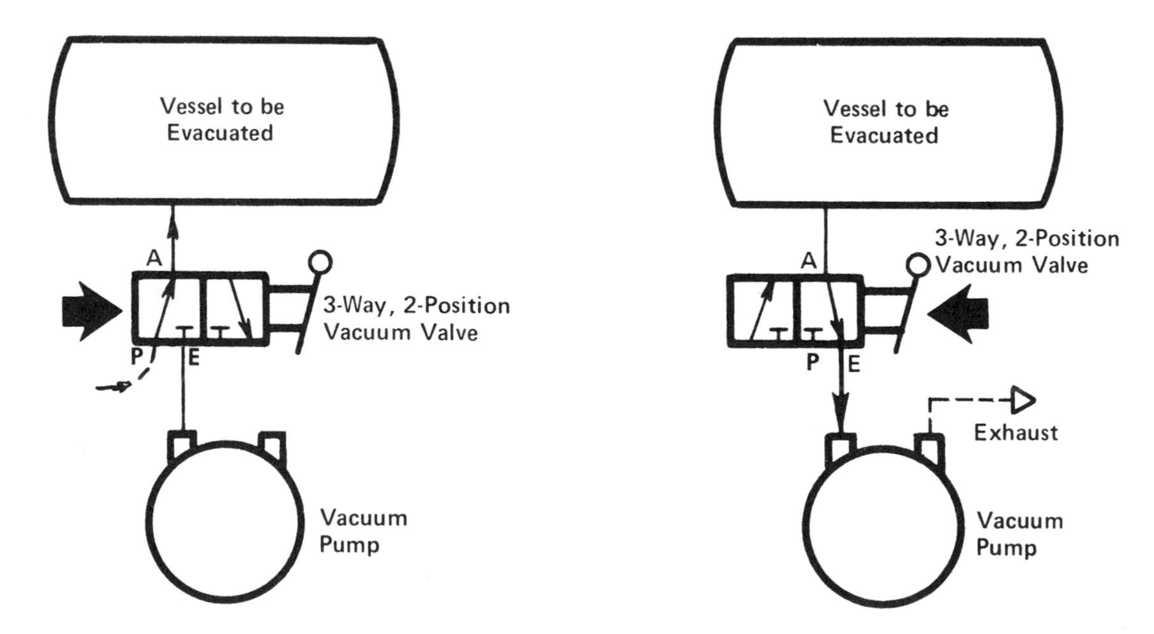

FIGURE 3-35. A 3-way directional air valve may be used to control vacuum circuits.

Vacuum System. Figure 3-35. In a vacuum application, atmospheric is the highest pressure in the system and should be connected to the pressure port of the control valve.

In the diagram on the left, the valve spool has been shifted to the left. The vacuum pump inlet is blocked and the vessel is vented to atmosphere.

In the diagram on the right, the valve spool has been shifted to the right. The vacuum pump is pulling a vacuum on the vessel and the atmospheric port, P, is blocked.

OTHER 3-WAY VALVE APPLICATIONS

The directional control applications on preceding pages are the more common ones for 3-way valves. On those applications, full pressure can be carried on inlet and outlet ports but the exhaust port must be vented. Valves designed for directional control usually have the connections marked P, A, and E or T. The E or T port is usually not rated for full pressure.

There are other 3-way valve applications for dual inlet or diverter service in which the valve must be capable of handling full pressure on any or all of its three ports. Some of these applications will be described in this section.

Note the official port markings for fluid power valves: inlet ports will be marked P, or P1 and P2 if there are two inlets; outlet ports will be marked with the first letters of the alphabet, A, B, C, etc.; exhaust ports for air valves will be marked E, or E1 and E2 on dual exhaust models; exhaust ports on hydraulic valves will be marked T for tank.

THREE-WAY DIVERTER VALVE APPLICATIONS

Diverter valve applications are those where a source of fluid can be directed to either of two outlets by shifting a valve spool. The spool valve in Figure 3-1 is a good example of diverter action. Any valve used for diverter service must be constructed so it can handle full inlet pressure on any port, or on any two ports at the same time.

Most diverter valve applications are hydraulic; normally there is no need on a compressed air system to divert air from one system to another since there is an unlimited supply of air stored in the receiver tank which can be used at will on several systems at the same time.

Graphic Symbol for Diverter Valve. Figure 3-36. The graphic symbol is constructed in the same way as shown in Figures 3-26 and 3-27 but the arrowheads show a flow direction which is different than for directional control action. The valve shown in this figure is a manual lever actuated model with spring return. For hydraulic service the only actuators normally used are manual lever or solenoid.

Connections to the rest of the circuit are usually drawn from the block nearest the return spring. Since a diverter valve has one inlet, two outlets, and no exhaust, ports are marked as shown on this illustration.

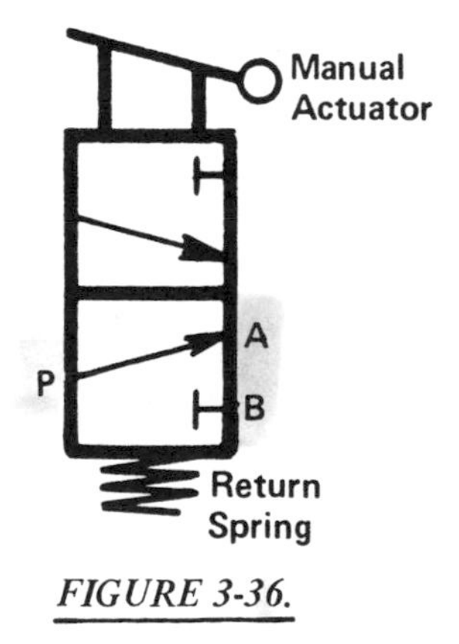

FIGURE 3-36.

Operation of Two Branch Circuits From One Pump. Figure 3-37. The purpose of a diverter valve is to route the oil from a hydraulic pump (or other source) to one of two branch circuits in a hydraulic system. A 3-way diverter valve can be used if cylinders in the branch circuits are single-acting (rams). To operate double-acting cylinders, either two diverter valves are required, one to divert the pressure line, the other the return line, or a special double section diverter valve is available which will divert both lines with one manual operation.

In Figure 3-37, either of two ram cylinders can be connected to the pump and relief valve with diverter Valve 3. A manual pump unloading valve, 4, can be opened to provide a free return path for pump oil when neither ram is being operated.

Outlet Ports A and B of the diverter valve feed into two branch circuits through check valves. The check valves prevent accidental dropping of a ram when unloading Valve 4 is opened by preventing

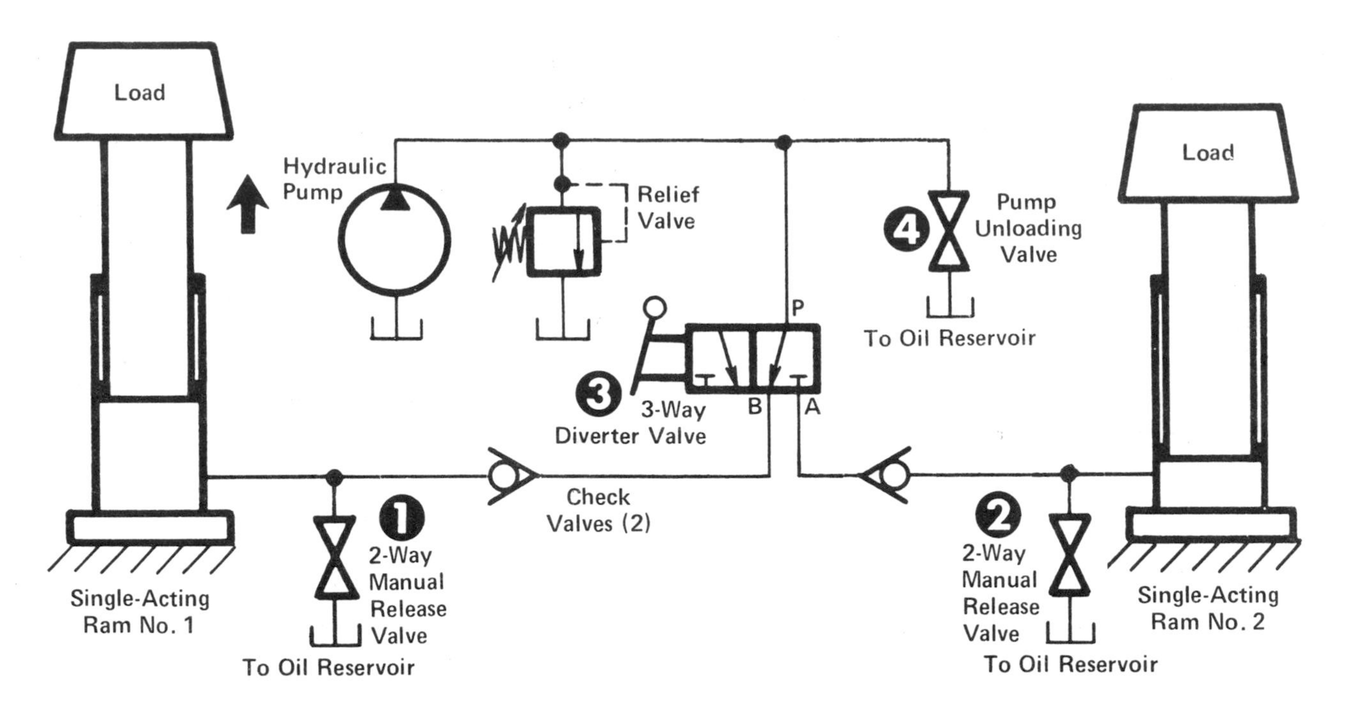

FIGURE 3-37. A 3-way diverter valve permits operation of two rams from one hydraulic pump.

back flow of the oil. Shut-off Valves 1 and 2 are connected across the two rams. When opened they will allow the rams to descend under weight of the load.

To operate the circuit, unloading Valve 4 should be opened prior to starting the pump. Diverter Valve 3 should be shifted to divert pump flow to the ram which is to be raised. For example, to raise Ram No. 1, shift Valve 3 to its right block, close release Valve 1, start the pump, then close unloading Valve 4. When the ram has extended to the desired height, open unloading Valve 4. The ram will stop and hold in this position, while the pump oil can flow unrestricted to tank. The pump may be stopped or used to extend Ram No. 2. To lower Ram No. 1, open release Valve 1, throttling with this valve to control rate of descent. Ram No. 2 can be raised and lowered in the same way. If the diverter valve has open porting in center crossover position, the pump oil can be split between the two branch circuits.

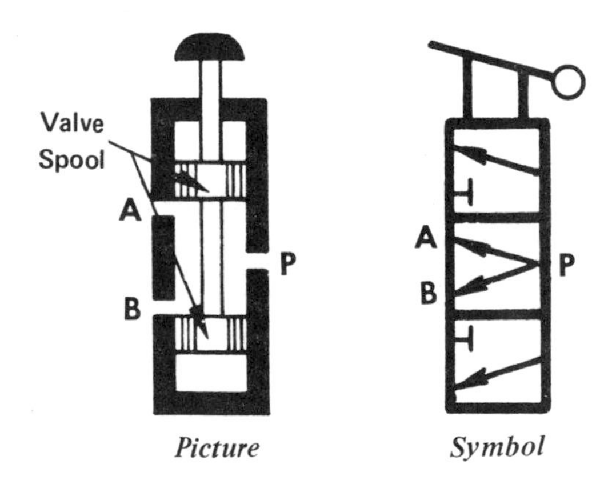

FIGURE 3-38. Open center diverter valve.

Open Center Diverter Valve. Figure 3-38. If the diverter valve has a working center neutral position in which all ports are open to each other (open center), then in the two-branch circuit above, an operator, by centering the valve, can move both rams at the same time. By throttling slightly to one side or the other of true center, the oil can be split in unequal proportion between the two branches. Or, an operator, by center throttling, can compensate for unequal loads on the two rams to keep them moving at the same rate of speed. Moving the valve spool all the way to one position, cuts oil off completely from the opposite ram.

DUAL INLET (SELECTOR) 3-WAY VALVE

A 3-way valve can also be used as a selector. On these applications two of the valve ports are used for inlets and the third port as the outlet. There is no exhaust connection on a selector valve. Inlets are marked P1 and P2; the outlet is marked A. Shifting the valve spool selects one or the other inlet to flow through to the outlet. Applications for oil hydraulics are few. Usually, the same fluid is connected to both inlets but at different pressures. Using different fluids may cause mixing of the fluids because of spool leakage. However, we are showing one application used on water hydraulics in a car wash where different liquids are used on the two inlets. Most applications involve compressed air, with a higher pressure on one inlet, a lower pressure on the other.

Graphic Symbol for Dual Inlet (Selector) Valve. Figure 3-39. The graphic symbol is constructed as shown in Figures 3-26 and 3-27 but the arrow heads show different flow directions than in those illustrations. In this example, a manual lever valve is illustrated with return spring. Dual inlet valves are used only occasionally with other than manual lever or solenoid actuators for either air or hydraulics.

Connections to the rest of the circuit are usually drawn from the block next to the spring. Since a selector valve has two inlets, one outlet, and no exhaust, ports are marked as shown here.

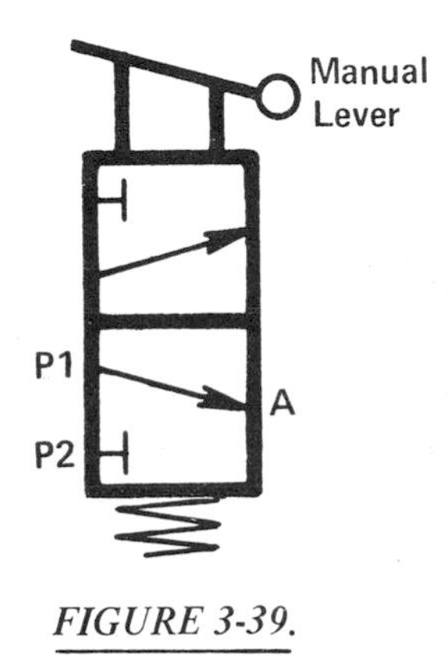

FIGURE 3-39.

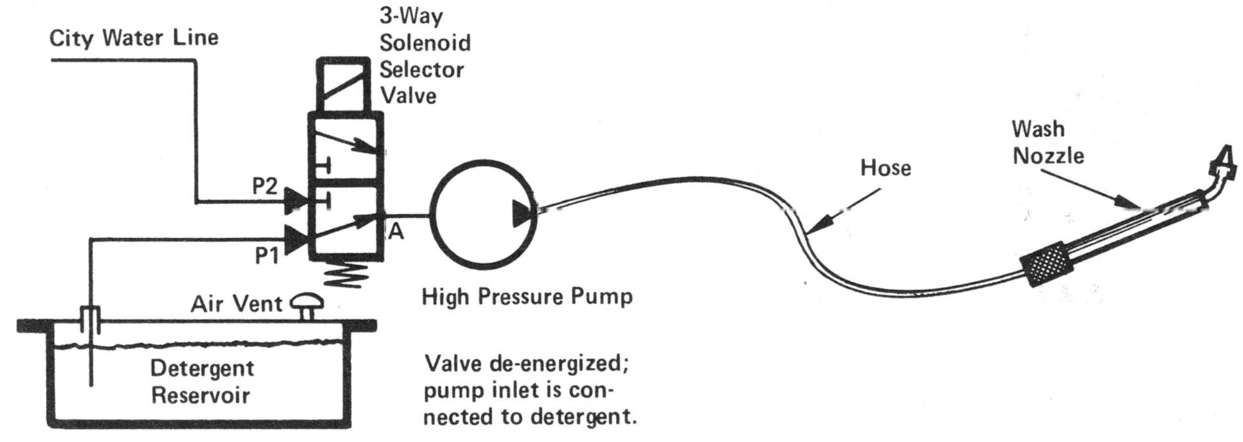

FIGURE 3-40.
Dual inlet 3-way valve used in car wash application.

Car Wash Application. Figure 3-40. A "do-it-yourself" car wash uses a piston pump to produce nozzle pressure. The pump can deliver a choice of clear water for rinsing, a detergent solution, and sometimes other liquids. A solenoid valve, when de-energized, connects the pump inlet to a detergent reservoir, and when energized, to the city water line for rinsing. The solenoid valve is controlled by a rotary selector switch. There are several ways a system like this can operate, this being one of them:

A coin (or coins) deposited by a customer starts an electric timer and starts the piston pump. The customer can change the selector switch for the desired fluid. When the timer stops, the pump also stops.

Where more than two fluid sources must be available, a series of 2-way normally closed valves must be used rather than one 3-way valve.

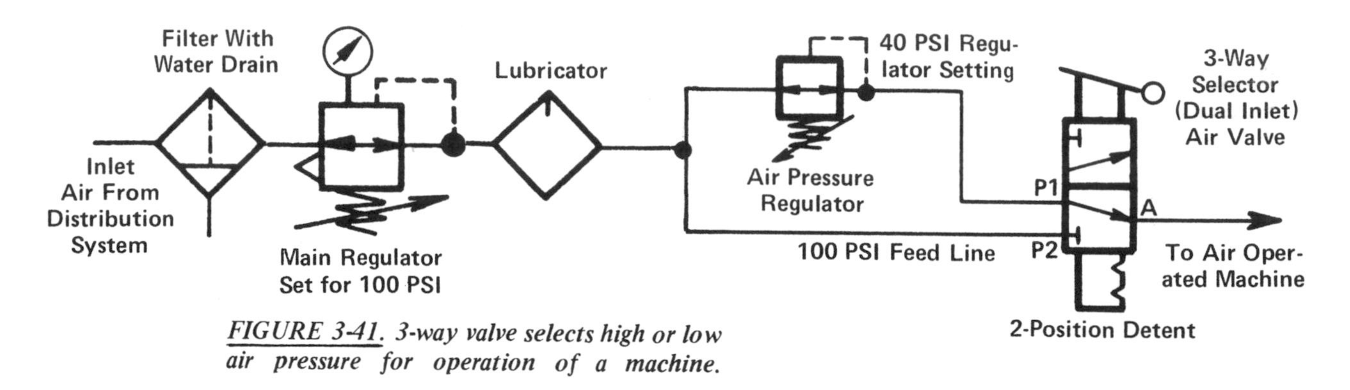

FIGURE 3-41. 3-way valve selects high or low air pressure for operation of a machine.

Dual Pressure Air Supply. Figure 3-41. A 3-way selector valve can be used to select either the full 100 PSI, for example, from the main regulator or the reduced pressure of 40 PSI, for example, from the downstream regulator. Lubricated air passing through a pressure regulator tends to precipitate the suspended oil. Therefore, it may be necessary to add a second lubricator in the 40 PSI line just ahead of P1 inlet on the selector valve.

A manual lever valve used as the selector should have detents on the spool to be certain the valve does not accidentally shift from low to high pressure under machine vibration.

For remote control of pressure level, a solenoid actuated selector valve may be employed. For fail-safe operation, in the event of a power failure, it should be connected so when the solenoid is de-energized, the lower pressure inlet is ported through to the outlet.

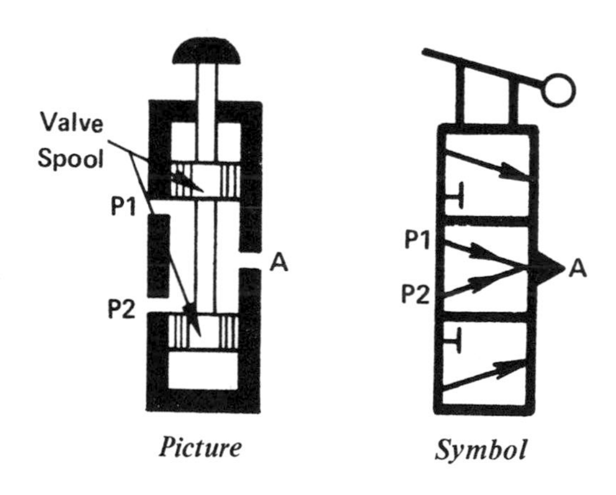

FIGURE 3-42. Dual inlet valve used for mixing.

Mixing Valve. Figure 3-42. A dual inlet selector valve may be used for proportional mixing of two liquids. The valve must have a center neutral position with open center porting (all ports connected together). In exact center position, equal volumes of the two liquids would be mixed provided the pressure in both liquids was the same. If under different pressures, the valve spool would have to be throttled slightly to one side of center to get equal mixing. Moving the spool to either extreme side position cuts off the flow of one liquid.

ACTUATOR SYMBOLS FOR 2-WAY AND 3-WAY VALVES

Actuators for Air Valves. Figure 3-43. A complete selection of actuators is not shown here, only those commonly used on 2-way and 3-way spool or poppet-type air valves with 2-position action. A more complete list including those used on 4-way directional control valves appears on Page 137.

Actuator symbols should be joined to the end of the valve symbol farthest from its "normal" position if it has one. This might be on either end depending on how flow arrows are drawn in the blocks.

Actuators for Hydraulic Valves. Of the actuators shown in Figure 3-43, only the manual lever and solenoid types are used to any great extent on hydraulic valves. In fact there are few valves designed and manufactured specfically for 2-way and 3-way high pressure hydraulic service in sizes larger than 1/4". Actuators for 4-way hydraulic valves are illustrated in the circuits on Pages 162 through 169.

Valve diagrams in Figure 3-43 show only the symbols for valve actuators, and do not show details of the symbol blocks with their flow arrows.

Part A, Figure 3-43. Solenoid. A solenoid shifts the spool or poppet in one direction and a spring returns it when the solenoid is de-energized. Double solenoid actuators, with a solenoid shift in both directions, are available but are seldom used on 2-way and 3-way valves.

Part B, Figure 3-43. Manual Lever. On spring return models the operator must hold the lever in its actuated position. When released, it allows the spool to return to its "normal" position. Spring centered models are available but seldom used on 2-way and 3-way valves.

Part C, Figure 3-43. Manual Lever, Detented. A detent is a click stop mechanism which holds the spool in the position to which it was last shifted. A spring loaded ball which drops into notches on the spool is a popular design. A valve with detents does not have a "normal" position. Two-position detents are standard; three-position detents are available but seldom used on 2-way and 3-way valves.

Part D, Figure 3-43. Palm Button. A plastic knob is attached directly to the end of the spool or poppet, and the valve is shifted by direct push-pull action. This actuator is limited to small valves because it lacks the lever advantage of a manual lever actuator.

Part E, Figure 3-43. Foot Operated. This hook symbol covers both pedal and treadle actuated valves. Pedals are actuated with the toe and are usually furnished with return spring. Treadle actuators, operated one way by the toe, the other way by the heel, are used more often on 4-way directional control valves.

Part F, Figure 3-43. Mechanical. The most popular mechanically actuated valve is the cam roller attached to the end of the spool or poppet, with spring return. Other mechanical actuators include a stem or clevis attached directly to the end of the spool.

Part G, Figure 3-43. Pilot-Operated. For spool-type valves only. The spool is shifted by fluid pressure working on the end of the spool, applied through a pilot port on the valve end cap. Double piloted actuators are available but seldom used on 2-way and 3-way valves.

A. Single Solenoid; Spring Return.

B. Manual Lever; Spring Return.

C. Manual Lever; 2-Position Detent.

D. Pushbutton; Spring Return.

E. Foot Operated; Spring Return.

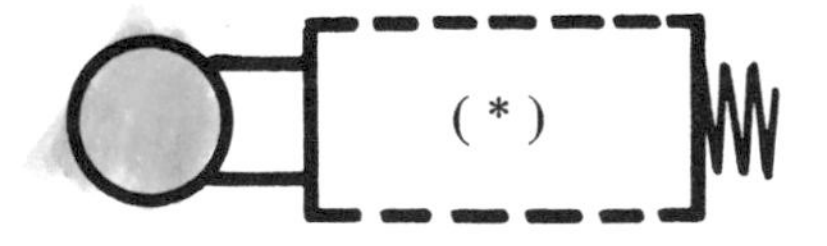

F. Mechanical; Spring Return.

G. Pilot Operated; Spring Return.

FIGURE 3-43. Common actuators for 2-way and 3-way spool-type air valves.

*Valve actuators, only, are shown. Porting may be any kind of 2-way or 3-way action shown previously in the chapter.

DOUBLE-ACTING HYDRAULIC CYLINDER CONTROLLED WITH 3-WAY VALVES

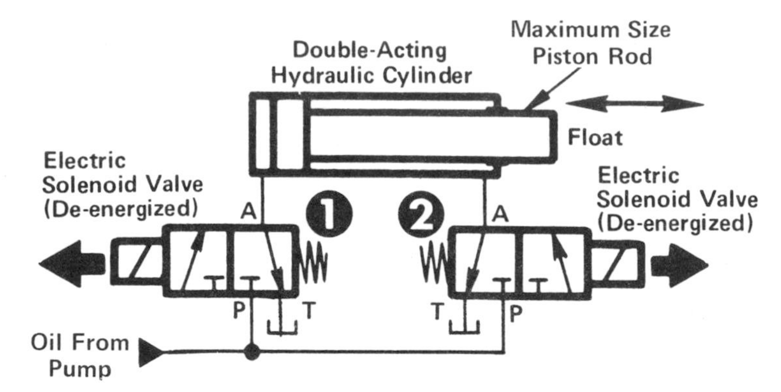

A. No current on either valve. Cylinder floating.

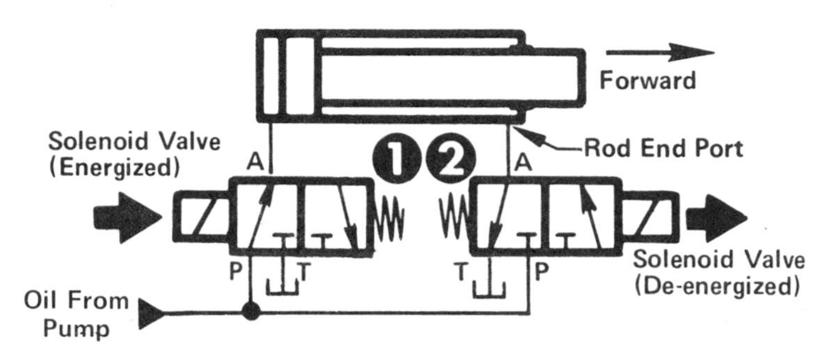

B. Current on Valve ❶ only. Cylinder extends.

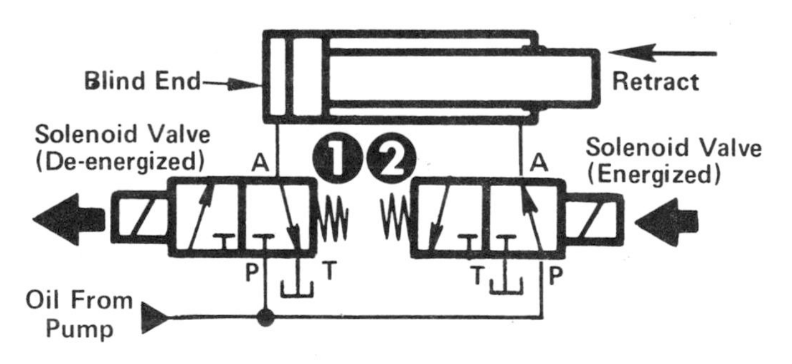

C. Current on Valve ❷ only. Cylinder retracts.

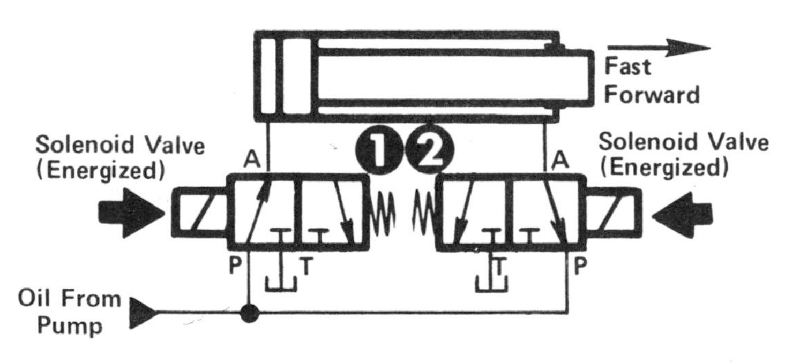

D. Current on both valves. Cylinder rapid forward.

FIGURE 3-44. Four Cylinder Actions.

Figure 3-44. On this page are shown four cylinder actions obtained from a double-acting hydraulic cylinder controlled with a pair of 3-way single solenoid, 2-position, directional control valves energized in four different combinations. The hydraulic cylinder has maximum size piston rod (2:1 ratio). Normally, one 4-way, 3-position valve (Chapter 4) would be used to control a double-acting cylinder but would give only three of these cylinder functions.

Part A. Both Valves De-energized.

With Valves ❶ and ❷ de-energized, both valve inlet ports are blocked, and both ports of the cylinder are open to tank (reservoir). This puts the cylinder in a "floating stop" position. It is free to follow the movements of the mechanism to which it is attached. Small cylinders can be positioned by hand when in a floating condition.

Part B. Valve ❶ Only, Energized.

If only Valve ❶ solenoid is energized, the cylinder extends at its normal speed since its rod end port is vented to tank through de-energized Valve ❷.

Part C. Valve ❷ Only, Energized.

If only Valve ❷ solenoid is energized, the cylinder retracts at its normal speed since its blind end port is vented to tank through de-energized Valve ❶.

Part D. Both Valves Energized.

Applying full system pressure to both ports of a cylinder which has unbalanced piston areas (single-end-rod type) produces rapid extension speed, but with a reduction of cylinder force in proportion to the increase in speed. This valving condition is called "regenerative" and is described in detail in Volume 2.

PRESSURE RELIEF VALVES FOR HYDRAULIC SYSTEMS

The pressure relief valve is the most important member of a group of pressure control valves. Other valves in this group are described in Volume 2. When a relief valve is connected across a pump pressure line (or any hydraulic pressure line), it acts as a "safety valve" to set a limit on the rise of pressure in that line. In valve terminology a relief valve is described as a 2-way, normally closed valve. Virtually every kind of positive displacement pump such as the gear, vane, and piston types described in Chapter 5, should be protected with a relief valve in every kind of circuit. However, a relief valve is not necessary on a centrifugal or impeller (non-positive displacement) pump.

Direct-Acting Relief Valve
Miniature Size

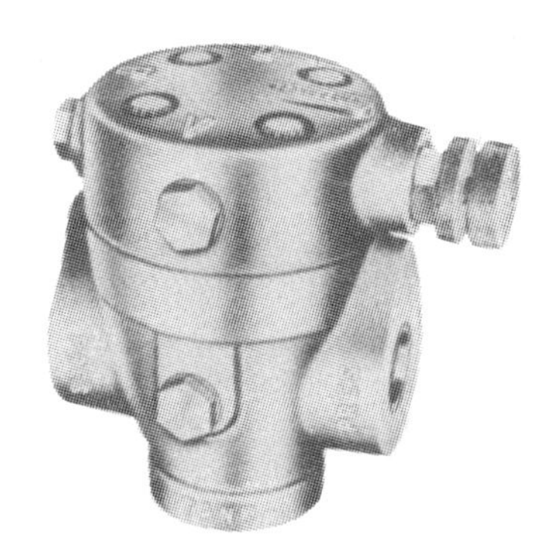

Pilot-Operated Relief Valve

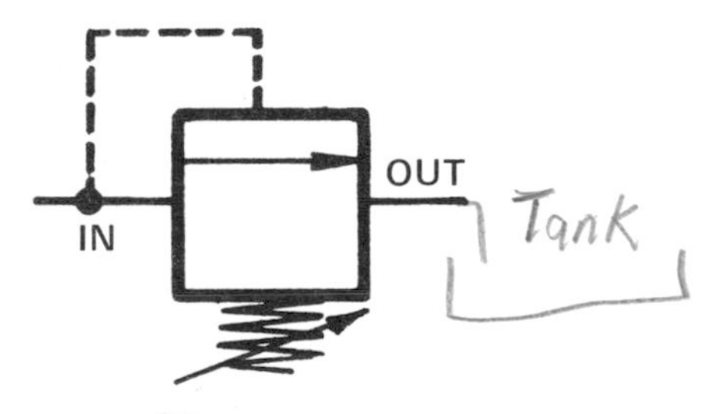

Graphic Symbol for
Use on Circuit Diagrams

The primary function of a positive displacement pump is to produce a flow of oil, either under pressure or un-pressurized as the situation requires. The oil will not be under pressure unless the flow becomes restricted. Then, the pump will produce whatever pressure level is required to maintain its rated flow. But it will produce a pressure no higher than necessary. For example, when the pump oil is used to extend a cylinder against load resistance, pressure in the pump line (which can be read on a gauge) will be directly proportional to the magnitude of the cylinder load. If the load is increased by 35%, pressure in the pump line would also rise by 35%. Therefore, the function of a relief valve is to protect the pump and other components against excessive pressure if the operator of a hydraulically-powered machine should try to move against a load which is greater than that for which the system was designed.

Note: Oil flowing across a relief valve generates heat which can be calculated with formula: *HP (heat) = PSI (across relief) x GPM (discharge) ÷ 1714.* BTU/hr = HP heat x 2545.

Types of Relief Valves. Relief valves fall into two categories – direct-acting or pilot-operated. Each type will be described in detail on the following pages.

Graphic Symbols. This symbol may be used for all kinds of relief valves. If the relief setting is adjustable, this can be shown with a slash arrow through the spring.

Direct-Acting Relief Valve. Figure 3-45. A "direct-acting" relief valve is one in which the poppet is held closed by direct force of a mechanical spring, usually adjustable. The spring force holding the poppet closed is opposed by hydraulic pressure, working against the exposed poppet area, tending to open the poppet. The "cracking pressure", then, is set by adjusting spring tension. Cracking pressure is the pressure at which there is an exact balance between spring force and hydraulic force. Any further increase in hydraulic pressure from the pump, if only a slight increase, will open the poppet a very small distance and allow a small part of the pump oil to escape. The pressure at which the poppet is

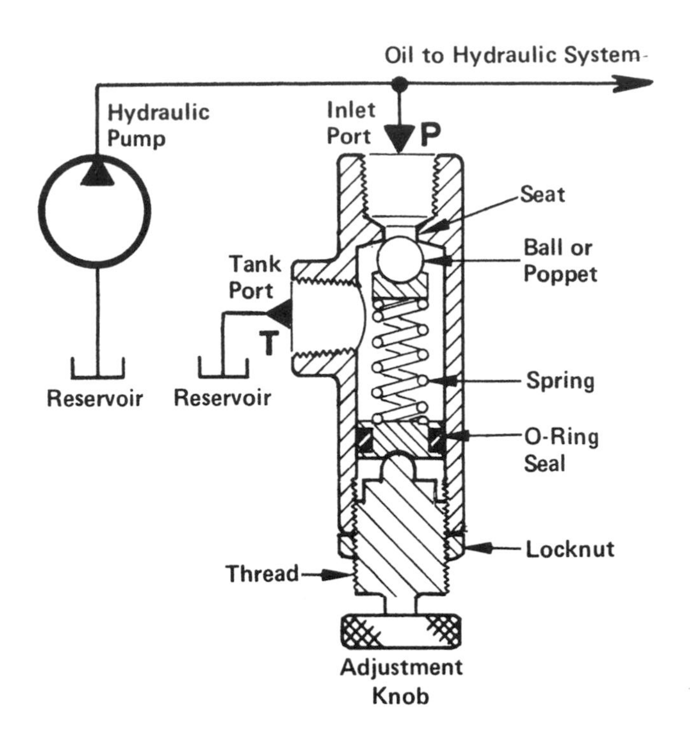

FIGURE 3-45. Direct-Acting Type of Relief Valve.

opened far enough to allow the entire flow to escape is called the "full relief valve pressure". It will be higher than the cracking pressure by some amount depending on the design of the spring and how much force is required to compress the spring far enough to create an opening large enough for the entire flow to pass to tank.

In Figure 3-45, spring tension is set on the knob to keep the poppet closed until pump pressure reaches the desired cracking pressure. Up to the cracking pressure, all pump flow is retained in the system. Between cracking and full relief pressure part of the oil is escaping and the remainder is still moving the cylinder. Finally, when pressure reaches full relief value, all oil escapes and the cylinder cannot continue to move.

Direct-acting relief valves are usually available only in relatively small sizes because it is difficult to design a sufficiently strong spring to keep the poppet closed at high pressure and high flow. They are used primarily in applications described later in this chapter. Other disadvantages are that they usually are not designed to cover as wide an adjustable range and have a larger differential between cracking and full relief pressure levels as compared to a pilot-operated relief valve.

Advantages of direct-acting relief valves are that they are less expensive than the pilot-operated types, and have a faster response when used to suppress a pressure spike. They are used as a pressure control element built into large pilot-operated relief valves as described in the next illustration.

Pilot-Operated Relief Valve. Figure 3-46. In this type relief valve a construction method is employed which makes it possible to handle higher pressures in combination with higher flows, and to be contained in a smaller frame than is usually possible with direct-acting relief valves built for the same flow and pressure rating. Adjustable hydraulic pressure, instead of an adjustable spring, holds the poppet on its seat up to the relieving pressure level.

Operating Principle. The valve is built in two stages. In Figure 3-46, the first stage includes the main poppet held in a normally closed position by a light, non-adjustable spring. This stage is built large enough to handle the maximum flow rating of the valve. The second stage is a small, direct-acting relief valve usually mounted as a crosshead on the main valve body, and includes a pilot relief poppet, a pilot spring, and an adjustment knob. This stage is very small in size, with the orifice being about 3/64 to 1/16" in diameter. The first stage handles full rated flow to tank; the second stage controls and limits pilot pressure level in the main spring chamber.

The first stage, with its main poppet is basically a 2-way, normally closed valve, held on its seat by the light main spring, which may have a cracking pressure of 25 to 75 PSI. The light main spring has little or nothing to do with maximum relieving pressure, but it does set a lower limit on the minimum pressure to which the main valve can be adjusted. Its main purpose is to keep the poppet in its normally seated position before the pump is started and until the relieving pressure level is reached. It permits the valve to be mounted upside down or in any position.

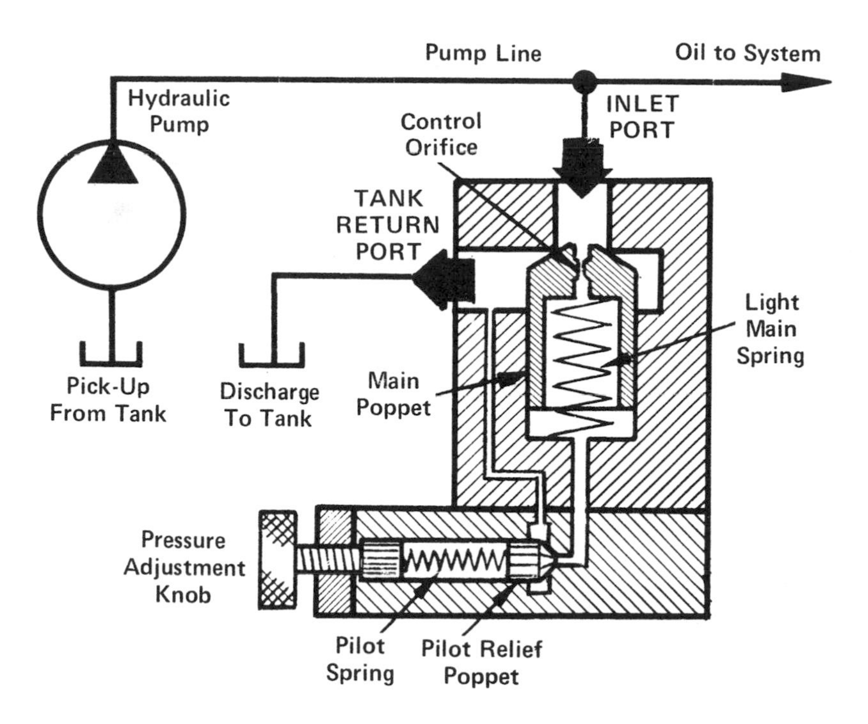

FIGURE 3-46. Pilot-Operated Type of Relief Valve.

Relieving action through the main poppet is as follows: As long as pump line pressure is less than relieving pressure set on the control knob, pressure in the main spring chamber is the same as pump line pressure because there is no flow through the control orifice and consequently there is no pressure drop from one side of it to the other. The light main spring exerts a force equivalent to 25 to 75 PSI to hold the main poppet seated. But if and when pump line pressure (also the pressure in the main spring chamber) rises higher than the adjustment set on the control knob, the pilot relief poppet becomes unseated. This starts an oil flow from the pump line, through the orifice, across the pilot relief poppet, and to tank. This restricted flow caused by the orifice creates a pressure difference between the pump line and the now lower pressure in the spring chamber. This pressure unbalance causes the main poppet to move off its seat just far enough (but no further) to discharge enough of the pump flow to prevent any further rise in the pump line pressure. On dead end condition or severe overload it will open fully if necessary. When pump line pressure drops below the control knob setting, the pilot relief poppet closes, flow through the orifice ceases, and the main spring can re-seat the main poppet.

APPLICATIONS FOR HYDRAULIC PRESSURE RELIEF VALVES

Hydraulic Pump Line. Figure 3-47. Relief valves can be used anywhere in the circuit where it is necessary to prevent pressure from exceeding a maximum level. The most common use is in the pump line to prevent the pump from developing excessively high pressure if the system should become overloaded.

Figure 3-47 shows the standard ANSI symbol for a relief valve. This symbol serves for both a driect-acting or a pilot-operated type of valve. The flow arrow inside the block is offset out of alignment with circuit connections to show the flow is normally blocked through the valve until inlet pressure flowing through the dash line becomes high enough to overpower the adjustable spring shown on the opposite side of the block. On most applications the flow size of the relief valve should be the same as the size of the pump pressure line. A relief valve is recommended for the pump line on every application where a positive displacement pump of the gear, vane, gerotor, or piston type, etc. is used.

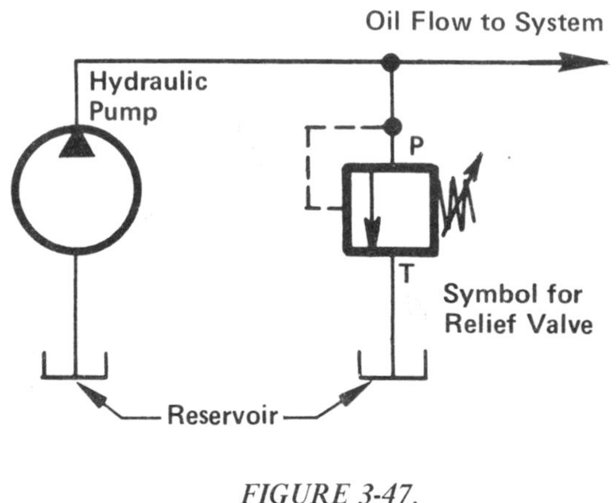

FIGURE 3-47.
Relief Valve Across Pump Line.

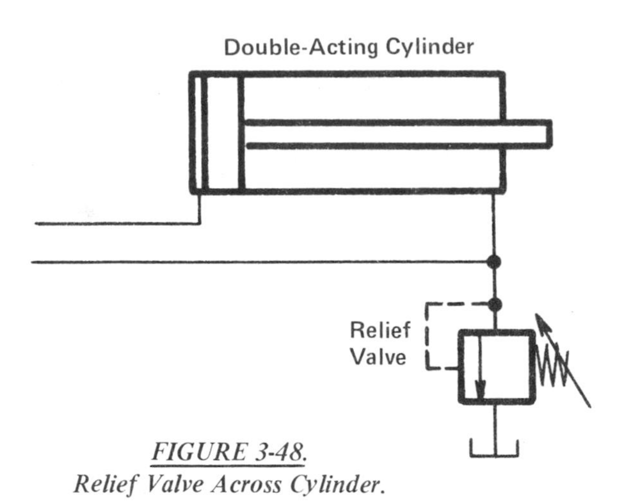

FIGURE 3-48.
Relief Valve Across Cylinder.

Cylinder Line. Figure 3-48. Sometimes a secondary relief valve is used across one of the ports of a cylinder (usually the rod end) to limit the pressure in one direction of movement to less than the main pump relief valve setting. Lower pressure on the rod end of a cylinder may be for the purpose of protecting the rod seal or the end cap from full system pressure.

Relief Valves for Compressed Air. In most industrial or shop air systems air is pumped and stored in a "receiver" tank for feeding into a distribution system which may serve many air-operated machines. There is no need for relief valves because maximum pressure is limited by stopping or unloading the air compressor when the desired maximum pressure is reached. At each air-operated machine, the optimum maximum pressure for that machine is set on a pressure regulator valve at the inlet to the machine. This valve, which is described in Chapter 5, is a pressure reducing valve, not a relief valve.

Although relief valves are almost never used to discharge a continuous flow of air to atmosphere for limiting the pressure level, a relief valve may occasionally be used as a safety valve, but is never intended to open except in a rare case of emergency. Examples are the safety reliefs built into the gas end of most hydraulic accumulators, and safety valves on vessels which are being pressure tested.

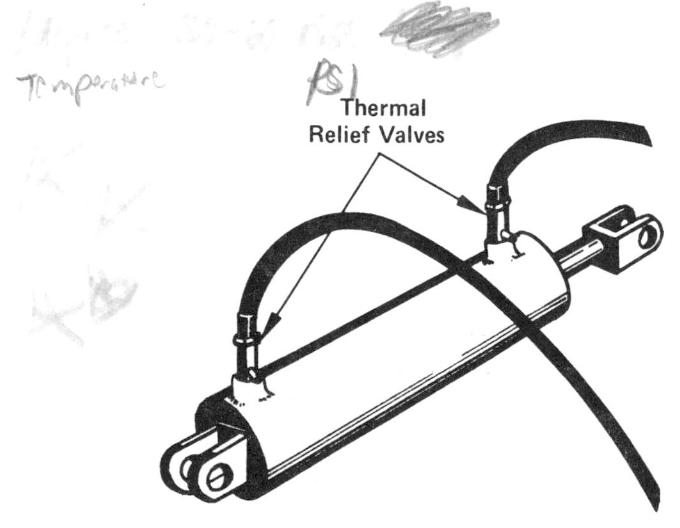

FIGURE 3-49. Thermal Relief Valves.

Thermal Relief Valve. Figure 3-49. During normal operation, most hydraulic systems will adjust to expansion of oil due to heating. But in a non-operating (idle) system with both cylinder ports blocked, a 1° rise in the oil temperature caused by an external heat source is said to cause a pressure rise of 50 to 60 PSI. For example, a mobile equipment cylinder, completely filled with oil and disconnected from a tractor hydraulic system by shut-off quick disconnect couplings is a locked system. If this cylinder is exposed to direct sunlight for several hours, a rise of 50° in oil temperature could cause a 2500 PSI rise above pressure already in the cylinder. The same thing could happen if a cylinder were uncoupled while cold then stored in a warm room. Cylinders to be stored should be partially drained.

A "thermal relief valve" installed in one or both oil ports will prevent a cylinder from bursting due to heat expansion. If the internal pressure builds up to the thermal relief setting, a drop or two of oil will be expelled to atmosphere through a small vent hole in the side of the valve. The term "thermal" does not imply any sensitivity to surrounding temperature. The action is strictly that of a pressure relief valve. However, thermal relief valves are non-adjustable and should be purchased for a relief pressure of 300 to 500 PSI higher than the maximum working pressure of the system so they will remain tightly closed under all normal operating conditions.

Thermal relief valves are used primarily on mobile or farm machinery cylinders; industrial cylinders are not usually subjected to a tightly blocked condition with a possible temperature rise.

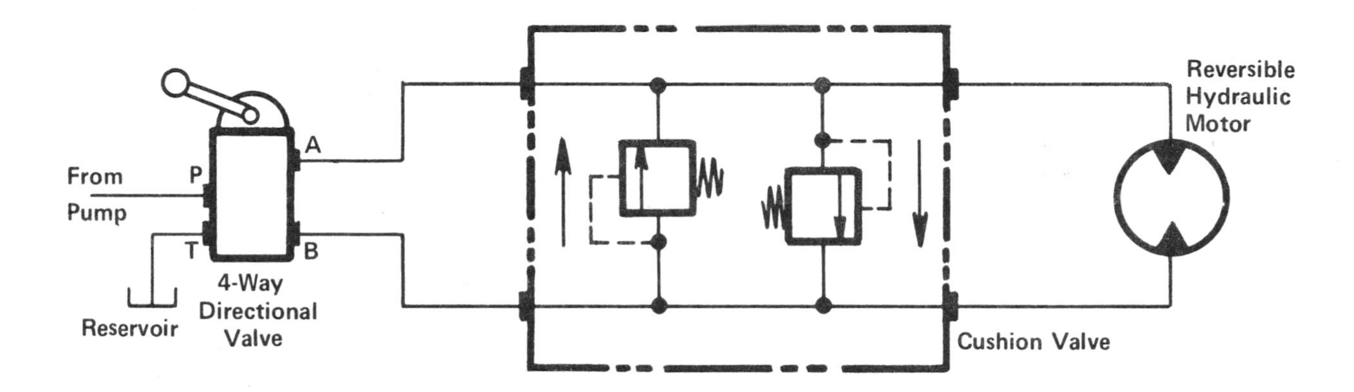

FIGURE 3-50. A cushion relief valve protects hydraulic motor against high pressure spikes.

Cushion Relief Valves. Figure 3-50. The cushion valve is also known as a "cross-over relief valve". It is strictly for hydraulic use, principally with hydraulic motors, occasionally with cylinders, to absorb momentary high pressure spikes which may pass through the system as the motor or cylinder is brought to a very abrupt stop while moving a high momentum load.

Two relief valves in the same housing face in opposite directions for protection in both directions of actuator movement. One relief valve in Figure 3-50 has its inlet on the top line and discharges into the bottom line. The other relief valve discharges from the bottom into the top line.

A pressure spike may be produced when a motor is suddenly stopped from a high rotational speed as the 4-way valve is shifted to neutral and the motor ports become blocked. During the brief stopping period, momentum energy from the motor and its load continues to drive the motor as if it were a pump. The flow of oil generated by this action becomes deadheaded against a blocked port in the 4-way valve. One of the relief valves in the cushion valve can discharge this oil flow to the opposite line, preventing an energy spike of a dangerously high level from being produced.

Direct-acting relief valves are employed because they usually respond faster than pilot-operated relief valves. The springs in these valves must be non-adjustable to prevent field tampering with their cracking pressure. The springs should have a cracking pressure from 300 to 500 PSI higher than the maximum working pressure of the system to prevent unwanted discharge while the system is working normally.

A cushion relief valve has four ports. Two ports connect to the control valve, the other two to the motor or cylinder. A cushion relief is recommended for all hydraulic motor applications where the motor operates above 500 RPM. Occasionally they are used with cylinders which are moving at high speed and carrying massive loads.

MISCELLANEOUS 2-WAY AND 3-WAY VALVES

The valves described in this section are classed as directional control valves because they do control the direction fluid is permitted to flow or not to flow. They have no control over the pressure level or the rate of flow. Port markings follow the recommended NFPA standards in which "P" is the inlet port and the first letters of the alphabet "A", "B", "C", etc. are the outlet ports.

Check Valves

A check valve has an internal poppet which is spring loaded toward a closed position. Fluid entering from one direction is completely blocked, but fluid entering from the opposite direction can push the poppet off its seat for free flow through the valve. By valve terminology it is a 2-way, normally closed valve. It carries the standard NFPA port markings or has a flow arrow stamped on its body to indicate direction of free flow. In the free flow direction there is a small loss of pressure of the amount which is required to compress the poppet spring. The internal spring is non-adjustable on all check valves.

For circuit drawings, the symbol shown in Figure 3-53 may be used. Port markings and flow arrow are usually omitted because of the clear nature of the symbol itself.

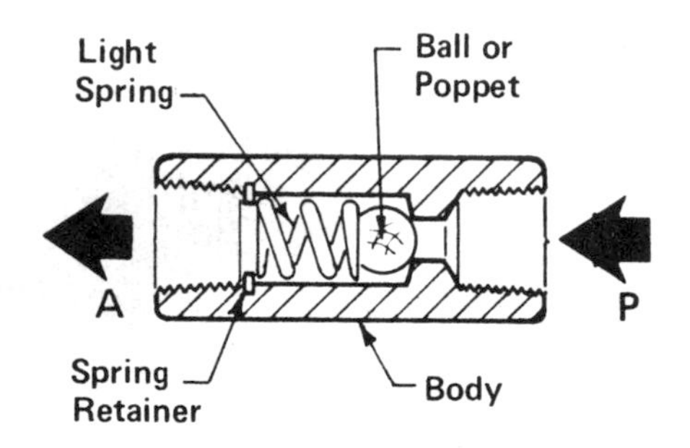

FIGURE 3-51. In-Line Check Valve. Typical Construction for Compressed Air.

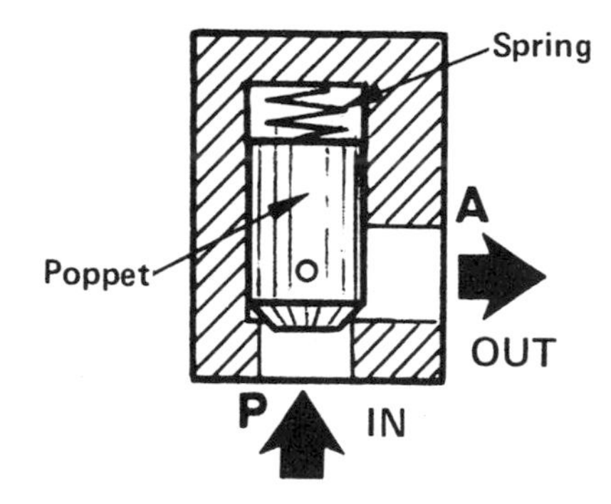

FIGURE 3-52. Angle-Type Check Valve. Typical for Hydraulic Use.

FIGURE 3-53. Check Valve Symbols.

Compressed Air Check Valves. Figure 3-51. The internal spring on an air check valve should be relatively light to keep the pressure loss to a minimum in the free flow direction. The free flow cracking pressure should be about 1 PSI but no more than 3 PSI. Most air check valves have a straight-through configuration with ports on opposite ends of a bar stock body. The preferred material of construction is brass because of presence of water in most compressed air lines.

Hydraulic Check Valves. Figure 3-52. Hydraulic check valves usually work at much higher pressures and a cracking pressure in the range of 10 to 20 PSI is acceptable. For oil service, valve bodies are usually cast iron with steel working parts. Angle construction as shown in this figure is more common than straight-through construction, probably because it is less expensive to manufacture.

Vacuum Check Valves. Figure 3-54. On a vacuum circuit it is most important to use a check valve with extremely low cracking pressure. Ordinary compressed air valves are not usually desirable, even those with 1 PSI cracking pressure. Most industrial vacuum systems operate on less than 20" Hg (10 PSI) vacuum. Using a check valve with 1 PSI cracking pressure would lose 10% of the working power.

Flapper or "swing" checks operate on very low pressure, using only the weight of the poppet to return the flapper to its normally closed position.

Flapper valves must be mounted vertically. If necessary to mount them horizontally, a very light spring should be used to keep them in the proper normal position. Some air check valves can be used for vacuum by removing the internal spring and mounting them vertically with poppet weight keeping them closed.

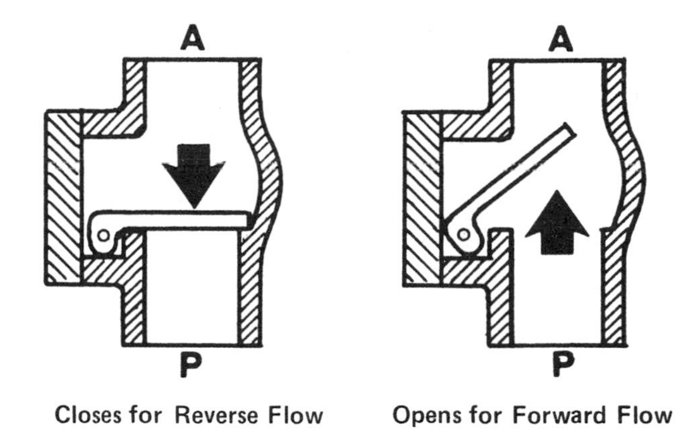

FIGURE 3-54. Flapper or Swing-Type Check Valve.

Check Valve Applications

Component By-Pass. Figure 3-55. Check valves are connected around other valves such as sequence, pressure reducing, flow control, counterbalance, etc., to permit fluid to flow freely in the return direction. These applications are typical in lines connecting a directional control valve with a cylinder where fluid must flow in both directions as the cylinder advances and retracts. Often the check valve is built into the same body as the component.

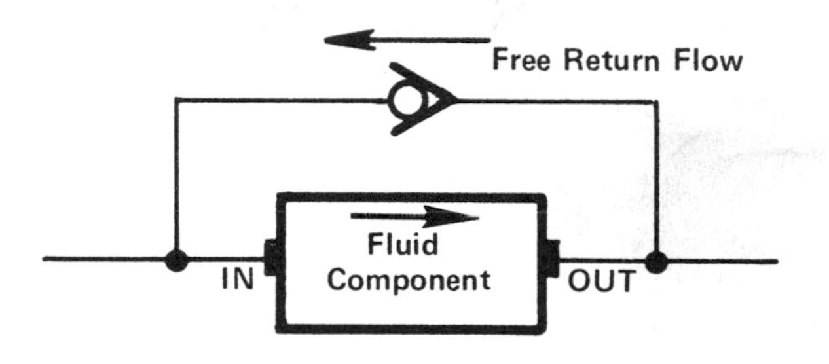

FIGURE 3-55. Component By-Pass.

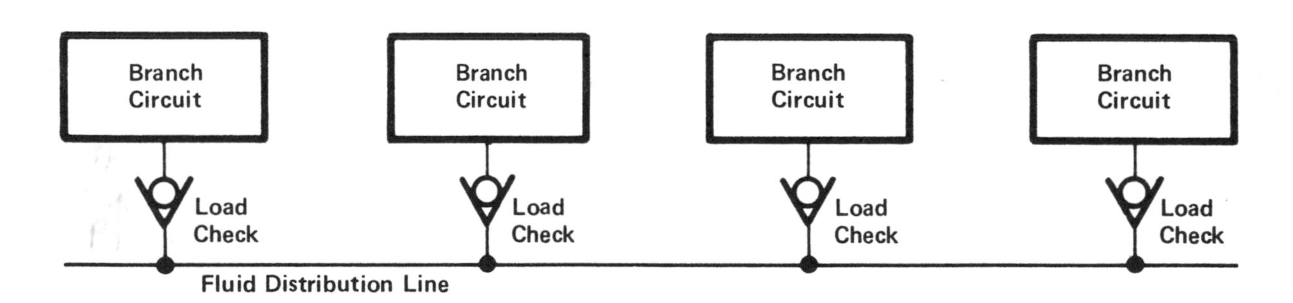

FIGURE 3-56. Check valves hold fluid in branch circuits to prevent "load dropping".

Load Holding. Figure 3-56. In hydraulic circuits having several branches feeding from the same pump, a check valve should be installed in the pressure line feeding the directional control valve in each branch to prevent oil which is already in one branch from backflowing into another branch which is less heavily loaded. Mobile circuits often use bank valves having "load holding" check valves built into the inlet of each valve section. Industrial components, especially solenoid valves, may not have these checks built into the valve. They can be added in the pressure feed line to each valve. Load holding check valves are not ordinarily used in compressed air applications.

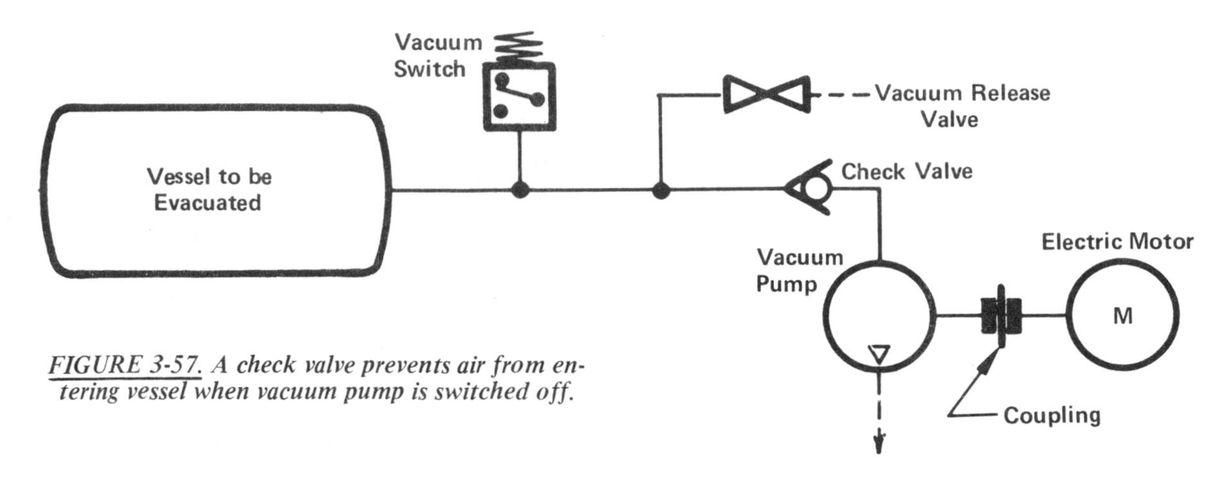

FIGURE 3-57. A check valve prevents air from entering vessel when vacuum pump is switched off.

Vacuum Storage System. Figure 3-57. The vacuum pump starts and stops in response to a vacuum switch wired to the driving electric motor. The check valve prevents loss of vacuum while the pump is stopped. Check valve cracking pressure reduces the maximum vacuum that can be produced, and represents a loss in the system. A flapper check valve with low cracking pressure should be used.

Other Check Valve Applications include the load holding check built into hand pumps, the free return check built into flow control valves, the pressure retaining check valve in series with an accumulator, and other applications where it is necessary to prevent backflow.

Quick Exhaust Valves

Used with single and double-acting cylinders to increase retraction speed by directly venting exhaust air to atmosphere at the cylinder ports rather than through a long connecting line back to the control valve. Since exhaust air restriction is eliminated, smaller control valves and lines can be used. Quick exhaust valves are obviously only for compressed air, not for hydraulics.

Operating Principle. Figure 3-58. Three main ports include an inlet which receives compressed air from a control valve; a cylinder port which connects with a close nipple to the cylinder port; and an exhaust port which vents directly to atmosphere, sometimes through a muffler.

Part A of Figure 3-58 shows the internal state of the valve when not activated. A rubber disc,

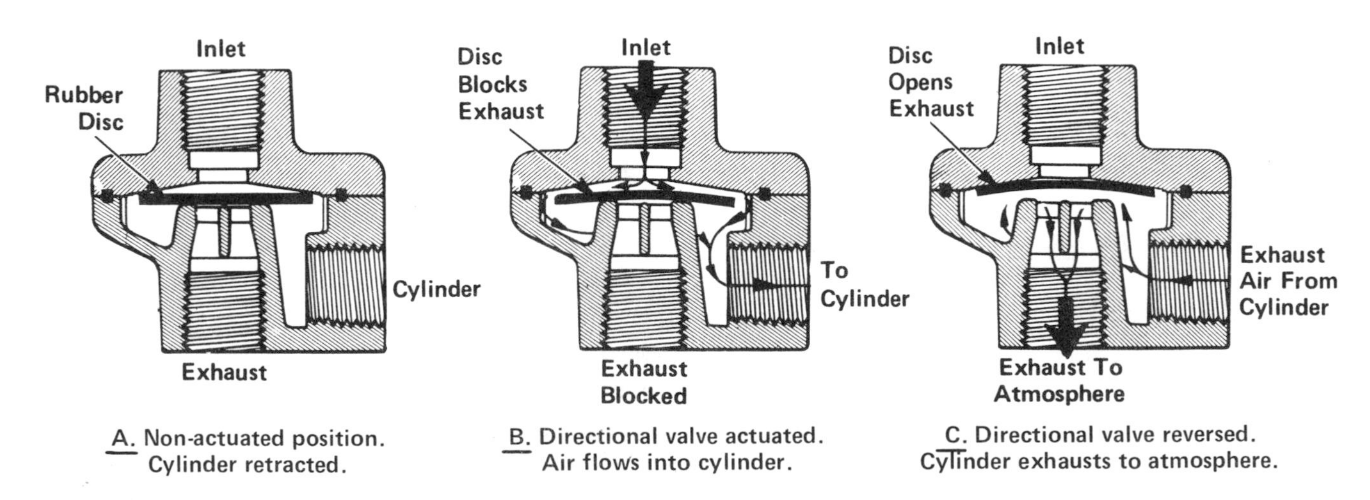

FIGURE 3-58. Quick exhaust valve shown in operating and non-operating states.

about 1" diameter, floats across and touches all three ports. Part B shows the valve state when air pressure is received from the control valve and pushes the rubber disc to make a tight seal across the exhaust port. This inlet air pressure bends the edges of the disc down to permit air to flow into the cylinder. In Part C the operator has released his control valve to retract the cylinder. When air pressure at the inlet of the quick exhaust valve has dropped a fraction of 1 PSI, trapped air in the cylinder unseats the rubber disc and vents to atmosphere.

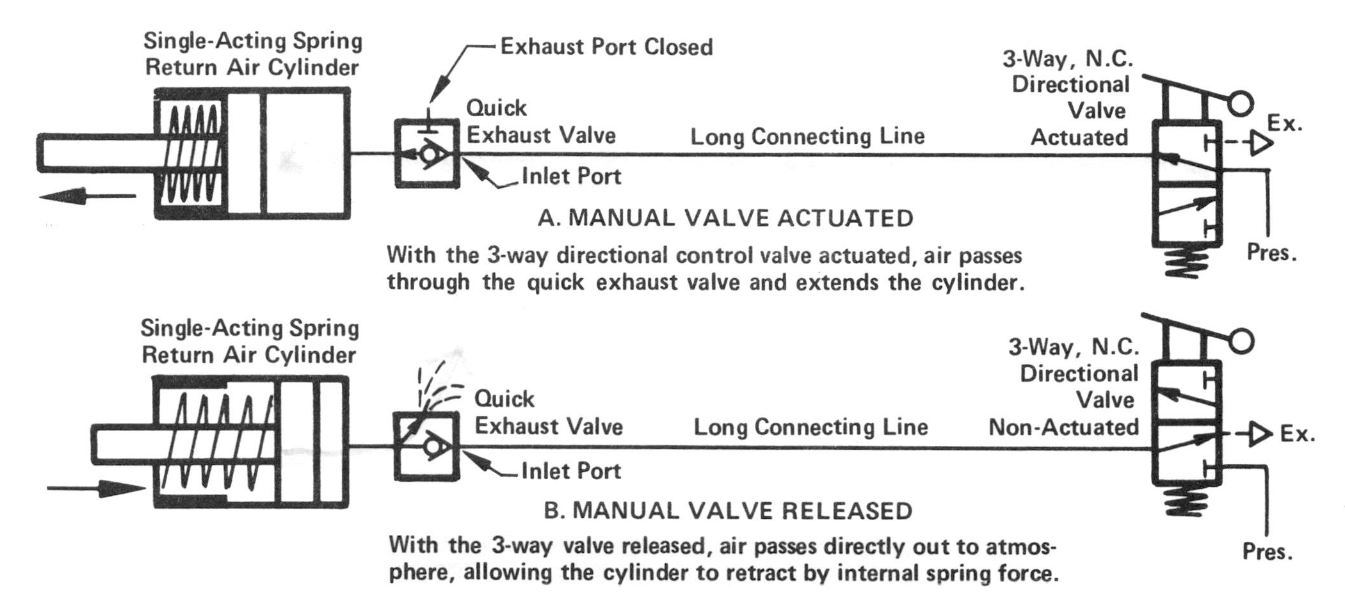

FIGURE 3-59. A quick exhaust valve permits rapid retraction of a single-acting air cylinder.

Quick Exhaust Valve on Single-Acting Cylinder. Figure 3-59. Retraction speed is increased. This circuit can also be used to control an air brake diaphragm, but instead of a spool-type 3-way directional valve for control, a pressure regulator type of brake control valve must be used, to be able to accurately vary exact braking force. A standard 3-way valve should not be used for air brake control.

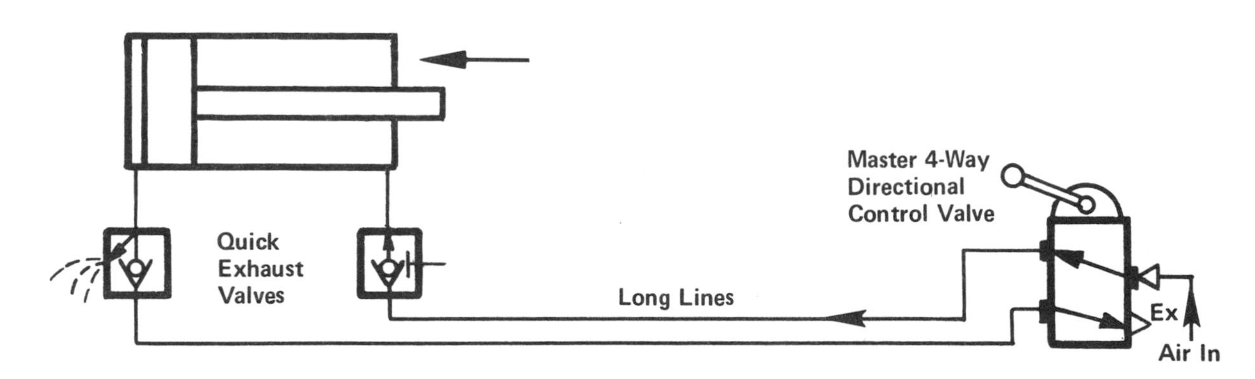

FIGURE 3-60. Quick exhaust valves speed up a double-acting air cylinder.

Quick Exhaust Valves on Double-Acting Cylinders. Figure 3-60. If lines connecting the control valve to the cylinder are unusually long or small in size, travel speed can be significantly increased by installing a quick exhaust valve directly at one or both cylinder ports, as needed. For in-plant operation, mufflers can be installed on the quick exhaust ports to reduce exhaust hiss.

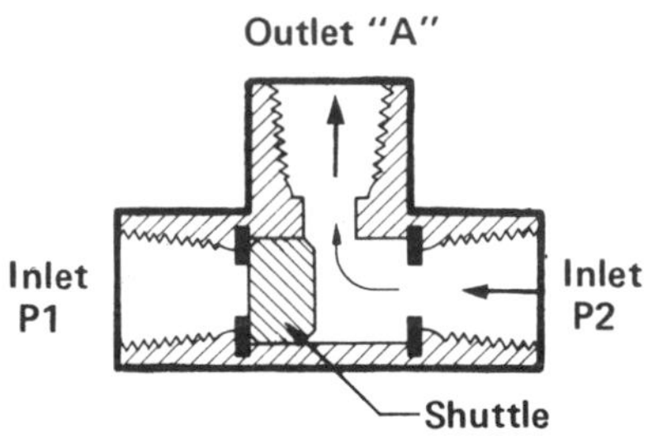

Part A. Cross Sectional View.

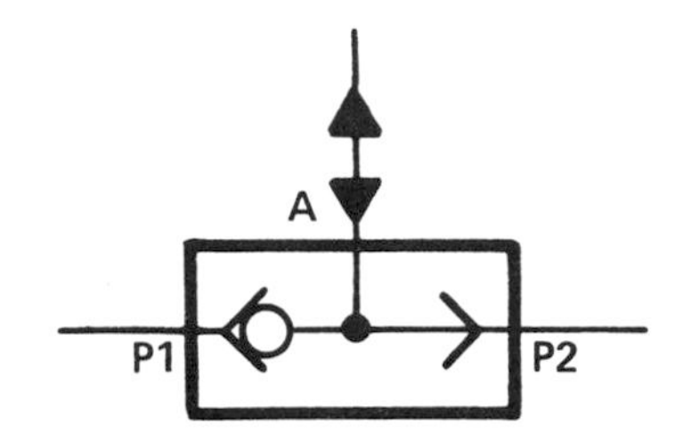

Part B. Graphic Symbol for Diagrams.

FIGURE 3-61. Shuttle Valve.

Shuttle Valve. Figure 3-61. This valve has the official designation "Double Check With Cross Bleed". Part A is a cross section picture; Part B is the ANSI symbol for schematic diagrams. The shuttle valve has two inlets, P1 and P2, and one outlet A. It can be classed as a 3-way valve with no normal position.

The purpose of a shuttle valve is to accept flow from either of two inlets, the one with higher pressure, and to pass it through to one outlet while keeping inlet fluids isolated from one another.

Shuttle valves are usually manufactured in small sizes, 1/4 or 3/8", since they are used primarily only for control functions handling a small flow of pilot fluid. If a larger size is required, as in the diagram below, a double piloted 3-way valve of any size can be used as a shuttle valve.

The floating poppet, or shuttle, is free to move back and forth, closing the inlet having the lower pressure, and giving preference to the inlet with higher pressure.

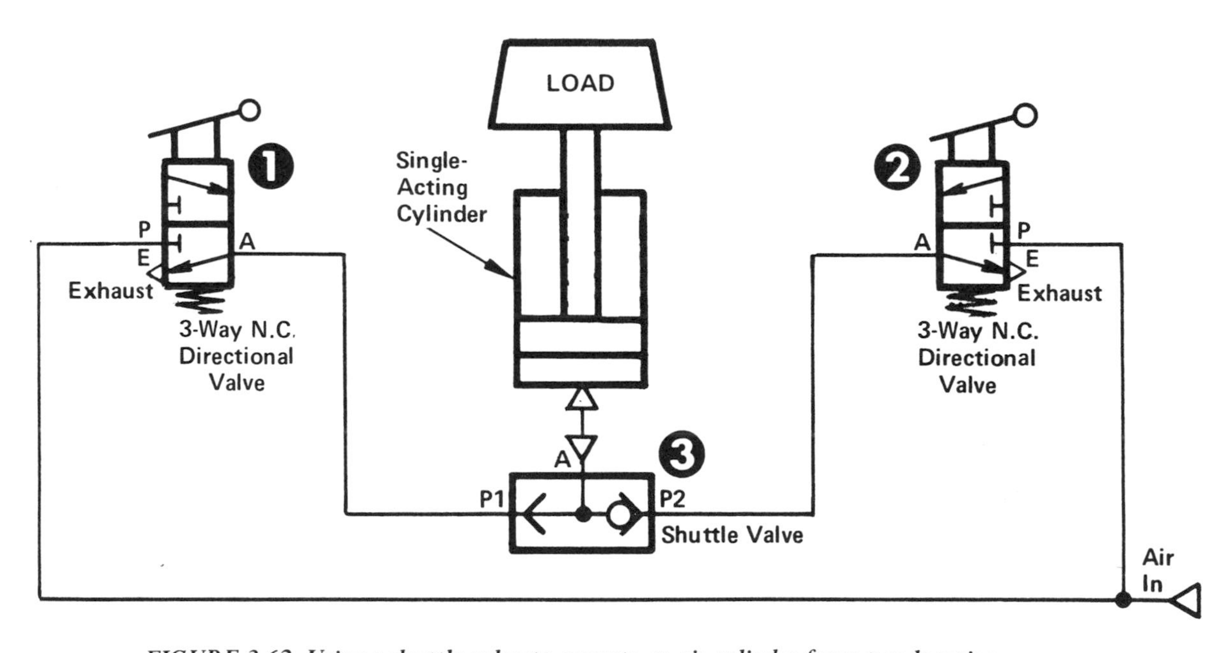

FIGURE 3-62. Using a shuttle valve to operate an air cylinder from two locations.

Shuttle Valve Application. Figure 3-62. This application is for controlling a single-acting cylinder from either of two control locations, using 3-way directional Valves 1 and 2.

If Valve 1 is actuated, air passing through this valve passes through Port P1 of the shuttle and into the cylinder. Shuttle action closes Port P1 to prevent loss of air through exhaust of Valve 2.

When Valve 1 is released, air returns through the shuttle and exhausts through Valve 1. The cylinder can be operated in like manner with control Valve 2. Standard check valves will not work in this circuit because air could not exhaust from the cylinder.

REVIEW QUESTIONS – CHAPTER 3

1. What is a "directional control" valve used for in a fluid power system? *– See Page 95.*

2. What is the difference in construction and action between a poppet valve and a spool valve? *–See Page 96.*

3. How many grooves are required on the spool of a 2-way valve? Of a 3-way valve? *– See Page 96.*

4. For controlling the maximum pressure in an air circuit, what is the name of the valve used? For controlling maximum hydraulic pressure? *– See Pages 97 and 98.*

5. What is the preferred body material for air shut-off valves? For water shut-off valves? For oil hydraulic shut-off valves? *– See Page 100.*

6. What is meant by the "normal" position of a valve and how is the normal position determined? *– See Pages 102 and 109.*

7. What is meant by the term "direct-acting" as applied to solenoid valves. Why are these valves limited in their size and pressure rating? *– See Page 105.*

8. How is a pilot-operated solenoid valve different in its action from a direct-acting solenoid valve? *– See Pages 105 and 106.*

9. What is the reason a pilot-operated type of solenoid valve cannot operate on low line pressures or on vacuum? *– See Page 107.*

10. How is a 3-way valve different from a 2-way valve? *– See Page 108.*

11. Which is the proper valve to use for controlling direction of a single-acting cylinder, a 2-way or a 3-way type? *– See Page 109.*

12. How many main portholes must a 3-way valve have? *– See Page 108.*

13. Give the NPFA standard port marking letters for a 3-way air valve. For a 3-way hydraulic valve. *– See Page 110.*

14. For directional control of a double-acting cylinder with 3-way valves, explain why TWO 3-way valves must be used. *– See Page 108.*

15. Explain the difference between a "direct-acting" and a "pilot-operated" relief valve. *– See Pages 121 and 122.*

16. Explain the function of the "control orifice" in a pilot-operated relief valve. *– See Page 123.*

17. Why is a pressure RELIEF valve not ordinarily used on a compressed air circuit? *– See Page 124.*

18. What is a "cushion relief valve and what is its purpose. *– See Page 125.*

19. Explain the flow action of a check valve. What is its purpose in an air or hydraulic circuit? *– See Page 126.*

20. Why is a LOW cracking pressure so important in a check valve used for air or vacuum? *– See Page 126.*

21. What is a "quick exhaust" valve and for what purpose is it used? *– See Page 128.*

22. What is a "shuttle valve". What is it used for. How is it different from two standard check valves placed back-to-back? *– See Page 130.*

23. What do the abbreviations N.O. and N.C. stand for? *– See Page 102.*

24. What kind of a valve is used to control the speed of a cylinder? *– See Page 97.*

25. What are the three general classifications of valves used for fluid power? *– See Page 95.*

26. Give the definition of "cracking pressure" for a hydraulic pressure relief valve. *– See Page 121.*

27. Name at least three advantages of a pilot-operated relief valve over a direct-acting type. *– See Page 122.*

CHAPTER 4

Control Valves– 4-Way and 5-Way

The simpler directional control valves, 2-way and 3-way types, were described in the preceding chapter. Spool-type models of those valves are designed with a single groove on the spool for handling only a single flow of fluid (Figure 3-1 for example). They can control direction on single-acting cylinders. The 4-way and 5-way valves described in this chapter have two (or more) grooves on the spool for handling two separate flows of the same fluid at the same time. They are capable of controlling direction on double-acting cylinders or reversible fluid motors.

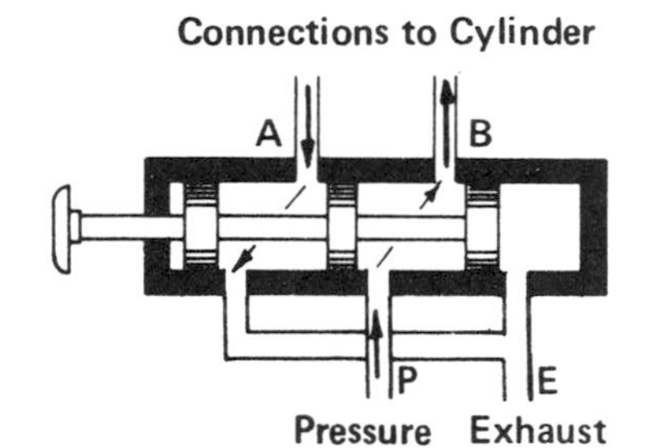

FIGURE 4-1(A). Four-Way Valve Action. Single-Exhaust Valve for Hydraulics.

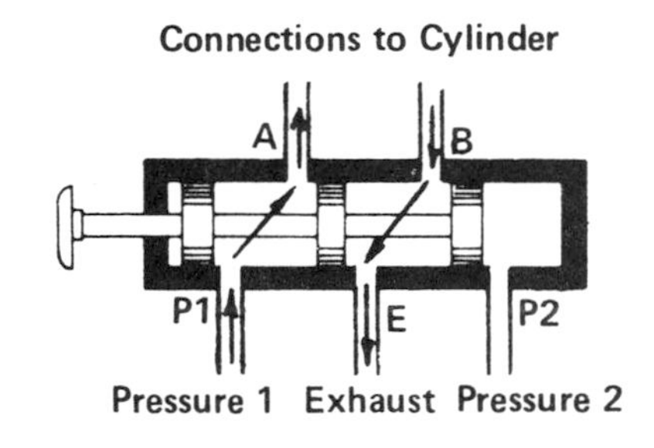

FIGURE 4-1(B). Five-Way Valve Action. Dual-Pressure Valve for Air.

Four-Way Valves. Figure 4-1(A). A 4-way valve is connected four ways into a circuit. It has four main ports which are: pressure inlet; two outlets or "cylinder" ports; and one exhaust (or tank) port. Some air valves may have two exhaust ports, both venting to atmosphere.

On some 4-way and 5-way valves, auxiliary ports, usually smaller in size than the main ports, may also be present for auxiliary functions such as vent or drain, external pilot pressure, bleed, or gauging.

Five-Way Valves. Figure 4-1(B). A 5-way valve is a special variation of a 4-way valve. It does have five main ports, all full size, and all serving separate functions. The main ports are: two pressure inlets; two outlets or "cylinder ports"; and one exhaust port. Five-way valves are suitable only for air operation. Their action will be described later in this chapter.

DIRECTIONAL CONTROL OF AIR CYLINDERS

Four-way valves built strictly to control air cylinders usually have one inlet marked "P", two outlet ports for connection to a double-acting cylinder and marked "A" and "B", and two atmospheric exhaust ports marked "EA" and "EB". This makes five main portholes but only four connections to the circuit since both exhaust ports serve the same purpose and could be teed together if desired. Valves with two exhaust ports are called "dual exhaust" valves. Exhaust Port EA (see Figure 4-3) handles exhaust flow discharged through cylinder Port A, and exhaust Port EB handles exhaust flow discharged through cylinder Port B.

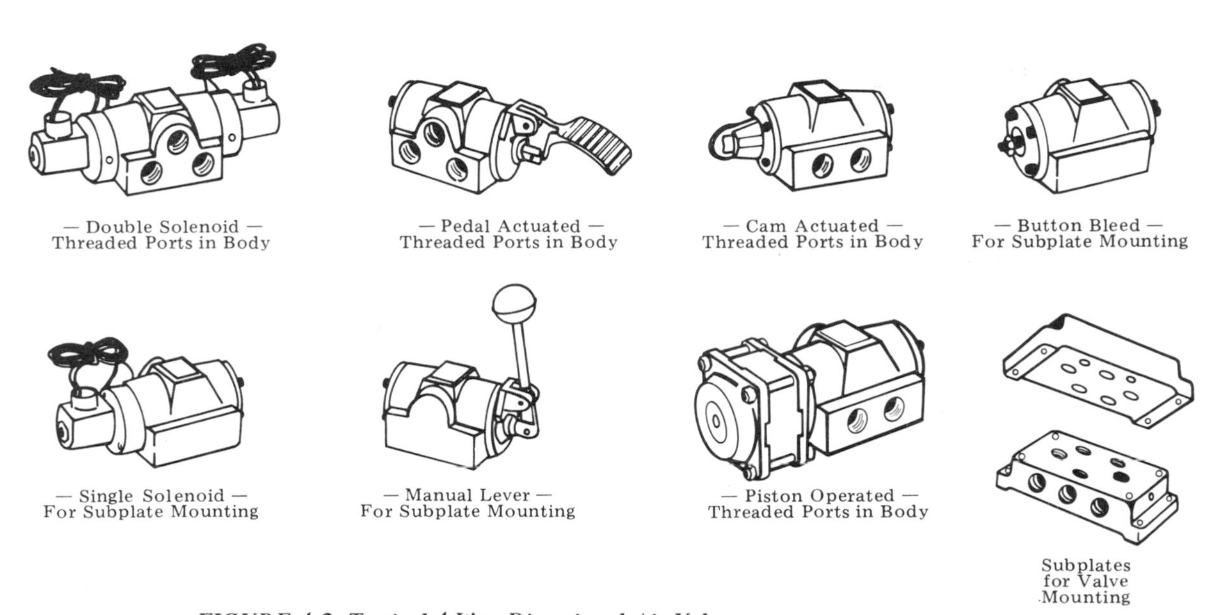

FIGURE 4-2. Typical 4-Way Directional Air Valves.

Typical Air Valves. Figure 4-2. Directional valves are offered in many shapes, sizes, and configurations. These illustrations are typical of some models available.

Manufacturers offer models with ports threaded in the body, also models with unthreaded portholes coming out the base of the valve. These are designed for bolting to a subplate. Circuit connections are made to threaded portholes in the subplate, either out the side or the bottom of the subplate. Examples are shown of each mounting arrangement.

Size range on air valves starts with sub-miniature 1/16" size, up to about 1" size. Few valves are available larger than 1". Very few applications require larger sizes.

Construction of Graphic Symbol. Figure 4-3. This diagram illustrates how graphic symbols for use on circuit diagrams are derived from a picture view of the valve. A 4-way dual exhaust valve is connected to operate a double-acting cylinder. On the left side of the diagram, in View 1, the valve is pictured with its spool shifted to its far left side position. The path of the air flow through the ports and across the spool grooves can easily be traced. On the right side of the diagram, in View 2, the same valve is pictured connected to the same cylinder, and with its spool shifted to its far right position. The air flow pattern through the valve has now changed, and can be traced in View 2.

The method of constructing a complete valve symbol is to use a small square box, Views 3 and 4, for each *working* position of the valve (two in this case) and draw a flow diagram to show air flow between ports, across the spool, when the valve has been shifted to that position. Arrowheads show direction of flow through the valve passages. (Some valves are designed to carry flow only in the direction of the arrowheads and may become inoperative if flow is carried in opposition to the arrowheads).

Most air valves are similar to the one pictured, with only two working positions and no neutral. The flow diagrams of Views 3 and 4 should be placed side-by-side to form a complete symbol as in View 5. Finally, appropriate actuators are drawn on one end (or both ends) of the complete symbol. There are a number of actuators available including, among others, manual lever, single or double solenoid, pedal, treadle, cam, etc. Figure 4-3 illustrates a valve with a manual lever and spring return.

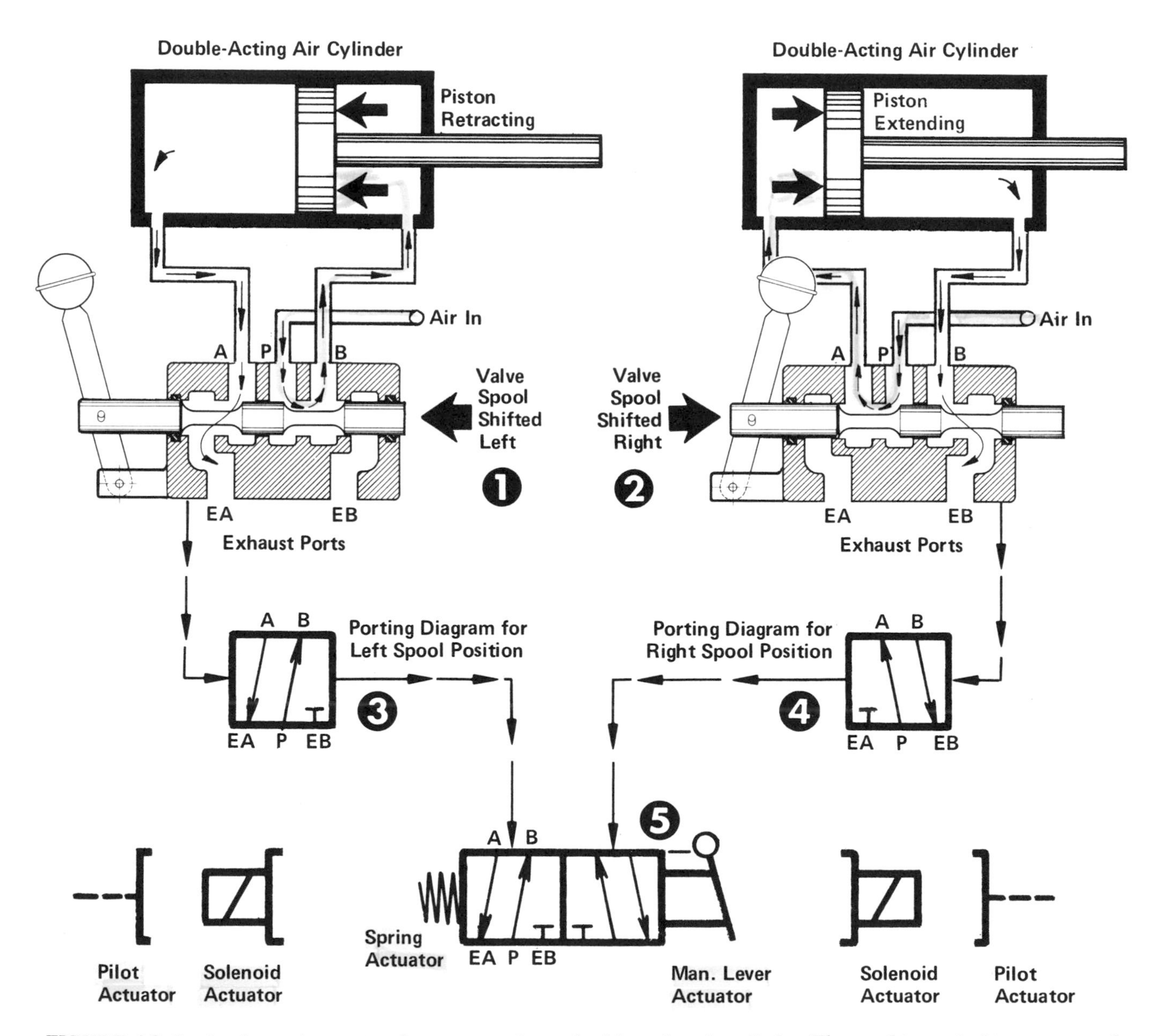

FIGURE 4-3. Dual-exhaust 4-way air valve connected to a double-acting air cylinder. The graphic symbol is constructed in the same way demonstrated on Page 103. The actuators shown above or those on Pages 136 to 138 may be used.

Schematic Diagram. Figure 4-4. Part A of this figure shows how the graphic symbol for a 4-way manual lever valve and cylinder should be drawn on a circuit diagram. The valve is shown in its

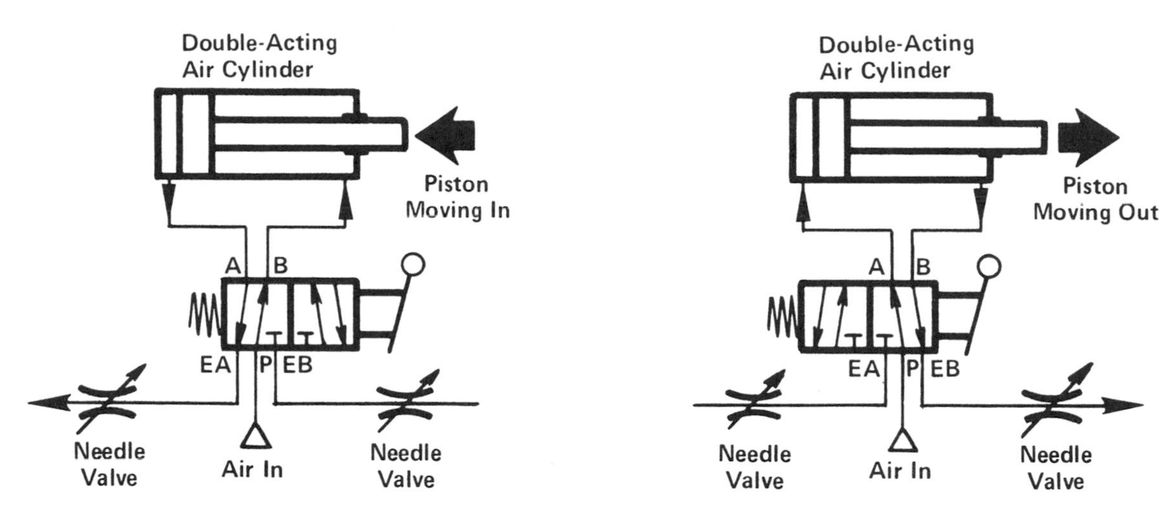

FIGURE 4-4. Demonstration of air cylinder operation with a dual-exhaust 4-way valve.

"normal" position, and the cylinder is connected so it will retract when the manual lever is released and the valve spool returns by spring force to its normal state. If desired, the cylinder connections could be reversed so the cylinder extends when the valve spool returns to its normal state.

Part B shows circuit flow when the valve has been shifted out of its normal state, causing the cylinder to extend. Unless otherwise indicated by a note on the drawing, the valve should be drawn in its un-actuated or normal state as shown in Part A of Figure 4-4.

This diagram also illustrates one reason why most air valves are built with two exhaust ports. An adjustable needle valve screwed into each exhaust port will permit the operator to trim the extension and retraction speeds as required by the job. Adjusting the left needle valve will change the retraction speed, while the right needle valve affects the extension speed. Adjustment of one needle does not change the speed established by the opposite needle valve.

Three-Position Air Valves. If an air cylinder must be stopped at a point between the ends of its stroke, a valve with a third, or neutral, position must be used. This is usually a closed center or a float center spool. These spools are illustrated on Page 155, Figure 4-25. Some valve manufacturers may offer special spools for special applications. Three-position air valves are usually available with all of the actuators shown on the following pages.

Good Design of Air Circuits. Because of the compressible nature of air, most air circuits work better if the cylinder does not stop at any intermediate point in its full stroke. If practical, the cylinder piston should be allowed to stall at both ends of the full stroke, either against its own end caps or against external positive stops. The piston can be held in a firm position against drift by keeping full pressure holding it against the stops. The 4-way valve should not be put in a center neutral position (if it has one) nor shifted into reverse until the cylinder is ready to move in the opposite direction.

It is difficult to stop an air cylinder in a precise mid-stroke position. When air is cut off, the piston continues to travel a short distance until a balance point is reached at which piston force is in balance with reactive resistance from the load. Although stopped, the piston will start up and move to a new balance position if the reactive load is increased or decreased. For these reasons, circuits using 3-position air valves should only rarely be used.

RULES FOR DRAWING ANSI GRAPHIC SYMBOLS FOR VALVES

These general rules apply to all 4-way and 5-way valves described in this chapter and to all 2-way and 3-way valves described in Chapter 3.

1. All connections between the valve symbol and the rest of the circuit must be made from just one of the blocks on the symbol. This seems obvious because the valve can be shifted to only one position at a time.

2. All connections from the valve symbol to the rest of the circuit should be made from the valve block which shows the flow state when the valve is un-actuated, or de-energized in the case of a solenoid valve. Valves which have an internal return spring or centering springs have a "normal" position to which the spool returns when the valve is un-actuated or de-energized. All actuators, of whatever type, can be visualized as having a "pushing" action against the end of the valve symbol blocks, pushing the entire block assembly one space so the next block will line up with circuit connections. This means that on spring return valves circuit connections should be made to the block nearest the spring, and on spring centered valves, to the center block. On valves which do not have internal springs, there is no "normal" position, and circuit connections can be made to any block. However it is customary to always draw circuit lines to the center position of all 3-position valves.

3. Arrowheads inside the porting blocks should, insofar as practical, show direction of flow when that particular block is moved into working position. In some circuits fluid may be free to flow in either direction at various times. This can be shown by using a double-headed arrow. In other cases a valve passage may be opened to pressure transfer but with no flow. This can be shown by using connecting lines between ports but omitting the arrowheads.

Refer to Part A of the preceding figure to see how these rules are applied. Any deviation from standard practice should be covered by a note on the drawing.

POPULAR 4-WAY VALVES FOR COMPRESSED AIR

Graphic symbols shown in this section show dual exhaust air valves – those with two exhaust ports. A single-exhaust air valve would be drawn in a similar manner to the hydraulic 4-way valves shown on Page 145, Figure 4-11, except with exhaust port marked "E" instead of "T". The following list of actuators covers the majority of those available from all manufacturers.

Four-way valves of some brands may be re-connected or may have unused ports plugged for use as 2-way, 3-way, or 5-way valves. These include those valves which use seals, like O-rings, which seal in both directions. Certain other brands including those with cup seals, if used in this way, might have the seals unseated or even washed out. It is best to consult factory literature before using a 4-way valve for any type service other than that for which it was intended.

These actuator symbols were taken from Specification ANSI Y32.10-1967, a copy of which, or equivalent, may be obtained from the National Fluid Power Association.

Solenoid Valves . . . To keep the spool in shifted position, current must be maintained on the solenoid of all valves which have spring returned or spring centered spools. Valves without internal springs will shift on a momentary impulse of current, one-half second or less, and will remain in shifted position until returned by an actuator on the opposite end (solenoid, pilot signal, manual lever, etc.).

• Single solenoid, 2-position, spring return. Valve spool springs back to original position when solenoid is de-energized. Dual exhaust model shown.

• Single solenoid, 2-position, no springs. Spool is returned by pilot air pressure from another valve after solenoid has been de-energized.

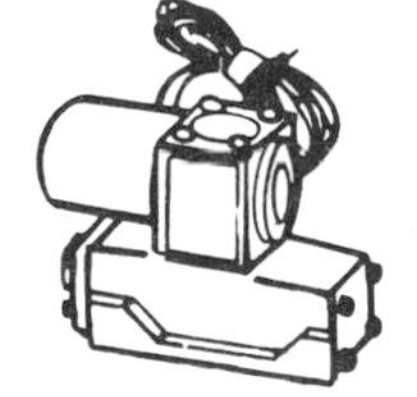

Single Solenoid Model

• Double solenoid, 2-position action. Valve shifts and remains shifted when one solenoid or the other is momentarily energized.

• Double solenoid, 3-position spool, spring centered. Porting in center neutral depends on spool selected. See air valve spools, Page 155.

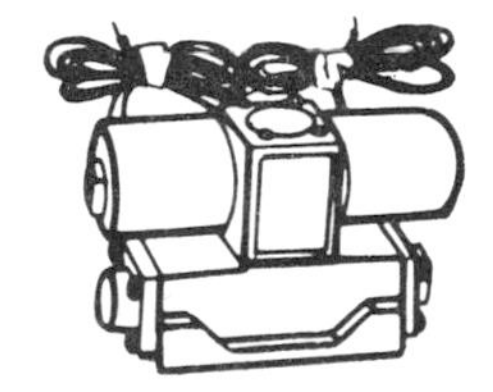

Double Solenoid Model

Button Bleed Valves . . . This type is constructed so the spool shifts when an end cap bleed button is pressed momentarily. This vents the end of the spool to atmosphere. Usually furnished only in 2-position double bleed models for air (only). A detailed treatment of these valves starts on Page 139.

• Button bleeder operated, 2-position. Spool shifts when one end cap or the other is vented to atmosphere through any small 2-way valve.

Button Bleeder Model

Manually Operated Valves . . . Includes all valves in which the spool is shifted manually by an operator. Direct manual control is used on valve sizes up to 1-inch. On larger sizes, due to the high force required, a 1/4" manual lever valve may be used to control a larger pilot-operated main valve.

Pushbutton or palm button actuators are not practical on valves larger than 3/8 inch. High shifting force may cause operator fatigue. Knobs are attached directly to the spool and have no lever advantage. Palm button valves are convenient for back-of-panel mounting, with knob extending through panel. Symbols shown are for dual exhaust construction.

• Friction Positioning, 2-position. Spool stays in shifted position by seal friction when manual lever or button is released by the operator.

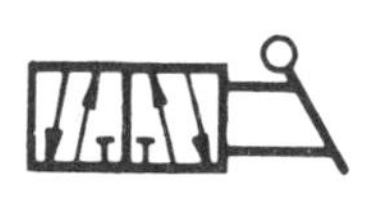

• Spring return, 2-position. Spool springs back to "normal" position when released.

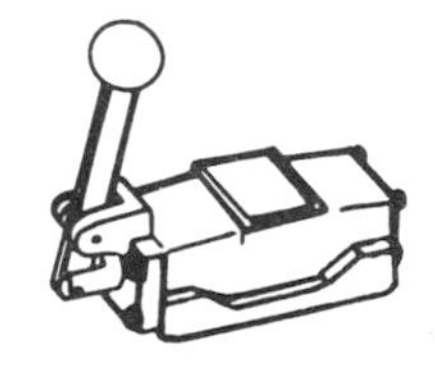

Manual Lever Model

• Spring centered, 3-position. Spool springs to center neutral when released. See Page 155 for center neutral spools for air valves.

• Click detent, 3-position. Spool stays in shifted position when lever is released. Detent protects against spool drift.

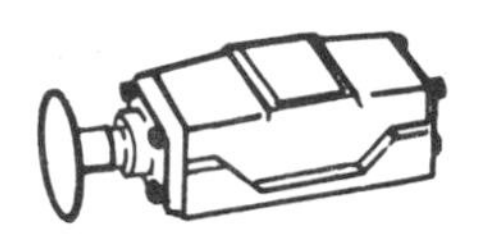

Palm Button Model

Foot Operated Valves . . . Includes pedal valves which are operated with toe, only, and treadle valves which can be operated with toe in one direction and heel in the other. Foot operated valves are usually available only in the smaller sizes, but can be used to pilot larger main valves.

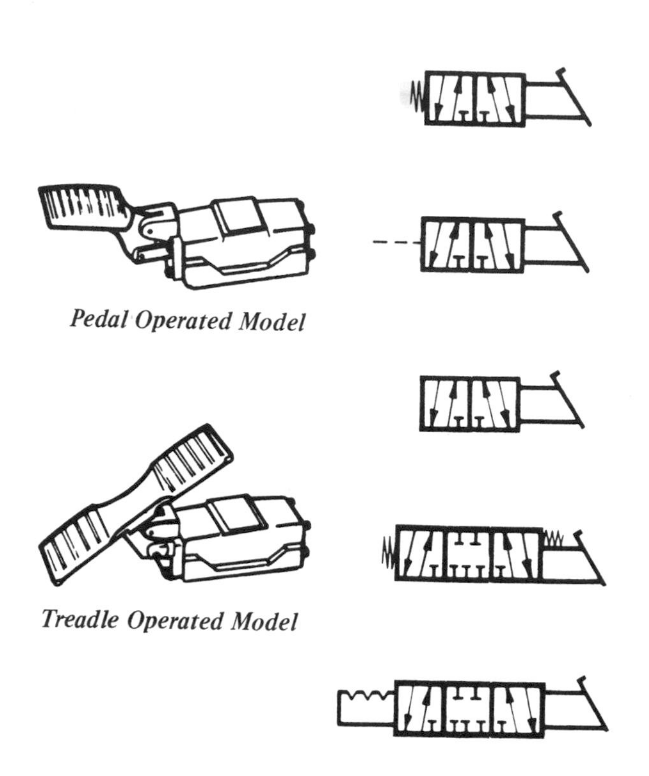

Pedal Operated Model

Treadle Operated Model

- Pedal actuated (treadle optional), 2-position spool is returned to "normal" position by spring force when pedal is released.
- Pedal actuated (treadle optional), 2-position spool returns to original position when a remote pilot signal, received from another source, is applied to end of spool.
- Treadle actuated 2-position spool held in each valving position by seal friction when treadle is un-actuated. Not suitable for pedal operation.
- Treadle actuated 3-position spring centered. Closed center spool shown. See Page 155 for air valve spool centers. Not suitable for pedal operation.
- Treadle actuated 3-position spool with click detents. Closed center spool shown. See Page 155 for air valve spool centers and Page 146 for detent information.

Pilot-Operated Valves . . . Includes all valves in which the main spool is shifted by application of air pressure to the end of the spool, either directly on the spool area or through a piston, diaphragm, or cylinder. A pilot-operated main valve may be controlled remotely by an operator using a miniature 3-way valve to pressurize and vent pilot lines.

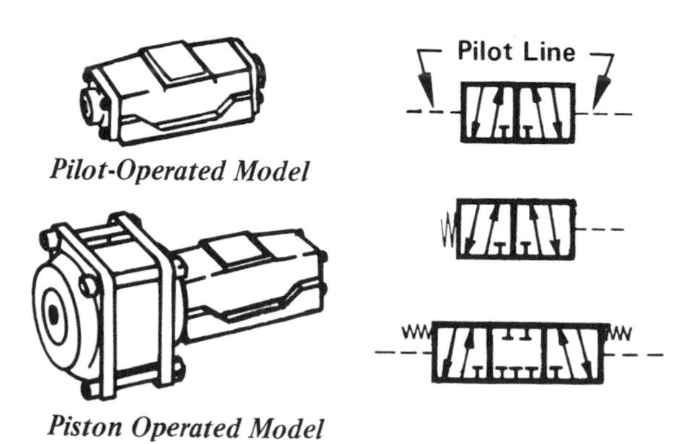

Pilot-Operated Model

Piston Operated Model

- Double pilot, 2-position, no springs. Spool stays shifted when pilot pressure is released.
- Spring return, 2-position, single pilot. Spool returns to "normal" position when pilot is vented.
- Spring centered, double pilot. Spool returns to center neutral when both pilot connections are vented. See Page 155 for air valve spool centers.

Mechanically Actuated Valves . . . These symbols may be used for any valve in which the spool is directly shifted by a moving member of the machine or by the cylinder, and includes cam roller, remote clevis, and stem actuated types. Mechanically actuated valves of small size can be used to pilot operate larger valves. Care should be used to mount these valves in a position where they will not be damaged if the actuating member should accidentally overtravel.

- Cam or mechanically actuated, 2-position, with spool spring returned to "normal" position.
- Cam or mechanically actuated, 2-position, with spool returned by remote pilot pressure signal from another source.

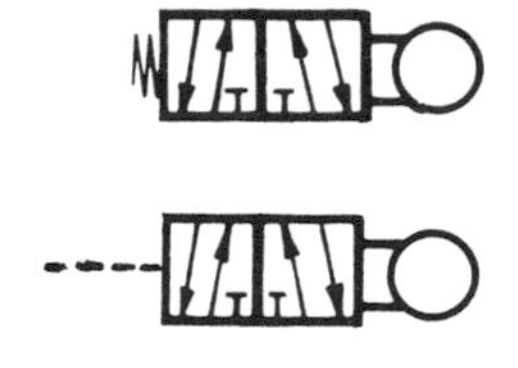

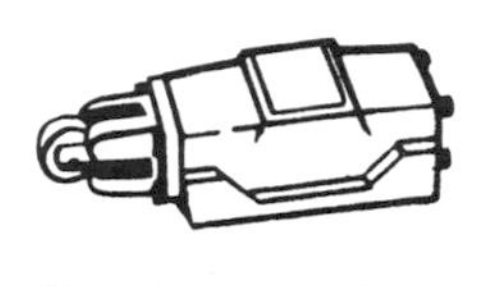

Cam Operated Model

BUTTON BLEED 4-WAY AIR VALVES

A button bleed valve is one of the standard 4-way valves available for air cylinder operation. The manner of shifting the spool makes it unsuitable for hydraulic operation, and it is seldom used on gases other than air because pilot air is discharged into the atmosphere. Its spool is shifted by air pressure by creating a pressure unbalance on opposite ends of the spool.

Typical 4-Way Button Bleed Valve. 1/2-Inch Port Size.

Button bleed 4-way valves give the simplest and most economical means of directional control of a double-acting air cylinder. Their operation is non-electrical, so they are particularly valuable in hazardous locations because they offer no fire or explosion hazard, and will operate reliably in ambient temperatures too high for solenoid valves.

As purchased, two miniature valves (referred to later as bleed buttons) are furnished screwed into the end caps of the valve body. These are miniature 2-way, normally closed valves which vent to atmosphere. Pressing one of the buttons will cause the valve spool to shift toward the button which was pressed. One or both bleed buttons can be unscrewed from the valve end caps, mounted on short hose extensions, and used as a control button by the operator, or can be mounted where they will be actuated by a moving part on the machine being controlled.

Button bleed main valves are usually furnished only as 2-position valves (no center neutral). Spring return and spring centered models sometimes may not operate reliably.

Shifting Principle. Figure 4-5. Shifting pressure is obtained internally from the main inlet port through small left and right orifices. When the valve is in a quiescent state there is no flow through these orifices and no pressure drop across them. Full inlet pressure is present on both sides of the

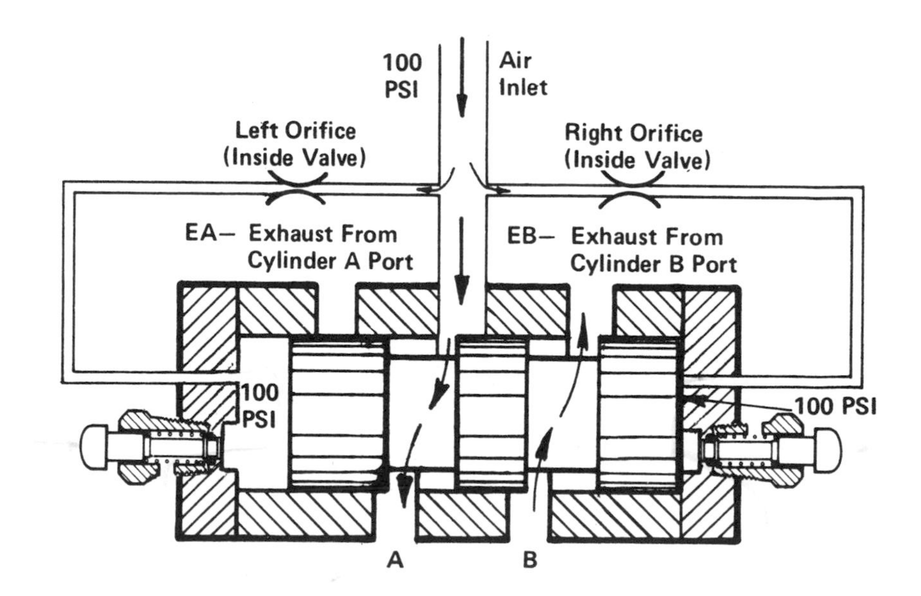

FIGURE 4-5. Schematic layout of a button bleed main valve and end cap bleed buttons, to illustrate spool shifting principle.

orifices and on both end areas of the spool, and the spool is in balance and will not move. To shift the spool this pressure balance must be upset.

In Figure 4-5, the spool has previously been shifted to the right side of its bore. To shift it to the left, the left bleed button must be pressed. This creates a small flow of air through the left orifice and out to atmosphere through the bleed button. The pressure drop thus created by air flowing through the left orifice causes the pressure to drop on the left end of the spool. Full pressure on the right end of the spool causes it to shift left.

After the left bleed button has been released, the flow of air to atmosphere ceases and in a matter of milliseconds full pressure is restored in the left end of the valve. The valve spool will continue to remain in its left position until the right bleed button has been actuated.

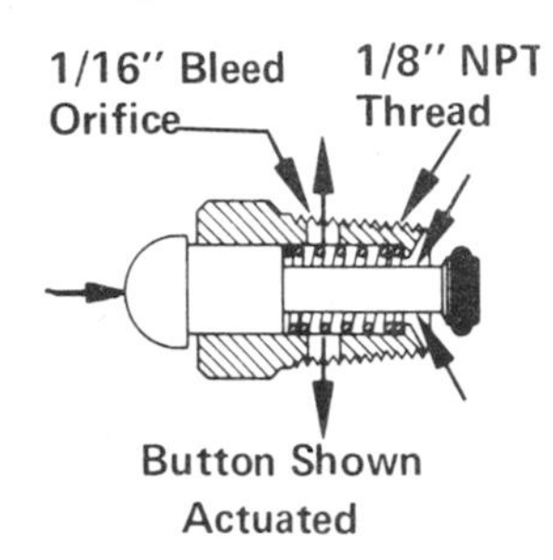

Bleed Button. Two bleed buttons are furnished as part of the master valve assembly. They are miniature poppet-type valves, 2-way normally closed, with discharge to atmosphere. Usually with 1/8" NPT pipe threads. Limited to about 150 PSI because the stem is not pressure balanced.

They can be removed from the master valve and mounted on hose extensions to operate the master valve remotely as shown in the next figure. Working circuits are shown later in this chapter.

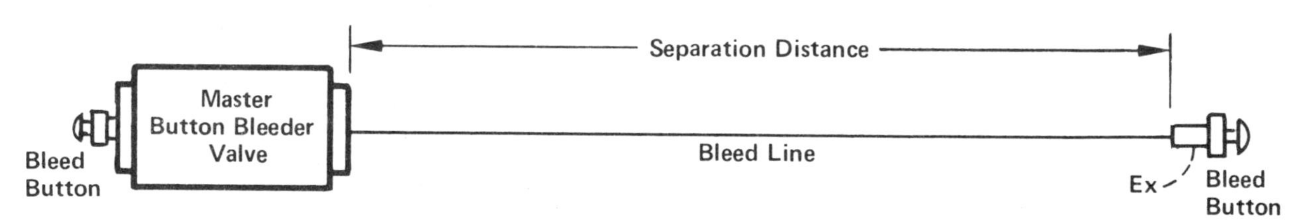

FIGURE 4-5. One or both bleed buttons may be located remotely for control of master valve.

Remote Operation. Figure 4-5. For remote shifting of the master valve the separation distance should not exceed 10 feet. Shifting speed becomes slow on greater separations.

Optimum diameter for connecting hoses seems to be 1/4" I.D. If line diameter is too large, shifting will be slow due to large air volume to be bled. If too small, shifting may be slow due to flow restrictions in the line. Hose up to a maximum of 3/8" I.D. is often used for bleed lines.

DIRECTIONAL CONTROL OF HYDRAULIC CYLINDERS

For most hydraulic circuits, a double-acting cylinder should be controlled with a 4-way valve having three working positions, including a neutral position for stopping the cylinder piston at any point along its stroke. (Some valves for mobile service may have a fourth position). Porting at each side position of the valve spool is standard 4-way action for extending and retracting the cylinder. The action is similar to that in 4-way air valves as already described. The neutral position is nearly always in the center of spool travel. The porting (the flow of oil between ports) in neutral can be one of several choices depending on the circuit action desired during the time the cylinder piston is stopped.

On ordinary hydraulic systems, the actuators for shifting the spool of a hydraulic valve are usually limited to only two types: manual lever actuators for mobile hydraulic systems and solenoid actuators on most indoor or industrial applications. There are, of course, some exceptions to this general practice.

X External Pilot Pressure Port

Y external drain port

The spool in a hydraulic valve is a metal-to-metal sliding fit in the valve body. At least two grooves are required so two independent flows can be handled at the same time without mixing. The high places between the grooves are called "lands". The spool must be a precision fit in the body to minimize leakage (called slippage) between ports. The only soft seals normally used on the spool are one seal at each end to prevent leakage to the outside. These seals will usually be rubber cups, O-rings, or quad rings. They are exposed to the pressure which exists in the tank port, and for this reason the tank pressure rating on many valves is less than the rating on the other ports.

Tank return flows come from both ends of the spool: from one end while the cylinder piston is extending and from the other end while the piston is retracting. These two flows, in a hydraulic valve, are combined inside the valve body and brought out through one tank port. On some valves the two internal tank return flows are combined through cored passages in the valve body. On other valves they may be combined through cross drilled holes in the body or spool.

Standard port markings for a hydraulic 4-way valve are: P for pressure inlet, T for tank return, and A and B for outlet ports which connect to a cylinder. Port markings for auxiliary functions may vary between manufacturers but many valves use the letter X for connection to an external pilot pressure source, and Y to indicate an external drain port. See Page 146 for more information.

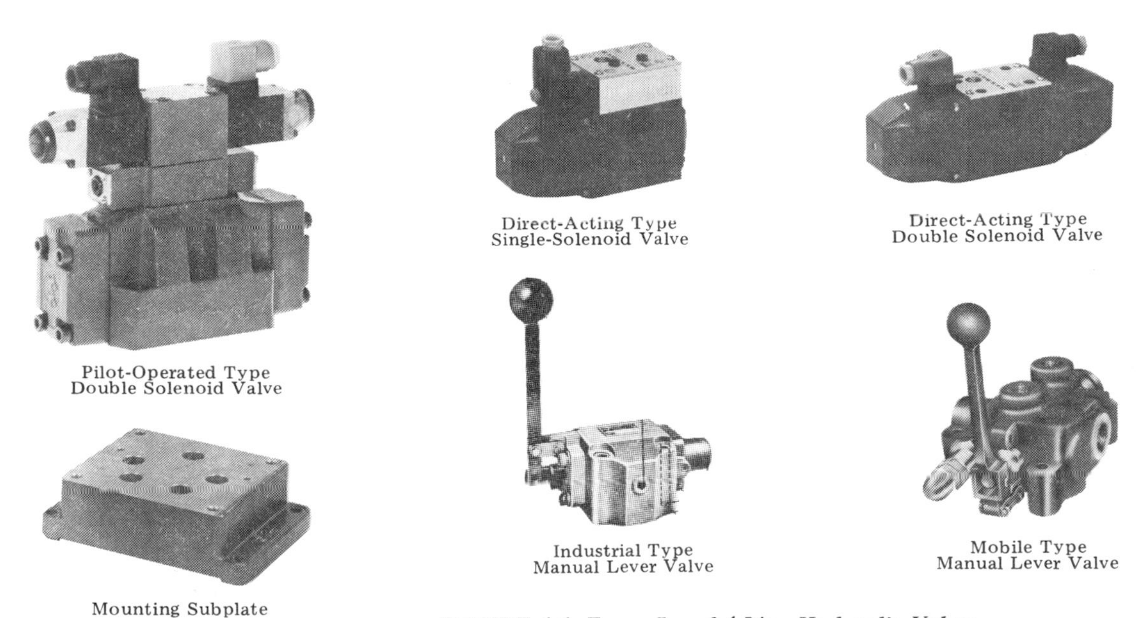

FIGURE 4-6. Examples of 4-Way Hydraulic Valves.

Typical Hydraulic Valves. Figure 4-6. Most of the solenoid and manual lever operated valves made for industrial hydraulic machines mount on subplates. Unthreaded portholes come through the base of the valve with O-ring seals, into matching holes in the subplate. Connections to the circuit are made to threaded portholes either in the sides or in the bottom of the subplate. In the past, connection portholes have been taper pipe threads, NPTF. To reduce thread leakage, the modern practice is to use straight thread portholes sealed with an O-ring.

Hydraulic 4-way valves for mobile equipment are mostly manual lever type, single spool or multiple spool bank valves. They usually have the connection portholes machined in the valve body, either NPTF taper pipe threads or straight threads with O-ring seal. Solenoid valves have been little used in the past because it was not possible to meter the flow with an ordinary solenoid. Solenoid valves are now available for mobile equipment which have "torque motor" type of solenoids which can be controlled with a rheostat. With these valves the operator can meter the flow as well as control direction.

Derivation of Graphic Symbol. Figure 4-7. This diagram shows a manual lever valve with its spool shifted into each of its three working positions. At 1, the spool is shifted to its extreme left position. A square block, 4, is used to represent the flow pattern through the valve in this spool position. At 3, the spool has been shifted to its extreme right position and another block, 6, shows internal flow pattern in this position. At 2, the spool has been shifted to a center neutral position in which flow is blocked between all ports. The flow pattern is shown at 5, and this spool is called a "closed center spool". Other spool types are shown in subsequent diagrams.

To make the complete graphic symbol for use on schematic diagrams, the square blocks for all working positions are gathered together into one assembly (shown at 7) and the appropriate actuator, or actuators, are drawn to one or both ends as appropriate. The actuators will be described later. Please refer to Page 136 for rules of construction and use of valve symbols.

The closed center spool shown in Figure 4-7 may also be used for compressed air valves, but only when a stop position is required along the length of the stroke. It should not be used on circuits where the cylinder stops only at extreme ends of its stroke.

Derivation of Graphic Symbol for a Closed Center 4-Way Valve

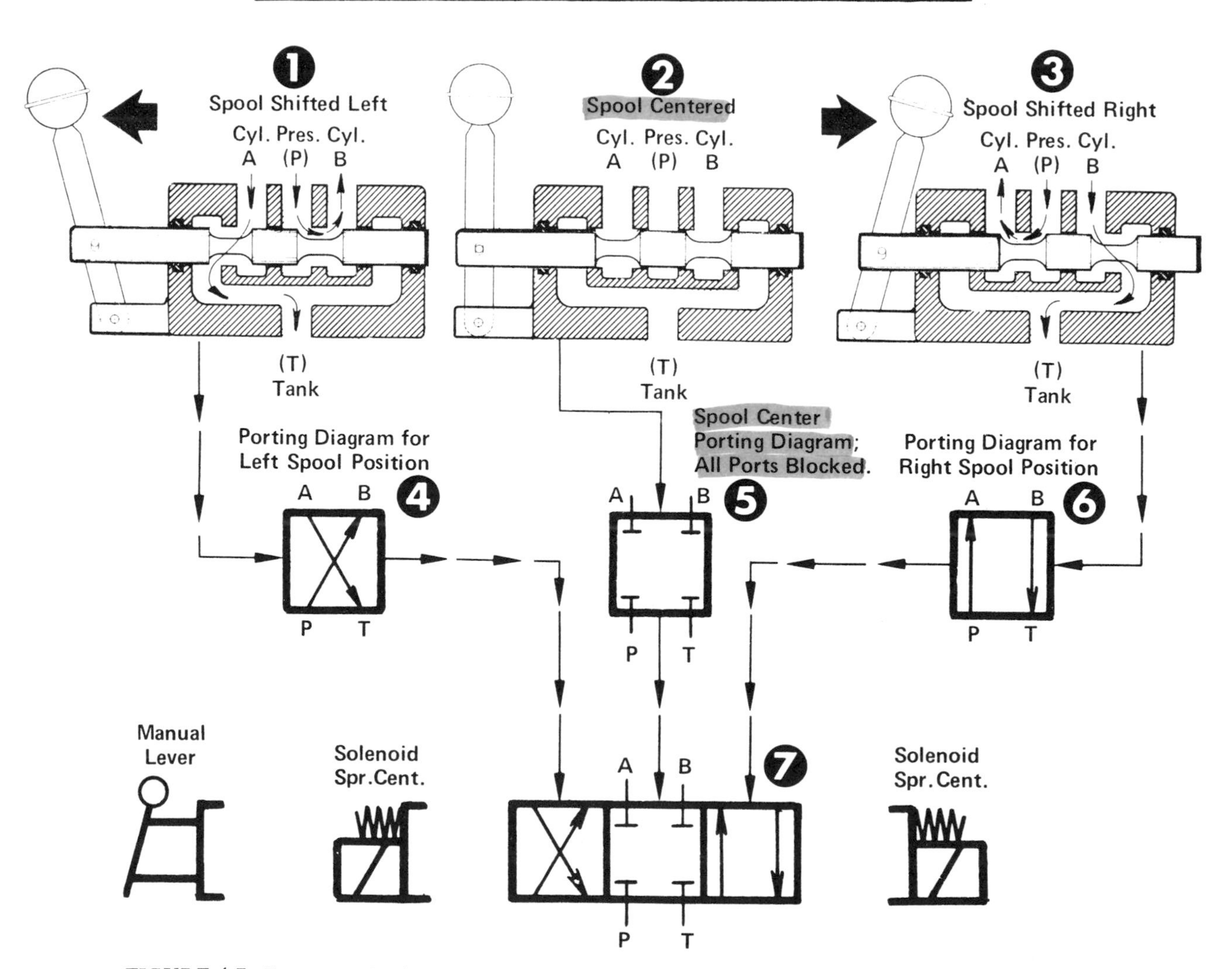

FIGURE 4-7. Example of a 4-way, 3-position valve used to control a double-acting hydraulic cylinder. The valve shown is a manual lever type with a closed center spool.

Complete Symbols. This is the way complete graphic symbols should be drawn on a schematic diagram. Circuit connections should normally be made to the center, or neutral, block. See Page 136 for rules on drawing valve symbols.

Other actuators shown previously for air valves, such as pedal, treadle, stem, cam, pilot-operated, etc., are sometimes available in miniature size. They are primarily used to control larger valves. Button bleeder valves are not used on hydraulics.

On sophisticated systems, electrically operated valves may be used in which a "torque motor" is used to modulate the spool position from its center neutral to provide speed control as well as directional control. These torque motors operate on low voltage D-C and are controlled by an operator using a rheostat.

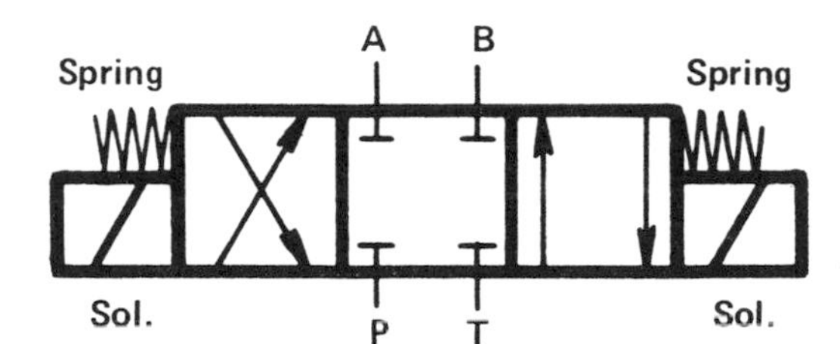

Symbol for double solenoid, spring centered valve with 3-position, closed center spool.

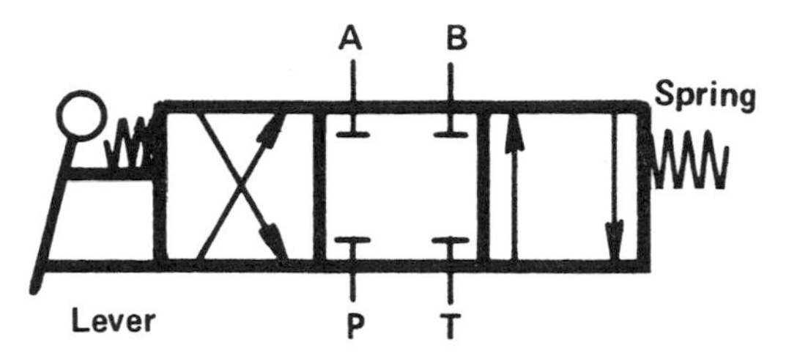

Symbol for spring centered manual lever valve with 3-position closed center spool.

Graphic Symbol for An Open Center 4-Way Valve

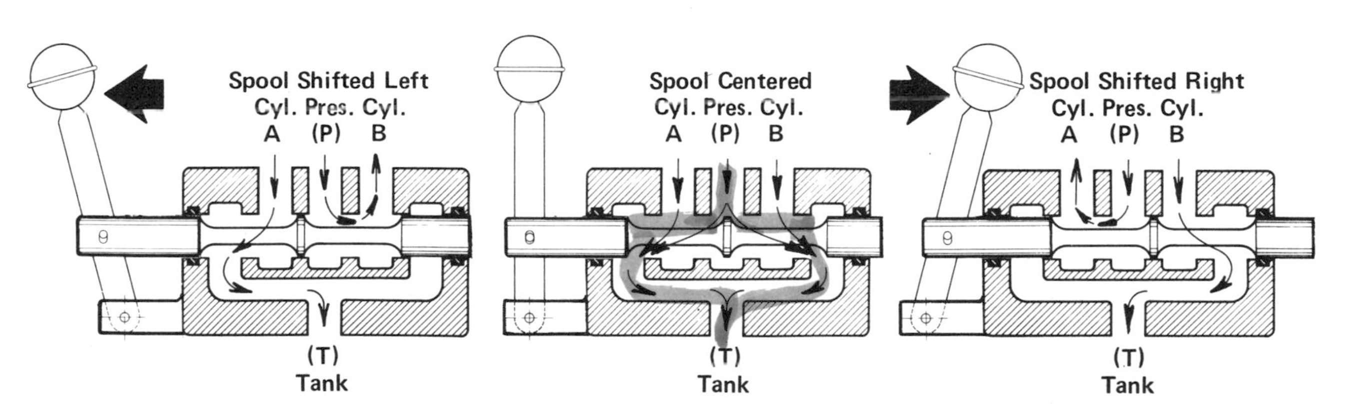

FIGURE 4-8. Example of a 4-way, 3-position valve used to control a double-acting hydraulic cylinder. The valve shown is a manual lever type with an open center spool.

Open Center Spool. Figure 4-8. In the two side working positions this valve operates identically to the closed center spool previously described. But in the center neutral position the inlet and outlet ports are open to tank.

This type spool is sometimes called a "motor" spool because it is frequently used to control a hydraulic motor in order to minimize circuit shock when the valve spool is centered. It is not usually recommended for control of cylinders because of the possibility of cylinder drift when the valve is in center position.

Symbol for double solenoid, spring centered valve with 3-position, open center spool.

The open center spool is almost never used on compressed air because its inlet port, P, becomes vented to atmosphere in neutral and the stored air in the compressor receiver tank would be lost.

Graphic Symbol for a Tandem Center 4-way Valve

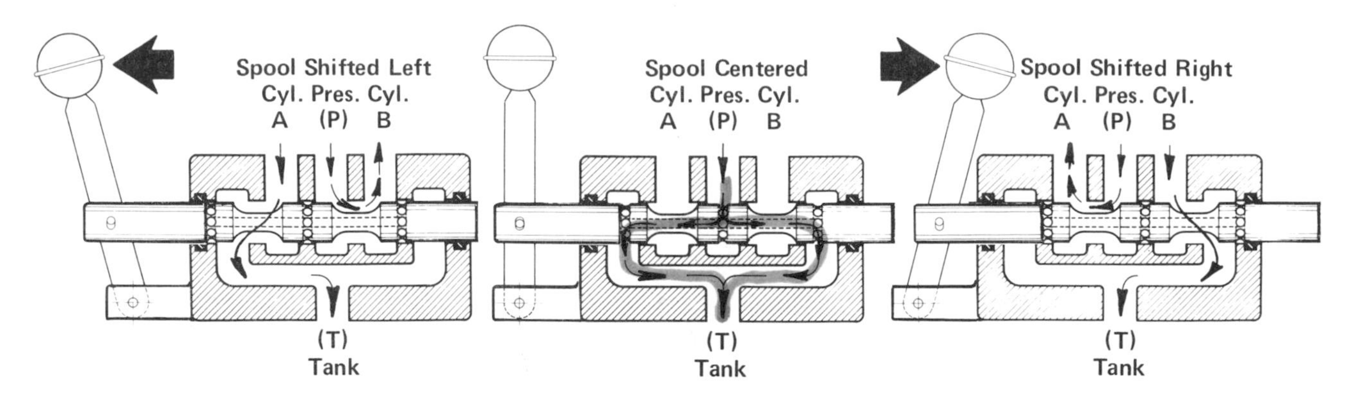

FIGURE 4-9. Example of a 4-way, 3-position valve used to control a double-acting hydraulic cylinder. The valve shown is a manual lever type with a tandem center spool.

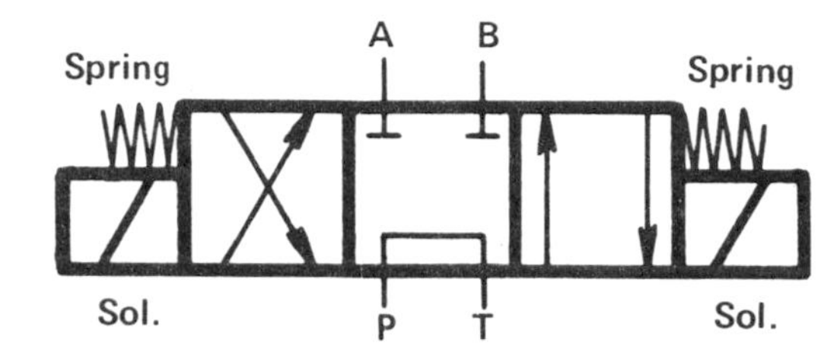

Symbol for double solenoid, spring centered valve with 3-position, tandem center spool.

Tandem Center Spool. Figure 4-9. This is a very popular spool for low power hydraulic systems because when shifted to neutral to stop cylinder motion it provides a free flow path for pump flow to tank. Although the pump shaft may continue to rotate, the hydraulic load is removed from the pump and the driving motor. This action is called "pump unloading". Also, cylinder ports are blocked in neutral which holds the cylinder against drift (except that caused by seal leakage in valves and piston leakage in cylinders).

On high power systems a closed center valve is usually preferred over a tandem center because of the shock generated by the shifting of a tandem center spool.

Tandem center valves are almost never used on compressed air because the pressure port is vented to tank (or to atmosphere) when the spool is in neutral. This would vent off all stored compressed air.

Graphic Symbol for a Float Center 4-Way Valve

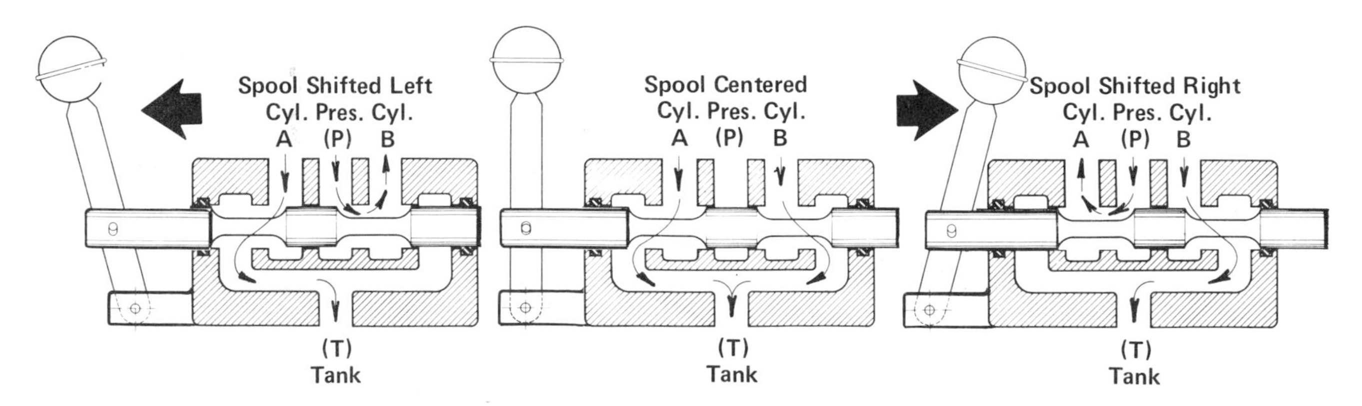

FIGURE 4-10. Example of a 4-way, 3-position valve used to control a double-acting hydraulic cylinder. The valve shown is a manual lever type with a float center spool.

Float Center Spool. Figure 4-10. The center neutral position of this spool blocks the valve inlet but vents both outlet ports, A and B, to tank, thus relieving any pressure block on either port of the cylinder. A cylinder connected to Ports A and B is free to "float" when the spool is centered, and its piston can be pulled or pushed by an external force. This spool is used to control cylinders on machines like scrapers and snow plows in which the cylinder must be permitted to float during parts of the machine cycle. It is a variation of the open center spool and is sometimes used for control of hydraulic motors. For example on hydraulic motor wheel drive of a vehicle, a float center valve would permit the vehicle to be towed.

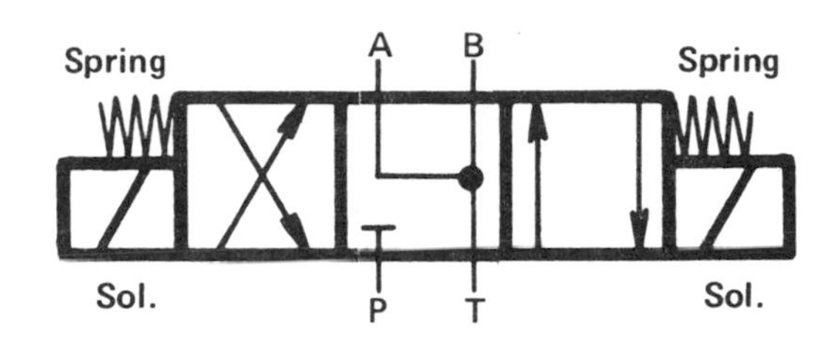

Symbol for double solenoid, spring centered valve with 3-position, float center spool.

This type spool is also available for air valves for similar applications in which the cylinder must be relieved of all pressure block during certain phases of its operation.

Two-Position Spool. Figure 4-11. This is a standard spool available from all hydraulic valve manufacturers. Although it is the spool most used for compressed air circuits, it is little used on hydraulics because most hydraulic circuits require a neutral or stop position for the cylinder.

For hydraulics it is usually available with single or double solenoid, manual lever, or pilot pressure actuators. It is seldom used for the main control but may be used in a branch circuit for auxiliary functions where a neutral position is not required.

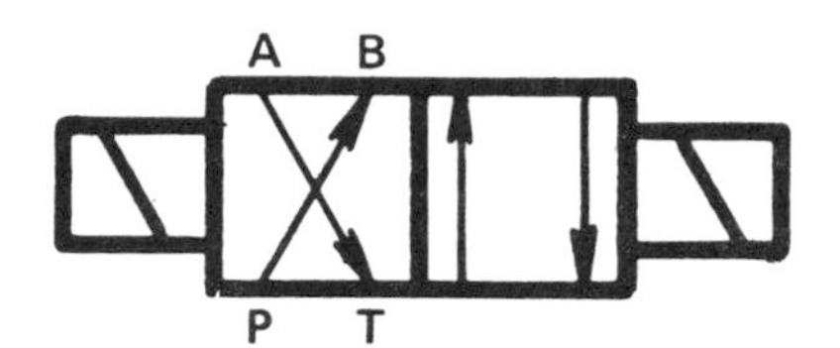

FIGURE 4-11. Double solenoid, two-position valve symbol.

Four-Position Spool. Figure 4-12. Three of the positions have standard 4-way, tandem center action with spring centering to a neutral position. The 4th position has open or float porting with a detent which holds it in this position until manually released by the operator. See next page for description of detent action. The 4th position may be used for "floating" a bulldozer cylinder, or to provide a towing position on vehicles with hydraulic motor wheel drive, for snow plow scraper cylinders, or for free fall of a winch powered with a hydraulic motor.

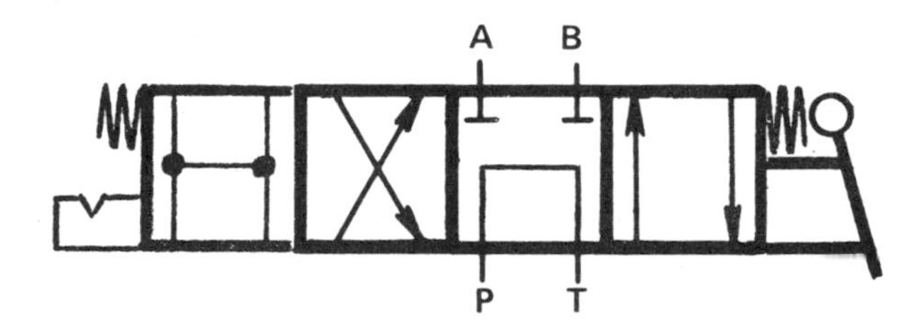

FIGURE 4-12. Example of 4-position, tandem center manual lever valve.

Valves with four positions are suitable only for manual lever actuation. Although available in single-spool models, they are primarily used in one or more sections of a multiple-spool bank valve.

A 4-position valve has no important application in compressed air circuits.

Other Valve Spools. Although the spools just described represent the majority of spool center neutral portings used in hydraulics, individual manufacturers may offer a number of variations which are used primarily for special circuitry.

In general, spool side positions have standard 4-way action, the special variation being in the center neutral position, and may include one of the cylinder ports, A or B, being vented to tank or connected to inlet pressure. These special variations are for hydraulic circuits and are rarely used on compressed air.

External Pilot and External Drain Ports. Please refer to the example of Figure 4-21 on Page 153. In addition to the four main ports of a 4-way valve, hydraulic solenoid valves of the pilot-operated type must be provided with a source of hydraulic pressure of 50 PSI or more for shifting the main spool. Usually, as shown in Figure 4-21, this shifting pressure is tapped internally from the pressure groove on the main spool. But on some applications this pressure can be obtained from an external source and connected to Port X on the valve subplate. When using external pilot pressure, the internal pressure line must be plugged inside the valve.

The pilot solenoid valve must have a drain connection to tank. Usually it is drained internally to the main spool tank groove. But on some applications it may be necessary to drain the solenoids to tank without combining with the main tank port. A separate drain line can be connected to Port Y on the valve subplate. When an external drain is used it is necessary to plug the internal drain line.

Spool Positioning Devices for Air and Hydraulic 4-Way Valves

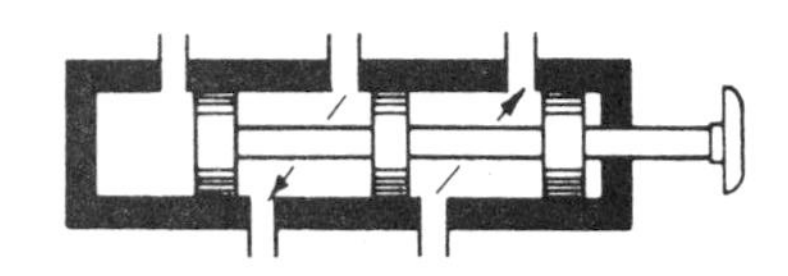

FIGURE 4-14(A). Friction Positioning.

Friction Positioning. Figure 4-14(A). On those valves which do not have internal springs to return the spool to a "normal" position when non-actuated, friction of the spool seals may be used to hold the spool in the last position to which it was shifted. While this positioning method is satisfactory on certain air valves which have high seal friction, it should never be used on hydraulic valves because of the danger of spool drift when the valve is handling excessive oil flow. Valves with this construction should be mounted with spool horizontal to avoid spool drift caused by machine vibration. Spool detents (see below) are a much safer method of holding a spool in a shifted position.

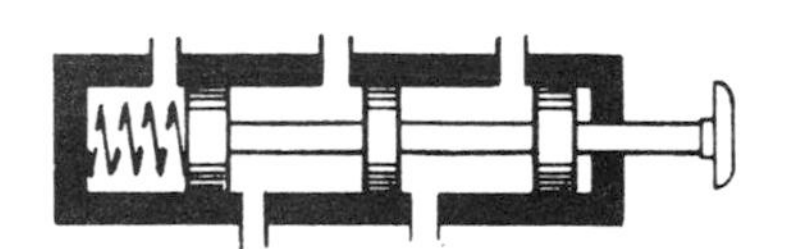

FIGURE 4-14(B). Spring Return.

Spring Return. Figure 4-14(B). Two-position valves may use a built-in spring to return the spool to its "normal position when it becomes un-actuated or de-energized. These valves do not have a neutral position. The "normal" spool position can be on either end of its travel depending on valve construction.

These valves are said to be "spring offset", "spring loaded", or "spring returned".

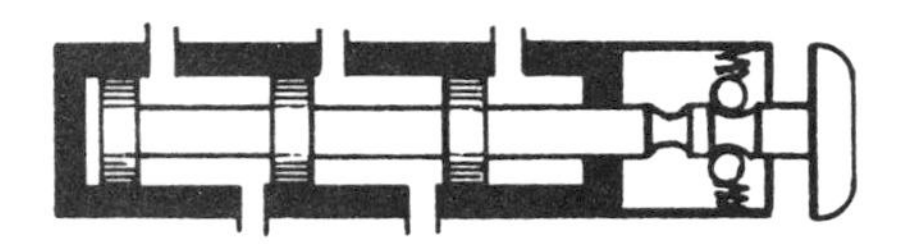

FIGURE 4-14(C). Click Detents.

Detents. Figure 4-14(C). A "click detent" is a mechanical arrangement for securing the spool in the position to which it has been shifted. For example, a spring loaded ball may drop into notches on an extension of the spool. Notches may be in the center neutral position only, or may be in the center and/or both side positions. Detenting is a much safer spool holding arrangement than depending on friction in the spool seals.

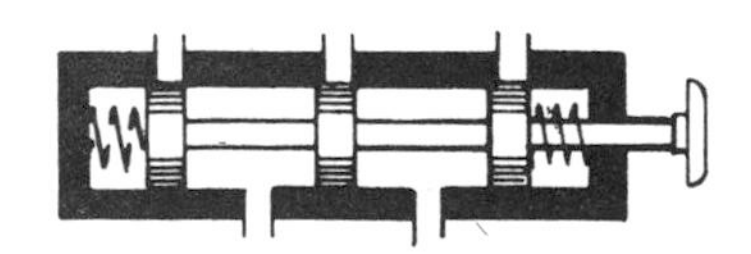

FIGURE 4-14(D). Spring Centering.

Spring Centering. Figure 4-14(D). Used primarily on hydraulic valves to obtain a 3rd working position which is usually a center neutral position. Essential on double solenoid valves. Optional on manual lever valves. Used less frequently in compressed air circuits.

ROTARY SHEAR-TYPE DIRECTIONAL CONTROL VALVES

Spool-type directional valves previously described appear to be more widely used than any other kind for compressed air and hydraulic oil service, but on some applications rotary shear valves may have some advantages over spool valves.

Rotary valves are more common as 4-way types but are also available with 2-way and 3-way action. They are not available as 5-way valves. Those with 4-way action have the same choice of center neutral porting previously described for spool valves – closed, open, float, and tandem center actions.

Models are available for the full pressure range covered by fluid power – up to 250 PSI on air and up to 10,000 PSI on hydraulics. Port sizes are available from 1/4" to 1", sometimes larger. Most rotary valves are operated mechanically with a rotary lever, but some manufacturers may offer solenoid operated models.

Figure 4-15. The body, containing the connection ports, is stationary. The handle is attached to a rotor which can be rotated through a 90° arc. Flow passages, equivalent to the grooves on a spool, are machined into the rotor. Turning the rotor causes various ports to be connected together. The center position is detented.

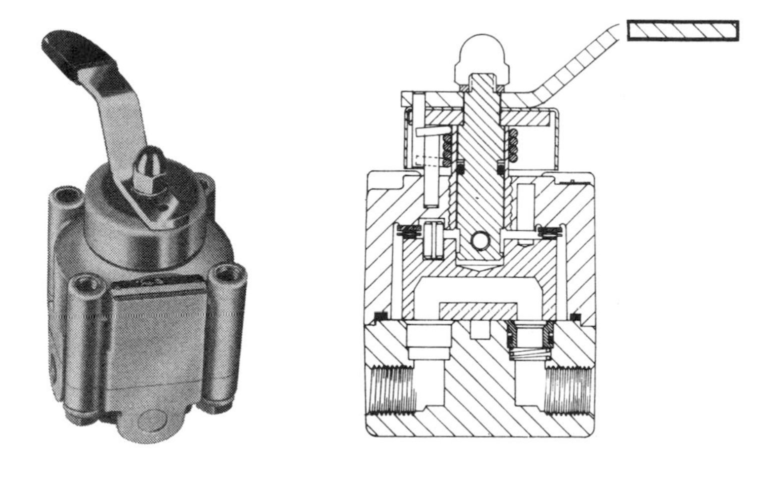

FIGURE 4-15. Rotary "shear-type" 4-way directional valve.

The action of a rotary valve is somewhat like poppet action, except that the rotor which is the poppet, does not lift off the port orifice; it slides off sidewise, shearing the fluid stream. Hence the name "shear-seal". Comparing the characteristics of a rotary valve with those of a spool valve of the same flow capacity:

(1). A rotary valve is usually more expensive than a spool valve of the same flow capacity. This may be a disadvantage, especially on mobile applications where initial cost of components is an important factor and where all components must be competitive in price.

(2). A rotary valve has better metering characteristics than most spool valves. This may make it better for some applications in spite of its higher cost.

(3). A rotary valve can only be partially pressure balanced even under static conditions. It requires more shifting effort as the pressure becomes higher. But it performs more smoothly than a spool valve at pressures of 3000 to 10,000 PSI. A spool tends to stick to the side of its bore and may lock up completely if the spool is allowed to remain shifted under high pressure for a short time.

(4). Since a rotary valve has no rubber moving seals, it can handle a wide variety of fluids.

(5). Although there may be a slight internal slippage between the ports of a shear-seal valve, it usually seals tighter than a spool-type valve. This makes it more efficient at higher pressures.

(6). The available range of sizes for rotary valves is more limited than for spool valves. Sizes larger than 1" may be difficult to obtain. The number of supply sources is also more limited.

(7). Some rotary valves may require an external drain line to tank on applications where there is a substantial back pressure on the tank port. In some cases this may also be true of spool valves.

FIVE-WAY DIRECTIONAL CONTROL VALVES

Continuing the description from Page 132, this valve is essentially a special variation of a 4-way valve, mostly for applications which require inlet pressures of two different levels. All 5-way applications could be handled with a standard 4-way valve by placing a pressure regulator in the appropriate cylinder line, by-passed with a check valve for flow in the reverse direction. A 5-way valve does have five main ports, all different, and is connected five ways into the circuit. Ports are: two pressure inlets, P1 and P2; two outlet or cylinder ports, A and B; and one common exhaust port, E, serving both cylinder ports. Most of the applications for 5-way valves are for compressed air.

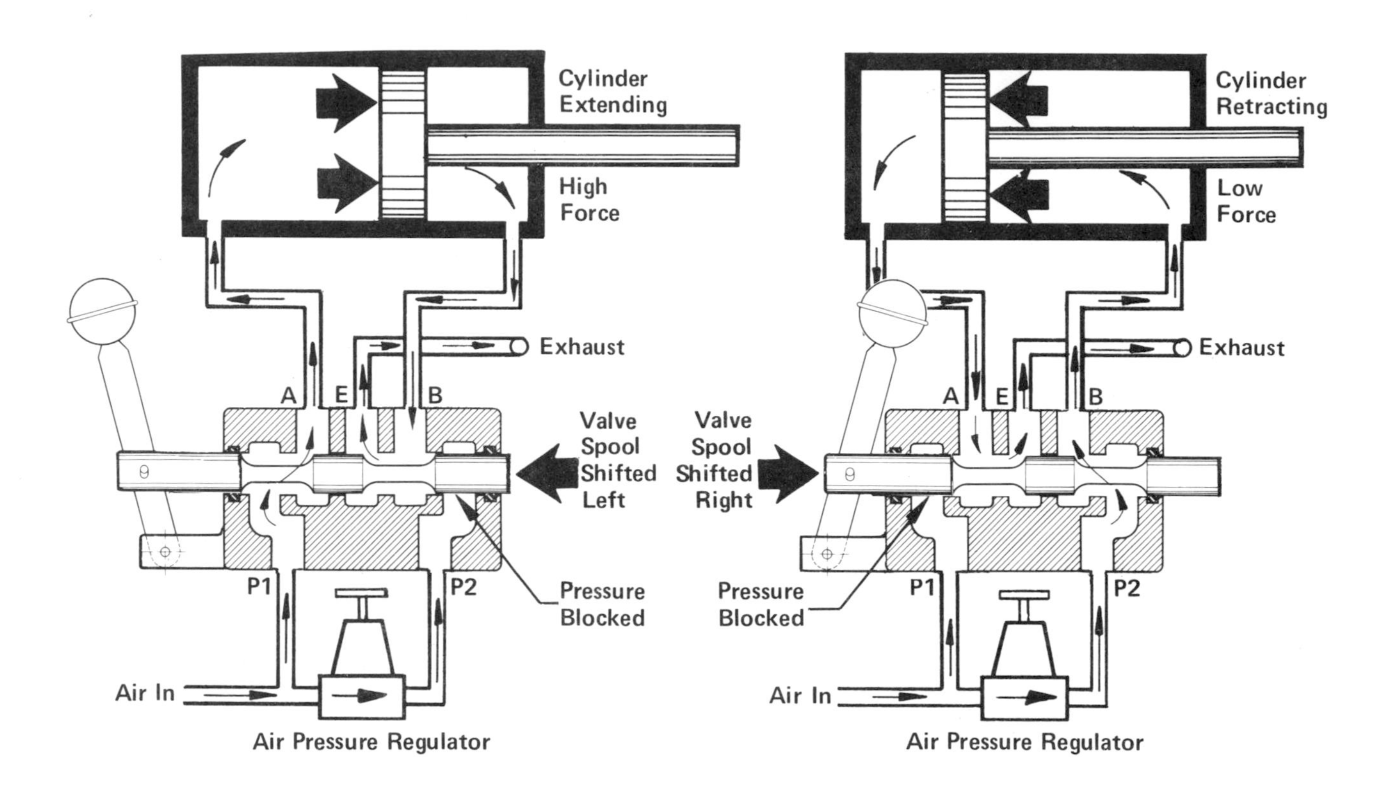

FIGURE 4-16. This cylinder is controlled with a manual lever, 5-way valve.

Action of a Five-Way Valve. Figure 4-16. A double-acting air cylinder is shown connected to a 5-way valve. Full air pressure is connected to inlet Port P1. Reduced pressure from a pressure regulator is connected to inlet Port P2. In the left view, the valve spool has been shifted to its far left position. Full pressure, entering inlet P1 passes out cylinder Port A and into the blind end of the cylinder. As the cylinder piston extends, exhaust air enters valve Port B and exhausts to atmosphere through Port E. Inlet Port P2 is inactive when the valve spool is in this position.

In the right view of Figure 4-16, the valve spool has been moved to its far right position. Inlet P2 is now active, supplying reduced pressure to retract the cylinder piston. Port P1 is inactive.

Operation of a double-acting cylinder with high pressure (high force) in one direction and reduced pressure in the other direction is one of the more important applications for a 5-way valve. A reduced cylinder force may be necessary to avoid damage to a workpiece, to prevent tool breakage, to avoid distorting or damaging a jig or fixture, or to reduce air consumption. A classic example is for operation of an air cylinder which drives the stem of a wedging-type gate valve up and down (open and closed). The "gate" has a slight taper, and when closed fits into a mating taper in the valve body to make a tight seal. The gate has a tendency to stick if forced closed with too much force. It may require 3 to 5 times the force to break the gate away as was used to close it. Incidentally, for this application, the connecting lines between valve and cylinder must be interchanged to obtain low closing force and high opening force.

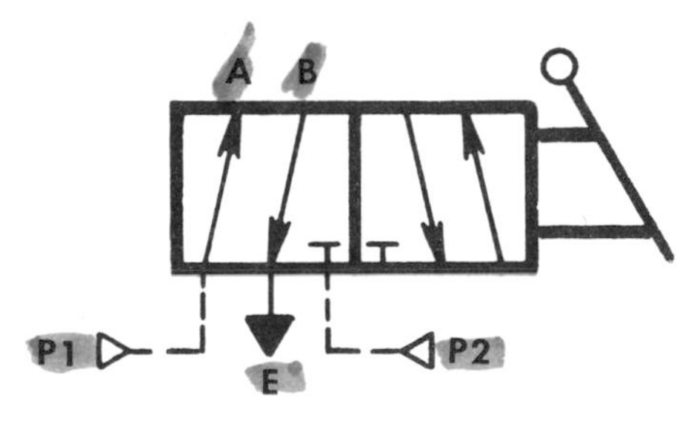

Graphic Symbol
Manual Lever 5-Way Air Valve

Air Saving Application. Another application is for saving air in one direction of cylinder travel, the free running or unloaded direction. In Figure 4-16 this is assumed to be the retraction direction. To control return speed a flow control valve would normally be placed in the line connecting valve Port A to the blind end of the cylinder, and would meter air out of the blind end. But with a flow control valve, when the cylinder reached stall at home position, air would continue to flow into the rod end of the cylinder up to full system pressure. This is unnecessary and is a waste of air. To save air the flow control valve should be removed and a regulator installed to feed low pressure air into valve inlet Port P2 as shown in Figure 4-16. The regulator should be used as a return speed control, and should be adjusted to the lowest pressure which gives satisfactory return speed, no higher. When the cylinder stalls at home position, only enough additional air would enter the cylinder to bring pressure up to the reduced level. On some applications, air consumption can be reduced by 15 to 20% for the overall cycle. A 5-way valve is ideal for this application.

The graphic symbol for a manually operated 5-way valve is shown above. For normal speed control in the forward direction in Figure 4-16, a flow control valve can be placed in a meter-out mode in the line connecting the rod end of the cylinder to valve Port B.

DIRECT-ACTING SOLENOID VALVES – 4-WAY

Figures 4-17 and 4-18. Direct-acting solenoid valves (as opposed to the pilot-operated type to be described next) are those in which the solenoid armature acts directly on the valve spool, and provides the physical force for shifting. Most modern solenoid valves use "push-type" solenoid action, with the solenoid armature extending through the center of the coil. In the de-energized state there is an air gap between the armature and the remainder of the magnetic structure. Energizing the solenoid coil causes the armature to pull in and seat against the iron core. The armature also pushes against the end

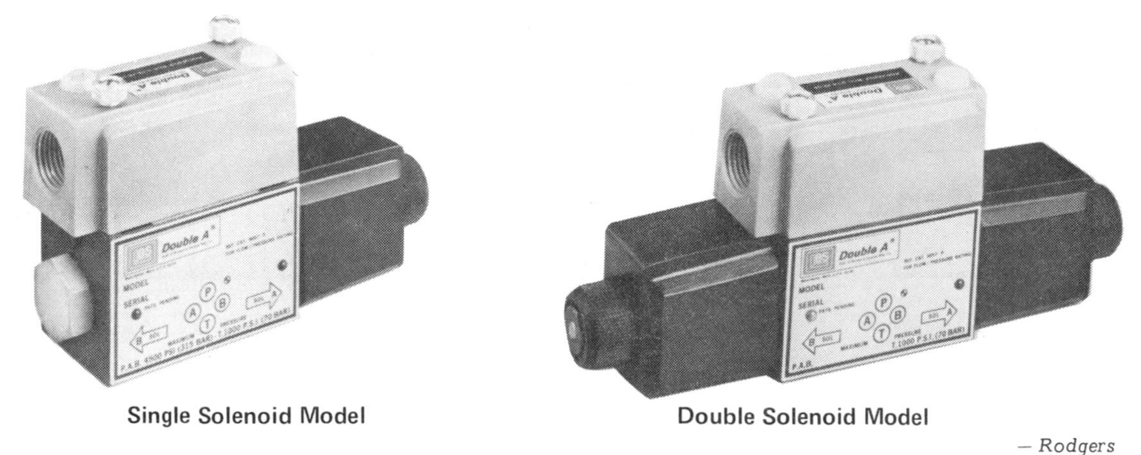

Single Solenoid Model Double Solenoid Model

– Rodgers

FIGURE 4-17. Direct-Acting Solenoid Hydraulic Valves – 4-Way

of the valve spool. A hydraulic model is shown in Figure 4-18, with standard 4-way action, and with the two exhausts (one on each end of the spool) combined internally and brought out in a single tank port. Double solenoid models are available with spring centered or 2-position action.

Low power electrical solenoids do not develop much force. Direct actuation is possible only on valve spools where the shifting force requirement does not exceed a few pounds. Direct-acting valves are, therefore, limited to low flow or low pressure applications. Few models are presently available in larger than 1/4 to 1/2'' size except for vacuum service. Some limitations on direct solenoid action are:

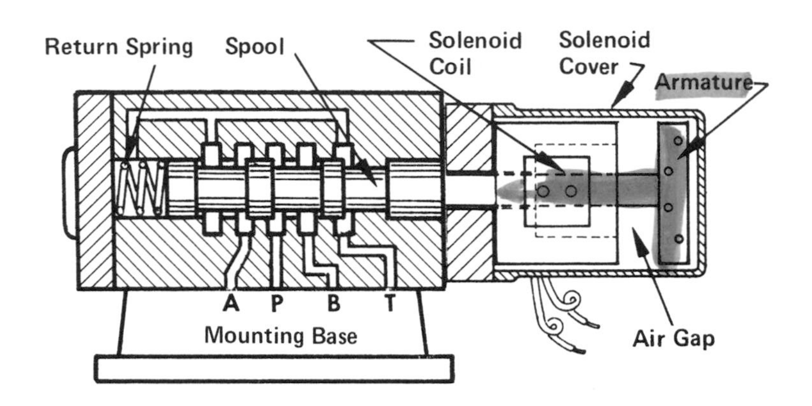

FIGURE 4-18. A direct-acting type of solenoid valve.

(1). An ''air gap'' solenoid has a gap in the magnetic core structure which closes when the armature seats. When first energized and until the moving armature can close this gap, the inrush current may be five to ten times normal steady state current. Any limit switches used in the solenoid circuit must be rated for this inrush current. Standard limit switches usually have contacts rated at 10 amperes at 115 volts, 60 Hz. This limits the size and power of the solenoid to a starting force of 5 to 10 pounds, increasing to 15 to 20 pounds when the gap is nearly closed. This is insufficient force to reliably shift the spool of a large valve.

(2). A solenoid should be allowed to close rapidly, to reduce current flow. Large solenoids tend to be self-destructive because of the high energy in the closing impact. Their life expectancy is usually less than that of smaller solenoids. A high impact also generates a high noise level near the solenoid.

(3). If two solenoids are coupled to opposite ends of the same valve spool, only one of them at a time can seat. If both solenoids were to be energized at the same time, the one which cannot seat will burn out in a short time from excessively high inrush current. In fact, even a momentary overlap in actuation where current is on both solenoids will burn out one of them in a matter of time.

(4). Because a solenoid armature should be allowed to seat as quickly as possible to reduce heating, the spool speed on a direct-acting solenoid hydraulic valve should not be reduced for the purpose of dampening shock. On pilot-operated solenoid valves, however, spool shifting speed may be controlled.

However, direct-acting solenoid valves do have several advantages over the pilot-operated type: They are simpler to use, usually cost less, and will work better on low pressure and vacuum. They do not require a source of pilot pressure for shifting. They usually have a faster shifting response, so are preferred on applications which require a high cycle rate or extremely fast response.

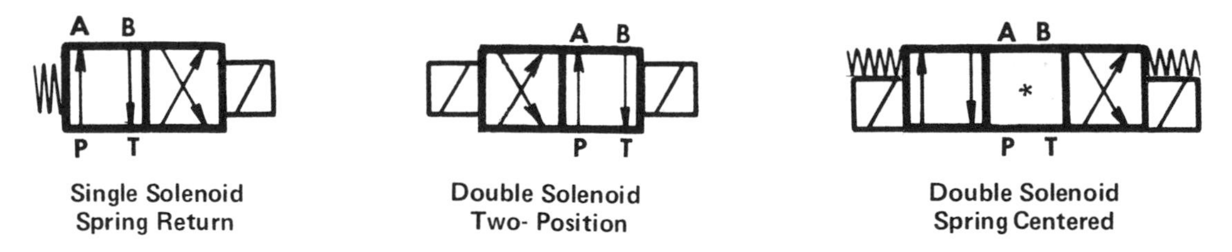

*Spool center is optional. Refer to Pages 142 to 145 for common spool center porting choices.

FIGURE 4-19. Graphic Symbols for Direct-Acting Solenoid Valves.

PILOT-OPERATED SOLENOID VALVES – 4-WAY

A "pilot-operated" solenoid valve is one in which the main spool is shifted, not directly by the solenoid itself, but by fluid pressure (air or oil) tapped from the inlet port. The solenoid opens and closes a miniature control valve orifice to apply fluid pressure to the end of the main spool. Thus, a miniature solenoid can control a very large main spool. Although shifting pressure is usually taken from the inlet port through an internal passage, it can also be brought in from an outside source through a port provided in the valve base. But in either case it is controlled and applied through miniature orifices opened and closed by the solenoid.

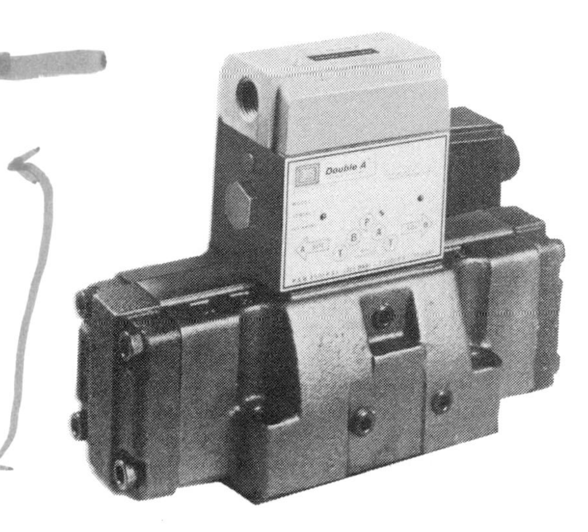

Single Solenoid 4-Way Hydraulic Valve, Pilot-Operated Type.

Single Solenoid Model. The general appearance of a hydraulic model is shown here. A typical pilot-operated air valve is pictured on Page 137. A schematic layout is shown below for an air valve. A hydraulic model works on the same principle but the main exhaust port, EA, and the pilot valve exhaust must be connected to the system reservoir. See Figure 4-21 for a hydraulic model.

The valve is built in two sections. The first section is the pilot section and includes a direct-acting 3-way or 4-way valve of miniature size, operated by a solenoid, and mounted "piggy-back" on top of a large main body which is sized for the rated flow. A mounting pad is machined on the top surface of the main body. The pilot section is bolted to this pad, with connections to the main body through matching holes sealed with O-rings.

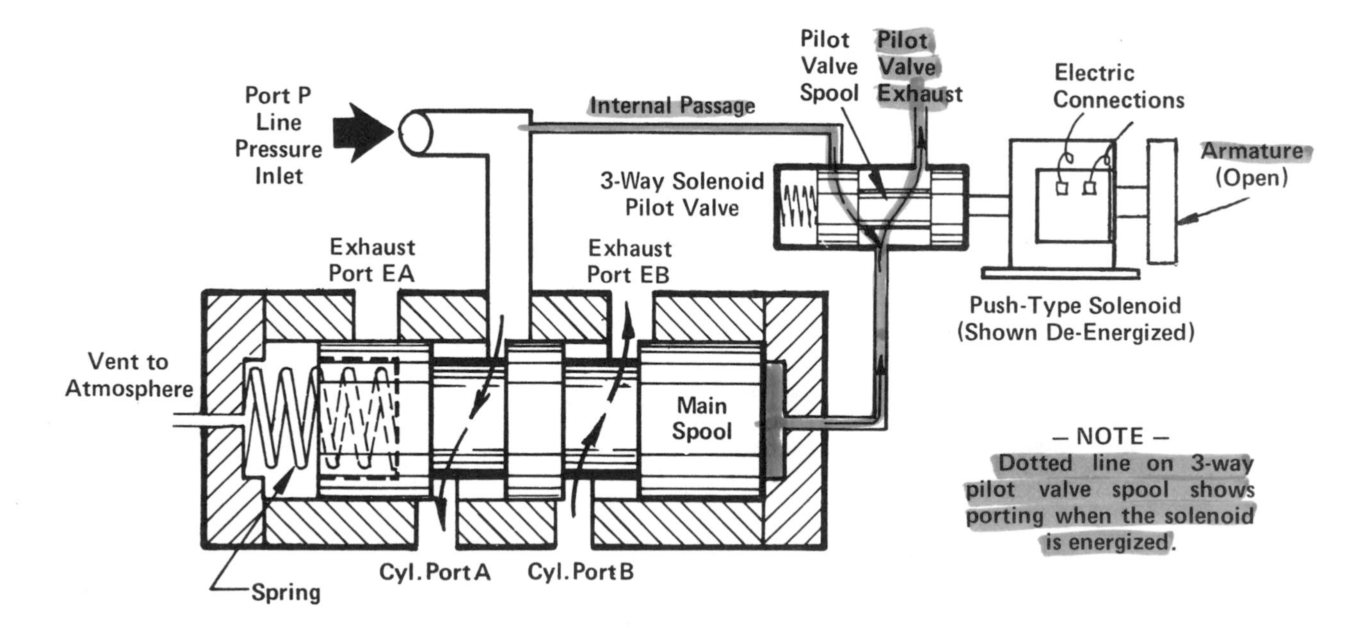

FIGURE 4-20. Schematic layout showing working parts of a pilot-operated single solenoid 4-way valve.

Working Principle. Figure 4-20. Working parts of this valve are laid out in schematic form for clarity of illustration. Working principle shown also applies to hydraulic valves.

The main body and spool are sized for handling the rated valve flow. In an air valve the main spool is shifted by air pressure obtained from the valve inlet through an internal passage. A hydraulic valve is usually shifted by hydraulic pressure. The spring end of the main spool must be vented to atmosphere (or to tank in the case of a hydraulic valve).

The solenoid opens and closes a miniature 3-way valve which, when open, will divert inlet air pressure to the right end of the main spool. This shifts the main spool to its left side position, and it will remain there as long as the solenoid is kept energized. When the solenoid is de-energized, inlet air is blocked at the pilot spool and the right end of the main spool becomes vented to atmosphere. The spool then returns to its "normal" position at the right side of its bore. Note: Energizing the solenoid will not cause the main spool to shift unless there is at least 50 PSI on the main inlet port.

Main spool shifting speed can be reduced by installing a flow control valve in the atmospheric vent on the left end cap. A flow valve in the pilot line to the right end cap will reduce spool return speed.

A minimum pressure must be available on inlet Port P for piloting. On air valves this must be 20 to 50 PSI, and on hydraulic valves 50 to 100 PSI depending on valve brand and size. Pilot-operated valves are not suitable for vacuum unless an external source of pilot pressure is available to use for piloting.

Power required on the solenoids varies from 8 to 10 watts for air valves to 50 to 100 watts for high pressure hydraulic valves. Ordinary limit switches can have a long and service-free life when handling these low powers.

Double Solenoid 4-Way Hydraulic Valve, Pilot-Operated Type, 2-Position and 3-Position Models.

Double Solenoid Valves. Figure 4-21. This hydraulic valve has a 3-position, spring centered, closed center, main spool. The small direct-acting "piggy-back" valve mounted on the main body controls shifting of the main spool. This pilot valve has a float center spool. With both solenoids de-energized, both ends of the main spool are vented to tank and centering springs bring the main spool to center neutral. Pilot pressure to both ends of the main spool is blocked when the pilot spool is centered.

Since both solenoids are coupled to the same pilot spool, energizing both solenoids at the same time would cause one armature to seat and the other armature would be held open. Its solenoid coil would burn out in a short time because of the high inrush current. When designing electrical circuits it is important that both coils of a double solenoid valve should never be energized at the same time.

Working Principle. With both solenoids de-energized in Figure 4-21, pilot spool and main spool are both in center neutral position. Energizing the solenoid on the right causes the pilot spool to move left, opening a passage for high pressure oil from main inlet Port P to flow across the pilot spool and into the right end of the main spool. The spool will shift left since its left end is vented to tank (or to atmosphere). It will remain there as long as the solenoid is kept energized. When both solenoids are de-energized, the centering springs will return both spools to their normal center positions. Note: The main spool will not shift out of center (when the solenoid is energized) unless there is at least 50 PSI available for pilot pressure. See Page 146 for external pilot pressure and drain ports.

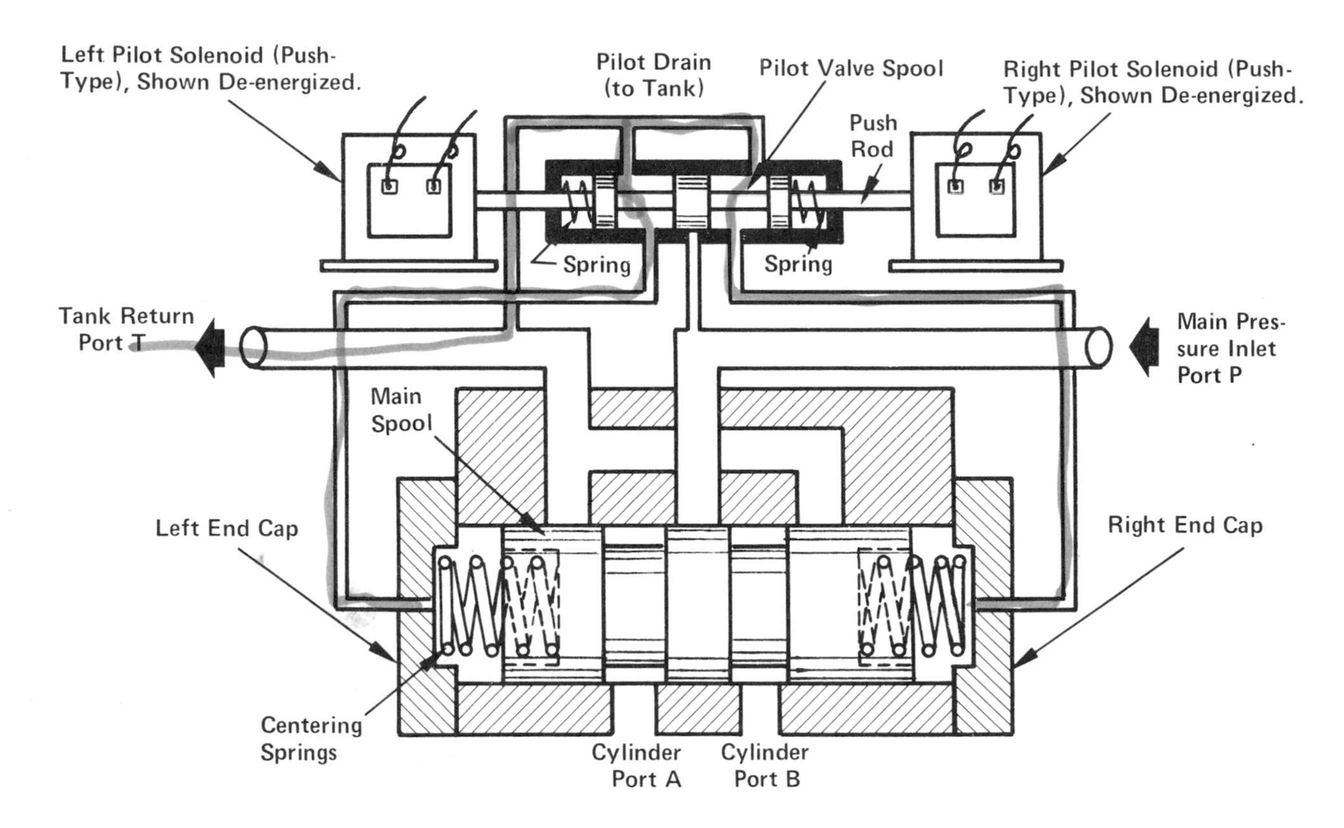

FIGURE 4-21. Double Solenoid, 3-Position, 4-Way Hydraulic Valve.

Two-Position, Double Solenoid, 4-Way Hydraulic Valves. Removing the centering springs from the main spool of the valve in Figure 4-21 will eliminate the center neutral position. Energizing one solenoid will cause the main spool to shift to the opposite side position, and it will remain in that position even after the solenoid has been de-energized, until the opposite solenoid is energized. The same "piggy-back" 3-position, float center pilot spool can be used as in the 3-position model of Figure 4-21.

To prevent the main spool from drifting out of position due to machine vibration or because of excessive flow, the main spool should be detented, and the valve mounted with spool horizontal.

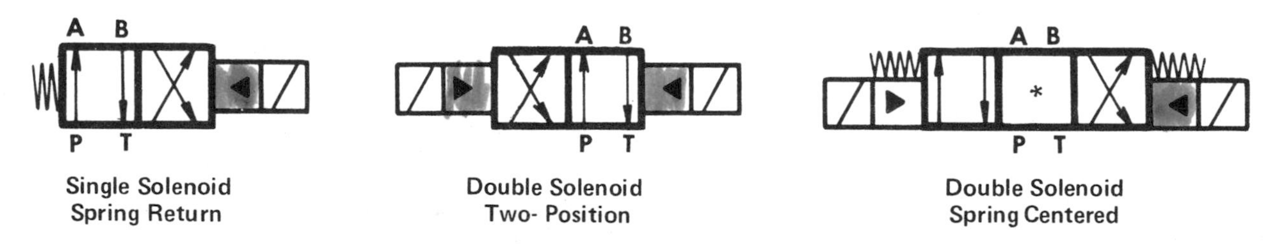

FIGURE 4-22. Graphic Symbols for Pilot-Operated Solenoid 4-Way Valves.

Graphic Symbols for Pilot-Operated Solenoid Valves. Figure 4-22. The main spool will not shift unless electric current is applied to a solenoid *and* at least 50 to 100 PSI fluid pressure is present for shifting power. Therefore, the symbol is drawn with both electric and pressure symbols as actuators, stacked one on top of the other. An open triangle indicates air piloting; a solid triangle indicates hydraulic piloting. This is logic symbolism for "AND" action.

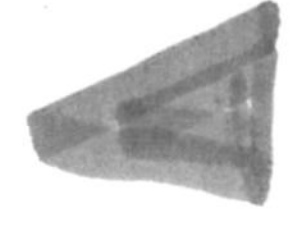

*Spool center is optional. Refer to Pages 142 to 145 for common spool center porting choices.

TRIANGLE Shows Air Pilot  

Hydraulic Pilot

DESIGNING COMPRESSED AIR CIRCUITS

Processing the Compressed Air. In most industrial plants raw air from the compressor and receiver tank is piped to those areas where air-operated machines are used. It is piped through a network of distribution plumbing. At the entry point to each machine, this "raw" air should be filtered and lubricated before being used in valves, cylinders, or air tools, to remove water and solid contaminants. Its pressure level should be reduced at each machine for the minimum level satisfactory for that machine. Every air circuit shown in this book, in fact all air circuits, should include a piece of air processing equipment sometimes called a "trio", although to keep the diagrams as simple and basic as possible, the trio assembly may not be shown on the diagram. A "trio" is a 3-part assembly including filter, pressure regulator, and lubricator, and usually with a pressure gauge. Each unit of this trio is described in Chapter 6.

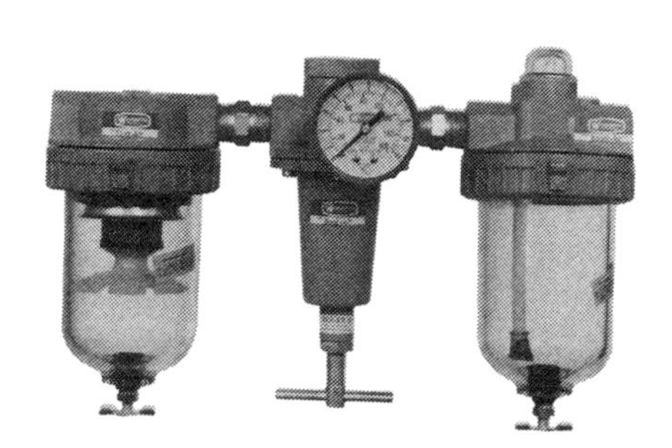

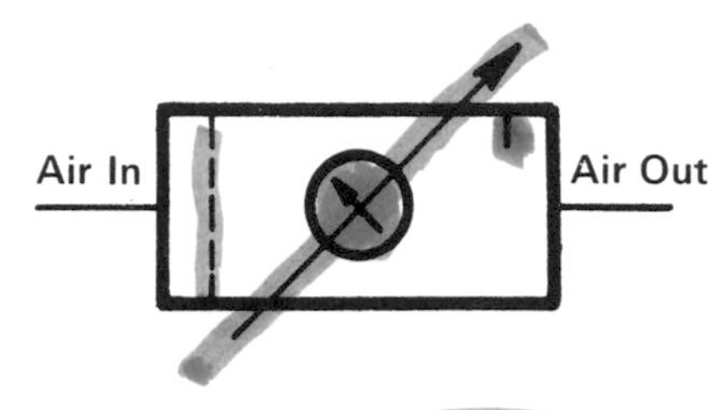

FIGURE 4-23. Simplified Symbol of a Trio Assembly for Use on Circuit diagrams.

Graphic Symbol for a Trio Assembly. Figure 4-23. If a trio assembly is included on a circuit diagram, the simplified symbol shown above is quick and easy to draw. Starting at the left, the vertical dash line represents the filter, the long slash arrow the adjustable regulator, the small circle a pressure gauge, and the short vertical line represents the lubricator. Air must be introduced first to the filter.

FIGURE 4-24. Exhaust Port Needle Valve Speed Control.

Speed Control of Air Cylinders. Control or limiting of speed in ***both*** directions is necessary in most air systems because air is coming from an almost limitless supply, and the flow resistance of the feed lines is the only factor which would limit maximum speed. An individual speed control valve must be provided for each direction because the load is usually quite different, and speed is directly affected by the load. Refer to Pages 61 to 66 for specific information on speed control methods. In most cases the use of a dual exhaust 4-way valve with a screw-in needle valve in each exhaust port is the most satisfactory as well as the cheapest type of bi-directional speed control.

Mid-Stroke Stopping of an Air Cylinder. Whenever possible the machine should be designed to avoid having to stop the piston anywhere along its stroke. It should be allowed to come against a positive stop at each end of its stroke. This stop may be the cylinder end caps or may be external stops. The cylinder should be allowed to stall against these stops with full air pressure remaining behind its piston. While stalled it will not consume air. It will not waste energy nor generate heat. Keeping it stalled under pressure holds it firmly in a fixed position so it cannot drift. For this kind of operation a center neutral or stop position on the 4-way control valve is not needed.

Stopping the piston of an air cylinder at a precise position along its stroke is very difficult. When the air supply is cut off, the piston will continue to travel for a short distance until a balance is achieved between the force of compressed air behind the piston and the reactive resistance of the load. The amount of overtravel is unpredictable, depending on type of load, travel speed, cylinder bore and stroke, size and length of connecting lines between 4-way valve and cylinder, and possibly other factors. Then, even after it does come to a stop it will be in an unstable condition. If the reactive load against the piston should increase or decrease, the piston would start up by itself seeking a new balance position. This could be a personnel hazard.

Solenoid Valve Control. For electrical control of an air cylinder in simple circuits, the preferred valve for most circuits is the double solenoid, 2-position model. It often eliminates the use of a holding relay in the electrical circuit which is usually required with a single solenoid valve.

FIGURE 4-25. These 3-position spools are available for most 4-way air valves.

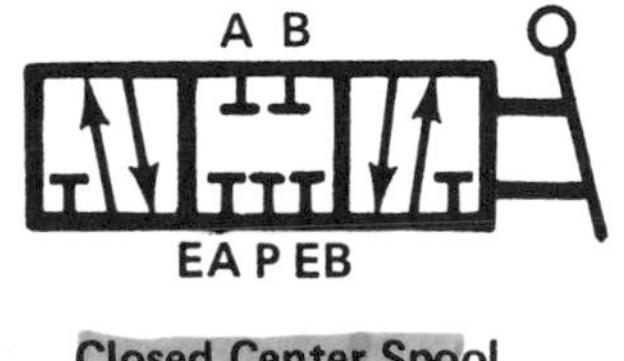

Closed Center Spool
With Dual Exhaust

Float Center Spool
With Dual Exhaust

Three-Position Spools for Air Valves. Figure 4-25. Although a center neutral position on a 4-way valve is seldom used for operating air cylinders, a neutral position is usually required when operating air motors. One of these spool types can be used, a full closed center or a float center. An air motor should not ordinarily be stalled and held against a positive stop because internal leakage would consume a sizeable amount of air resulting in a waste of energy.

Choosing Valve and Plumbing Size. A rule-of-thumb for valve sizing is to select one with the same port size as the port on the cylinder to be controlled. Cylinders up through 3" bore have 1/4" pipe ports, so a 1/4" 4-way valve is usually as large as needed even for cylinders larger than 3" bore. If several cylinders receive air through the same valve, then a larger valve size may be needed to get sufficient speed. Plumbing size is selected the same way. The inside area of 1/4" Schedule 40 pipe will handle cylinders at least up to 3" bore. Of course the use of larger valves and plumbing may be beneficial by reducing pressure losses.

Air Line Shut-Off Valve. Figure 4-26. A shut-off and vent valve should be installed in the air feeder line to each air-operated machine, just ahead of the trio assembly. Refer to Figure 3-29, Page 111 for details.

The valve used must not only shut off the air supply but must vent the pressure trapped in the machine. If a 3-way valve of the right size is not readily available, a 4-way standard dual exhaust air valve may be used by plugging the ports which will not be used, the A and EA ports in this example.

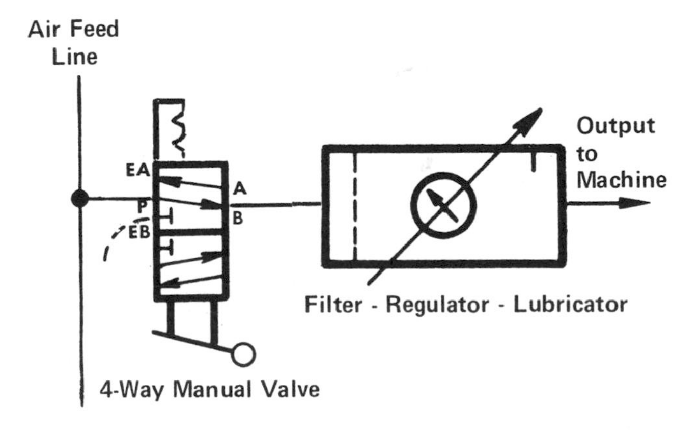

FIGURE 4-26. 4-Way Shut-Off and Vent Valve.

AIR CYLINDER CONTROL WITH 4-WAY VALVES

Compressed air for operation of these circuits should be obtained from the shop air distribution system through a 3-way shut-off valve and trio assembly although, for simplicity, these items have been omitted from the diagrams. The pressure regulator in the trio assembly should be adjusted to the lowest pressure which will produce the required force on the cylinder piston.

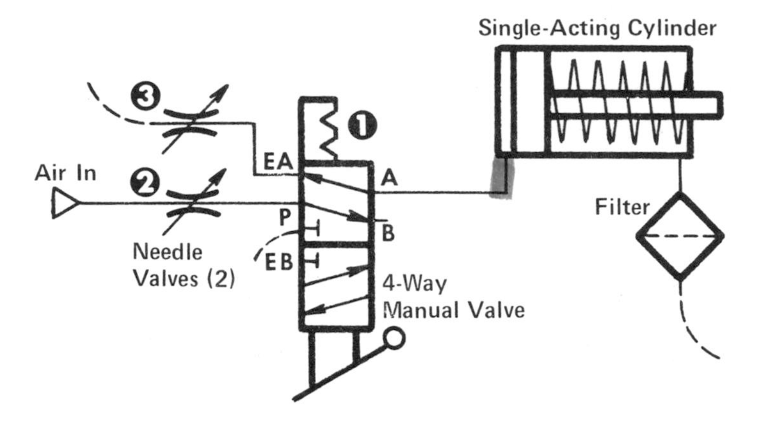

FIGURE 4-27. Manual Lever Control of Single-Acting Cylinder.

Air Vise or Clamp. Figure 4-27.

A single-acting cylinder, with spring return, is controlled with a manual lever, dual exhaust 4-way valve, 1. The valve spool has a 2-position detent to hold it in a shifted position. A spring returned valve might be more appropriate for some applications.

Cylinder extension (or vise closing) speed can be controlled with needle Valve 2 placed in the air inlet. If it should be necessary to limit retraction speed this can be done with needle Valve 3 placed in exhaust Port EA. Exhaust Port EB and cylinder Port B are not used for this application and should be plugged to keep dirt out of the valve.

If possible, a filter should be installed in the cylinder rod end port to keep dirt from being sucked in as the cylinder retracts. This port should never be plugged. Some single-acting cylinders simply have an unthreaded vent hole on the rod end; they will not accept a filter connection.

Instead of the 4-way valve, 1, a 3-way control valve will give the same action described. Also, a treadle valve can be used in place of the manual lever valve.

Instead of a single-acting cylinder, a standard double-acting cylinder can be used. Connect it as shown in some of the following diagrams.

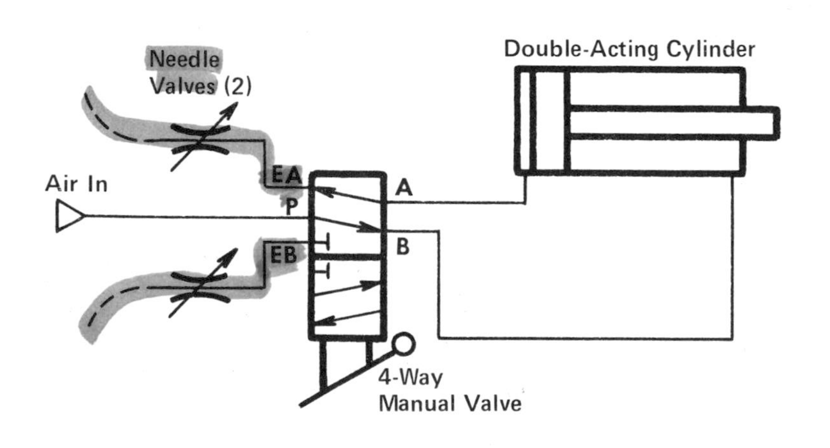

FIGURE 4-28. Manual Lever Control of Double-Acting Cylinder.

Manual Lever Control. Figure 4-28.

This is the basic circuit for control of a double-acting air cylinder. A dual exhaust, manual lever valve is shown. A pedal, treadle, stem, knob, or solenoid actuated valve can be substituted. Friction positioning of the spool is shown. For pedal or solenoid actuation a spring return valve should be used.

Screw-in needle valves in valve exhaust ports give adjustable speed control in both directions of movement. If desired, two flow control valves, installed in lines connecting valve to cylinder, could also be used. Screw-in mufflers can be installed in the exhaust lines to reduce air hiss.

Button Bleed Control. Figure 4-29. A button bleed 4-way valve of suitable size for the application is used for this circuit. Two bleed buttons are furnished, screwed into the main valve end caps. Fingertip actuation of these buttons will shift the 4-way valve, or the buttons can be removed and installed remotely, connected to the valve end caps with short lengths of hose, and used as pushbuttons. Button 1 causes the cylinder to extend; Button 2 causes it to retract.

Note: The action of a button bleed valve, as far as the schematic is concerned, is that the valve blocks move *away* from the button which has been actuated. Physically, this is opposite to the way the valve works, since actuation of a bleed button causes the spool to move *toward* the end which has been bled.

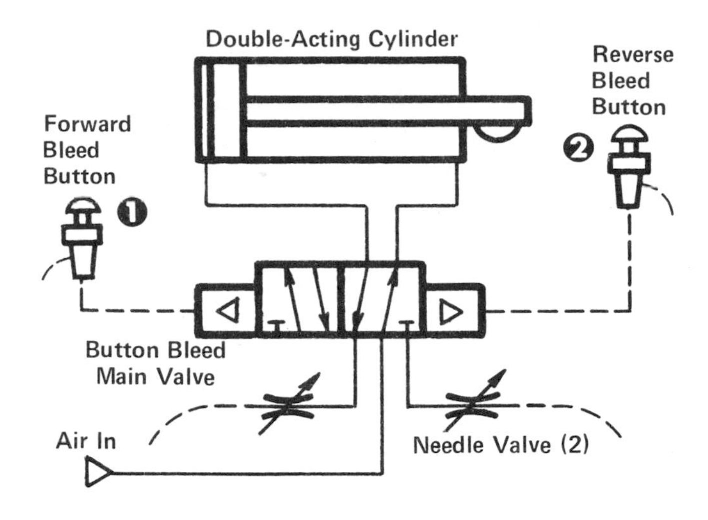

FIGURE 4-29. Button Bleed Air Cylinder Control.

Spring return or 3-position spring centered button bleed valves may not operate reliably. We recommend using only the double bleed, no-spring models.

For speed control, screw-in needle valves installed in the dual exhaust ports are shown. If preferred, these needle valves may be replaced with a pair of flow control valves installed in the cylinder lines.

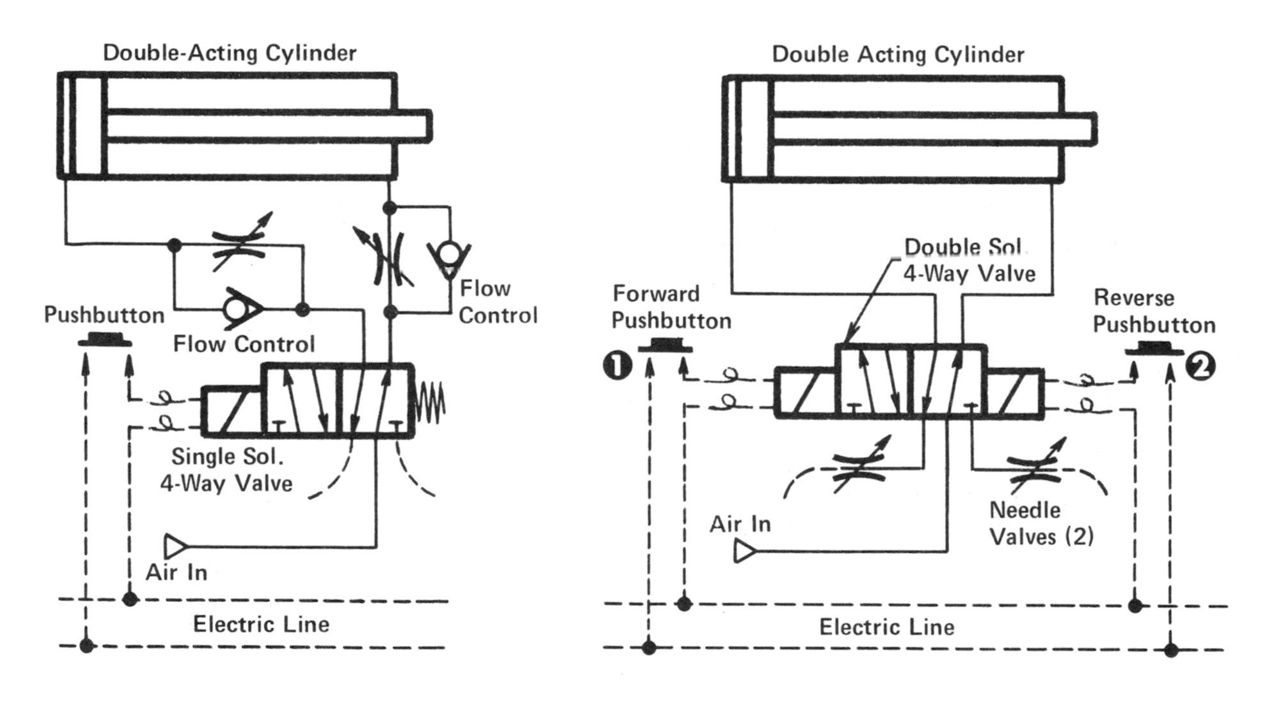

PART A. Single Solenoid Valve Control.

PART B. Double Solenoid Valve Control.

FIGURE 4-30. Double-Acting Air Cylinder Controlled With Solenoid 4-Way Valves.

Solenoid Valve Control. Figure 4-30. Most air circuits use one of two types of solenoid valves: the single solenoid valve (Part A) with spring returned spool, and the double solenoid valve (Part B), which does not have a return spring or centering springs. Choice between these types will depend on the kind of electrical starting signal which will be supplied. The single solenoid valve must have current maintained on its coil to keep the spool in a shifted position. The double solenoid valve will shift on a momentary electrical signal (½ second or less) and its spool will remain in the position to which it was last shifted without maintaining current on either coil.

Valves may be either direct-acting or pilot-operated models. Caution! Do not use air pressure higher than the valve rating. On some air valves excessive pressure will cause the spool to shift unexpectedly without current being applied to either coil. This could cause personnel injury.

In Part A of Figure 4-30, a single solenoid valve is shown. Its solenoid is energized from an electrical source through a normally open (N.O.) pushbutton. The "fail-safe" condition of this circuit is with the cylinder retracted when the valve solenoid is de-energized. If electric current should fail, the cylinder would retract. By interchanging the connections between valve and cylinder, the cylinder would go to its full extension if electric current should fail. On the diagram the electric circuit is shown in dash lines.

Pressing the pushbutton will shift the solenoid valve and the cylinder piston will extend. It will continue to extend and will hold in a stalled condition at the end of its stroke as long as the solenoid coil is kept energized. Releasing the pushbutton will de-energize the valve and the cylinder will retract.

Speed control is by means of two flow control valves placed in connecting lines between valve and cylinder in a meter-out mode. Exhaust needle valve speed control, as shown in Part B could be used with the same results.

In Part B of Figure 4-30, a double solenoid valve is shown. As connected, when Pushbutton 1 is pressed, the 4-way valve spool will shift and the cylinder piston will extend. It will remain stalled at the forward end of its stroke even though the pushbutton is released. To retract the piston, Pushbutton 2 must be pressed momentarily. Dash lines on the diagram show the electrical circuit.

A 3-position 4-way valve can be used if necessary to stop the piston along its stroke, although this is rarely necessary or desirable on most air operated machines.

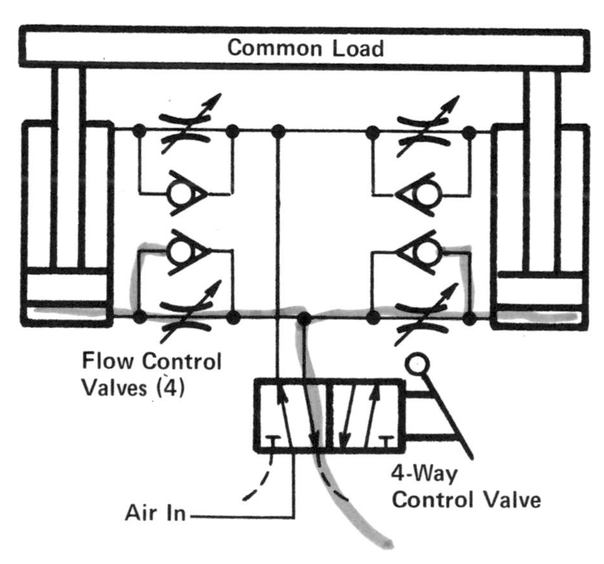

FIGURE 4-31. Synchronizing Two Air Cylinders.

Several Cylinders Moving in Synchronization. Figure 4-31. Two or more air cylinders attached to a common load will tend to travel at different speeds. The cylinder with the larger share of the load will travel slower. To get them to move together, flow control valves must be installed in each cylinder line. They must be experimentally adjusted until the cylinders move together in both directions. From time to time slight adjustments will have to be made on the flow control valves.

If the proportion of load on each cylinder should change, all control valves will have to be readjusted. Whenever possible, on a platform lift for example, it is better to use one larger cylinder in the center rather than a smaller cylinder at each corner. This eliminates the synchronizing problem. A synchronizing circuit for air cylinders is shown in the "Air-Over-Oil" chapter of Volume 2 textbook.

AUTOMATIC RETRACTION OF AIR CYLINDERS

Circuits shown under this heading will cause the cylinder to retract automatically at the end of the forward stroke without additional action by the operator, after the cylinder has been set in motion by a very brief signal from the operator or from another machine or process.

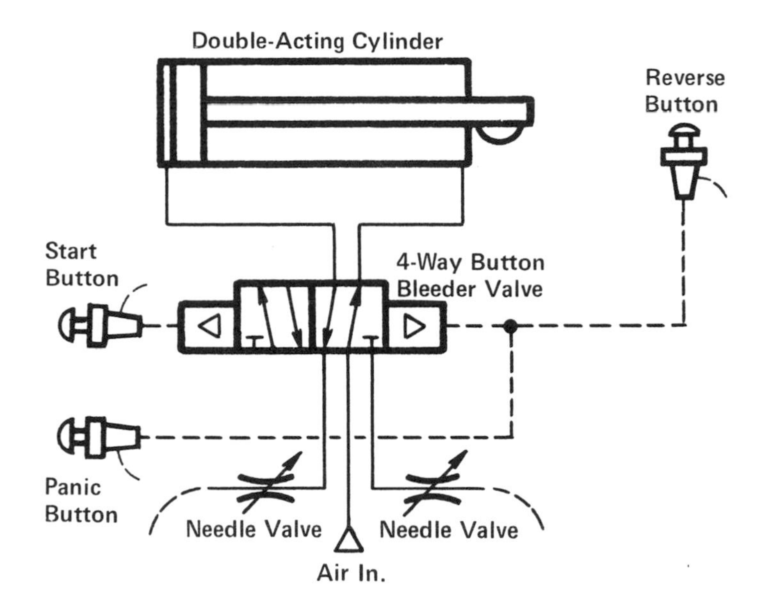

FIGURE 4-32. Button Bleed Automatic Retraction.

Bleed Button Automatic Retraction. Figure 4-32. The main valve is a 4-way button bleed model of suitable size for the application. It is shown as a dual exhaust model with needle valves installed in both exhaust ports for cylinder speed control. Alternatively, two flow control valves can be installed in the cylinder lines, and air mufflers installed in the valve exhaust ports.

The two bleed buttons which are usually supplied with a 4-way button bleed main valve can be unscrewed from the valve end caps. One of them, Button 1, can be remotely mounted as a start pushbutton. The other one, Button 2. can be removed and mounted where the cylinder or a moving part of the machine will actuate it at the end of cylinder travel. A third one, Button 3, must be purchased separately. It can be mounted near the operator as a "panic" button to be used in an emergency if he must quickly retract the cylinder before it completes its forward stroke. It is connected in parallel with retract Button 2, and does the same thing.

To operate the cylinder, the operator momentarily presses bleed Button 1. This shifts the 4-way valve and the cylinder starts forward. It continues to travel even though the start button has been released. At the end of its forward stroke, bleed Button 2 is actuated. This returns the 4-way valve to its original position, causing the cylinder to retract to home position, stop, and remain stalled. The operator can then initiate another cycle.

Solenoid Valve Automatic Retraction. Figure 4-33. The main valve is a 4-way, double solenoid, 2-position model of suitable size for the application. It has no return or centering springs.

The cycle starts when the operator momentarily presses Pushbutton 1. This energizes the left solenoid of the 4-way valve. The valve spool shifts and the cylinder starts its extension. Forward speed can be adjusted with needle Valve 3. At the forward end of the stroke the cylinder or a moving member of the machine actuates the limit switch. This causes the right solenoid of the 4-way valve to become energized. The valve spool returns to its original position and the cylinder retracts and stalls at home position until the operator initiates the next cycle. Retraction speed can be adjusted with needle Valve 4.

A "panic" pushbutton should be placed near the operator so he can quickly retract the cylinder at any time during its forward stroke. The pushbutton is wired in parallel with the limit switch and gives the same action as if the limit switch had been actuated.

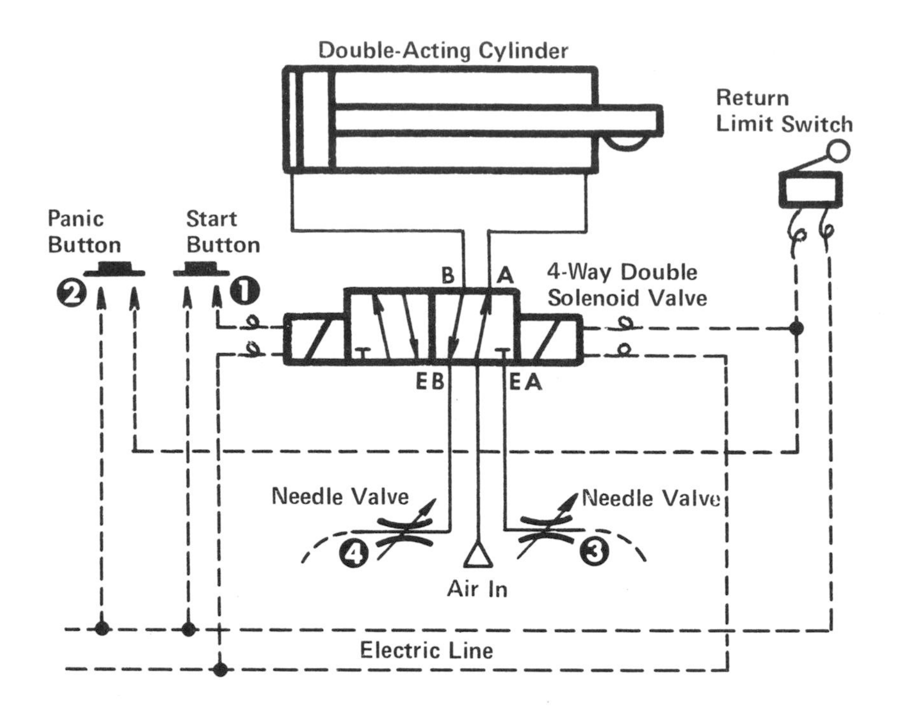

FIGURE 4-33. Solenoid Valve Automatic Retraction.

Note: If the limit switch is placed so it will reverse the cylinder before the end of the complete forward stroke, be sure to mount it so it will not be damaged if the cylinder should accidentally overrun.

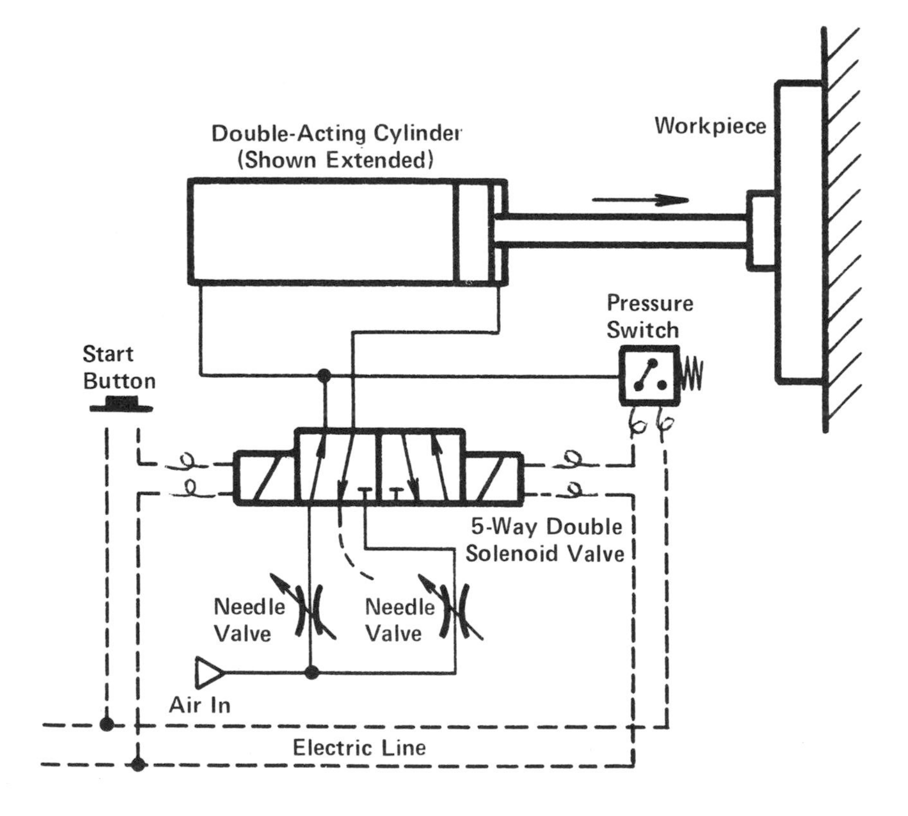

FIGURE 4-34. Pressure Switch Automatic Retraction.

Pressure Switch Automatic Retraction. Figure 4-34. A double solenoid 5-way valve of suitable capacity for the application is used. It has two working positions (no return or centering springs). The return actuator is a pressure switch, which unlike a bleed button or a limit switch, can cause retraction to start at any point during the forward stroke if the cylinder stalls. This circuit is useful if workpieces of varying thickness are being handled, where a limit switch or bleed button would not work. The pressure switch should be adjusted to "make" (close its contacts) at a pressure just under the pressure level supplied to the circuit. For example, on a supply pressure of 100 PSI, the switch should be adjusted to close contacts when pressure reaches 90 to 95 PSI.

The operator starts the cycle by pressing the pushbutton. This energizes the right solenoid and starts the cylinder forward. The cylinder continues to travel until it stalls against the workpiece, at whatever point in its stroke this may occur. Pressure then builds up behind the cylinder piston. When it reaches the pressure switch setting, the switch trips, the left solenoid becomes energized, and the piston retracts.

A 5-way instead of a 4-way valve is used so that meter-in speed control can be used by placing two needle valves in the two pressure inlets of the 4-way valve. Exhaust needle valves and meter-out flow control valves do not work well with a pressure switch because a relatively high pressure exists behind the piston throughout its forward stroke and the pressure switch might trip prematurely. Meter-in speed control with a 5-way valve is more reliable in this case.

CONTINUOUS RECIPROCATION OF AIR CYLINDERS

Circuits shown under this heading will cause an air cylinder to reciprocate continuously back and forth between forward and reverse limit points which may be the total stroke of the cylinder or could be less than maximum stroke with reversal points determined by the placement of bleed buttons, limit switches, or pressure switches.

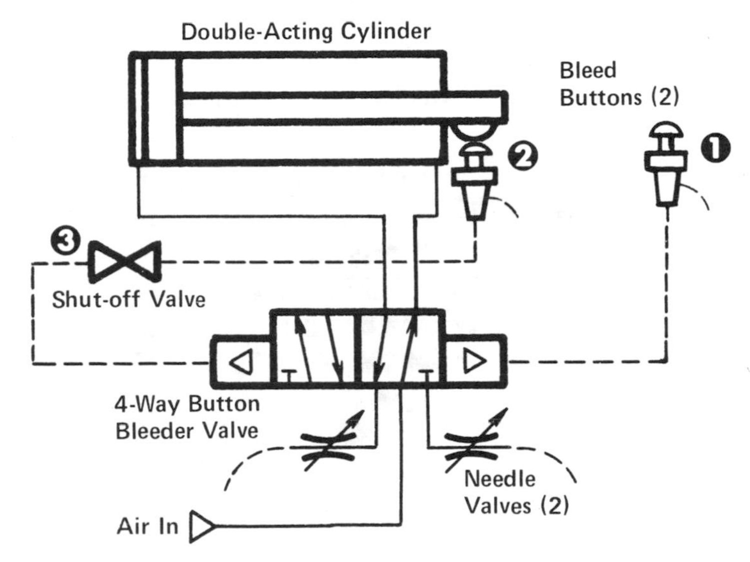

FIGURE 4-35. Reciprocation With Button Bleed Valves.

Bleed Button Reciprocation. Figure 4-35. The 4-way valve used in this circuit is a double bleed, 2-position, 4-way valve of a suitable size for the application. Two bleed buttons are furnished as part of the main valve. They must be unscrewed from the main valve end caps and mounted on hose extensions, no longer than 10 feet, at positions where they can be actuated by the cylinder or by a moving member of the machine at each end of the desired stroke.

Valve 3 starts the action. But be- this valve is actuated, the cylinder is at home position, standing on button Valve 2. Opening Valve 3 bleeds the left end of the main valve and starts the cylinder forward. It will continue to travel back and fortth between Buttons 1 and 2 until Valve 3 is closed. Then, it will continue to travel until it reaches home position before it stops.

If Valve 3 is placed in the line to bleed Button 1, the cylinder will stop in a fully extended position.

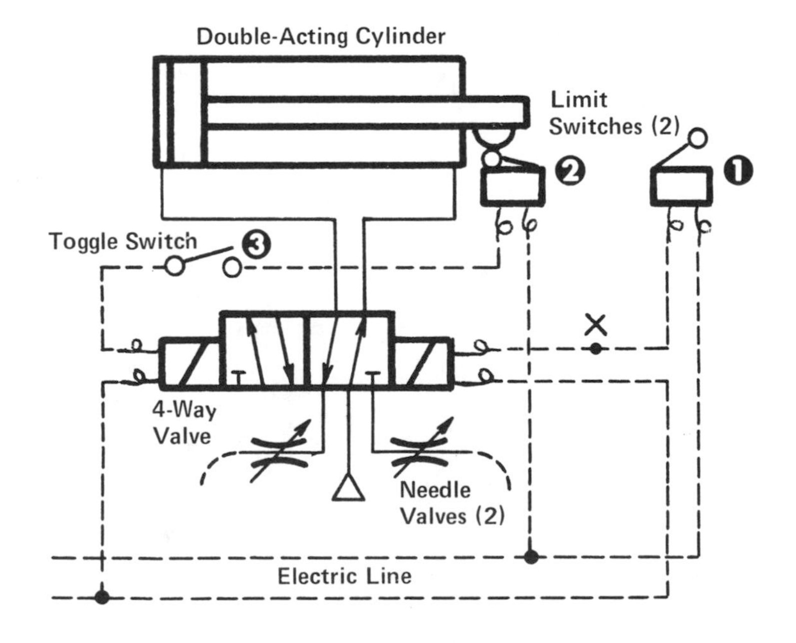

FIGURE 4-36. Reciprocation With Solenoid Valves.

Limit Switch Reciprocation. Figure 4-36. A limit switch is placed at each end of the desired stroke, which can be less than the maximum cylinder stroke if desired.

In Figure 4-34, dash lines show electrical wiring. Switch 3 starts the action. Before this switch is closed, the cylinder is at home position (retracted) and standing on limit Switch 2. Closure of Switch 3 energizes the left solenoid coil of the 4-way valve and starts the cylinder forward. It will travel back and forth between the two limit switches until Switch 3 is opened. The cylinder may not stop immediately; it will continue until it reaches home position. Moving the location of Switch 3 to Point X will cause the cylinder to stop in a fully extended position. Note: If the limit switches are not located at extreme ends of the cylinder stroke, the cylinder may ride past them if electric current fails, if a solenoid coil should burn out, and certainly every time the cylinder is stopped. Mount the limit switches so the cylinder, when it rides past them, will keep them actuated, and so the cylinder can ride past them in the opposite direction without damage.

HYDRAULIC CYLINDER CONTROL WITH DIRECTIONAL VALVES

Those circuits previously shown for air cylinders are usually not well suited for hydraulic cylinders because of basic differences between the two media, especially because air is highly compressible while hydraulic oil is not. For example, air cylinders are usually operated with 2-position valves having no neutral position. The cylinder is brought up against a positive stop and held there under full pressure to prevent drift. Hydraulic cylinders, on the other hand, should be stopped by shifting the control valve to a neutral position. Holding active pressure against them while they are stopped would in most cases waste power and generate heat as the pump oil was forced to discharge across a relief valve. Pressure compensated pumps described in Chapter 5 are exceptions. While the cylinder is stopped, it is important to remove the hydraulic load against the pump. This is called "pump unloading" and is discussed further in Chapter 5. The neutral position of the control valve is often used to "unload" the pump as will be shown in circuits to follow.

When a hydraulic cylinder is stopped, the pump which supplies oil to it is usually allowed to continue running but with the hydraulic load removed. Low horsepower systems may be exceptions to this rule, as on a truck tailgate lift, where the electric motor is switched off to stop the lift cylinder.

Another difference between air and hydraulic systems is the power level. Industrial air systems are designed to operate at pressures up to 100 to 150 PSI while hydraulic systems usually operate at much higher pressures.

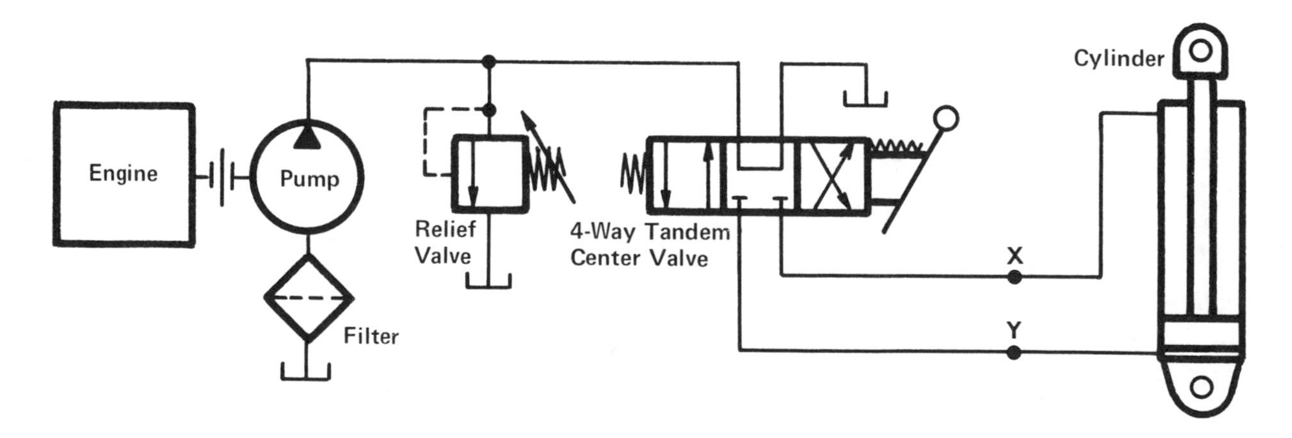

FIGURE 4-37. Manual Control of a Double-Acting Hydraulic Cylinder.

Basic Hydraulic Circuit. Figure 4-37. A double-acting cylinder is controlled with a manual lever, 4-way tandem center valve. This valve provides not only directional control for the cylinder, it also provides speed control and unloading of the pump while the cylinder is stopped. Driving power comes from an engine or electric motor. The pump is one of the positive displacement types described in Chapter 5. A strainer or filter should be used at the pump inlet. A typical filter is a stainless steel 100-mesh type. A relief valve should be teed into the pump outlet line, discharging to tank.

This is a general purpose circuit, useful on many jobs. A typical use is on a log splitter which usually requires a 3 or 3½" bore cylinder operating at 1000 to 1500 PSI pressure. Pump flow capacity depends on the speed required. Refer to horsepower tables on Page 277 for motor HP required according to the flow rating of the pump to be used.

Speed control valves may be unnecessary when a manual lever 4-way valve is used. The valve can be throttled to control speed. However, if desired, a flow control valve can be added at *either* Point X or Point Y. Only one speed control valve should be needed; the pump volume will limit maximum speed in the other direction. If cylinder speed is too fast in both directions, pump shaft speed should be reduced or the pump replaced with one of lower volume. Speed control valves use energy and create heat, thereby reducing circuit efficiency; they should be used only when necessary.

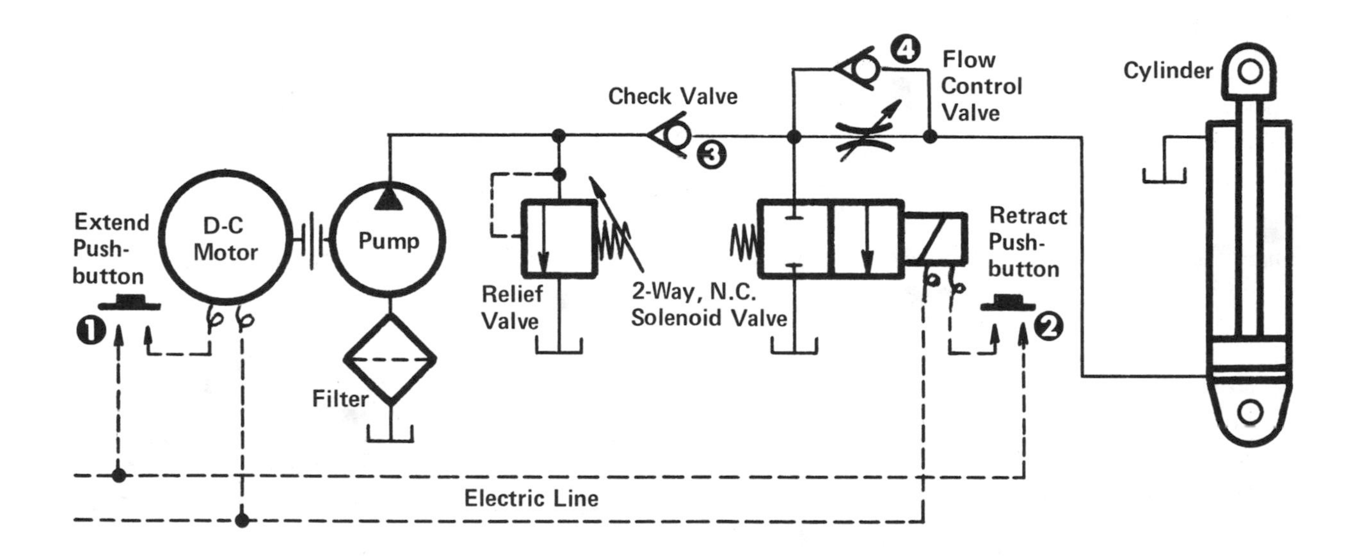

FIGURE 4-38. Tailgate Lift Circuit for a Hydraulic Cylinder.

Tailgate Lift. Figure 4-38. This circuit is designed for a single-acting cylinder with gravity return, typical of a tailgate lift. The hydraulic pump, usually a small gear pump, is driven with a D-C electric motor. A 100-mesh strainer should be used on the pump inlet. Check Valve 3 prevents backflow of oil through the pump while the lift is holding.

To raise the lift cylinder, Pushbutton 1 is pressed to start the electric motor and pump. To lower the cylinder, Pushbutton 2 is pressed to energize the 2-way solenoid valve. Flow control Valve 4 should be adjusted to limit down speed. No speed control is needed for the up speed. If heavy loads must be lowered very slowly, the solenoid valve and the flow control valve should be replaced with a manual valve which has good throttling characteristics.

Note: Starting and stopping an electric motor to control cylinder action should only be used on low power applications where usage is intermittent. Frequent starting of a large motor will cause it to overheat because of high inrush current and will cause possible damage.

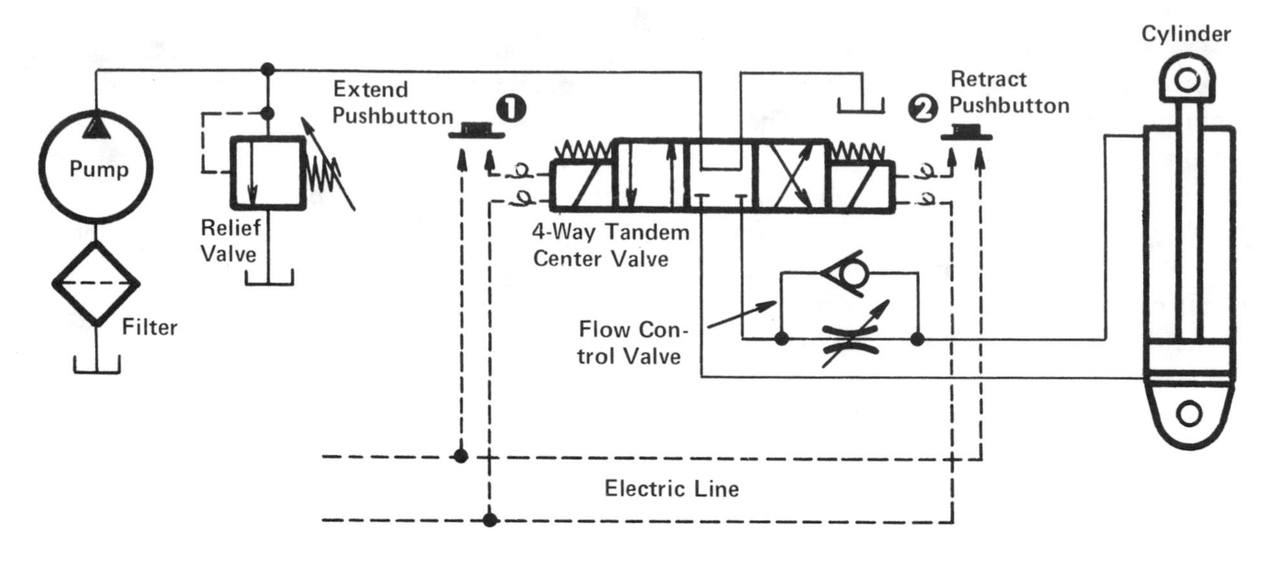

FIGURE 4-39. Solenoid Valve Control of a Hydraulic Cylinder.

Basic Electrical Control. Figure 4-39. For controlling one double-acting hydraulic cylinder on systems of less than 50 HP, a 4-way double solenoid valve with tandem center, spring centered spool is most popular. On higher power systems a tandem center valve may generate too much shifting shock, and other valving arrangements may be better. These are described in Volume 2.

In this figure, for illustration of basic control, a pushbutton is wired in series with each solenoid coil. Pressing, and holding, Pushbutton 1 will cause the cylinder to extend. Pressing, and holding, Pushbutton 2 will cause it to retract. When neither pushbutton is pressed, the valve spool returns to its center neutral position. In neutral position, oil flow from the pump can pass through the valve and

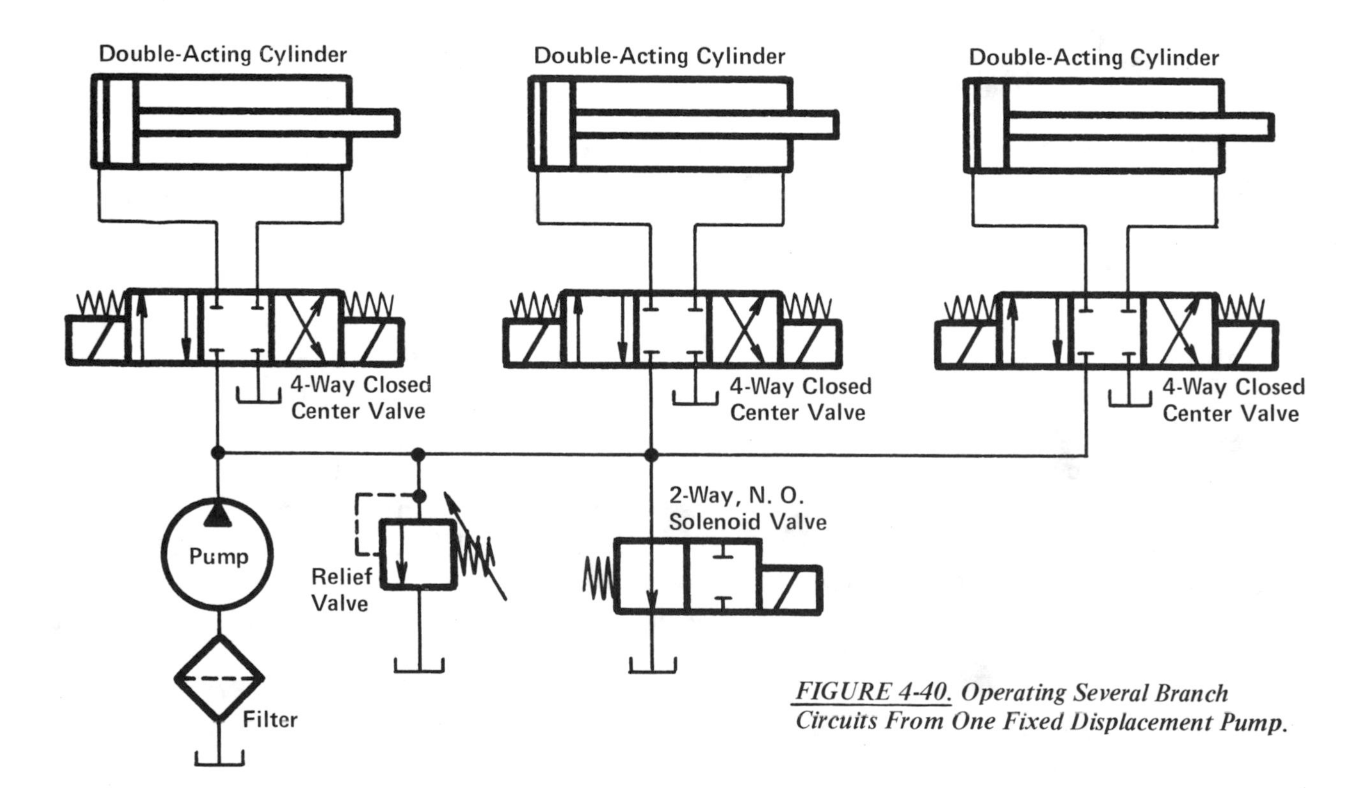

FIGURE 4-40. Operating Several Branch Circuits From One Fixed Displacement Pump.

flow freely to tank. Since the pump is working against very little flow resistance when the valve is in neutral, it is said to be in an "unloaded" condition.

If a hydraulic cylinder is allowed to stall at either end of its stroke, pump pressure will quickly rise and the entire pump flow will discharge across the relief valve to tank at full relief valve pressure setting.

In practice, the electrical circuit will be more involved than in this simple example. One or more limit switches and holding relays may be necessary. For more information on electrical control circuits for solenoid valves, see Volume 2 and also the Womack book "Practical Fluid Power Control".

Several Branch Cylinders on One Pump. Figure 4-40. If two or more branch cylinders must work at the same time, and if full pump pressure must be available to all working cylinders, tandem center valves should not be used because the available pump pressure must divide between the cylinders since tandem center valves place the cylinders in series across the pressure. Branch circuit 4-way valves should have closed inlet ports (when in neutral). This places the cylinders in parallel across the pump pressure.

In Figure 4-40, if all 4-way valves are allowed to center at the same time, this will throw a deadhead load against the pump. Since the pump flow cannot go into any of the branch circuits, and if there is no provision for unloading the pump, it must build up sufficient pressure to discharge the flow across the relief valve to tank. This will convert the full input power into heat, and the entire system will soon be in an overheated state. The 2-way solenoid valve has been added for unloading the pump when its flow is not needed for any of the branch circuits. The electrical circuit must be designed in such a way that when none of the 4-way valve solenoids is energized, the 2-way valve will also be de-energized and the pump flow can pass freely to tank. But if either solenoid of any valve is energized, the 2-way valve solenoid will also be energized. This will block the free flow path of the oil and force it to flow to the valve whose solenoid has been energized. The 2-way solenoid valve is a N. O. (normally open) model and should be able to pass full pump flow at a pressure drop of no more than 25 to 75 PSI.

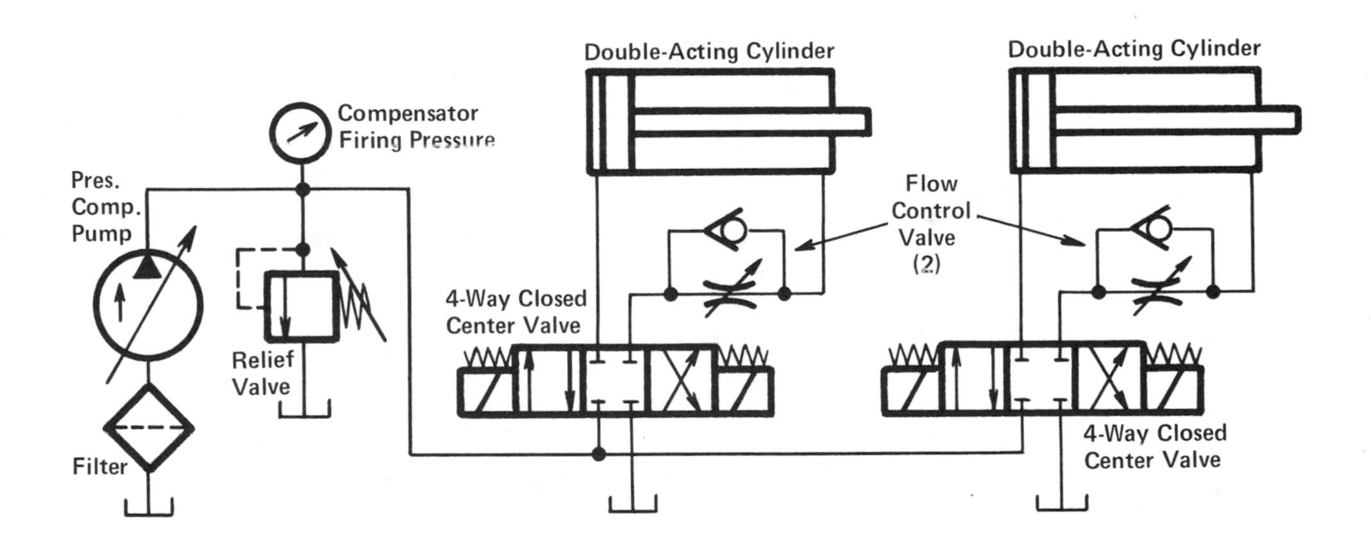

FIGURE 4-41. Pressure Compensated Pump Operation of Several Branch Circuits From One Pump.

Pressure Compensated Pump. Figure 4-41. The use of this type pump is another way of operating several cylinder branch circuits from one pump. The graphic symbol for a pressure compensated pump is a circle with large slash arrow indicating that its displacement (pumping volume) can be varied, and with a small vertical arrow inside the circle showing that its displacement is varied by a built-in pressure compensator mechanism.

The action of this kind of pump is described in Chapter 5. Briefly, it is constructed in such a way that the length of piston stroke or the eccentricity of its cam ring (in the case of a variable vane pump) can be varied to reduce its flow rate if and when pump outlet pressure builds up to the "firing" pressure which has been pre-adjusted on its compensator. The compensator does this automatically without attention from the operator. The compensator normally permits the pump to work at its maximum displacement, and only reduces displacement if pump line pressure rises too high.

The action of the pressure compensated pump in Figure 4-41 can be described as follows. When all 4-way valves have been centered, a deadhead load is placed on the pump. Pressure at the pump outlet port will rise until the compensator "firing" pressure is reached. The compensator will then reduce pumping volume to a lower rate (to zero if necessary) to prevent a further rise in pressure. The pump will continue to run and will maintain high pressure in the pump line but will produce no flow from its outlet. Of course there will be a small internal mechanical and fluid power loss in the pump and the pump will continue to pump only enough flow to satisfy its internal leakage. So the only power consumed from the driving motor will be for pump losses and the only heat produced in the oil will be from pump internal leakage.

To understand the advantages of a pressure compensated pump it is important to remember that even if the pump continues to maintain full high pressure but if it produces no flow, there is no power being produced or delivered.

The 4-way valves must have spools in which the inlet port is closed in neutral position. All valve inlet ports must be connected in parallel to the pump pressure line. Full pressure is always available to all 4-way valves, but if two valves are shifted at the same time the two circuits must divide the pump flow.

CONTINUOUS RECIPROCATION OF A HYDRAULIC CYLINDER

Figure 4-42. Hydraulic applications for cylinders to automatically retract at the forward end of their stroke, or to continuously reciprocate back and forth until stopped by an operator, nearly

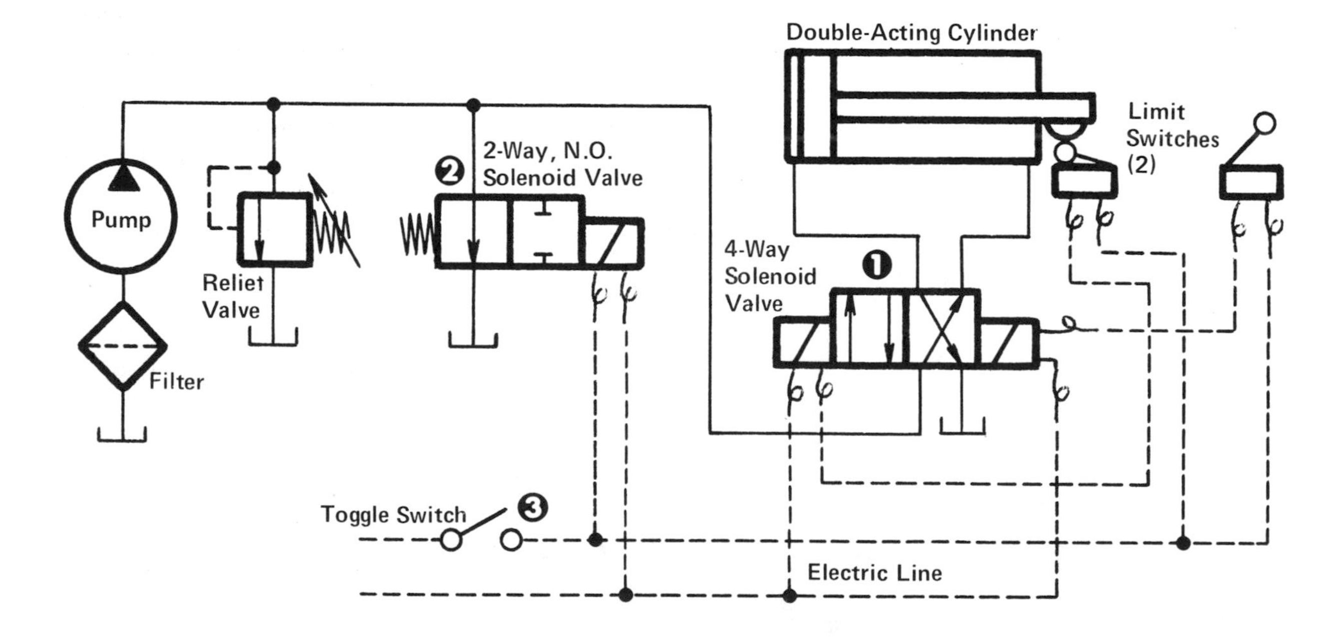

FIGURE 4-42. Automatic Reciprocation of a Hydraulic Cylinder – Electrical Control.

always use solenoid type 4-way valves, with limit switches to define the points of reversal, both forward and reverse, and these points may or may not be at the ends of the full cylinder stroke.

In this diagram a double solenoid valve, 1, of a size large enough to handle pump flow is used. Its spool will shift and will remain in shifted position when the appropriate solenoid coil is momentarily energized. As with most all hydraulic circuits, a means should be included to unload the pump when the cylinder has been stopped, and for this purpose Valve 2 is provided. It is a 2-way, normally open valve of sufficient size to pass the full pump flow to tank at a pressure drop of no higher than 25 PSI.

The circuit operates as follows: Switch 3 starts the action. But before this switch is closed, the pump is running and its full flow is being discharged to tank through Valve 2 at no more than 25 PSI back pressure on the pump line. The cylinder is standing in the position where it was last stopped. This could be in a partly extended position or at one end of its stroke. Closure of Switch 3 energizes Valve 2, blocking the pump discharge and causing pressure to build up in the pump line. Switch 3 may or may not also energize a solenoid on Valve 1, depending on whether the cylinder cam is standing on one of the limit switches. The cylinder starts up and if partly extended moves in the direction it was going when last stopped. It will continue to travel (reciprocate) back and forth between the two limit switches until Switch 3 is opened. It will immediately stop at whatever position it may be in, as pressure is lost in the pump line by the opening of Valve 2. Switch 3 can be jogged to get the cylinder back to one end of its stroke. Note: When Valve 2 is de-energized, the cylinder could drift if supporting a reactive load, or one requiring less than 25 PSI, because the cylinder ports on Valve 1 are not blocked.

Caution! If either limit switch is placed to actuate before the end of the full cylinder stroke, be sure to mount it so it will not be damaged if the cylinder should accidentally run past it.

For the design of more elaborate electrical control circuits, refer to the Womack book "Practical Fluid Power Control". See inside rear cover of this book for description.

BANK VALVES FOR CYLINDER CONTROL

A bank valve includes two or more individual manual lever 3-way or 4-way hydraulic valve spools in a 1-piece iron casting assembly. A pressure relief valve to serve the pump line is usually built into the inlet end cap of a bank valve.

A bank valve is a compact way of building a number of directional control valves in a small, space-saving assembly for use on manually operated hydraulic circuits where a number of branch circuits are to be operated from one pump.

In addition to the advantage of a compact control package, a bank valve is just about the only practical way of operating several branch circuits from one pump and (1), being able to have full pressure at the same time on all branch circuits whose control levers are shifted, and (2), being able to unload the pump to tank when all control handles have been moved to neutral position. Individual 4-way valves connected together with close pipe nipples cannot give this kind of performance,

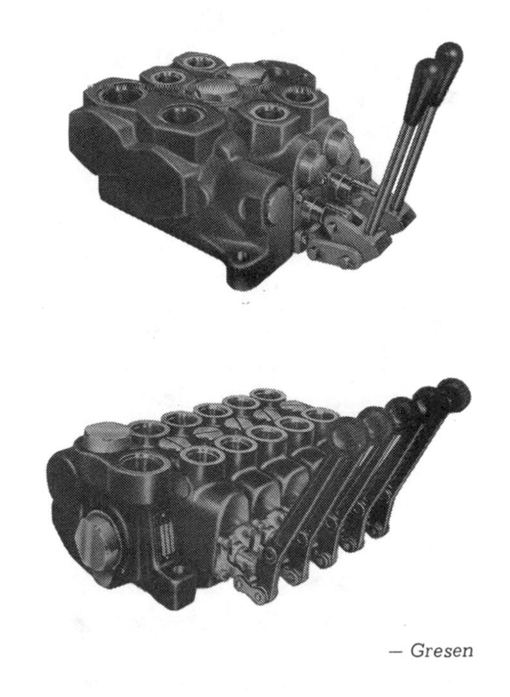

– Gresen

FIGURE 4-43. Examples of Bank Valves.

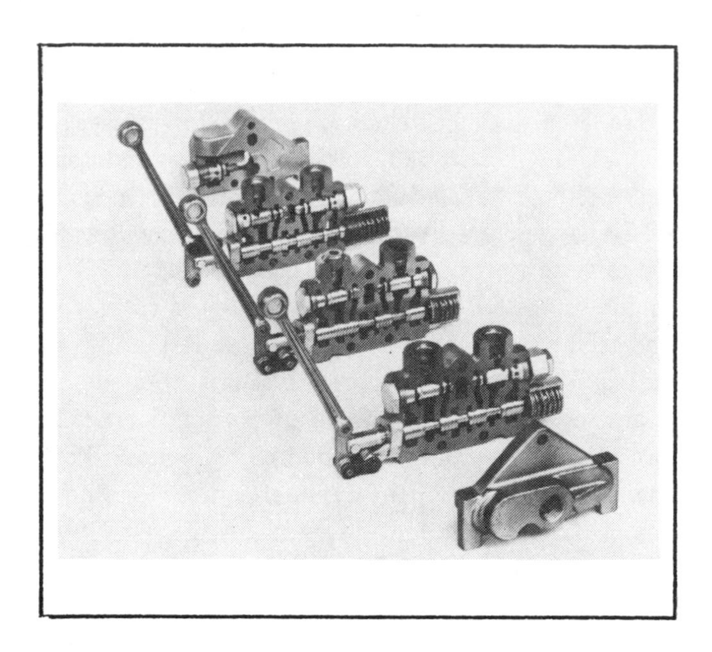

FIGURE 4-44. Sectional-Type Bank Valve.

because bank valves have internal passages which connect between sections to give valving actions which are impossible with a group of individual valves. Bank valves can also reduce the number of external plumbing lines.

Figure 4-44. A "stack" or "sectional" valve is another form of bank valve and used for the same circuit action. It is assembled from individual sections bolted together and enclosed with a front and rear end cover. A pump relief valve is usually contained in the front end cover.

The advantage of a stack valve is that a section can be replaced with one of a different kind without replacing the entire valve. They are popular with fluid power stocking distributors who can assemble an entire valve to a customers specification from individual sections carried in stock.

Bank valves are strictly for hydraulic circuits; they have no application on air. Due to the nature of air, the problems that bank valves were developed to solve do not exist in a compressed air circuit.

Figure 4-43 shows examples of a 2-section and a 5-section stack valve. They are available from 1/2" up to 1½" size, perhaps larger from a few manufacturers.

Bank Valve Action. Figure 4-45. A 3-branch circuit is shown with 4-way valve spools for operation of double-acting hydraulic cylinders. Three-way sections are available for operation of single-acting cylinders, and 4-way sections with open center porting are available for operation of hydraulic motors. Various manufacturers have other sections available for special circuitry.

In Figure 4-45, all valve spools are in neutral position. Pump oil can flow through open passages in all sections and be discharged to tank without placing a heavy hydraulic load on the pump. If one spool is shifted, Spool 1 for example, the free flow center porting becomes blocked and the pump flow is diverted to Cylinder 1. Return oil from Cylinder 1, after passing through Section 1 spool, dumps into the valve housing and is discharged directly to the tank port without returning through downstream spools. If two spools are shifted at the same time, Spools 1 and 2 for example, Cylinders 1 and 2 are both connected in parallel to the pump line and both can receive full pump pressure although they will have to divide the flow.

On most bank valves, to eliminate additional external connections, a pump relief valve is built into or ahead of the first section. Other optional features are available such as load holding checks, pilot-operated check valves on cylinder ports, etc.

Speed control valves are not often needed with manual bank valves. Cylinder speed can be controlled by metering with the valve spool. Spool metering can also be used to divide the pump flow in any ratio between two cylinders.

Power Beyond. This is an optional feature available so additional 4-way valving can be added downstream for operating additional branch circuits from the same pump. Refer to Volume 2 for a description of this feature.

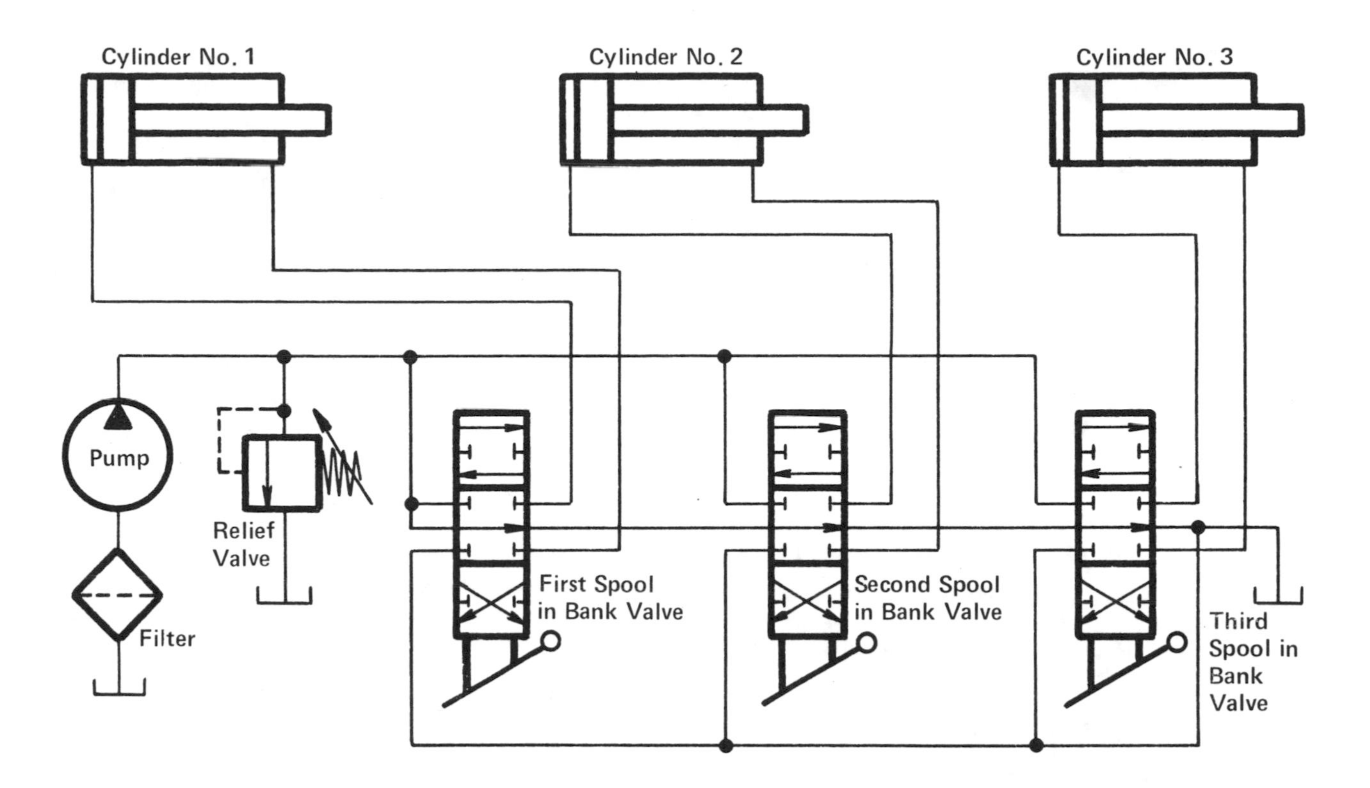

FIGURE 4-45. Internal Circuit of a 3-Spool Bank Valve.

Solenoid Operated Bank Valves. These have been available from a few sources in the past but they have not been popular because of the abrupt shifting action of a solenoid. The ability to meter the oil or to divide it between two branch circuits is one of the outstanding features of manually operated bank valves. These functions are not possible with solenoid operation.

Recently, several manufacturers have developed bank valves operated with solenoid "torque motors". These are low voltage (usually 12 or 24 volts) direct current coils, and using servo principles, a valve spool can be metered to any position by controlling current on the torque motor with a rheostat. These valves are expected to become popular, especially on mobile equipment where spool metering and flow dividing are very important capabilities.

REVIEW QUESTIONS – CHAPTER 4

1. Explain the difference between a 4-way valve and a 5-way valve. *– See Pages 132 and 148.*

2. What is a "detent" on a directional valve? *– See Page 146.*

3. Name the valve ports which are identified by these symbols: A, B, P, T, E, EA, EB, P1, and P2. *– See Pages 133 and 141.*

4. On what kind of applications would a button bleeder 4-way valve be preferred over a solenoid valve? *– See Page 139.*

5. Explain the shifting principle of a button bleeder valve. Why can this valve not be used on hydraulics? *– See Page 139.*

6. On button bleeder valves, what is the maximum recommended separation distance between main body and bleed buttons? What effect is the result of too long a separation distance? *– See Page 140.*

7. What is the basic difference between a direct-acting and a pilot-operated solenoid valve? *– See Pages 150 and 151.*

8. Why is it not practical to build direct-acting solenoid valves in sizes to handle high flow at high pressure? *– See Page 150.*

9. Name at least one advantage of a DIRECT-ACTING over a pilot-operated solenoid valve. *– See Page 150.*

10. What is usually the minimum line pressure which must be maintained for the shifting of a pilot-operated solenoid valve? *– See Page 152.*

11. Why is it better design on most air circuits not to stop the cylinder in a mid-stroke position? *– See Pages 135 and 162.*

12. What is the advantage (in some circuits) of a dual exhaust 4-way air valve instead of a single exhaust valve of the same kind? *– See Page 135.*

13. Describe the porting or the flow pattern through a tandem center hydraulic valve when the spool is in center neutral position. Describe also the center porting of an open center, a closed center, and a float center hydraulic valve. *– See Pages 142 to 144.*

14. Which of the spool types mentioned in Question 13 could be used on compressed air? Of those which could or should not, give the reason. *– See Pages 142 to 144.*

15. Which of those spool types mentioned in Question 13 is sometimes called a "motor spool"? *–See Page 143.*

16. On what kind of applications (name two) would a 5-way air valve offer advantages over a 4-way valve? *– See Pages 148 and 149.*

17. While many types of actuators are available for 4-way air valves, only two of these are commonly used in hydraulic circuits. Name these two. *– See Page 140.*

18. On what kind of application would a pressure switch be better than a limit switch for reversing a cylinder at the forward end of its stroke? *– See Page 160.*

19. Three kinds of 4-way solenoid valves are used: single solenoid, double solenoid with two positions, and double solenoid with three positions. Which of these types must have current maintained on the coil to keep the spool in shifted position? *– See Page 136.*

20. Which of the types mentioned in Question 19 is most often used on hydraulic circuits? Which one is most often used on air circuits? *– See Pages 135 and 145.*

21. How many main ports are there on a 4-way valve? On a 5-way valve? *– 4 ports; 5 ports.*

22. On what kind of applications are float center hydraulic 4-way valves used? *– See Page 145.*

23. Although a rotary shear-type 4-way valve may be more expensive than an equivalent size spool valve, it has several advantages. Name at least two advantages. *– See Page 147.*

26. What is the simplest way to synchronize two air cylinders operating through the same 4-way valve, to cause them to move at the same speed? *– See Page 158.*

27. Although more expensive, what advantage does a pressure compensated pump offer when two or more branch circuits must operate from one pump at the same time? *– See Page 165.*

28. Aside from a more compact assembly, what important advantage does a bank valve have over a number of single valves connected together with pipe nipples? *– See Page 167.*

29. What is the advantage of a stack-type bank valve over one made from a solid casting? *– See Page 168.*

Pumps do not pump pressure, they pump fluid

Pressure is a resistance to flow.

CHAPTER 5

Air and Hydraulic Pumps

The technology of fluid power is basically the transmission of power from one location to another. The advantage on some applications of transmitting power through pipes rather than electrically through wires or mechanically through gears and chains, is the ability to finely and precisely control both force and speed. Fluid power transmission always starts with a source of mechanical power which may be from an engine or an electric motor.

The first step in fluid power transmission is to convert mechanical power input into fluid power by means of an air compressor or a hydraulic pump. The fluid power is then pushed through a pipe to the point of usage and must then be converted back into mechanical power by means of an air or hydraulic cylinder, motor, or rotary oscillator. While still in its fluid state its force and speed can be regulated by means of various kinds of valving. An important point to remember is that power, whether mechanical, electrical, or fluid, must always consist of both force and speed. In the case of fluid power, force is directly proportional to the gauge pressure above atmospheric, and speed is directly proportional to the rate of fluid flow. No matter how much pressure may show on a gauge, power is being transmitted only when the fluid is in motion.

For fluid systems operating indoors the most common source of mechanical input power is from an electric motor. For systems operating outdoors, the most common source is from an engine.

AIR COMPRESSORS

It is not within the scope of this textbook to describe the many kinds of air compressors used in dozens of industries nor to cover the wide range of air applications from 10 PSIG instrument air to more than 2500 PSIG in applications like a wind tunnel. Our brief discussion is limited to basic information on air compression in the usual pressure range, from 60 to 125 PSIG, which is used in industrial plants for operation of cylinders, air motors, and rotary oscillators.

Basically, an air compressor is the device which takes input mechanical power and delivers air fluid power for transfer through pipes to the point of use. Conversion of mechanical power into air power is inherently an inefficient process because of the very nature of air. According to Charles' Law, air becomes hot when compressed. Most of this heat of compression, which represents input energy, escapes by radiation, so this reduces conversion efficiency. To reach a high pressure, air must be compressed in steps or stages. It can be safely compressed only a limited amount each step. Then it must be cooled before being compressed to a higher pressure. If compressed too much on each stage, not only is the compression less efficient but there is the hazard of an explosion if any hydrocarbon material, such as a lubricant, should come into contact with the overheated air. For practical usage, compression ratios are usually limited to 4 or 5 to 1 for each stage to stay well below the "dieseling" point. This limits the practical pressure output of a single-stage compressor to 50 to 75 PSI, starting from atmospheric pressure intake. To generate higher pressure, the air must then be cooled, at least to some extent, before being compressed to a higher pressure. In a multi-stage compressor the range is usually held to 3 to 1 or even less per stage. Typically, air is compressed in the first stage to about 40 PSIG. After cooling it is used to supercharge the inlet of a second stage. Interstage cooling is very important. If air were to enter the second stage at the same temperature it left the first stage, much of the advantage of multi-staging would be lost. For interstage cooling on small compressors, the hot air from the first stage is conducted to the second stage through finned tubing which is exposed to the air stream from the flywheel fan. On larger compressors the cylinder jackets are cooled with circulating water, and water can also be used for interstage cooling. To reach a pressure of 2500 PSIG, as many as six or seven stages may be required, with interstage cooling between each stage.

FIGURE 5-1. A typical small air compressor used in many industrial plants and service stations.

Motor-Driven Air Compressors. Figure 5-1. In an industrial plant most air compressors receive their mechanical input from electric motors. One large compressor, say 100 HP capacity, may serve many individual machines working in the 1/8 to 1½ HP range. For outdoor use, especially on portable air compressors, diesel or gasoline engines usually furnish the mechanical power input. For common use in the 90 to 125 PSIG pressure range, a 2-stage compressor usually works out to be most economical and efficient. At these pressures, each 1 HP of mechanical input will produce about 4 to 6 SCFM (standard cubic feet per minute) of air. Power input required on various compressors is shown in the table on Page 276. Information on Page 263 shows how to estimate air consumption for reciprocating cylinders.

Figure 5-1 shows a typical 2-stage, piston-type compressor driven by a 25 HP electric motor. Motor and compressor are usually mounted on top of the storage tank, called a "receiver". Water will collect in the bottom of the tank and should be drained at least once every day. Manual or automatic draining may be used.

The receiver tank holds the "raw" compressed air which is fed into the distribution system serving the plant. But before using it in air cylinders, valves, or air tools, it should be filtered and lubricated, and its pressure level adjusted for each particular machine. These operations are done with a "trio" assembly at the entrance to every machine. Each unit of the trio assembly is described in Chapter 6. Sometimes the water vapor must also be removed with an air dryer.

FIGURE 5-2. Miniature Air Compressors, Up to 1½ HP.

Miniature Air Compressors. Figure 5-2. A variety of small air compressors is available, up to 1½ HP for use on applications requiring a small amount of air. These compressors are intended to be used on a 1 to 1 basis – an individual compressor for each machine. A packaging machine, for example, may have a small compressor incorporated as an integral part of the machine and permanently installed as a part of the machine.

COMPRESSOR TERMINOLOGY

RECIPROCATING COMPRESSOR. One in which the compressing element is a piston having a reciprocating motion in a cylinder barrel.

ROTARY COMPRESSOR. A machine in which air is compressed by the positive action of rotating elements, as for example, vanes or impellers.

ONE-STAGE COMPRESSOR. A machine in which air or gas is compressed in each cylinder from initial intake pressure to a final discharge pressure.

TWO-STAGE COMPRESSOR. A machine in which air or gas is compressed from initial pressure to an intermediate pressure in one or more cylinders or casings, and thence to a final discharge pressure in one or more cylinders or casings.

MULTI-STAGE COMPRESSOR. A machine employing two or more stages.

VACUUM PUMP. A compressor which operates with initial intake below atmospheric and discharges at atmospheric pressure or slightly higher.

AIR-COOLED COMPRESSOR. A machine cooled by atmospheric air circulation around cylinders or casings.

WATER-COOLED COMPRESSOR. A machine cooled by water circulated through jackets surrounding the cylinders or casings.

INTERCOOLER. A heat exchanger for removing heat of compression between stages.

AFTERCOOLER. A heat exchanger for cooling air or gas discharged from a compressor. They provide an effective means for removing moisture from air or gas.

AIR RECEIVER. A tank for storage of air discharged from a compressor.

The compressors usually have an integral electric motor mounted with them in the same frame. There are also belt-driven models and air storage (receiver) tanks of various sizes available.

Rotary vane compressors deliver a relatively high volume of air but at low pressure, 10 to 15 PSI. Piston compressors deliver a smaller volume of air but at high pressure – up to 175 PSI. Diaphragm compressors deliver a small amount of air at low to moderate pressure.

Cycling an Air Compressor. As a rule, on compressors of 25 HP or less, the electric motor is stopped between cycles. A pressure switch teed into the receiver tank cuts off the motor when pressure in the tank reaches a pre-set "cut-off" pressure. It starts the motor when air usage has caused the tank pressure to drop to a pre-set "cut-in" level.

On larger compressors or those driven with an engine, it is impractical to stop the motor or engine between cycles. When tank pressure comes up to the cut-off level, the motor or engine is allowed to remain running but the compressor outlet is vented to atmosphere. This is called "unloading". A check valve in the receiver tank holds air in the tank when the compressor unloads. Unloading can be done with a pilot-operated dump valve which receives a pilot pressure signal from the receiver tank. When tank pressure falls to the cut-in level, the dump valve closes. Unloading can also be done with a solenoid valve operated from a pressure switch, or by a magnetic clutch which stops the compressor while allowing the motor or engine to continue running.

The air compressor should have sufficient capacity so that it can maintain system pressure by running only about 75% of the time. When installing a new plant air supply it is well to provide generous capacity for the future addition of more air-operated machines.

POSITIVE DISPLACEMENT HYDRAULIC PUMPS

A "positive displacement pump" is any kind of pump in which the internal working elements make such a close fit together that there is very little leakage or slippage between them. The general categories of such pumps are gear, vane, and piston types, but there are several variations of each type. The volume of the internal cavities which increases and decreases during each shaft revolution is called the "displacement" and is usually expressed in C.I.R. (cubic inches per revolution). The flow volume of such pumps is directly proportional to shaft speed, and is catalog rated as GPM (gallons per minute) at a certain RPM. (Note: 231 cubic inches = 1 gallon). Because of the tight fit between the pumping elements, these pumps can produce very high pressure in the pumped fluid. The tighter the fit, the higher the pressure that can be pumped efficiently. That is why piston pumps usually carry a higher pressure rating than other types – it is practical to build these pumps with a closer internal fit.

In contrast to the positive displacement types, there is the impeller or centrifugal type. Its impeller does not make a tight fit with the housing so there is a great deal of internal slippage. Therefore, these pumps are limited to producing maximum pressures which are too low to be of much practical use in hydraulic systems. They are used mostly for secondary chores, for moving a high volume of oil but at a maximum pressure of 20 to 30 PSIG, and they are rarely used to produce the main flow of fluid power. They are described later in this chapter.

Purpose of a Hydraulic Pump. A pump is a device for changing mechanical input power into the same amount of fluid power, minus friction and flow loss in the pump itself. Its primary purpose is not to produce ***pressure***, but to produce ***flow***. The flow output is directly proportional to shaft speed; if speed is increased by 50%, the flow output will increase by the same amount. It will produce its rated flow at zero pressure and will be fulfilling its primary purpose. It will only produce pressure if its flow should become restricted, as by putting a load against a cylinder piston which is being moved by the oil flow. In fact, the pressure produced in the pump flow is directly proportional to the resistance placed aginst the flow; in other words, by the magnitude of the load placed on the cylinder. It produces just enough pressure to accomplish its primary objective – of keeping the flow coming at a constant rate.

On a tight-fitting positive displacement pump, if the oil flow is deadheaded (blocked), the pump will continue to build up pressure to the limit of the horsepower available for driving its shaft, or until something breaks – a line, a valve, or the pump casting. For this reason, a pump must be "unloaded" rather than deadheaded during those times in the machine cycle when its flow is not needed. The subject of pump unloading is described later in this chapter.

In the study of various kinds of pumps to follow, the student will see that the same principle is involved in various ways – the principle of opening up internal cavities during one-half of the shaft revolution, then closing these cavities to a smaller volume during the other half revolution. There are various ways of valving the oil in and out of the cavities.

Lubrication of Hydraulic Pumps. No external lubricating system is required for hydraulic pumps since the fluid being pumped circulates around the shaft bearings. It not only keeps bearings lubricated, it carries away heat of bearing friction. The fluid must have lubricating qualities. This is no problem when pumping hydraulic oil but must be considered when pumping other fluids. Plain water should not be pumped with any kind of pump designed for mobile or industrial hydraulic service. Fire resistant fluids including phosphate ester and water with lubricating oil emulsion can usually be handled by most pumps, but to get a satisfactory life it may be necessary to operate them at pressures below the maximum rating for oil. Check with catalog information or contact the pump manufacturer before operating at full pressure rating on fluids other than oil.

Filtration for Hydraulic Pumps. Some pumps are much more sensitive to contamination in the oil than are others. For example, piston pumps used in hydraulic transmissions must have extremely well filtered oil or they can be destroyed in a few minutes operation. On the other hand, gear pumps can tolerate a fair amount of dirt with only a small reduction in life expectancy.

For all pumps, regardless of kind, we recommend a strainer or filter to protect the inlet. Then additional filtering can be added in the pressure or return line for those pumps requiring finer filtration. See comments under each pump type.

Pump Life Expectancy. Pump bearing life depends on two factors, shaft speed and operating pressure. Life is inversely proportional to rotational speed. A pump running at 1200 RPM will have a life expectancy about 50% greater than if run at 1800 RPM. Life is inversely proportional to the cube of pressure. Stated another way, a pump operating at 1000 PSI will have 8 times the life expectancy of the same pump operating at 2000 PSI. (Pressure ratio is 2:1, and the cube of 2 is 8).

Pump life varies with manufacturer and with type of application. Pumps intended for industrial service are usually designed with larger bearings for longer life than are pumps designed for mobile service. Using a mobile pump for heavy-duty industrial service may result in unsatisfactory life.

The above factors are only for bearing life. Actual life depends also on factors which are the responsibility of the user: System filtration and oil cleanliness, amount of side or end loading permitted on the shaft, oil temperature, degree of cavitation at the pump inlet, and misalignment of pump and motor shafts. The pump manufacturer has no control over these factors.

Pump life is calculated on the number of hours during which the pump is operating at maximum system pressure; unloading time is not counted even though the pump remains running. There is no exact answer as to how long a pump will last. On applications requiring longest possible pump life, discuss details of the application with the engineering department of the pump supplier.

GRAPHIC SYMBOLS FOR HYDRAULIC PUMPS

The graphic symbols shown here are for use on fluid schematic drawings, and serve for any kind of positive displacement pump or air compressor. They are the official symbols recommended by the NFPA (National Fluid Power Association) and the ANSI (American National Standards Institute).

5-3(A). Fixed Displacement Pump Symbol

5-3(B). Variable Displacement Pump Symbol

5-3(C). Pressure Compensated, Variable Displacement Pump Symbol

5-3(D). Over-Center Variable Displacement Pump Symbol

FIGURE 5-3 Pump Symbols.

Fixed Displacement Pump. Figure 5-3(A). A circle is the basic symbol for all pumps, both air and hydraulic. The triangle points toward the pump outlet; a solid triangle indicates a hydraulic pump, an open center triangle indicates an air compressor. Hydraulic and air motors use a similar symbol but with the triangle at the inlet port pointing inward. This symbol can also be used for impeller pumps.

Variable Displacement Pump. Figure 5-3(B). Certain kinds of positive displacement pumps can be constructed so the displacement can be increased or decreased. A long slash arrow drawn through the basic circle indicates the variable displacement feature.

On some models the displacement can be changed from outside the pump by means of a screw or wrench adjustment or by a handwheel. On other models the displacement change must be made from inside the pump.

When displacement is reduced, the pumping rate decreases the same amount (at a constant shaft speed). The required drive HP also changes in proportion to change in displacement (at a constant pressure level).

Pressure Compensated, Variable Displacement Pump. Figure 5-3(C). These pumps also are built so the internal displacement can be changed. The small vertical arrow inside the circle indicates that displacement is automatically reduced, without operator assistance, by a built-in mechanism called a pressure compensator. The compensator keeps the pump at maximum displacement until the pressure in the pump outlet line tends to build into an overload level (because of overload on the system cylinder). The compensator will then reduce pump displacement enough (to zero displacement if necessary), to keep pressure from rising higher. Compensators are described later in this chapter.

Over-Center Variable Displacement Pump. Figure 5-3(D). This type of pump is used in most hydraulic transmissions. The displacement can be varied by operator control or automatically by a compensator. In addition, the pumping elements can be displaced across center (zero displacement position), causing the pump to resume its pumping. But because the valve timing changes by 180° on the other side of center, the pump flow is in the opposite direction – the previous inlet port becomes the outlet and the previous outlet now becomes the inlet. This change in direction of flow is accomplished without reversing direction of shaft rotation.

To indicate this type pump, two triangles are used along with the slash arrow. Most transmission pumps also have a pressure compensator which is indicated with the small vertical arrow.

HYDRAULIC GEAR PUMPS

Figure 5-4. Gear pumps have for many years been popular on all kinds of hydraulic systems but especially on mobile equipment because they are less expensive and have a greater tolerance for dirt than any other kind. Until recently, though, they had a shorter life expectancy because of the extremely heavy side load which the gears impose on shaft bearings. This also caused them to be rated at lower maximum pressures. In recent years, due to improvements in manufacturing techniques, gear pumps are now available for high pressure operation – up to 3500 PSIG for certain sizes from some manufacturers. They are available in a very wide range of flow capacities, starting at a fraction of 1 GPM up to 80 GPM for single pumps and 160 GPM for double pumps. A few manufacturers may offer even larger capacities.

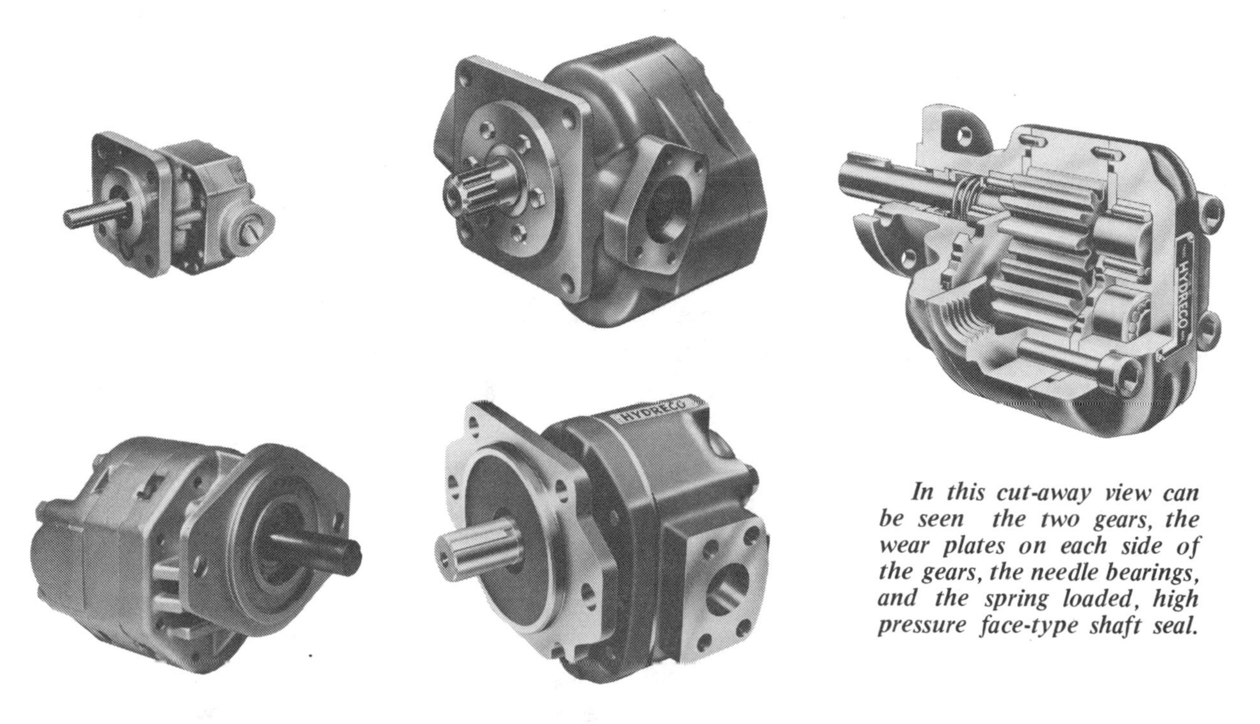

In this cut-away view can be seen the two gears, the wear plates on each side of the gears, the needle bearings, and the spring loaded, high pressure face-type shaft seal.

FIGURE 5-4. Appearance of Typical External Gear Pumps.

External Gear Pump Construction. Figure 5-5. An "external" gear pump is one in which each gear is external to the other, as contrasted with internal gear pumps in which one gear is inside the other.

Two steel gears, usually spur gears, rotate inside a cast iron or zinc alloy housing. The "drive" gear is keyed to the shaft. It meshes with and turns the "driven" gear. Gear teeth make a tight sliding fit in the housing. As the gears revolve, a flow path is created around the outside of each one. Oil, trapped in the slots between teeth is carried around and discharged into the cavity with the outlet port. It is forced from the outlet port by the oil coming in behind it. Meshing of the teeth in the center of the pump seals outlet port from inlet port. A suction is created at the inlet port as oil is carried away by the gear teeth. This suction draws oil into the pump from the system reservoir.

Helical instead of spur gears are sometimes used because they give quieter operation and produce a smoother flow. But they do impose an end load on the shaft. The ultimate in external gear construction is the use of herringbone gears which produce a smooth quiet flow without end thrust on the shaft.

Bearing failure is probably the major cause for gear pump failure because all types of gears produce very heavy side loading on the bearings. In most cases of pump failure, a bearing fails first. This leads

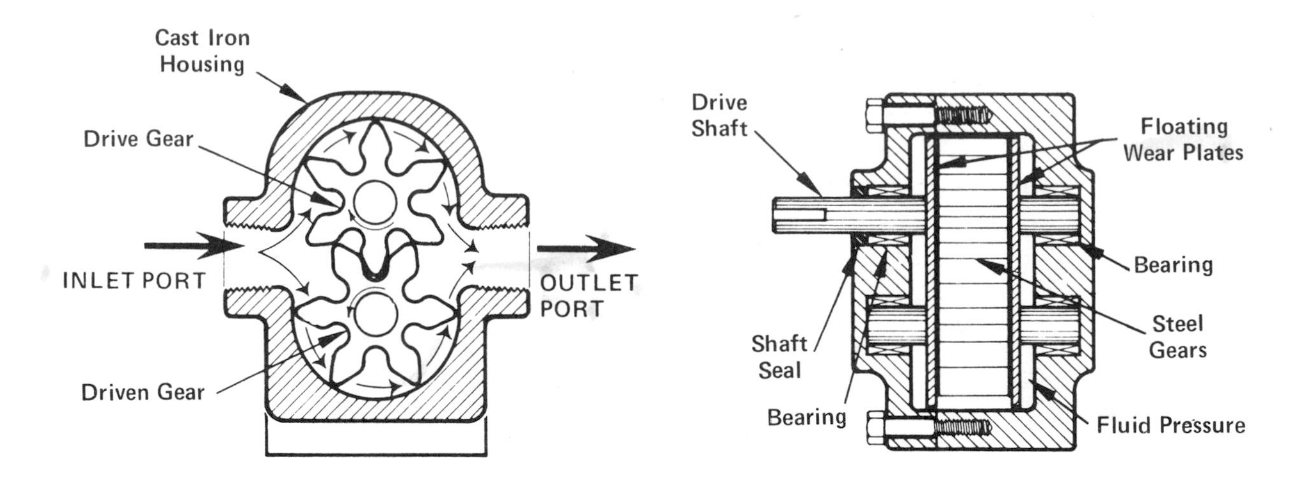

FIGURE 5-5. Typical Construction of an External Gear Pump.

to other damage such as key or shaft breakage, heavy scoring of the housing or wear plates, and sometimes to breakage of the pump housing. Pump life could, in many cases, be greatly extended if bearings were replaced on a routine basis about every 2000 hours of operation.

Displacement is fixed on a gear pump at the time of manufacture. There is no practical way to change the displacement. However, output flow can be varied by changing shaft speed.

Wear Plates. Those gear pumps built for high pressure usually have floating "wear plates" on each side of the two gears. These plates can be seen in Figure 5-5. They have an oval shape to fit the bore in the housing, and are held snugly against each side of the gears by internal fluid pressure obtained from the pump outlet through small drilled passages. The plates take up clearance space at the sides of the gears, thus compensating for manufacturing tolerances in fitting the gears into the housing. They improve pump efficiency, particularly at higher pressures by reducing internal slippage.

Direction of Rotation. Unlike gear *motors* which are built for reversible rotation, gear *pumps* are usually built for single-direction rotation, either clockwise (CW) or counter-clockwise (CCW) when looking into the shaft end of pump. If run in the wrong rotation, pump flow would also be in the wrong direction. The inlet port would become the outlet port. High pressure slippage oil would back up through the internal drain passage, collect behind the seal and blow it out.

Gear pumps can be built for bi-rotational operation in one of these ways: (1), By using a high pressure face type shaft seal instead of a low pressure lip seal. The seal would not blow out when the pump was operated within its maximum pressure rating. The problem with this seal is that it may at times "weep" slightly, so it is used primarily on pumps built for mobile equipment where a slight seal leakage may be permissible. (2), An external drain port can be provided which connects into the area just behind the shaft seal. A separate line must be run from this port to tank. And (3), The pump can be constructed with a set of four internal check valves as illustrated in Figure 5-36. These check valves provide a free flow passage for slippage oil to return to the inlet port. Methods (2) and (3) prevent pressure from building up behind the shaft seal and blowing it out.

Shaft Bearings. Some gear pumps are built with needle bearings, others with sleeve or bushing type bearings. Sleeve bearings are less expensive and work well with electric motor drive where full shaft RPM is always maintained. They can outlast needle or roller bearings provided the oil is kept well filtered. Pumps for mobile equipment should always have needle bearings because they are more dirt tolerant and less likely to freeze when operated at full pressure at reduced speed (engine idle).

GEAR PUMP VARIATIONS

Pumps shown under this heading are variations of standard gear pump design but the same basic pumping principle is used; that is, two mating elements, one of them keyed to the shaft, rotate together, opening up and closing down internal cavities on each revolution of the shaft. The method of feeding oil into and out of the elements varies somewhat from standard external gear pumps. They are less popular than external gear pumps, being more costly and available from fewer sources and in a more limited range of sizes and pressure ratings.

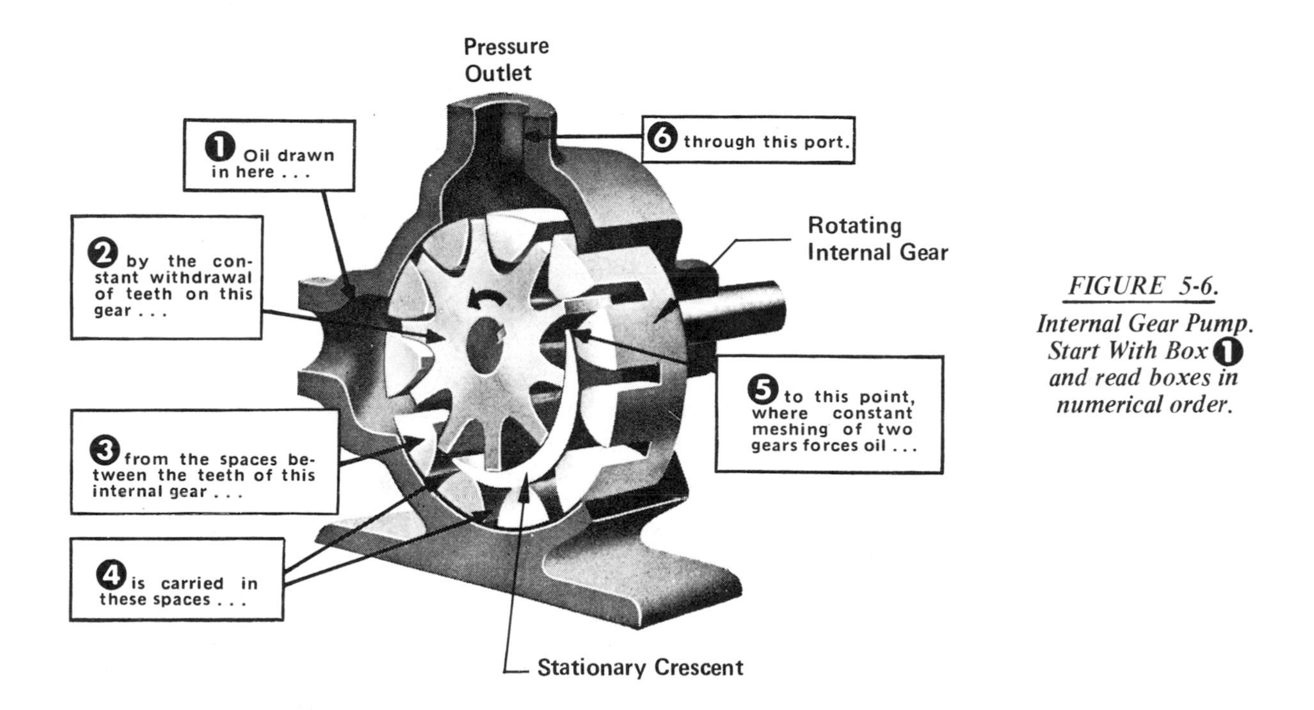

FIGURE 5-6.
Internal Gear Pump. Start With Box 1 and read boxes in numerical order.

Internal Gear Pumps. Figure 5-6. An internal gear pump is one in which the two gears or pumping elements are one inside the other. In this example the inner gear is keyed to the shaft. It meshes with and carries the outer gear with it. Oil moves through the pump by the opening of cavities between the gear teeth on the inlet side and meshing of the teeth on the discharge side. A stationary crescent-shaped divider separates the suction and discharge portions of the fluid. As with most gear pumps, shaft rotation must be in a specified direction to avoid blowing out the shaft seal.

Advantages claimed for this design are smooth and almost pulseless flow and slightly more horsepower capacity in the same physical size. Disadvantages are higher cost, a very limited range of sizes, low to only moderate pressure ratings, and few sources of manufacture.

Gerotor Pumps. Figure 5-7. This is another type of internal gear pump with one gear or "gerotor" inside the other. Oil flow is produced by opening and closing of internal cavities on each revolution of the shaft.

The inner gerotor is the drive element and is keyed to the shaft. It meshes with the outer gerotor and both elements rotate inside a bore in the housing. The inner element has one less tooth. This causes a rotational progression in the relative position of one element to the other. For example, a certain tooth on the inner element advances into the next slot on each shaft revolution.

Fluid enters and leaves tooth spaces through kidney shaped openings in the side plates. These openings are connected to inlet and outlet ports. They are shown in dash lines on the drawing.

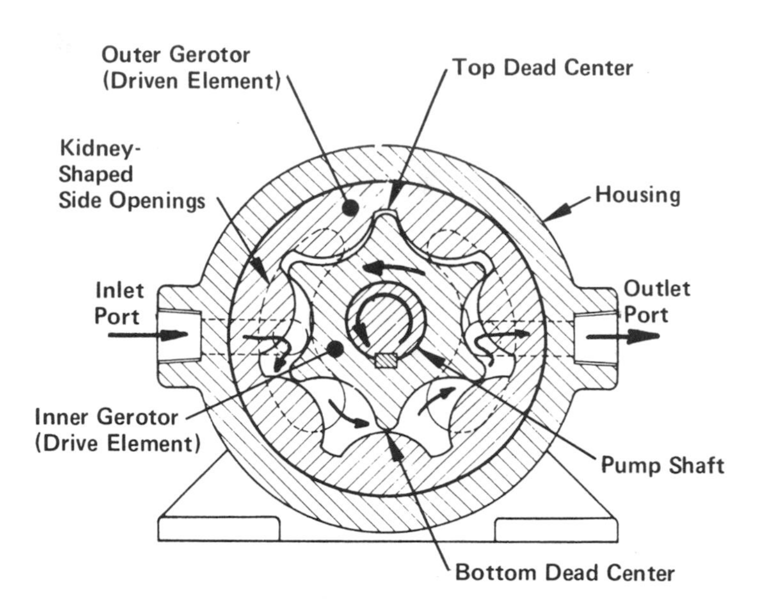

FIGURE 5-7. Gerotor-Type Internal Gear Pump.

Each tooth of the inner element is in sliding contact with the surface of the outer element and never loses contact with it, thus keeping a tight oil seal between the two elements and preventing backward slippage as oil is moved through the pump.

As the elements rotate, a vacuum is created as the teeth start to separate at top dead center position. This vacuum draws oil from the inlet port through the kidney shaped openings in the side plates to fill the cavities which are opening up. When bottom dead center position is reached, the teeth start to engage, closing down the cavities between the elements and forcing the oil to the outlet port through the other kidney shaped side openings.

The surface contact speed between the elements is relatively slow, minimizing surface wear. The contact points make one complete cycle for every 6 to 9 shaft revolutions (depending on the number of teeth in the elements). The opening and closing of cavities between teeth is gradual, across long ports, minimizing shock and turbulence.

Advantages of gerotor-type pumps include very smooth and quiet operation. Disadvantages are that they are more sensitive to dirt than are external gear pumps. They must have well filtered oil, say 25μm, because of the large area of sliding fit between outer element and housing. The pump can be quickly destroyed by dirty oil which might do little damage in an external gear pump. Manufacturers recommendations on filtering should be carefully followed. A further disadvantage is that they are mechanically unbalanced and impose a heavy side load on shaft bearings. If rotated at excessive speed, centrifugal force may produce severe vibration and high bearing loads. Gerotor-type pumps are available from a much more limited number of sources than are standard gear pumps, in a more limited range of sizes, and for low to only moderately high pressures.

HYDRAULIC VANE PUMPS

Vane Pump Principle. Figure 5-8. The rotor is keyed to the shaft and is rotated by it. Vanes are a slip fit in the slots in the rotor. They are carried around by the rotor, and are kept in continuous contact with the cam surface by centrifugal force or other means. In this simple example the cam surfaced is machined in the housing, eccentric (off-center) to the shaft centerline. This causes the cavities between adjacent vanes to expand during the one-half revolution they are exposed to the inlet port, and to diminish during the one-half revolution they are exposed to the outlet port. Oil is ported into and out of the cavities through semi-circular slots in the side plates which are connected to the inlet and outlet ports. When a vane passes top dead center, for example, the cavity between it and the preceding vane starts closing in volume, forcing oil which is carried in the cavity to discharge into the outlet port. At bottom dead center, the cavities start opening up. The vacuum which is produced draws oil in through the inlet port.

Oil flow produced is directly proportional to the displacement and shaft speed. Displacement can be expressed in C.I.R. (cubic inches of volume change during one revolution). Shaft speed is usually rated in RPM. As with all pumps, no pressure is produced unless there is resistance to flow. Then the pressure, usually expressed in PSIG, is directly proportional to the load (or flow) resistance.

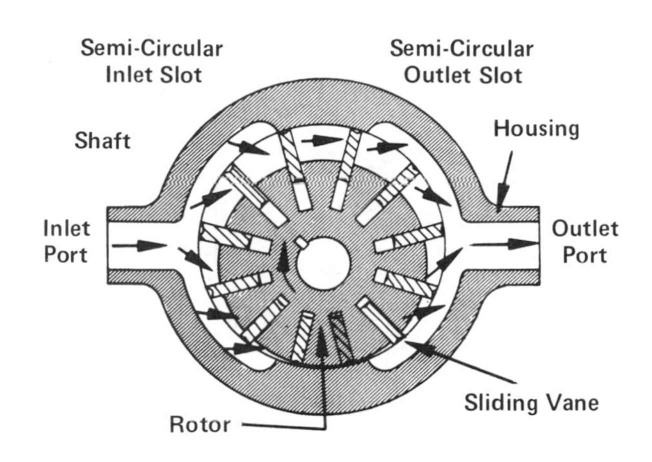

FIGURE 5-8. Vane Pump Principle.

The construction shown in this figure is a single-lobe pump in which the cam surface has only one lobe. It is circular in form but offset with respect to the shaft. Each vane extends and retracts once during each shaft revolution. The construction is somewhat typical of a rotary air compressor. Hydraulic pumps of single-lobe design are usually built with a pressure compensator (see next figure).

Vane Pump With Pressure Compensator. Figure 5-9. This is a variable displacement, pressure compensated pump of single-lobe design using the vane principle illustrated above.

A pressure compensator on any pump is a device which, when the pump outlet pressure has increased up to the "firing pressure" due to increase in load resistance, will automatically reduce pump flow output (to zero if necessary) to prevent any further rise in pressure. On Figure 5-9 the compensator can be seen projecting from the right side of the pump case.

FIGURE 5-9. Variable Displacement, Pressure Compensated Vane Pump.

On most pumps the pressure at which the compensator "fires" is adjustable and has been pre-set to the maximum pressure which the system is designed to handle. The adjustment is locked and the operator does not have access to it. It works automatically to override the operator in the sense that if the operator should try to move a load which would require a pressure beyond the capacity of the system, the compensator will not permit overloading to a higher pressure. Instead, it will reduce the pump displacement to zero and the system will simply stall until the overload is removed.

A compensator, then, protects the pump and the entire system from being damaged by excessive pressure; it protects the driving motor or engine from being overloaded and stalled, and prevents overheating of the system.

Due to the unbalanced nature of single-lobe pumps, they are more limited in both pressure and flow than the balanced, double-lobe pumps to be described later. For example, maximum ratings on the pump illustrated is 21 GPM at a shaft speed of 1750 RPM, and with a pressure limitation of 1500 PSI. High speed and/or high pressure will produce excessive vibration.

Pressure Compensated Pump will give enough flow to maintain pressure

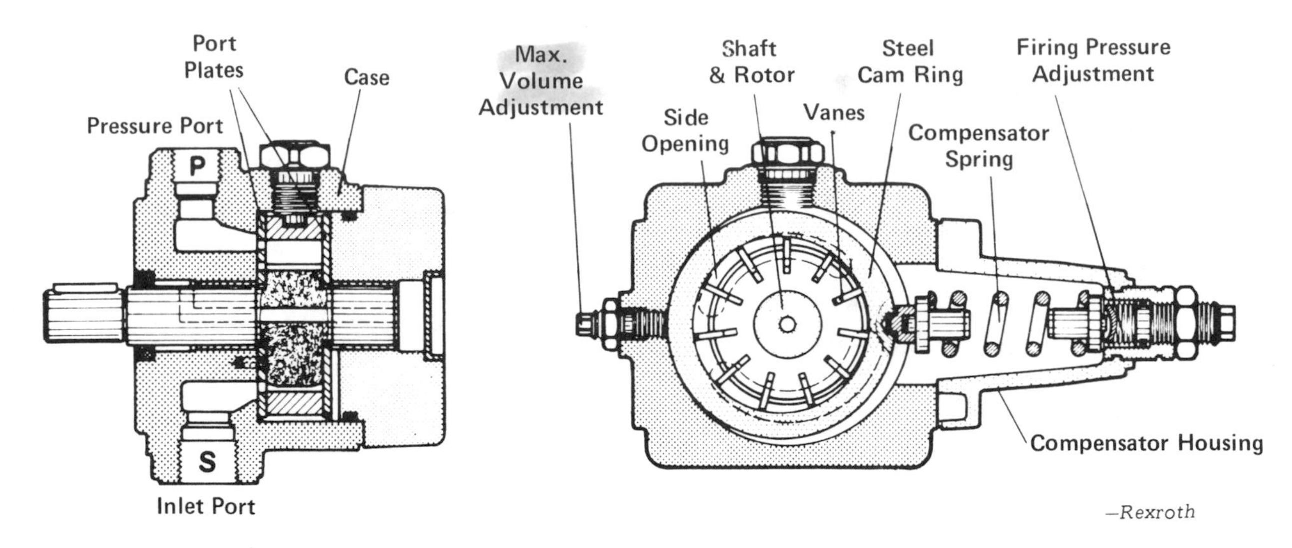

FIGURE 5-10. Internal View, Pressure Compensated Single-Lobe Vane Pump.

How the Pressure Compensator Works. Figure 5-10. Inside the pump can be seen the rotor which is keyed to the shaft. The vanes can slide up and down in the slots in the rotor. They are kept in contact with the cam surface by springs under the vanes or by centrifugal force. The cam surface in this pump is a steel ring which is held in its far left eccentric position by the compensator spring. In this position the pump is at maximum displacement and will produce maximum flow. Hydraulic pressure which is developed by the pump, working against the cam ring, is trying to move the ring to the right, to a concentric position with the shaft, at which position the pump output will be reduced to zero.

Tension on the compensator spring can be adjusted to establish the "firing pressure", the pressure at which hydraulic pressure working against the cam ring can overcome the spring tension and compensate the pump to zero flow.

When the pressure overload is removed from the system, the compensator spring can again put the pump on full stroke.

The maximum volume screw can be adjusted to reduce the maximum flow from the pump when it is on full stroke. This may be used when the driving motor or engine has insufficient horsepower to run the pump at full displacement.

BALANCED VANE PUMPS

A balanced vane pump is one having two lobes on the cam surface on opposite sides of the shaft. The cam surface, instead of being circular is roughly elliptical, so each vane makes two strokes on each revolution of the shaft. The two pumping chambers mechanically oppose each other so the side load produced by one chamber is exactly balanced by an equal side load from the other chamber. Thus, bearing loads from internal pressure are virtually zero, and the only bearing loads are those from external side and end loading.

FIGURE 5-11. Balanced Vane Pump.

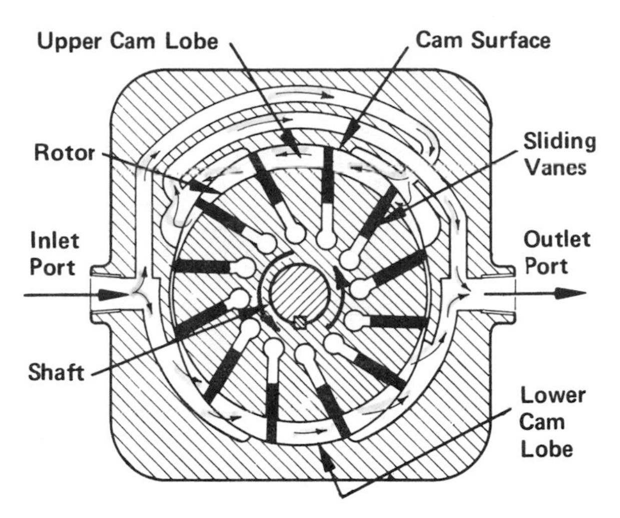

FIGURE 5-12. Working Principle – Balanced Vane Pump.

Pumping Action. Figure 5-12. Inlet oil is connected to both cam lobes through internal passages. Outlet oil is collected from both lobes through other passages and combined at the outlet port.

Pumping volume is directly proportional to shaft RPM and to the displacement; displacement being the change in total volume of all cavities between all vanes during one shaft revolution.

The vanes must remain in tight contact with the cam surface at all times. Some pumps rely solely on centrifugal force to extend the vanes. These pumps do not perform well at low RPM. Other pumps supplement the centrifugal force with a set of springs under each vane. They perform well at low RPM but have a problem of spring fatigue. Still other pumps use hydraulic pressure under each vane, obtained from the outlet through internally drilled passages. Extension force on vanes is proportional to outlet pressure which, in turn, is proportional to hydraulic load. When outlet pressure is low, extension force is also low and vane wear is minimized. But when outlet pressure is high, vane extension force increases to maintain a tight seal between vane and cam surface.

Balanced vane pumps, due to symmetrical construction, are difficult to build with variable displacement. Vane pumps cost a little more than gear pumps of comparable quality. Bearing loads are low, and when operated on clean oil they have exceptionally long life. Common service problems, other than shaft seal replacement, may include occasional vane or vane spring replacement, or replacement of cam ring due to washboarding caused by erosion at points of high pressure and high velocity.

Vane pumps are only moderately tolerant of a small amount of oil contamination but are more sensitive to dirt than are gear pumps. In addition to a 150μm inlet strainer they should be protected by at least 25 to 40μm filtration in the pump pressure or tank return line of the system.

PISTON PUMPS IN GENERAL

Piston pumps can be manufactured with closer internal fits than can other pumps. This means internal slippage can be less so they can operate with reasonable efficiency at pressures both too high or too low for satisfactory operation of other pumps.

All piston pumps have working elements enclosed in a case. Pressure in the case must be kept low to avoid over-pressuring and blowing out a shaft seal. External case drain connections are provided which should be drained directly to reservoir without combining with other tank return lines. They should never be plugged. Back pressure in case drain lines must be kept very low. Before starting up a new system with a piston pump, the case should be filled with oil to avoid dry running before internal slippage can fill the case.

BENT-AXIS PISTON PUMPS

Figure 5-13. The cylinder block is offset at an angle to the shaft axis. This causes the pistons to stroke as the cylinder block rotates with the shaft.

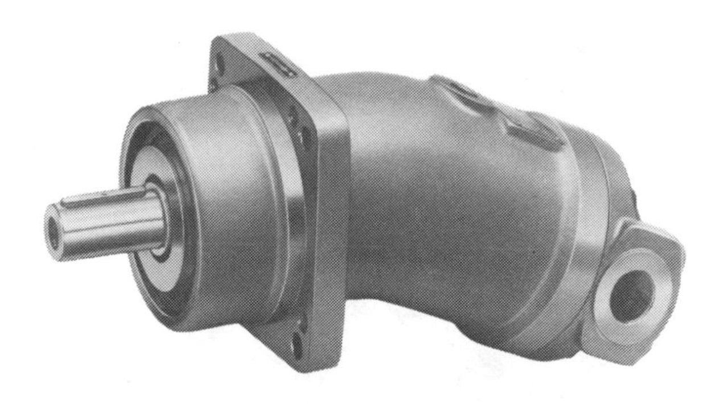

FIGURE 5-13. Bent-Axis Piston Pump.

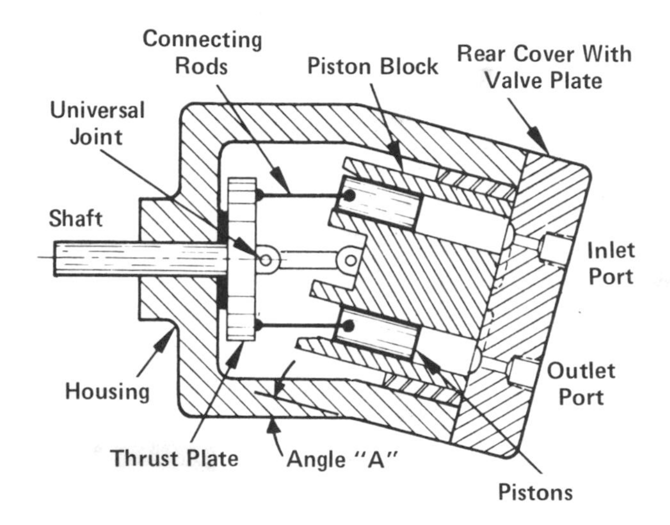

FIGURE 5-14. Working Parts of a Bent Axis Pump.

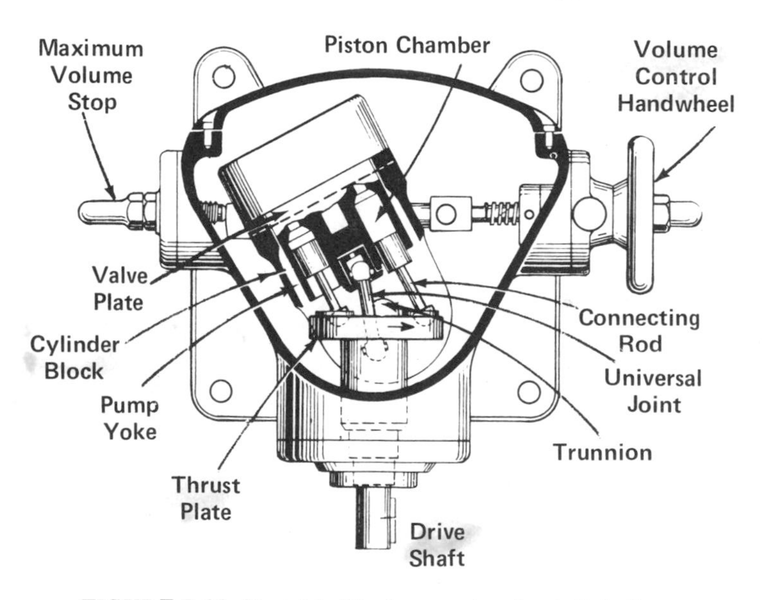

FIGURE 5-15. Variable Displacement – Bent Axis Pump.

How the Pump Works. Figure 5-14. As the shaft turns it carries the thrust plate with it. A universal joint connection to the piston block causes the block and the pistons to rotate with the shaft and thrust plate. Connecting rods from each piston provide the linear motion for the piston stroke.

Portholes in the rear cover connect to semi-circular feed grooves either in the rear cover or in a valving plate in the rear cover. Position of these grooves is rotationally timed so oil is fed into each piston while it is on its suction stroke and oil is connected to the outlet port when the piston is on its discharge stroke.

Volume of flow is directly proportional to number of pistons, bore and stroke of the pistons, and to shaft RPM. Since displacement cannot be changed, the only way to vary the flow volume is to vary the shaft speed.

Variable Displacement Bent Axis Pump. Figure 5-15. Working parts are similar to the fixed displacement pump, but the cylinder block is hinged around a trunnion or pintle on each side. The angle between cylinder block and shaft axis can be mechanically changed with the volume control handwheel. When the angle is reduced to zero, the pump flow becomes zero. On some pumps the cylinder block can be moved across center to the opposite angle with the shaft axis. In this case, flow through the pump reverses, with the previous inlet becoming the outlet, etc. The maximum volume stop can be adjusted to limit maximum flow.

A valve plate with semi-circular grooves is timed to port oil into and out of the piston chambers at the proper time. Trunnions which provide hinge supports for the cylinder block are hollow. Oil is ported in and out of the pump through these hollow joints.

STRAIGHT-AXIS PISTON PUMPS

Swash Plate Type. Figure 5-16. In this pump the cylinder barrel containing the pistons is parallel to the shaft axis. The stroking device is a non-rotating swash plate mounted at an angle to the barrel. The angle of tilt on the swash plate determines the piston stroke and therefore, the pump displacement and flow.

How the Pump Works. Figure 5-17. The cylinder block containing the pistons is rotated by the shaft. A shoe on the end of each piston rides the surface of the swash plate, imparting a reciprocating motion to the pistons. On the suction stroke the pistons are pulled back by a spring loaded retractor ring.

Oil is fed into the pistons from the inlet port through semi-circular valving grooves in the rear cover or in a separate valving plate inside the rear cover. Oil is discharged to the outlet port through another similar valving groove.

Variable Displacement Swash Plate Pump. Figure 5-18. This is the type of swash plate pump used in most hydrostatic transmission drives. The swash plate is hinged on trunnions or pintles, one on either side of the cylinder block. These trunnions can be hollow and can serve as porting passages to bring oil into and out of the pump.

Wobble Plate Pump. Figure 5-19. This is another version of a straight-axis piston pump. In this version the cylinder with its pistons remains fixed while the motivating element, a cam or wobble plate is attached to the shaft and rotates with it. Return springs keep each piston riding the surface of the wobble plate.

Valving of the oil in and out of this pump is through a set of two check valves for each piston. Inlet check valves

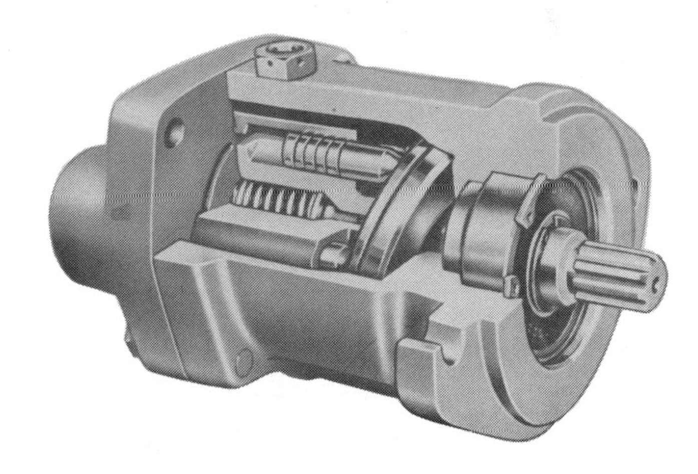

FIGURE 5-16. Swash Plate Piston Pump.

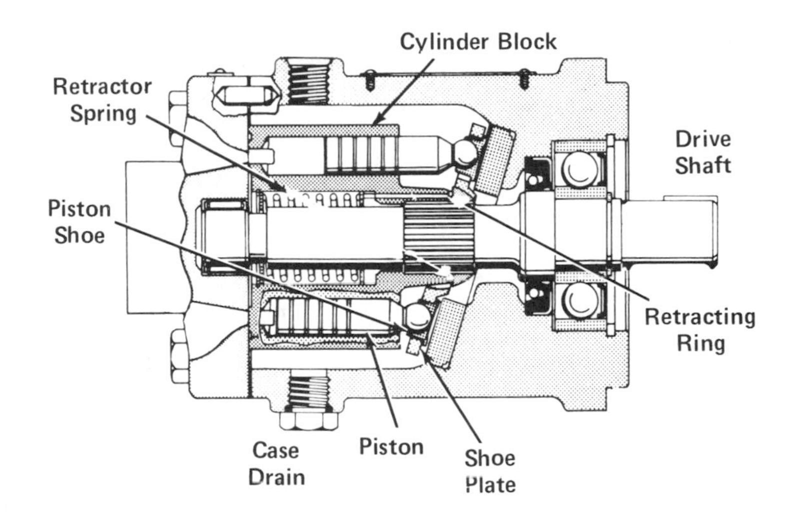

FIGURE 5-17. Cutaway View – Swash Plate Piston Pump.

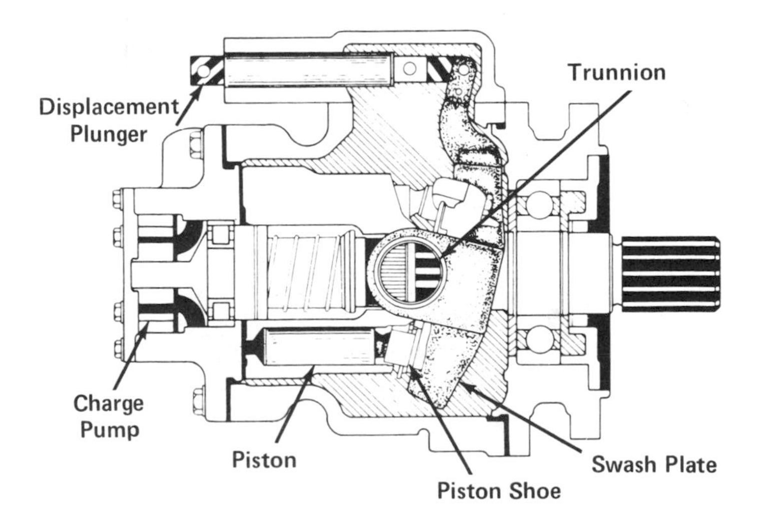

FIGURE 5-18. Variable Displacement, Swash Plate Piston Pump.

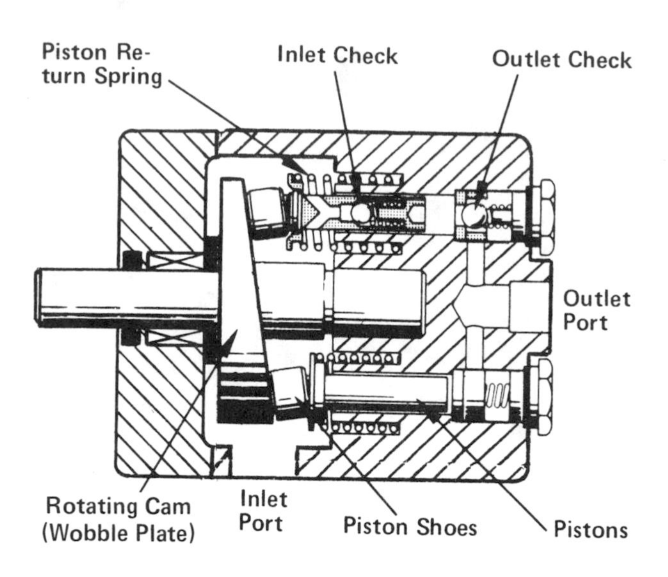

FIGURE 5-19. Wobble Plate Piston Pump.

are in the hollow center of each piston and suck oil in from the pump case. Outlet check valves are located in the back of the case or in a rear cover. They discharge into a common pressure manifold which leads to the outlet port.

To make a variable displacement version, since the angle on the wobble plate cannot be varied, the piston stroke must be shortened to reduce pumping volume. This is done with a mechanism which limits the distance the pistons can be retracted by their return springs. This reduces the volume of the cavity which can suck in intake oil and reduces the output flow.

This pump is usually available in low volume sizes for pressures to 10,000 PSI.

RADIAL PISTON PUMPS

Figure 5-20. Radial pumps are designed so the pistons stroke in a direction at right angles to the shaft axis. Not only this kind but all piston pumps are designed with an odd number of pistons – 3, 5, 7, or 9, to reduce the pressure ripple in the output. Less torque is required to start the pump against a deadhead load.

In Figure 5-20, an eccentric cam is mounted on the shaft. Each piston has a shoe which rides on the surface of the cam. On the suction stroke of each piston oil is drawn in through a check valve from the case. On the discharge stroke, the oil is discharged through another check valve into a common manifold which connects to the outlet port. This construction is limited to fixed displacement operation.

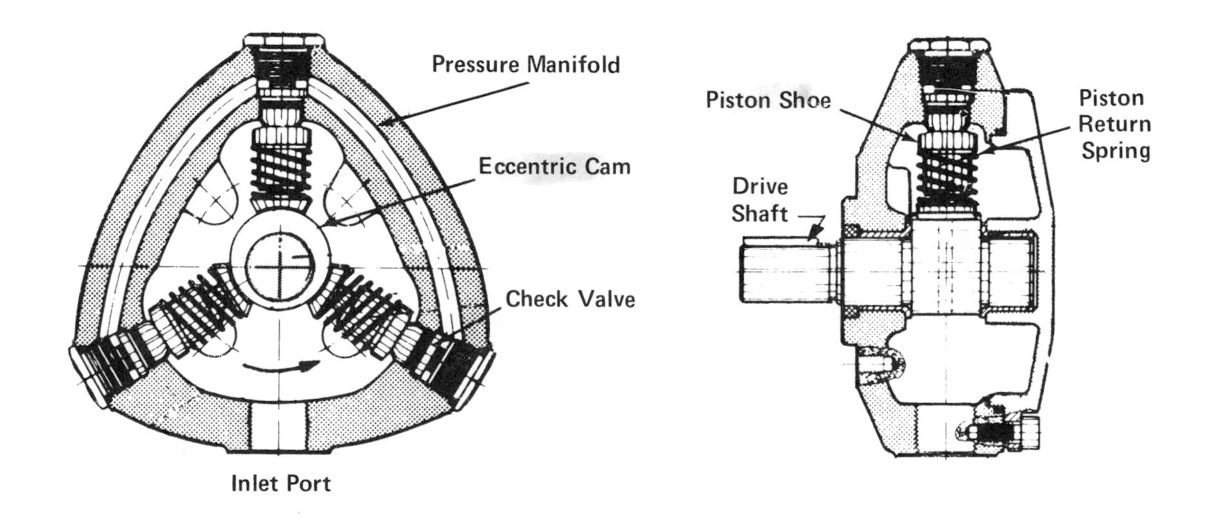

FIGURE 5-20. Working Principle – Check Valve Type, Radial Piston Pump.

Figure 5-21. In this construction the cylinder block carrying the pistons is keyed to the shaft. The reactor or cam ring is stationary but can be shifted sidewise to change its eccentricity to the shaft axis which changes the pump displacement. The pintle is a stationary valving element which is in line with the shaft but is stationary; it valves the inlet oil to the pistons on their suction stroke and accepts discharge oil on the compression stroke. "Out" and "In" passages in the pintle area connect through a commutator-type valve to inlet and outlet ports in the pump case.

The pistons are forced inward by the reactor ring; they are returned usually by spring force which holds them in contact with the reactor ring surface as they stroke.

The reactor ring is shown in its far left shifted position for maximum displacement and maximum flow. Using the handwheel the operator can move the reactor ring to the right, reducing pump flow. When the ring is concentric with the shaft axis the pistons are on zero stroke and there is no output flow. If the reactor ring is moved over center to the right, the pump resumes pumping but flow is in the opposite direction.

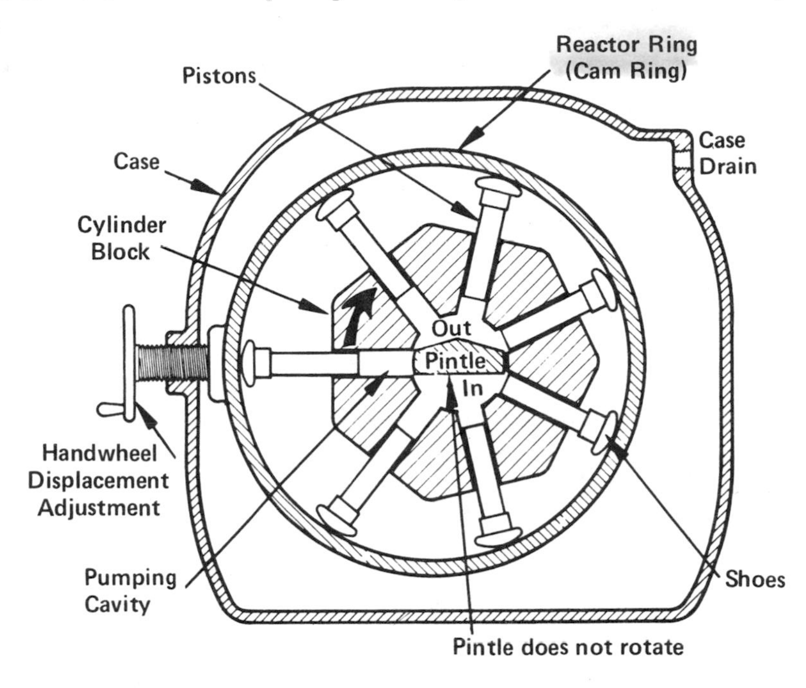

FIGURE 5-21. Variable Displacement Radial Piston Pump.

PRESSURE COMPENSATORS FOR HYDRAULIC PUMPS

A pressure compensator is a mechanism which can be built into certain pumps for the purpose of automatically reducing pump displacement to a lower flow (to zero if necessary), to prevent any further rise in pressure if the hydraulic system should be overloaded. A compensator is pre-adjusted to act (or fire) when the pressure has increased up to the maximum design level. It will act automatically without the necessity of operator attention. In fact it will override the operator to prevent him from placing an overload on the system.

The action of a pressure compensator on a single-lobe vane pump has been described on Page 182. Please note that variable displacement action (or a pressure compensator) cannot be built into a balanced vane pump.

On the graphic symbol for a hydraulic pump the presence of a compensator is shown by drawing a small vertical arrow inside the circular symbol. See Page 176.

Any of the types of piston pumps previously described which are available with variable displacement can be furnished with a pressure compensator.

On a sudden severe overload such as a cylinder stalling against a positive stop, a high pressure spike can be generated by the pump before the compensator, being a mechanical device, has time to act. To protect the pump and other components against the severity of such a spike, a small non-adjustable relief valve should be connected across the pump outlet line of most pressure compensated pumps. The relieving pressure should be about 300 to 500 PSI higher than the compensator firing pressure.

NON-POSITIVE DISPLACEMENT HYDRAULIC PUMPS

Figure 5-22. In contrast to the positive displacement gear, vane, and piston pumps described earlier in this chapter, these pumps have an impeller which does not make a tight sliding fit with the housing. Therefore, there is a high rate of slippage – fluid from the outlet leaking back to the inlet. For this reason these pumps cannot produce the high pressure required in an industrial hydraulic system. Energy is carried by the mass of the fluid moving at high velocity.

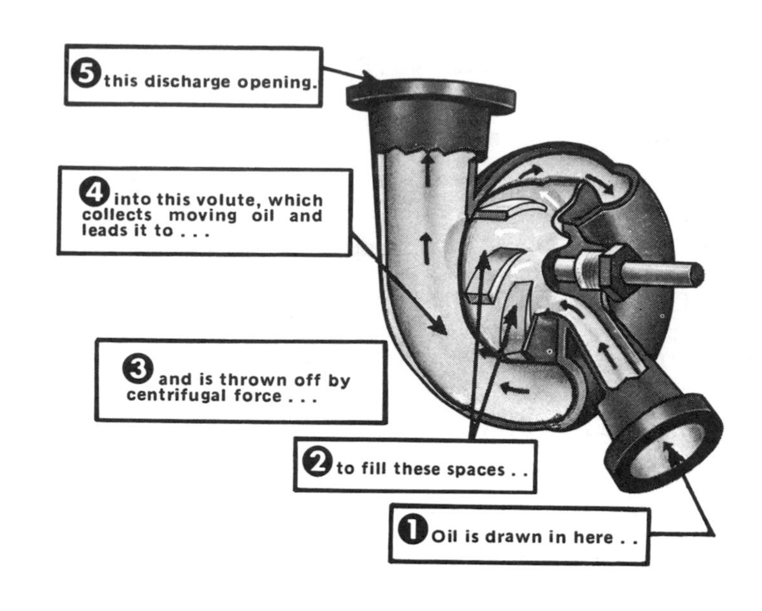

FIGURE 5-22. Impeller Type, Non-Positive Displacement Pump. Start with Box 1 and read boxes in sequence.

Although impeller pumps are not suitable for generating the main flow of fluid power, they can perform less important chores such as transferring fluids from one location to another, like filling a hydraulic tank from a barrel of oil, or supercharging the inlet of a high pressure pump.

Because there is no metal-to-metal contact between impeller and housing and because they do not depend on the pumped fluid for bearing lubrication, they can pump liquids which cannot be handled by conventional hydraulic pumps. They can pump fluids which are highly contaminated, those which have no lubricity, or those which may be corrosive to other pumps. However, they work best on low viscosity fluids such as water, kerosene, light mineral oil, etc.

A single-stage impeller pump is limited to 20 to 50 PSIG pressure. Some foundries and rubber molding companies generate higher pressures by using multi-stage pumps in which the outlet pressure from one stage supercharges the inlet of the next stage, etc.

The output flow from an impeller pump is not constant. It is greatest at little or no back pressure, and decreases as back pressure increases. For this reason a relief valve is not needed to protect the pump, as the entire flow can backslip internally if back pressure becomes too high.

HYDRAULIC HAND PUMPS

Hand pumps are used where a source of input power is not available or where the extra expense of a power pump is not warranted. For example, they are used on shop presses and other portable equipment, and always to power a ram-type cylinder for squeezing, press fitting, straightening, and many other uses. They serve as standby pumps on hydraulic systems, for emergency power if the main power input should fail. They are capable of developing pressure just as high as power driven pumps although at a much slower rate. They give excellent control of oil flow on applications like pressing small bushings on or off a shaft.

Hand pumps are always of the piston type, and are usually constructed with a piston (occasionally with two pistons in push-pull) working between two check valves. When the handle is raised for the suction stroke, oil is drawn into the cylinder cavity through the inlet check valve either from a self-contained or from an external reservoir. When the handle is lowered for the discharge stroke, oil in the cylinder cavity, not being able to backflow through the inlet check valve, is forced through the outlet check valve into the working circuit, a hydraulic ram.

Most hand pumps include an oil reservoir, one (or two) pistons, two check valves, one release valve for return flow, and a pressure relief valve. Single-acting pumps work in one direction with an idle handle return. Double-acting pumps work in both directions of the handle.

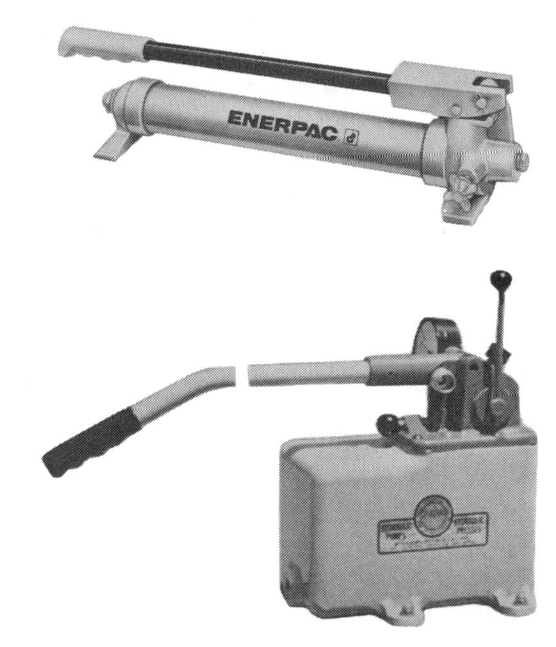

FIGURE 5-23. Typical Hand Pumps.

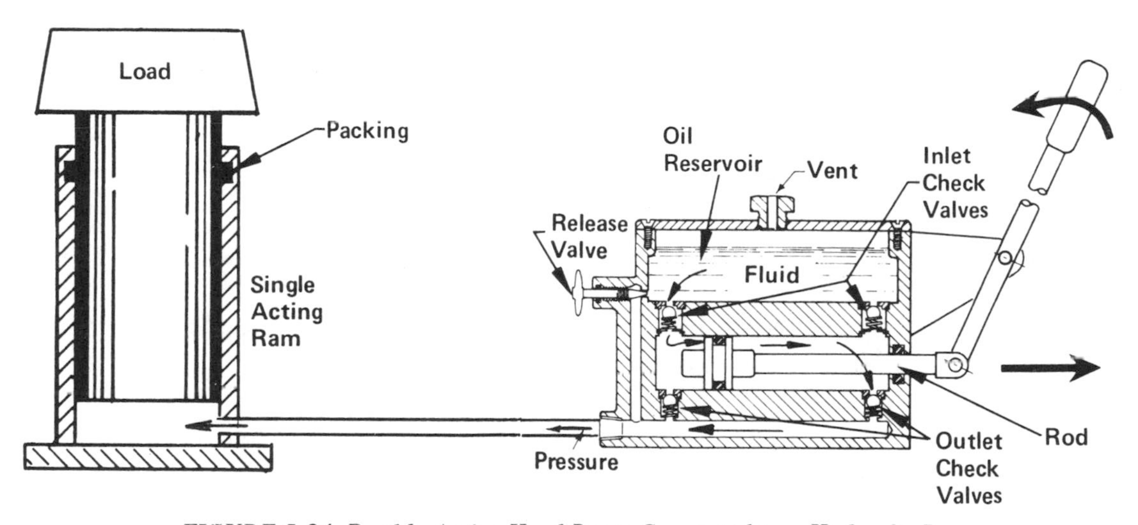

FIGURE 5-24. Double-Acting Hand Pump Connected to a Hydraulic Ram.

Double-Acting Hand Pump. Figure 5-24. In a double-acting hand pump there are either two single-acting pistons or one double-acting piston as in this figure. There are two sets of check valves, an oil reservoir, a release valve, and usually a pressure relief valve.

This figure shows a pump connected to a hydraulic ram. To raise the ram, the release valve on the hand pump must be closed. Stroking the pump handle draws oil from the reservoir, through the check valves and into the ram. The ram will stop and hold at any height when the pumping is stopped. To lower the ram, the release valve can be opened, draining the oil back into the reservoir.

Calculating Handle Strokes. On any hand pump, to calculate the number of handle strokes to raise the ram to a desired height, calculate the cubic inch volume to be filled in the ram according to its bore and the height to which it will be raised. Divide this by the cubic inch displacement per stroke of the hand pump. This rating will be found in the hand pump catalog.

ACTION OF A POSITIVE DISPLACEMENT HYDRAULIC PUMP

A positive displacement pump is one of the gear, vane, or piston pumps (or variation thereof) described earlier in this chapter. Due to the tight fit of the pumping elements, these pumps, when driven at a constant shaft speed, will deliver a relatively constant flow of fluid except for a small internal leakage which increases as pressure builds up.

We emphasize again that the primary purpose of a hydraulic pump is to produce a flow of fluid. It will, if possible, do this without building up pressure. Pressure build-up by a pump is a secondary function, and occurs only when a resistance is placed against the flow of fluid. Then, the pump will build up only enough pressure to maintain its flow against the flow resistance. Therefore, pressure read on a gauge placed on the pump line will be exactly proportional to the resistance which the pump is working against. Power drawn from electric motor or engine prime mover will also be directly proportional to the flow resistance in the pump line. In some hydraulic systems, because of light loading against the cylinder, a gauge in the pump line may not rise to relief valve pressure until the cylinder bottoms out.

To demonstrate the pumping action of a positive displacement pump, a hydraulic test stand could be set up and operated under conditions described below. See also diagrams on opposite page.

Figure 5-25. A hydraulic pump, driven by an electric motor, is picking up fluid from the reservoir and producing a flow of 20 GPM. Since the manual load valve is wide open there is practically no resistance to flow. The pump can move its rated flow through the load valve and back to reservoir without building up appreciable pressure. None of the fluid is escaping across the relief valve which is adjusted to open at 1000 PSI. A pressure gauge verifies that there is no pressure build-up in the system.

Figure 5-26. Now, the load valve has been partially closed, thus creating a resistance to flow. The pump will now generate enough pressure, but only enough, to maintain its full flow. The pressure which is necessary, for example 500 PSI, to push the entire 20 GPM of the pump through the partially closed valve can be read on the pressure gauge. The electric motor or engine must supply an equivalent amount of power to the pump. None of the pump flow is escaping across the relief valve because system pressure is less than the relief valve setting.

Power to force the fluid through the restriction of the valve generates heat, the amount of which can be calculated from the flow of 20 GPM and the gauge pressure of 500 PSI. This heat is carried by the flow of fluid into the reservoir to increase its temperature.

Figure 5-27. Finally, the load valve is completely closed. The pump continues to run at the same speed and to produce the same flow providing its driving motor has sufficient power. However, it must generate increased pressure to maintain its flow. Without the relief valve, pump pressure would continue to increase until the driving motor stalled, or until some part of the system ruptured from high pressure. The relief valve provides a maximum limit on pump pressure. The entire flow of 20 GPM must discharge across the relief valve back to reservoir. Allowing a pump to discharge across a relief valve for an extended period is highly undesirable in a working system. Since fluid power produced by the pump is not being converted into mechanical power, the entire input power to the system is being converted into heat. Not only is efficiency of the system reduced to 0% during these periods, but the wasted power will overheat the system in a very short time.

On hydraulic systems larger than fractional horsepower, it is not practical to stop the electric motor or engine every time the cylinder is stopped. The pump should remain running but some means should be provided in the circuit for pump fluid to flow without restriction during idle periods. This action is called "pump unloading", and several methods of unloading are shown in following pages.

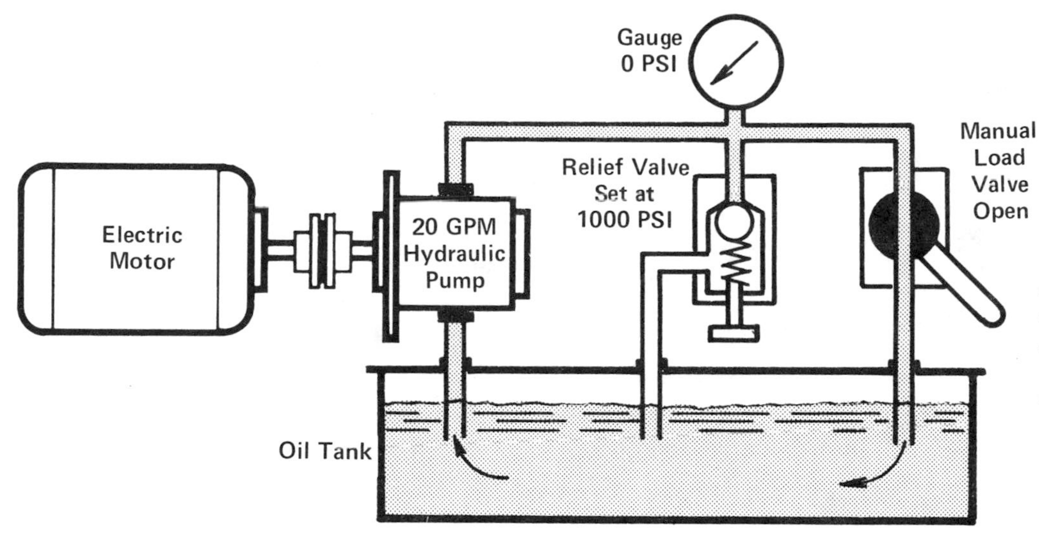

FIGURE 5-25.

Hydraulic test stand demonstrating free flow of hydraulic oil back to reservoir when the load valve is wide open. See description across page.

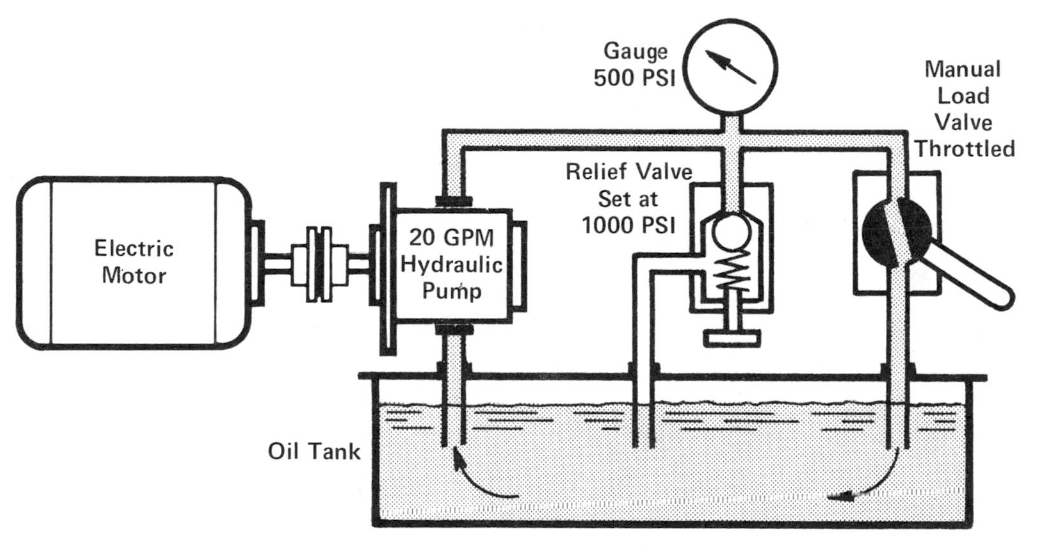

FIGURE 5-26.

When the load valve has been partially closed, the pump must build up enough pressure to overcome flow resistance through the valve, thus generating heat in the oil. The flow volume of a positive displacement pump remains constant except for a small amount of oil leakage in the working elements due to increased pressure.

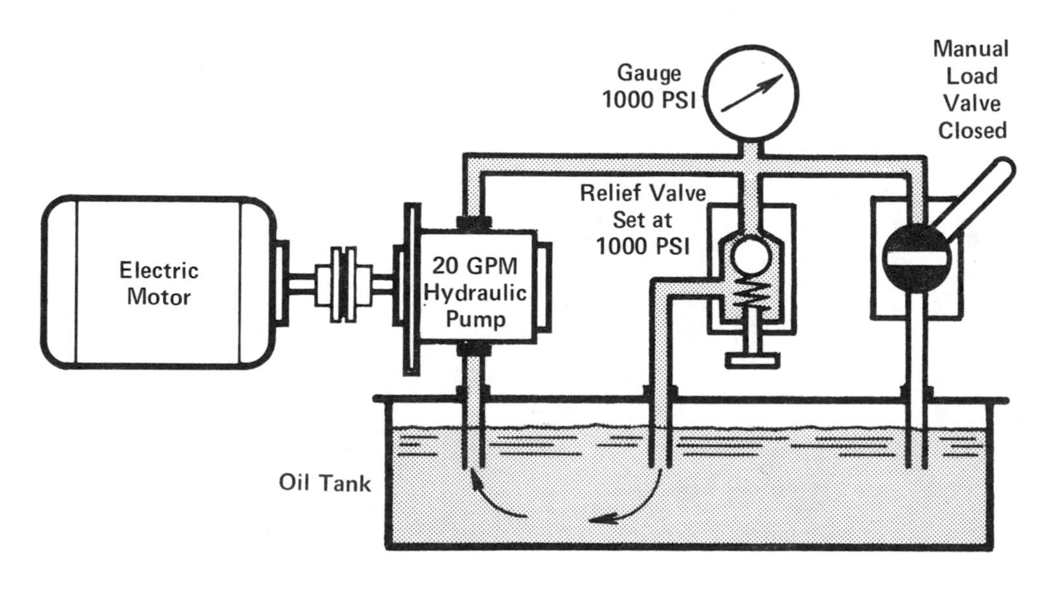

FIGURE 5-27.

If the load valve is completely closed, the pump must build up sufficient pressure to force the oil to discharge to reservoir over the relief valve. This can overheat some systems very quickly.

HP = PSI6 x GPM

UNLOADING A HYDRAULIC PUMP

In the matter of unloading, air systems act quite differently from hydraulic systems. On compressed air, when the operator stops the cylinder, the inlet flow is simply blocked; there is no power being used and no heating. But on a hydraulic system, using a gear pump for example, when the operator stops the cylinder, the hydraulic pump is usually allowed to continue running, and a free path is provided for the pumped fluid to return to the reservoir. If the oil flow were to be blocked, as in an air system, the pump would have to generate enough pressure (and power) to discharge the fluid across a relief valve. Heat generated by this discharge would be carried to the reservoir by the oil flow.

Since one horsepower of energy is equal to 746 watts, every 1 HP of fluid energy going across a relief valve will put the same amount of heat into the oil as if a 746 watt electric heating element were immersed in the reservoir. To avoid overheating the oil and to reduce power waste, all hydraulic systems of, say, 1 HP and over, should have a provision for unloading the pump during periods when the cylinder is stopped. Only waste power turns into heat; fluid energy converted into mechanical power and used against a load is not converted into heat.

Methods of Pump Unloading. Fluid horsepower is made up of two components – pressure and flow. HP = Pressure x Flow. To reduce HP to zero, as for unloading a pump, can be done in one of two ways: (1), The pressure can be reduced to zero, and, regardless of the volume of flow, HP in the flowing oil will be zero. Or, (2), the flow can be reduced to zero, and regardless of what a pressure gauge may read, the power will be zero.

Pressure Unloading, Fixed Displacement Pump. Figure 5-28. A fixed displacement pump has no adjustment nor any provision for an operator to change its pumping rate. When using a pump such as a gear pump, a tandem center 4-way control valve can be used, which provides a free unloading path for the full flow of pump oil when the valve spool is centered. By removing resistance to flow, the gauge pressure is reduced to zero and no power input is required to keep the pump running (except a small amount due to flow resistance through the pump line and 4-way valve passages). When the valve is shifted to either side position, oil is diverted into the cylinder, and pressure will build up in direct proportion to load resistance.

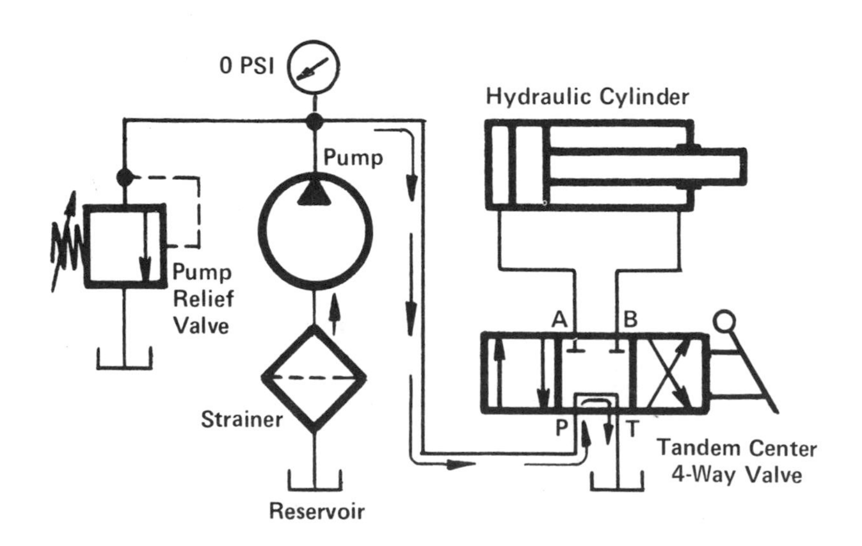

FIGURE 5-28. Standard circuit for operation of a double-acting hydraulic cylinder from a fixed displacement (gear) pump.

Multiple Branch Circuits With Tandem Center Valves. Figure 5-29. When two or more branch circuits are to receive oil from the same pump, the 4-way tandem center valves can be connected in series, or tandem. When both valve spools are centered, the pump is unloaded. Either valve spool can be shifted to either side position and oil will be diverted into that branch. Caution: Tandem (or series) circuits cannot be used if both branches must receive oil at the same time and if full pump

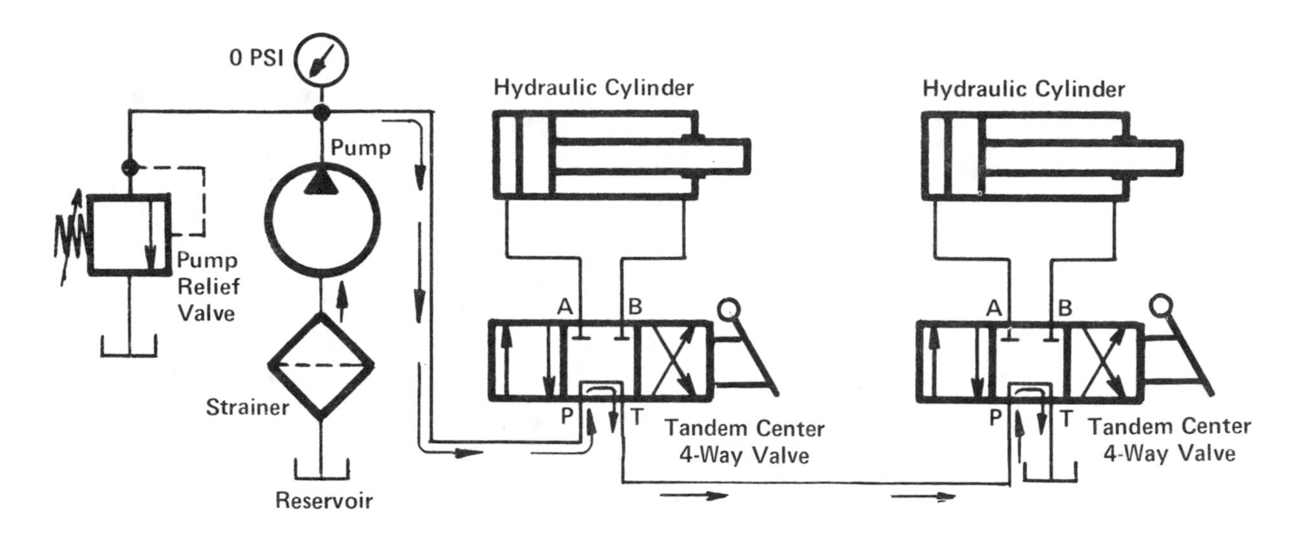

FIGURE 5-29. Multiple branch circuits from one pump using tandem center 4-way valves.

pressure must be available to both. Tandem circuits will divide the available pump pressure in direct proportion to the ratio of loads which are being carried. Full pressure will not be available to either branch.

Unloading by Blocking the Flow. Figure 5-30. Where two or more branch circuits must operate from one pump, and if full pressure must be available to two or more branch circuits at the same time, a variable displacement pump with pressure compensator is often employed. Closed center valves must be used so the pump will become deadheaded when all branch circuit 4-way valves are centered. When the pump becomes deadheaded, a pressure gauge on the pump line will show full compensator firing pressure, but the pump will be producing no power because it has reduced its own displacement to zero and is pumping no flow (except for a small amount to make up the leakage losses in the pump itself).

Review the action of a variable displacement, pressure compensated pump on Page 182. The compensator automatically stops the pump from pumping when the pump outlet becomes deadheaded. The optional pressure relief valve shown in Figure 5-30 is recommended for all pressure compensated pumps. On a very sudden deadhead created by shifting the 4-way valves to neutral, there may be a

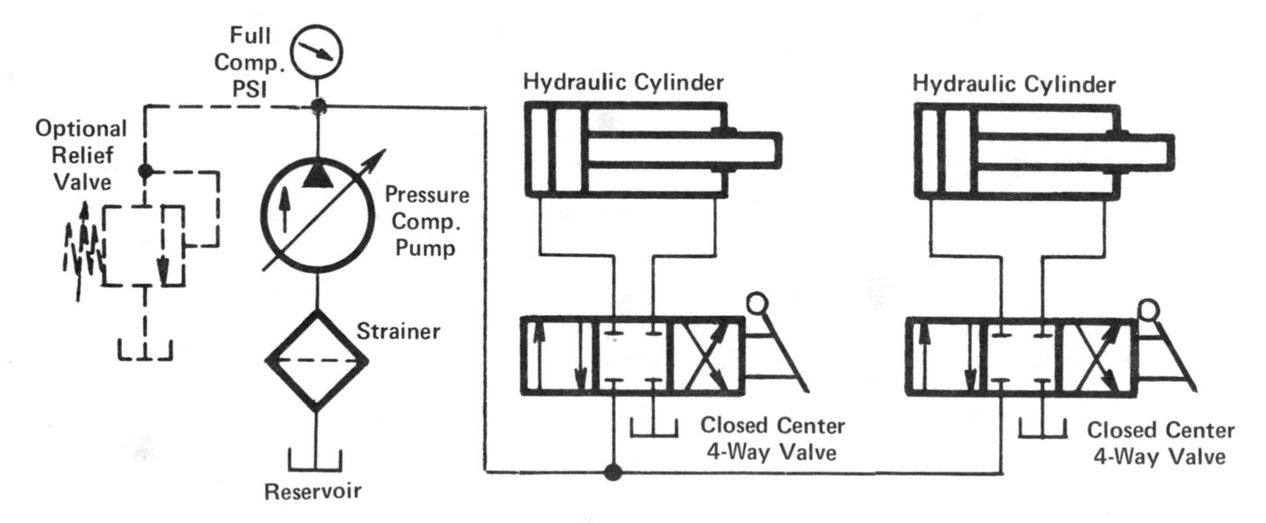

FIGURE 5-30. Multiple branch circuits from one variable displacement, pressure compensated pump.

brief but very high pressure transient pass through the system during the few milliseconds which it takes for the pressure compensator to act. Usually, a very small, direct-acting relief valve set to a pressure about 300 to 500 PSI higher than the compensator firing pressure, will prevent the transient from rising to a dangerous level.

In Figure 5-30, although all branch circuits can receive full pressure at the same time, they must divide the available pump flow when more than one branch is in operation.

Bank Valves for Unloading. For manually controlled hydraulic systems, especially on mobile equipment, bank valves provide an excellent means of operating several branch circuits from one pump while still unloading the pump when none of the branches is working. Bank valve action can be reviewed on Page 168.

INPUT HORSEPOWER TO A HYDRAULIC PUMP

Flowing power in the fluid stream is made up of a combination of gauge pressure and volume of flow. Pressure is usually measured in PSI and flow in GPM (Chapter 1). Both of these conditions must exist before any power is present. Pressure by itself can transmit force from one location to another but at a zero power level; that is, no work can be done. Flow by itself can transmit fluid from one location to another but unless it is under pressure it has no capability for transmitting power or doing work at the remote location. The amount of flowing power is in direct proportion to each of these factors. If the flow volume is doubled (at the same pressure), twice the power will flow. If the pressure is doubled (at the same flow), twice the power will also flow.

To increase the horsepower transmitted the rate of flow can be increased either by using a pump which has a greater displacement or by increasing the shaft RPM of the existing pump. Or, to increase the horsepower the pressure level can be raised by increasing the setting of the pressure relief valve on the pump line. But remember that increasing the flow of horsepower by either of these methods requires a corresponding increase in driving power to the pump from an electric motor or engine. For instance, when replacing a broken pump with a new one, it is important that the flow rate of the new one be the same as, or less, than the old one. If the flow is greater, be sure there is sufficient driving power available to run it at the operating pressure required in the system.

Input Power to Drive a Hydraulic Pump. For "ball parking" the horsepower required, use the "rule-of-1500" given in the box on the opposite page. A more exact calculation can be made later.

Horsepower Table. The table on the opposite page is a sample of the format used to determine the input horsepower required to produce a given combination of GPM (left column) and PSI (top line). A more complete chart will be found in the Appendix on Page 277. Figures in the chart are based on a pump efficiency of 85% for conversion of mechanical power into fluid power. Most pumps will fall into the range of 80 to 90% efficiency, so the chart should be accurate to within ±5% for almost any pump. The chart was calculated from the formula: $HP = PSI \times GPM \div 1714 \div 0.85$. If actual pump efficiency is known, it can be used in this formula in place of 0.85.

Using the Table. The range of 500 to 5000 PSI covers the usual range of hydraulic pressures, but values for other pressures can be determined by adding values from the table. For example, at 3500 PSI pressure, add the values in the 3000 and 500 PSI columns. For systems operating at less than 500 PSI, horsepower calculations become inaccurate because pump losses become a greater percentage of the converted power.

Electric Motor Drive. When operated under normal environmental conditions, a 3-phase squirrel cage induction motor (the kind used for most systems), has a service factor of 10 to 15% and can be overloaded to this extent without damage to insulation. If the fluid system does not require full horsepower continuously, but only for short duration peaks, and with intervals between peaks at less than full horsepower, the motor will usually stand an additional overload. We recommend a maximum overloading of not more than 25% above nameplate horsepower and for not more than 10% of total running time. We do not recommend overloading single-phase motors beyond their normal service factor.

If motor is to be operated at unfavorable conditions such as unusually high or low room temperature, high altitude, etc., the motor manufacturer should be consulted about overloading. The motor, in some cases may actually have to be de-rated.

Engine Drive. It is almost needless to say that an engine cannot be overloaded beyond its rated torque without stalling. Engine driven hydraulic systems must be carefully designed to avoid overloading.

RULE-OF-1500 FOR ESTIMATING PUMP DRIVE HORSEPOWER

This rule is handy for rough estimating and is easy to remember. It states that 1 HP of drive is required for each 1 GPM of flow at a pressure of 1500 PSI, or any equivalent of these numbers.

For example, a 5 GPM pump operating at 1500 PSI would need 5 HP, or at 3000 PSI would need 10 HP. A 10 GPM pump at 1000 PSI would need 6-2/3 HP, or the same pump operating at 1500 PSI would need 10 HP, etc.

Another handy rule-of-thumb is that it takes about 5% of the full rated horsepower to idle a pump when it is running "unloaded"; that is, when it is re-circulating oil at near zero pressure. This amount of power is needed to take care of flow losses through the pump plus mechanical friction losses in bearings and pumping elements.

INPUT HORSEPOWER REQUIRED FOR A HYDRAULIC PUMP

Figures in the body of this chart show input horsepower needed to produce flows in the first column at the pressures shown along the top of the chart. This is part of a more complete chart which appears on Page 277. (Pump efficiency is assumed to be 85%).

GPM	500 PSI	750 PSI	1000 PSI	1250 PSI	1500 PSI	1750 PSI	2000 PSI	2500 PSI	3000 PSI	3500 PSI	4000 PSI	5000 PSI
5	1.72	2.57	3.43	4.29	5.15	6.00	6.86	8.58	10.3	12.0	13.7	17.2
10	3.43	5.15	6.86	8.58	10.3	12.0	13.7	17.2	20.6	24.0	27.5	34.3
15	5.15	7.72	10.3	12.9	15.4	18.0	20.6	25.7	30.9	36.0	41.2	52.5
20	6.86	10.3	13.7	17.2	20.6	24.0	27.5	34.3	41.2	48.0	54.9	68.6
25	8.58	12.9	17.2	21.4	25.7	30.0	34.3	42.9	51.5	60.1	68.6	85.8
30	10.3	15.4	20.6	25.7	30.9	36.0	41.2	51.5	61.8	72.1	82.4	103
40	13.7	20.6	27.5	34.3	41.2	48.0	54.9	68.6	82.4	96.1	110	137
50	17.2	25.7	34.3	42.9	51.5	60.1	68.6	85.8	103	120	137	172
60	20.6	30.9	41.2	51.5	61.8	72.1	82.4	103	124	144	165	206
70	24.0	36.0	48.0	60.1	72.1	84.1	96.1	120	144	168	192	240
80	27.5	41.2	54.9	68.7	82.4	96.1	110	137	165	192	220	275
100	34.3	51.5	68.6	85.8	103	120	137	172	206	240	275	343

PUMP AND MOTOR SHAFT ALIGNMENT

If a hydraulic pump is separately mounted, as on a foot bracket, and coupled directly into the shaft of an engine or electric motor, both pump and motor should be mounted on a very rigid common base which is sufficiently rigid that it will not flex under torque surges between the two units. Pump and motor shafts should be very carefully aligned. This applies not only to hydraulic pumps but also to hydraulic motors, air motors, and air compressors. Even a small amount of misalignment places a side load on bearings which they were not designed to support. Bearing life will inevitably be shortened.

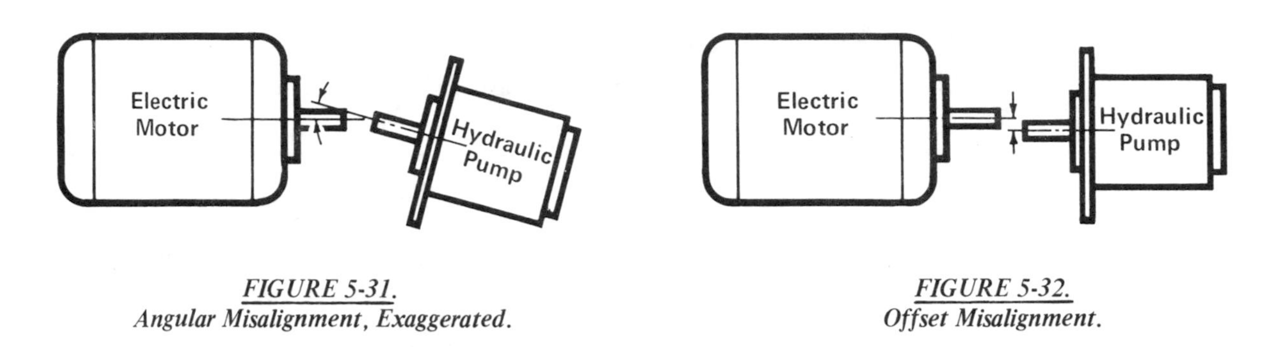

FIGURE 5-31.
Angular Misalignment, Exaggerated.

FIGURE 5-32.
Offset Misalignment.

Misalignment is of two types. Figure 5-31 shows an exaggerated case of angular misalignment in which the two shafts may be the same height from the baseline but are not parallel. Figure 5-32 shows offset misalignment in which the shafts may be parallel but are shifted sidewise or are not the same height from the mounting base. This type of alignment may be caused by machining tolerance in the electric motor shaft to base dimension, which could be as much as 1/32" short of nominal dimension. It may be necessary to place shims under the motor feet to bring its shaft height to within a few thousandths of an inch of the pump shaft height. Both angular and offset misalignment can be present on the same installation.

Note: When replacing a foot mounted hydraulic pump, its foot mounting bracket should not be replaced unless it has been damaged. If the old pump was in good alignment with the motor shaft, a new pump of the same kind will usually be in perfect alignment when mounted on the original foot bracket. However, the alignment should always be checked because it may not have been accurate on the first pump. In fact this might have been a contributing cause to the original pump failure.

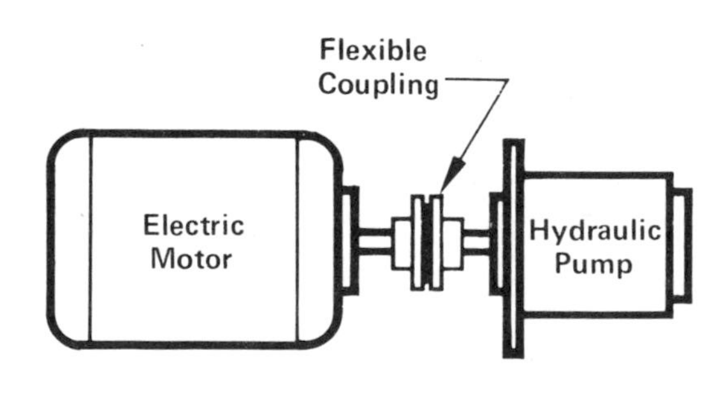

FIGURE 5-33. Flexible Coupling.

When the two shafts have been accurately aligned, a flexible coupling should be used to join them. The purpose of this coupling is to correct small alignment errors which may remain. But if the basic alignment was poor, the coupling itself will wear out prematurely. Good practice dictates that shaft alignment should be as near perfect as possible before the coupling is installed. On some couplings, such as the roller chain type, alignment error can be detected after the coupling has been installed, by rotating the shafts about 30° at a time while checking for binding in the coupling.

SIDE DRIVE OF A PUMP WITH BELT OR GEARS

Any kind of side drive directly applied to a pump shaft such as belt, gear, or chain, places a side load on pump bearings and bends the pump shaft. Unless the pump is specifically designed for side drive, this invites shaft breakage and premature bearing failure. Some hydraulic pumps and motors, operating at speeds less than 1000 RPM, are designed with large shafts and bearings to withstand the extra strain of side drive. However, most high speed pumps and motors are built for direct drive and their working life is always shortened by side loading. Some of them may even carry catalog ratings specifying the amount of side loading which is acceptable, but nevertheless their life expectancy is always greater when used on applications where there is no side or end loading of the shaft.

Figure 5-34. If side drive must be used, the sheave or gear should be mounted as closely as possible to the pump bearing and with the hub facing away from the pump as shown here. By keeping the drive axis as nearly over the bearing as possible, shaft deflection will be kept to a minimum. Side loading can also be kept to a minimum by choosing diameter of sheaves, sprockets, or gears as large as practical. Side thrust is reduced in the same ratio that diameter is increased. By observing these rules for good design and by reducing pressure and speed to less than maximum catalog ratings, pump life can be prolonged.

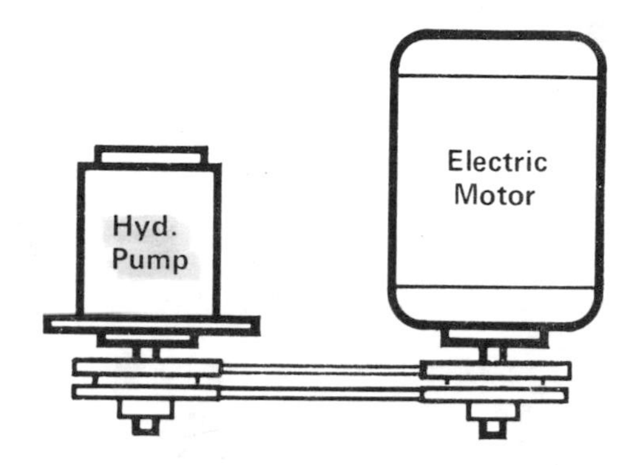

FIGURE 5-34. Side drive places an extra load on pump bearings and shaft.

Figure 5-35. Any side loading placed directly on the pump shaft always reduces the life expectancy of any pump. To have the convenience of side drive without placing a side load on the pump shaft, the best arrangement is to mount the sheave or gear on a jackshaft (idler shaft) which is supported on pillow block bearings. The pump, after being carefully aligned with the jackshaft, can be coupled to it with a flexible coupling. The jackshaft can be side driven by an electric motor or engine. Bearings and shafts on these components are usually quite adequate for side loading.

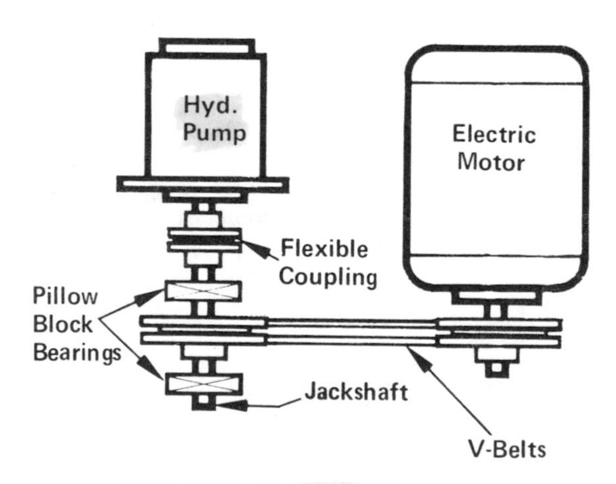

FIGURE 5-35. Side load is removed if pump is driven from a jackshaft.

DIRECTION OF PUMP ROTATION

Specifying Pump Rotation. It is imperative with most pumps that they be rotated in the correct direction. Check valve piston pumps in Figures 5-19 and 5-20 are exceptions, as they can be rotated in either direction and will always deliver oil from their outlet port. With other pumps the direction of oil flow will reverse if direction of rotation is reversed, or they may not pump at all.

When ordering a new or replacement pump, its rotation should be specified with a notation such as,

"CW (clockwise) rotation when viewing shaft end of pump" or "CCW (counter-clockwise) rotation when viewing end opposite shaft" etc. The standard designation for shaft rotation on a pump, CW or CCW, is specified when viewing the shaft end. This is opposite to the determination normally used for electric motors. When installing a pump, look at the rotation marking on it either on the nameplate, on a separate plate, or look at the markings on the ports.

Uni-Directional Pumps. These must be rotated in the direction marked. Some models can be changed to the opposite rotation in the field by shifting a drain plug from one hole to another, turning over a cam ring, or rotating the ring 90 degrees. Other uni-directional pumps cannot be reversed in the field.

Pumps like the internal gear, external gear, vane, and their variations, will very likely blow a shaft seal if rotated under pressure in the wrong direction. If this happens the pump will pump air back through the inlet strainer into the reservoir.

Bi-Directional Pumps. This group of pumps can be rotated at will in either direction without making a field modification. Four kinds of pumps fall into this group:

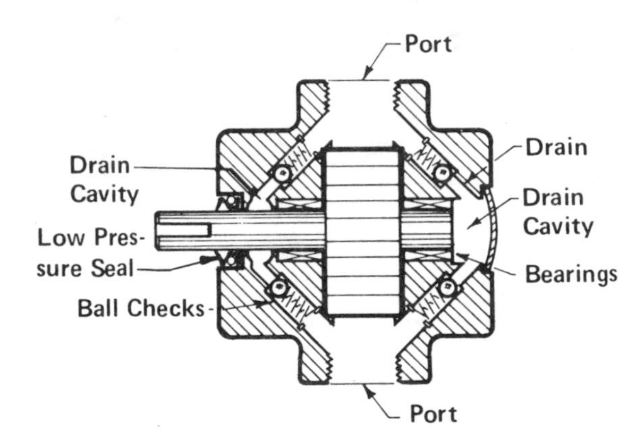

FIGURE 5-36. This pump is made bi-directional by a set of check valves for draining slippage oil.

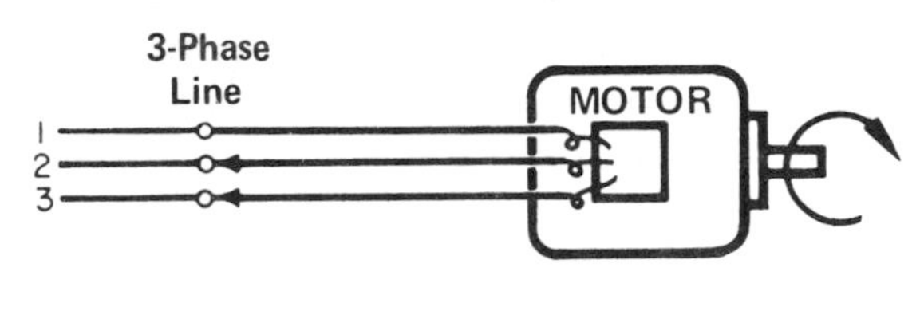

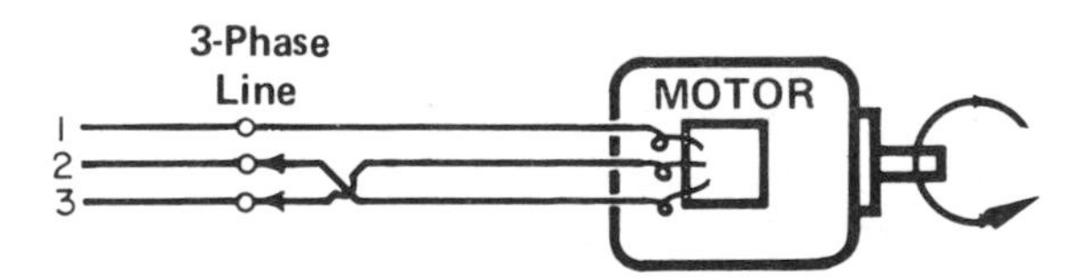

FIGURE 5-37. A 3-phase electric motor may be reversed by interchanging any two line wires.

(1). Figure 5-36. Those built with a set of internal check valves as shown in this figure. Leakage oil which seeps into the bearing and seal cavities can be drawn by suction back to the inlet port. Since direction of flow changes when shaft rotation is reversed, for CW rotation one port is the inlet and for CCW the other port becomes the inlet. Check valves prevent oil from the outlet port from backflowing to the inlet port through cavity drain holes.

(2). Those pumps which have a high pressure (face-type) shaft seal are capable of rotation in either direction. Since the oil flow reverses when the shaft rotation is reversed, the inlet and outlet port connections must be interchanged if shaft rotation is reversed.

(3). Check valve piston pumps shown in figures 5-19 and 5-20 may be rotated in either direction. Oil flow does not reverse when shaft rotation is reversed.

(4). Some pumps which have an external drain port may be run in either rotation. Internal leakage is brought out the drain port, and this port should always be connected to reservoir, never plugged. Oil flow reverses when shaft rotation is reversed, so inlet and outlet connections must be interchanged.

Checking Direction of Rotation. After installing a pump, test it for correct rotation by very briefly jogging the electric motor. If motor and pump shafts are enclosed and cannot be seen, rotation can be checked by viewing the oil in the reservoir through the filler or cleanout opening, or by listening through one of these openings. A pump running backward will pump air into the reservoir and the turbulence can be seen and heard.

Pumping assemblies disconnected from the electric power line and moved to a new location should always be tested for rotation when connected back into the electric line. On 3-phase electric circuits the motor will run in reverse if the line wires happen to be re-connected in a different sequence.

Reversing an Electric Motor. Figure 5-37. A 3-phase motor can easily be reversed by simply interchanging any two of the three connecting wires.

A single-phase electric motor cannot be reversed simply by reversing the plug in the electric socket. The starting winding must be reversed in relation to the running winding, and this is done by interchanging wires inside the motor junction box. Instructions will be found either inside the junction box or on the motor nameplate. A single-phase motor, once wired for the correct rotation will always run correctly when moved to another location.

PUMP INLET DESIGN

Suction Strainers. Figures 5-38 and 5-39. Because of the close metal-to-metal clearances inside a positive displacement pump, the oil entering the inlet should be filtered to remove solid particles large enough to jam between working elements. Abrasive particles, even though not large enough to jam the pump will cause excessive wear in pump, valves, and cylinder. If possible, these particles should be prevented from entering the power stream. The most practical way to remove large particle contamination is to install a stainless steel wire mesh strainer in the pump inlet line either below the oil level in the reservoir or external to the reservoir near the pump inlet port.

Most pumps create a slight suction at their inlet port which draws oil into the inlet by force of atmospheric pressure. Although gear pumps, for example, are capable of creating a suction of nearly 20" Hg, this would harm them. Suction vacuum should not be permitted to exceed 5" Hg on a gear pump

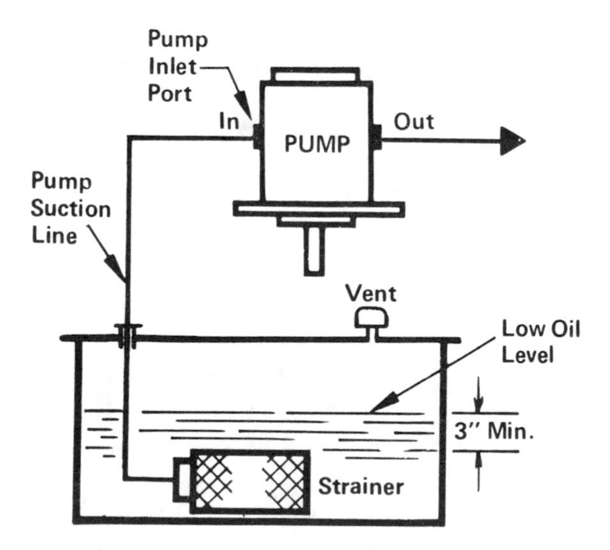

FIGURE 5-38. Inlet strainer immersed in the reservoir below oil level.

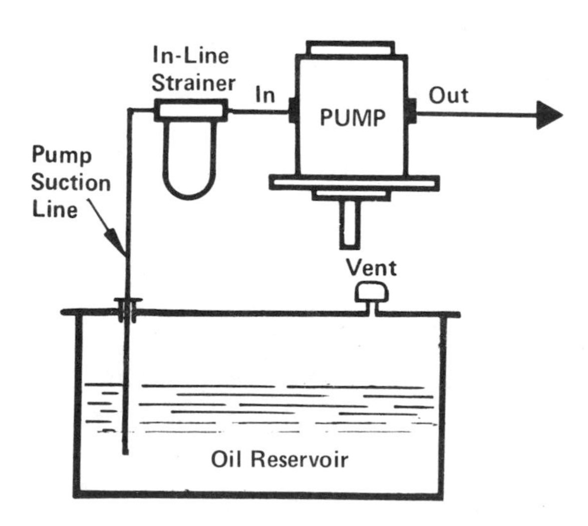

FIGURE 5-39. Inlet strainer in the pump inlet line, external to the reservoir.

and even less on a vane or piston pump. To help keep the inlet vacuum to a minimum, the inlet strainer should be very generous in size with no more than 1 to 2" Hg flow restriction when clean.

For petroleum hydraulic oil, the recommended inlet filter is a 100-mesh (100 wires to the inch) screen. This gives a filtration of about 150μm (micro-metres) which is about as fine as practical without over-restricting flow into the pump. All other fluids, having a higher specific gravity than oil, flow with more restriction. They should be passed through a coarser filter – about 60-mesh wire with a 260μm rating. As a rule-of-thumb, an area of 10 square inches of filtering area should be provided for each 1 GPM of flow of oil of average viscosity. For oils with over 500 SSU viscosity, for fluids with greater specific gravity, and for cold weather start-ups, a strainer with 2 to 3 times this area is recommended.

Finer strainers, or even micronic filters can be used in the pump inlet but sizes large enough to keep flow restriction to less than 2" Hg, are very large and very costly. If used, they should be sized to create no more than 1 to 2" Hg flow restriction.

Much finer filtration is required in some systems using certain types of pumps. An inlet strainer should be used as recommended above, supplemented with a micronic-type filter in the pump pressure line or in the main tank return line. In these locations there is sufficient pressure available for moving the oil through the filter, and somewhat higher flow losses are acceptable

All plumbing joints in the pump inlet line should be made up air-tight. Even small air leaks will cause erratic operation. Strainers installed in the reservoir (Figure 5-38) should be submerged at least 3 inches below the minimum oil level. If not submerged deep enough, vortexing may occur, drawing air into the pump.

Note: All plumbing in the pump inlet should be much larger than in the rest of the system; it should be large enough to keep flow velocity in the range of 2 to 4 feet per second.

PUMP CAVITATION

A pump is said to be "cavitated" when inlet oil, for whatever reason, is not entirely filling the cavities which open up for it on the intake part of the pumping cycle.

Cavitation may cause damage in the pump bearings if the flow of oil is not sufficient to carry away the heat produced by mechanical friction. Lack of cushioning between pumping elements may also cause impact damage.

If not properly designed, or if not properly maintained, any system can have or can develop cavitation. More pump and system failures occur from cavitation than from any other single cause.

Vacuum Cavitation. Figure 5-40. Voids may occur between pumping elements if sufficient oil cannot flow into the pump because of restrictions such as clogged inlet strainer, undersize plumbing, too great a suction head, too long a plumbing run, oil too viscous (perhaps because of cold temperature), or if a hose is used for pump inlet connection, the hose may have collapsed or become restricted at some point.

Partial vacuum at the pump inlet will cause vaporization of some of the oil which will cause air to come out of solution with the oil. Although most of the oil and air vapor will go back into solution when the oil is pressurized, the immediate result is cavitation, causing increased pump wear and vibration.

Air Leak Cavitation. Figure 5-41. This is called "aeration" and is a form of cavitation and should be considered as such. Air can enter the oil stream as a result of one or more of these causes:

(a). Low oil level in the reservoir may have uncovered a part of the inlet strainer. Vortexing may be drawing in air even though the oil level is above the top of the strainer.

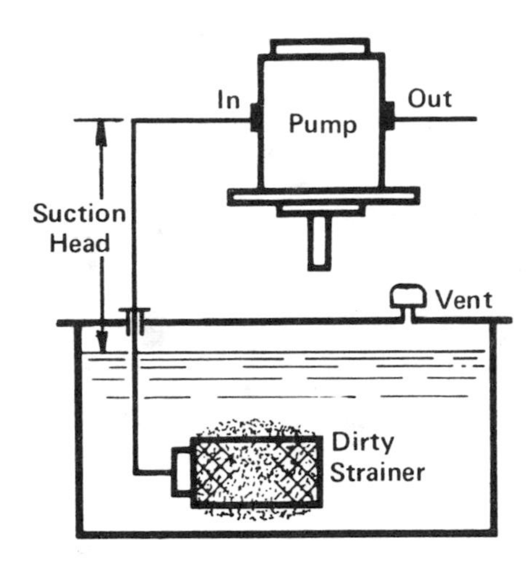

Vacuum cavitation caused by a dirty strainer.

FIGURE 5-40. Vacuum cavitation is the inability of a pump to pull in a full charge of oil by suction.

Air leaks also cause cavitation.

FIGURE 5-41. Aeration or the entry of air into the oil system is another form of cavitation.

(b). Leaks in the inlet plumbing line, especially at unions. A leak may also be present at the pump inlet port if, in replacing a pump, a taper pipe thread fitting was screwed into a straight thread port. Note: straight thread ports have a machined surface around the thread for seating of the O-ring seal.

(c). Worn out pump shaft seal. If there is a vacuum on the inside of the shaft seal, air may be leaking into the pump. To test shaft seal for air intake, squirt a little oil around shaft near seal while pump is running. If oil is sucked in, an air leak is present. (Note: if pump is badly worn, oil may be leaking into the seal cavity faster than it can be carried away by the internal drain. This usually blows the seal).

(d). On some installations air may be entering through a worn piston rod seal in the cylinder. If cylinder is mounted with rod extending upward, a partial vacuum may be created in the rod end as the cylinder retracts under a heavy load.

Cavitation Check List. Cavitation should be suspected if pump operates with an increased noise level which greatly increases as load is applied, or if the pump case becomes very hot around the shaft. Sometimes the pump is unable to build the pressure high enough to pick up the full load or to produce the maximum force. The cylinder may move erratically or may stall before full pressure is reached. These symptoms also may indicate a badly worn pump instead of cavitation. If cavitation is suspected, the following points should be considered:

(1). The inlet strainer or filter may be too small for the pump flow.

(2). The inlet strainer may be partly stopped up with dirt. All hydraulic pumps should have an inlet strainer. If one is not visible in the pump inlet line it is probably immersed in the reservoir. To test for cavitation, some users temporarily remove the strainer, or take the element out of an in-line strainer, leaving only the housing. The pump should not be run long with the strainer removed. If the symptoms clear up with the strainer removed, the problem is restriction in the inlet line or a dirty strainer.

(3). The inlet plumbing line may be too long or too small in size.

(4). The suction head may be too great. See Figure 5-42.

(5). Air may be leaking into the oil for reasons given on the preceding page.

(6). Oil level in the reservoir may be low, exposing part of the inlet strainer. Remember that oil level rises and falls as cylinders retract and extend. Check reservoir oil level while all cylinders are fully extended.

(7). The baffle inside the reservoir may be restricting free flow one side to the other, causing air pick-up at the inlet strainer.

(8). The oil viscosity may be too high either because of low temperature or from the wrong oil having been added to the reservoir.

(9). If the pump is belt-driven, sheave ratio may have been changed, causing pump to run faster than was originally intended. Inlet strainer size should be increased.

(10). The oil may have an excessive amount of foam in suspension after the system has been in operation for an hour or two. This may be caused by turbulence created by tank return oil discharging at too high a velocity into the reservoir. Enlarge the diameter of the return line for a few feet before it enters the reservoir.

Foam may also be generated by return oil discharging above oil level in the reservoir, or from the reservoir capacity being too small for the pump flow, or from use of the wrong oil. For example, motor oil may not have the foam suppressant additive which is required in hydraulic oil.

INLET CONDITIONS FOR A HYDRAULIC PUMP

To avoid cavitation damage in a hydraulic pump, it should not be run with its inlet subjected to excessive vacuum. Some piston pumps should not be run with more than 1 to 2" Hg vacuum, others can run with slightly higher vacuum. Vane pumps should not have more than 4 to 6" Hg, and gear pumps with no more than 6 to 10" Hg vacuum. Each pump manufacturer has his own recommended limitations on the degree of inlet vacuum for his own pumps.

(1). Suction Head. Figure 5-42. If the pump is mounted at a higher elevation than the oil reservoir, the height from the oil surface to the pump inlet is called the "suction head". This should be kept to a minimum because each foot of head will produce about 1" Hg vacuum on the pump inlet if pump suction is used to elevate the oil. This is in addition to the vacuum (loss of pressure) due to flow resistance in the line.

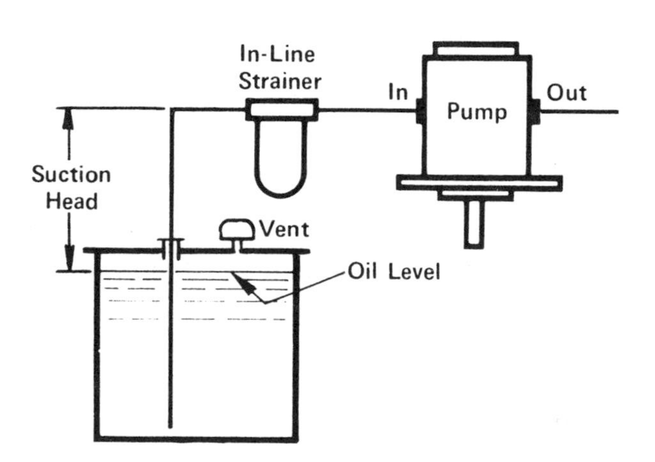

FIGURE 5-42. Suction head on a pump is the height to which the oil must be elevated by pump suction.

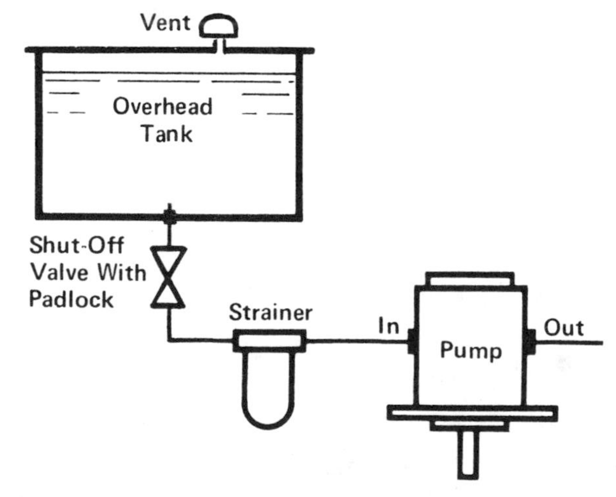

FIGURE 5-43. A flooded suction (pressurized inlet) may be obtained with an elevated reservoir.

(2). Flooded Suction. Figure 5-43. By placing the reservoir at a higher elevation than the pump, the inlet vacuum (due to suction head) can be eliminated, and replaced with a positive inlet pressure which will help oil to flow into the pump. However, the loss of pressure due to flow resistance still remains, so inlet piping should be large enough to reduce oil velocity to 2 to 4 feet per second.

Most pumps will tolerate only a small amount of positive inlet pressure, usually less than 15 PSI, so elevation of the reservoir should not exceed 10 to 15 feet. Too high an inlet pressure will blow out the shaft seal of most pumps.

A shut-off valve is usually required on an elevated tank so the oil can be shut off for servicing of the pump and inlet strainer. To prevent accidental cavitation of the pump if someone should start the pump with the valve closed, the valve should be padlocked in the open position.

(3). Horizontal Separation. Horizontal separation of pump and reservoir should be kept to a minimum because additional vacuum is required to flow oil around corners and through pipe. We recommend that pipe diameter be enlarged, on runs longer than 5 to 10 feet, so flow velocity will not exceed 1 foot per second.

(4). Priming a New Pump. When a pump is first installed, pumping cavities are filled with air. This air must be expelled before the pump can perform properly. Oil must be drawn into the cavities to replace the air. This is called "pump priming". This is no problem if the reservoir is at a higher elevation than the pump, but if the suction head is too great it may be difficult to get the pump primed.

A check valve piston pumps is one of the hardest to get primed because of the cracking pressure of the inlet check valves. A gear pump usually primes readily.

The most effective way to prime a pump is to first slightly crack a fitting in the outlet line. Then turn the pump and motor shafts slowly by hand until oil begins to come out the fitting. On large pumps which cannot be turned by hand, jog very briefly with the electric motor, then let the motor and pump come to a complete stop before jogging again. Jog until oil seeps out the cracked fitting. Often a pump can be primed in this way which will not prime by continuous running of its electric motor at high speed.

Should more severe measures be necessary, a low air pressure can be applied to the top of the oil in the reservoir. Caution! do not apply more than 1 PSI pressure obtained through a low range pressure regulator. This is enough pressure to lift the oil at least 2 feet. Air must be applied with caution because some large tanks could be damaged by more than 1 to 3 PSI.

Many piston pumps draw intake oil from their own case. On all piston pumps, the case should be filled with oil through one of the case drain ports before the pump is started.

Some piston pumps (except those of the check valve type) can be primed by connecting a hose to the outlet port, immersing the hose in a bucket of oil, then turning the shaft backward. This will suck in oil to fill the pumping cavities. But this may not fill the case, which should be done separately.

(5). Losing Pump Prime. Most positive displacement pumps are self-priming. This means they will develop enough suction to purge themselves of air and pull in a charge of oil. The only time a pump should have to prime itself is on its initial installation, or after it has been drained for cleaning, repair, or overhaul. Normally the inlet line will remain full of oil when the pump is stopped even though the reservoir is at a lower elevation. If oil drops down out of the inlet line when the pump is shut down, this may be caused by an air leak in the plumbing or around the pump shaft seal. A check valve type piston pump may lose prime if any one of its inlet check valves develops a leak.

(6). Supercharging a Pump. Figure 5-44. A pump is said to be supercharged if oil is supplied to its inlet under pressure, usually from another pump. Most gear and vane pumps and some piston pumps,

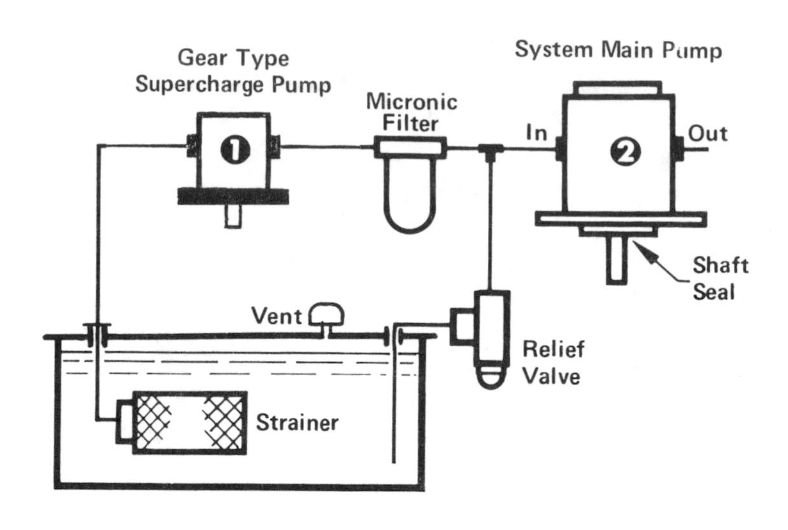

FIGURE 5-44. A low pressure, high volume gear pump is used to supercharge the inlet of a high pressure main pump.

when operated under normal conditions do not need to be supercharged. But when operated with extra long inlet lines, or run at higher than normal speed, or with too great a suction head, their inlet suction may be insufficient to keep them from cavitating. Certain types of piston pumps have low suction or none at all, and must be supercharged under all operating conditions.

Supercharging is usually done with a low pressure auxiliary pump, a gear type or even an impeller type. The supercharge pump must have slightly greater displacement than the main pump.

In the diagram, Pump 1 is the supercharge pump. Excess oil which cannot be accepted into the main pump must be discharged to the reservoir through a low pressure relief valve set for a pressure of 5 PSI for most pumps, or up to 300 PSI when supercharging a closed loop hydrostatic system using piston pumps. When supercharging any pump it is well to find out manufacturers maximum rating on inlet pressure. Too much pressure may damage or blow out the shaft seal.

If micronic filtering is to be included, the best place to locate a micronic filter is in the supercharge pump line as shown in Figure 5-44.

Supercharging may also be done by locating the reservoir at a higher elevation. Figure about 1 PSI of supercharge pressure for each 3 feet of elevation. To obtain 5 PSI supercharge pressure the reservoir would have to be elevated about 15 feet above the pump inlet.

NOTES ON PUMP OPERATION

Case Drains. All piston pumps have case drain connections, sometimes as many as 3 or 4. Slippage oil which by-passes the pistons accumulates in the case and lubricates the moving parts. Some pumps pick up suction oil from their own case. A drain line must be run to reservoir with minimum restriction, and without being combined with any other tank return lines. Plug all case drain ports except the one used, which should be the one nearest the top of the case. Never plug all drain connections; this would cause pressure to build up in the case and blow out the shaft seal. It is very important to prevent case pressure from exceeding 15 PSI for most pumps.

Other pumps – gear and vane – do not have a case in the same sense as do piston pumps. Seal drain ports are seldom seen on these pumps. Leakage of oil into the seal cavities is handled in one of the ways previously described in this chapter.

Storing a Pump. If a pump removed from a system is to be out of service for a considerable time, say longer than 6 months, its case should be filled with oil and its ports plugged before it is stored. A small air pocket should be left to allow for expansion of the oil if the pump should be subjected to heat during storage. This will prevent bursting of the housing. The oil used should preferably be a rust preventive type, as some petroleum oils may not give complete rust protection. Do not use detergent or other oils which are not compatible with the rubber seals. Your oil supplier can advise on type of oil to use.

Using Fluids Other Than Petroleum Oil. All other hydraulic fluids have less lubricity than petroleum oil. Since the pumped fluid must also lubricate the pump bearings, the pump must be used at lower pressure on other fluids if the same bearing life is to be obtained. Water has no lubricity and should not be used in pumps designed for pumping oil. Water should be pumped by impeller pumps at low pressures, and by suitable piston pumps at high pressures.

Fluids (even petroleum oil) of much lower or higher viscosity are not pumped as efficiently as petroleum oil of the correct viscosity. Pumps which have automatic take-up for wear, such as vanes in a vane pump or wear plates in a gear pump, are usually more efficient in pumping thin fluids if the fluids have sufficient lubricity. Refer to Pages 39 and 271 for effect of viscosity on Pump performance. Fluids with higher specific gravity than oil (and this includes all other hydraulic fluids) cause greater cavitation, and the pump may have to be supercharged or run at a lower speed.

Pump seals, particularly the shaft seal, must be compatible with the pumped fluid. Incompatibility is evidenced by swelling of the rubber and rapid deterioration. For example, synthetic rubber seals which work well on petroleum oil are not compatible with synthetic fire resistant fluids.

Pump case material must also be compatible with the pumped fluid. Corrosive fluids have to be pumped with stainless steel pumps. Water pumps should be constructed of brass or stainless steel or should be internally plated.

Maximum Pump Speed. Maximum horsepower per dollar and per pound of weight is obtained by operating a small pump at high speed instead of a large pump at low speed. However, any pump should not be operated above maximum catalog rated speed because it may cavitate if inlet passages are too small to carry the increased flow at high RPM. Vibration increases with speed because of increased centrifugal force on unbalanced rotating elements. It can become destructive not only to the pump but to the structure on which the pump is mounted. Both of these conditions, cavitation and vibration, produce excessive noise and shorten pump life. Some pumps may gall inside because of centrifugal force squeezing out oil at high rotating speeds.

Minimum Pump Speed. Since GPM flow decreases in direct proportion to decrease in RPM, a pump for engine drive should be selected for adequate flow at the lowest engine speed where flow will be required. A pump becomes less efficient at lower speeds because internal slippage loss (which is proportional to pressure, not to speed) represents a greater proportion of input horsepower.

Pumps which have sleeve bearings may seize if the pump is operated at high pressure while it is running at low RPM. Overheating may occur at low speed because pump flow may not be sufficient to carry away heat generated by friction losses and internal slippage. Rubber shaft seals can be damaged by excess heat. Overheating can occur on either roller bearing or sleeve bearing pumps.

Those vane pumps which depend solely on centrifugal force to keep the vanes in tight contact with the cam surface should not be operated, under pressure, below 1000 RPM.

The acceptable minimum speed will vary with pressure and oil viscosity. The pump manufacturer should be consulted on any application where the pump will be operated under heavy load at speeds less than 600 RPM.

Cold Weather Operation. Hydraulic oil "thickens" as its temperature decreases, and becomes more difficult for the pump to pick up by suction. If the equipment must be started in cold weather, the hydraulic oil must be carefully selected for the lowest viscosity which will still give good performance after the equipment warms up to normal operating temperature. The V.I. (viscosity index) rating of the oil is of major importance on machines subject to cold starts. See information on Page 40.

On some jobs it may be necessary to keep the oil from becoming excessively cold during the night or during long periods when the system is shut down. It may be desirable to pre-heat the oil before start-up. If electric power is available, an immersion heating element can be installed inside the reservoir below oil level and connected to a thermostat also installed below oil level. A small impeller type pump can be installed to keep the reservoir oil in gentle circulation, but a pump is usually not needed if the heating element is properly chosen.

Caution! Other petroleum products such as gasoline, kerosene, distillate, jet fuel, etc. should never be added to hydraulic oil for the purpose of thinning it. These products may not chemically mix with the oil and will eventually evaporate leaving a residue. But in the meantime they reduce lubricity and offset the effectiveness of the oil additives which suppress foam and inhibit rust.

Electric Heating Elements. Immersion heating elements for water are designed for 20 watts per square inch of element surface. Since oil is a poor heat conductor, these elements should not be used for heating hydraulic oil, except as later described, unless a circulating pump is used. The oil in direct contact with the element surface will carbonize excessively. Electrical supply houses can supply heating elements for oil which are rated at 10 watts per square inch, and these can be used in the reservoir without a circulating pump.

Water heating elements can be used for oil heating if operated at half their rated voltage. For example, a 5 kW, 440-volt element can be operated on 220 volts and will produce 1/4th its wattage rating, in this case 1.25 kW. It will operate at a surface temperature of 5 watts per square inch which is a safe condition without the necessity of using a circulating pump.

Cleanliness. Dirt is one of the worst enemies of a hydraulic system. If the system is conservatively designed and properly maintained it should give good service for many years. But a system which is poorly designed and/or neglected may develop service problems in a very short time. All reasonable precautions should be taken to keep dirt from entering fluid lines and reservoir, but some of the dirt is internally generated by mechanical wear of moving parts, overheating, packing deterioration, and other causes. To have a trouble-free and long-lasting system, it should be designed conservatively in the first place including suitable strainers and filters, and properly sealed reservoir. Then, it should be serviced at regular intervals including a check of oil level, cleaning wire mesh inlet strainers, replacing micronic filter elements, and cleaning the outside of the reservoir to help prevent dirt from accidentally getting in when the filler cap is removed. These points are particularly important:

(1). Be careful of the oil added to the reservoir. Add the same brand and kind that is already in the system. It is true that many oils are compatible with one another, but some oils are not. Use only new oil taken from original factory containers. But even new oil may contain dirt, so pour it through a strainer using a clean funnel or hose. Never add used, old, or unknown oil to the system.

(2). Keep filler opening sealed at all times except when adding new oil.

(3). All air vents on the reservoir, including the filler opening, should be protected with good quality breathers having a micronic rating.

(4). One main factor in generating internal contamination is excessive heat in the oil. Industrial systems should, if necessary, be protected with oil coolers. It may not be possible to add coolers to moving equipment, so the oil should be frequently changed.

(5). Every pump should be protected with a wire mesh strainer on its intake, then additional fine filtering should be added to systems on which a dirt sensitive pump is used, and usually on systems which operate continuously (more than two or three hours every day). Filtering circuits are shown in Chapter 6.

REVIEW QUESTIONS — CHAPTER 5

1. What are some advantages of fluid power over electrical or mechanical power transmission on certain applications? — *See Page 171.*

2. Why does air have to be compressed in several stages, usually no more than a 3:1 compression ratio per stage, to reach a high pressure level? — *See Page 172.*

3. Give the definition for a "positive displacement" pump. — *See Page 174.*

4. What does the abbreviation C.I.R. mean in relation to a hydraulic pump or motor? — *See Page 174.*

5. How are the bearings lubricated on most hydraulic pumps? — *See Page 175.*

6. Name at least three conditions in a hydraulic system which will shorten pump life. — *See Page 175.*

7. The basic graphic symbol of a pump is a circle. If there is a long slash line through the circle, what does this tell about the kind of pump? — *See Page 176.*

8. What creates the "suction" at the inlet of a hydraulic gear pump? — *See Page 177.*

9. What are the purposes of the wear plates in a gear pump? — *See Page 178.*

10. What may happen if a pump shaft is rotated in the wrong direction while the pump is under pressure? — *See Page 178.*

11. On those pumps which have a pressure compensator, what is the purpose of the compensator? — *See Page 181.*

12. Which types of pumps can be built with variable displacement so they can use a compensator? Which types cannot? — *See Pages 181 and 187.*

13. Explain "balanced" construction in a vane pump. What, inside the pump, is balanced? — *See Page 182.*

14. What is the difference in operating principle between a bent-axis and a swash-plate axial piston pump? — *See Pages 184 and 185.*

15. Why are piston pumps designed with an odd number of pistons, 3, 5, 7, or 9, etc? — *See Page 186.*

16. What is the reason an impeller (non-positive displacement) pump cannot produce high pressure? — *See Page 188.*

17. What is the primary function of any hydraulic pump? What is its secondary function? — *See Page 190.*

18. What is meant by "pump unloading" and why does a pump have to be unloaded during periods when no mechanical work is delivered from the system? — *See Page 192.*

19. What are the two ways in which a pump can be unloaded? — *See Page 192.*

20. Why should a relief valve be used with a pressure compensated pump? — *See Page 193.*

21. What is a good efficiency rating for most positive displacement pumps? — *See Page 194.*

22. What is the "rule-of-1500"? — *See Page 195.*

23. Under normal conditions how much can an induction electric motor be overloaded on a continuous basis? — *See Page 195.*

24. What are the two ways of possible misalignment between pump and motor shafts? — *See Page 196.*

25. What kind of damage can be caused by running a pump shaft backward? — *See Page 198.*

26. If a 3-phase induction electric motor runs in the wrong direction, how can its rotation be reversed? — *See Page 198.*

27. What is the recommended mesh size of an inlet strainer on petroleum hydraulic oil? For water base or other fire resistant fluid? — *See Page 200.*

28. What is "pump cavitation"? — *See Page 200.*

29. What is the most common cause of pump cavitation? — *See Page 201.*

30. What is the most common cause for air getting into the oil? — *See Page 201.*

31. What is the "suction head" of a hydraulic pump? — *See Page 202.*

32. What is meant by "supercharging" a hydraulic pump? — *See Page 204.*

33. What are some of the dangers of operating a hydraulic pump above its maximum rated speed? — *See Page 205.*

CHAPTER 6

Other Fluid Power Components

CONDITIONING OF COMPRESSED AIR

"Raw" compressed air directly from an air compressor should not be used in fluid power equipment, air tools, or instrumentation, until it has been cleaned, its pressure level adjusted, and if necessary, lubricated. On some applications it may also have to be de-humidified before it can be safely used.

Of course, every air compressor should have an intake filter and this filter will remove a great deal of dust particles of larger size. However, it will not remove fine dust nor moisture vapor. Since an intake of about 8 cubic feet of free air will be compressed to about 1 cubic foot of compressed air, the relative humidity will also be increased to 8 times the humidity of the intake air. Usually then, the compressed air in the receiver tank will be saturated, that is, will be at 100% relative humidity and the remaining vapor will condense into liquid water and will collect in the receiver tank where it can be drained off. Fine contaminants which pass through the intake filter will also be concentrated 8 times.

Charles' Law states that when air is compressed its temperature will increase. Actually the temperature may momentarily reach 450° F in the compressor cylinder. This high temperature plus lube oil in the compressor plus dirt particles from the air, plus condensed water will produce many unwanted chemical compounds which are carried into the receiver tank. While most of these abrasive compounds, varnishes and sludge, may precipitate in the receiver tank, some of them will pass into the distribution system. To prevent these unwanted by-products from entering valves, cylinders, and air tools, all compressed air should be processed at the entry point of every piece of equipment in which it is used.

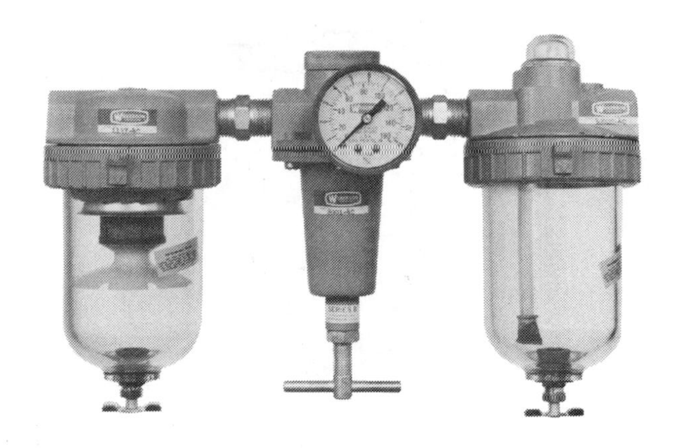

FIGURE 6-1. *Trio Assembly for Air Processing.*

Trio Assembly. Figure 6-1. This is a 3-part assembly consisting of a filter, pressure regulator with gauge, and lubricator. The three units are assembled with close nipples. It is generally accepted practice to install a separate trio assembly for each air-operated machine. It should be installed as close as practical to the first component in the machine which is usually the directional control valve. It is very poor practice to attempt to use one large trio assembly to serve several machines.

Duo Assembly. Sometimes equipment is being served which should not be lubricated. In this case a duo assembly consisting of filter and regulator should be used. Equipment which should not be lubricated includes air dryers, certain types of instrumentation, air used in the processing of food and beverages, air for medical and dental use, and certain pressure intensifiers and cylinders which have been pre-lubricated with moly grease.

Order of Installation. Trio and duo assemblies must be assembled in the proper order, with filter first, regulator next, and lubricator always last in line. Each unit of the assembly must be plumbed correctly with air entering the inlet port so air flow will agree with the arrow usually stamped or cast into the housing. Any unit connected in backward will not function.

Trio Sizes Available. Standard trio assemblies are manufactured in pipe sizes from 1/4" through 1"; miniature units in 1/8" and 1/4" pipe size. For larger than 1" air lines the individual filter, regulator, and lubricator units are offered individually rather than in assemblies.

CAUTION ON USE OF PLASTIC BOWLS

Filter and lubricator bowls are made from a clear polycarbonate plastic. This material is the best yet found for normal conditions of service where pressure does not exceed 250 PSI nor the temperature 150 degrees F. But like all plastics it is not compatible with certain chemicals. Plastic bowls should never be installed in a location where they will be exposed to fumes such as paint thinner, for example. Nor should outside or inside of the bowl ever be wiped with a rag which has been used in a solvent. If cleaning is necessary, clean cotton waste should be used. Or, they may be washed with non-detergent hand soap or in kerosene. Even a momentary exposure to harmful chemicals will start chemical deterioration which will develop into "craze" marks, and the bowl may soon explode under pressure. Any bowl starting to show craze marks should be immediately removed from service.

For locations where fumes may be present or where temperature or pressure may exceed the above values, filters and lubricators may be ordered with metal bowls.

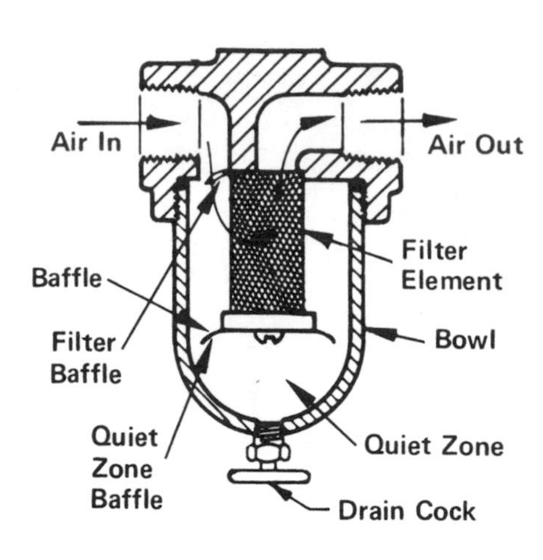

FIGURE 6-2. Air Line Filter.

Air Line Filter

Figure 6-2. First, air entering the filter is passed through a circular baffle plate which imparts a high velocity cyclonic motion which causes water and heavier particles of dirt to be hurled out by centrifugal force against the sides of the bowl. These contaminants drain down the sides of the bowl into the quiet zone below the baffle. Here they remain until expelled either by a manually operated drain cock or by a float-type automatic drain.

Next, the air is passed through a very fine filter element to remove the small particles still remaining. Various filtering materials are used such as cotton (cellulose), felt, sintered bronze, etc. The modern trend is to filter down to a 5μm particle size. Even microscopic dirt particles will definitely cause excessive wear and early failure of cylinders, valves, air tools, and other components.

Draining the Filter. Condensate (with entrained dirt) should be drained at least once a day or oftener if necessary. It should never be allowed to rise above the quiet zone under the baffle. If it should, the dirt would very quickly stop up the fine filter element and stop air flow. Filters can be purchased with built-in float-type automatic drainer or one can be attached externally to the drain cock.

Note: The filter described is mechanical in action. While it will very efficiently remove solids and water in liquid form, it will not reduce the humidity (water vapor content). Therefore, water may condense downstream of the filter unless a dryer is used to reduce humidity.

Air Pressure Regulator

Figures 6-3 and 6-4. The purpose of a pressure regulator is to reduce the higher pressure in the plant air distribution system to a lower and more suitable level for each machine in which the air is used. It is the counterpart of a pressure reducing valve in a hydraulic system.

A regulator is much more than a simple orifice to reduce flow. Such an orifice would permit pressure to climb to the distribution level if air flow were deadended as by a cylinder stalling against its end cap. It is not a relief valve. Such a valve would waste a great deal of air by discharging to atmosphere to maintain a constant pressure level. A regulator can be described as a 2-way, normally open, poppet valve which is closed by excessive pressure on its outlet working against a diaphragm.

Importance of a Pressure Regulator. All plant air systems require a separate pressure regulator (usually a part of a trio assembly) to be used on each unrelated machine in the plant which operates from the air distribution system. It permits the operator of each machine to individually set the pressure on his machine according to the precise level needed, thereby providing consistent and repeatable performance at the desired force and speed level. It further serves to reduce air waste by permitting an operator to adjust pressure on his machine to no more than actually required.

Non-Relieving Pressure Regulator. Figure 6-3. This type is so named because it will not permit overpressure on its outlet port to escape to atmosphere. Therefore, this is the kind, rather than the self-relieving kind, which must be used for handling gases or liquids which should not be allowed to escape to atmosphere. The self-relieving kind, to be described later, is preferred for handling compressed air.

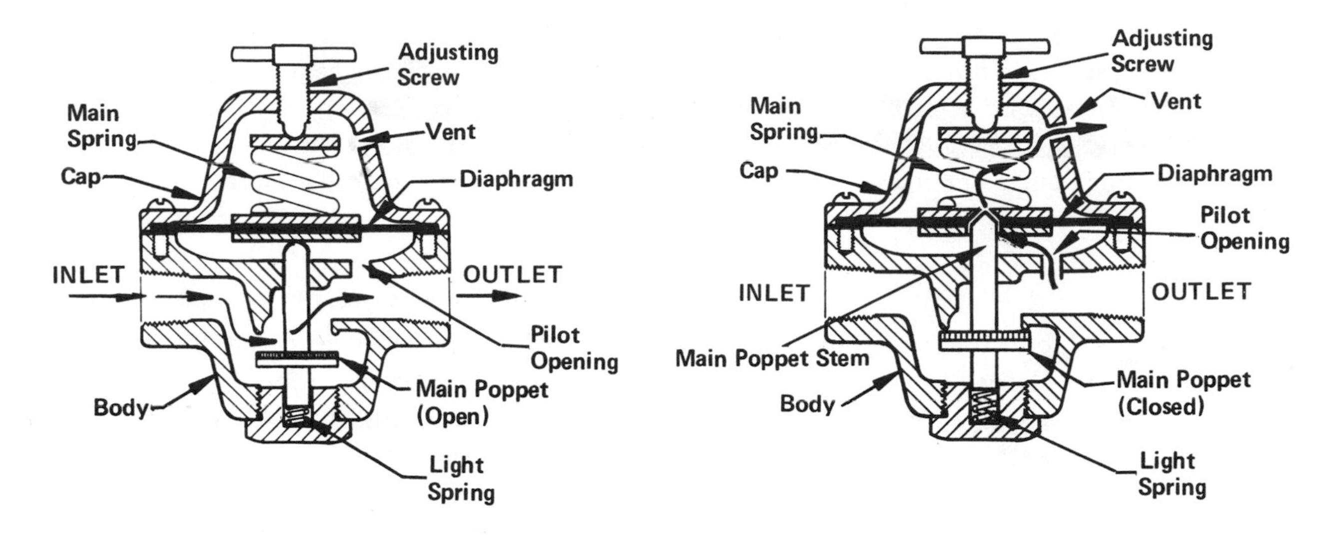

FIGURE 6-3. Non-relieving air pressure regulator, shown in a metering position.

FIGURE 6-4. Self-relieving air pressure regulator, shown in a relieving position.

The main poppet controls flow of air and reduces or cuts off flow, when necessary, to prevent outlet pressure from rising above the level which has been adjusted. Opening and closing of the main poppet is controlled by a diaphragm. When the diaphragm moves up or down, the main poppet moves with it. The diaphragm is in a state of force balance between outlet air pressure, obtained through the pilot opening, working under it and trying to push it up, and mechanical spring force working on top of it and trying to push it down. The outlet pressure which is high enough to balance the spring and still keep the main poppet open is the maximum pressure the regulator can deliver. Any higher pressure will move the main poppet to restrict flow, as necessary, to prevent outlet pressure from building any higher. To obtain a higher pressure from the regulator the spring must be tightened. Then, outlet pressure can rise higher before the balance point is reached which will cut down the air flow.

If the cylinder which is receiving air through the regulator should reach the end of its stroke and stall, this "deadend" condition would cause outlet pressure to rise until the main poppet completely closed at the pressure level set on the adjusting screw of the regulator.

Note: The light spring is only for the purpose of keeping the main poppet stem in contact with the diaphragm; it has almost no effect on the outlet pressure level.

Self-Relieving Regulator. Figure 6-4. Any excess of pressure which might appear on the regulator outlet due to unusual circuit conditions can vent off to atmosphere on this type regulator. For this reason it is preferred on most compressed air circuits, but should not be used on gases where a possible atmospheric venting would be undesirable.

The action is very similar to that described for the non-relieving regulator except that there is a small vent hole in the center of the diaphragm. As long as the system is working properly this hole remains covered by the end of the poppet stem, and no air escapes. But if excessive pressure should back up on the regulator outlet port, the diaphragm can move away from the stem, uncovering the vent hole, allowing excess pressure to escape through another vent opening in the regulator cap. After excess pressure is relieved, the diaphragm will again seat against the end of the stem.

On air operated machines, the circuit condition which would cause excessive pressure to appear on the regulator outlet is this: Suppose a steady pressure is being held, with no air flow, on a deadend circuit, as by a cylinder stalled against its end cap. Then, suppose the operator should decide to lower the setting on the regulator. This would leave a pressure trapped in the cylinder which was higher than the new setting on the regulator. On a self-relieving regulator this excess pressure could vent to atmosphere; on a non-relieving regulator it could not vent until the next cycle of the cylinder.

A self-relieving regulator is considered safer to use on deadend circuits – those applications where a steady pressure, with no flow, must be held for a relatively long period. Any leakage of the main poppet could not build up excessive pressure on the outlet, but on a non-relieving regulator, it could.

Air Line Lubricators

Figure 6-5. A lubricator is the 3rd unit in a trio assembly. By the time air reaches the lubricator it should have been cleaned, its water removed, and its pressure level adjusted for the machine which it serves. The purpose of a lubricator is to inject a finely divided oil mist into the air stream to lubricate components downstream including valve spools, cylinder pistons, and air tools. Once installed and adjusted it requires no further attention except for replenishment of its oil reservoir from time to time. It works automatically to start the oil feed as soon as air starts to flow and to stop feeding oil when the air flow stops. The rate of oil feed is proportional to the volume of air flow.

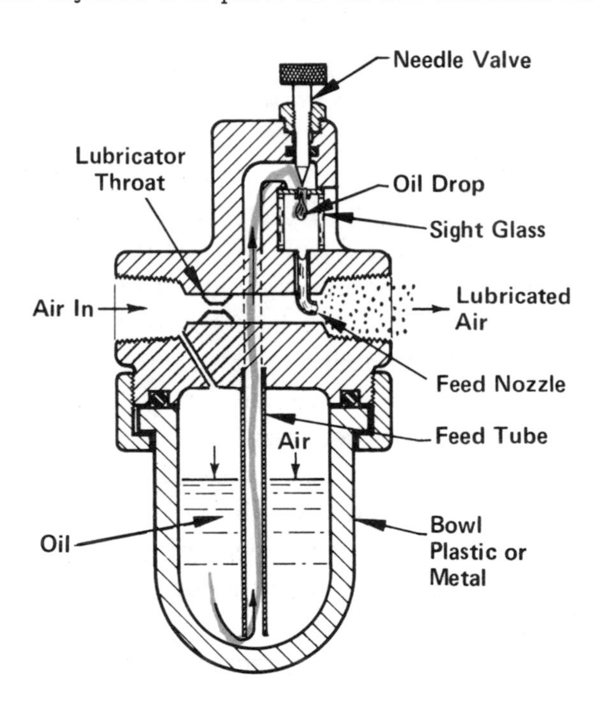

FIGURE 6-5. Simple Air Line Lubricator.

The lubricator shown in Figure 6-5 we call a "simple" or "standard" lubricator to distinguish it from a more sophisticated type to be described later and called an "atomizer" type.

The rate of feed can be adjusted with a built-in needle valve. Oil which is fed into the air stream is broken up into a fine mist by high velocity air flow and carried downstream suspended in the air. It will carry a short distance but will eventually precipitate in the lines and on working surfaces of components. Feed rate should be very low – just enough to keep a light film of oil on working surfaces. Excessive oil feed may, over a period of time, gum up working parts of components, and may create a breathing hazard for people working in the area if it is discharged into the atmosphere. Oil precipitators are available to collect excess mist if this should be a problem.

Rate of oil feed will have to be determined experimentally. When installing a new machine the lubricator should be adjusted to near its minimum feed rate. It should be observed, under working conditions, for a few days to be sure a small amount of oil is being consumed. Usually it is better to under-lubricate than to over-lubricate.

Principle of Automatic Feed. Air pressure from the regulator inlet forces oil to be fed into the air stream. All lubricators work on the automatic feed principle illustrated here.

A small pressure drop must be produced in the throat of the lubricator to be used for feed pressure. It can be produced in various ways – by a fixed or variable orifice, a restrictor disc or flapper, or by any other means. When air is flowing, the outlet port must always be at a slightly lower pressure than the inlet, by 1 to 5 PSI according to the volume of air flow.

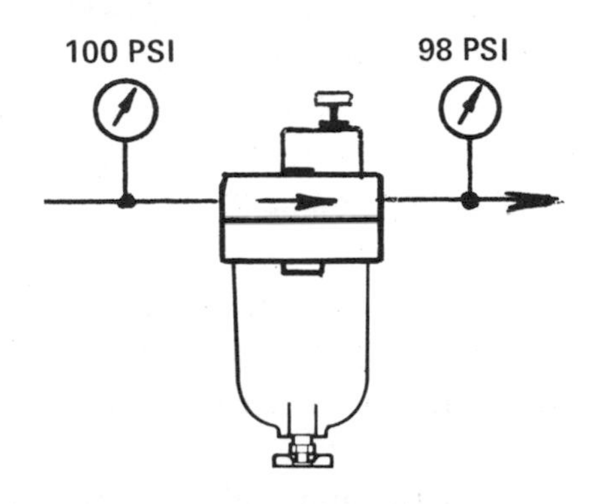

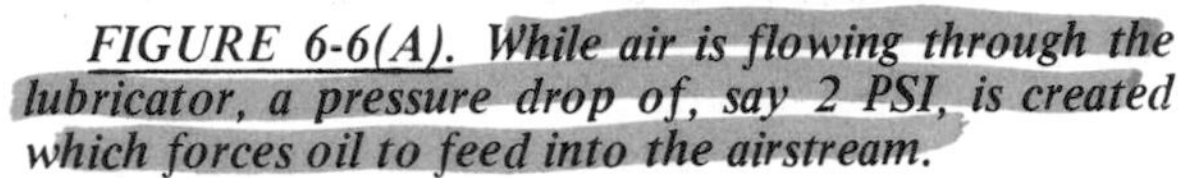

FIGURE 6-6(A). While air is flowing through the lubricator, a pressure drop of, say 2 PSI, is created which forces oil to feed into the airstream.

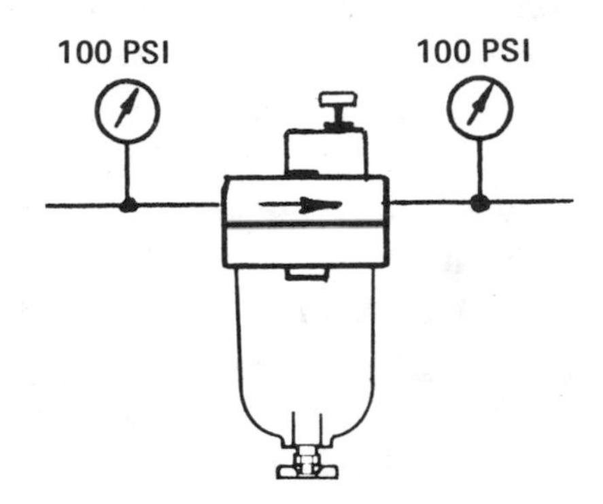

FIGURE 6-6(B). When air flow stops, the 2 PSI pressure drop disappears and both ports are at the same pressure – 100 PSI. Oil feed stops.

In Figure 6-6(A), the air is flowing. The inlet pressure might be 100 PSI, for example. Pressure loss through the throat of the regulator might be 2 PSI. The inlet pressure of 100 PSI is working on the surface of the oil (see Figure 6-5) to force it up through the feed tube, through the adjustable needle valve, and discharge it through the feed nozzle into the outlet port area where the pressure is only 98 PSI. As the oil leaves the feed nozzle, velocity of the air breaks it up into a fine mist which is carried downstream suspended in the air.

Oil will feed at a certain rate when the pressure drop through the lubricator throat is 2 PSI. If air flow is increased, the pressure drop might increase to 3 PSI which would cause a greater volume of oil to be fed.

In Part (B) of Figure 6-6, air flow has stopped. Since there is no flow through the regulator throat there is no pressure drop across it. The outlet port is at the same pressure as the inlet so there is no air pressure difference to produce an oil flow.

Oil Feed Rate. Most lubricators have a needle valve for adjusting the rate of feed and a sight glass through which the drip rate can be observed. It is hard to set an exact rate of feed but a starting point for a new machine, using the kind of lubricator shown in Figure 6-5 in a 1/4" size, one drop every 3 minutes is suggested. Observe the oil level for a few days to be sure a small amount of oil is being consumed. If there is evidence of oil blowing out the exhaust ports of the 4-way control valve, cut the drip rate to one-half. The oil observed in the sight glass is all going into the air system.

The air stream breaks the oil into fine particles of various sizes. The smaller particles may remain suspended for quite a distance in the air stream but larger particles may precipate in only a few feet of travel. The lubricator should be installed very close, as close as practical, to the inlet of the 4-way valve, but in any case should not be separated by more than 10 feet.

Lubricating Oil. Use a plain petroleum oil of very light weight, equivalent to SAE 5W or 10W. There is a possibility that detergents in the oil could cause damage to the clear plastic bowl. Adding anti-freeze to the oil is not recommended except with the consent of the lubricator manufacturer who will know what chemicals may cause damage to the bowls.

Atomizer-Type Air Line Lubricators

This is a more sophisticated version of the simple lubricator shown in Figure 6-5. It operates on the same principle of automatic feed but breaks the oil into much finer particles, 2μm (microns) or less, which will remain suspended in the air for much greater distances, for hundreds of feet.

In operation, a small amount of the inlet air is used to project the feed oil at high velocity against a baffle plate. Particles larger than about 2μm are allowed to precipate back to the oil reservoir for re-processing while the smaller particles enter the air stream. Oil feed is observed through a sight glass but only about one out of every 20 to 30 drops seen in the sight glass actually goes into the air line. So the drip rate on these lubricators should be set for 20 to 30 times the rate which would be used on a simple lubricator.

Atomizer-type lubricators should be used where oil must remain in suspension for a long distance, and for circuits using small valves and cylinders where oil feed should be quite low. They can be adjusted accurately for very low oil feed and yet can provide sufficient lubrication for cylinders of large size. Their main disadvantage is their slightly higher cost.

Standard (simple) lubricators are preferred where cost is a factor and for applications where a heavy oil feed is required and the oil does not have to remain suspended for more than 10 feet.

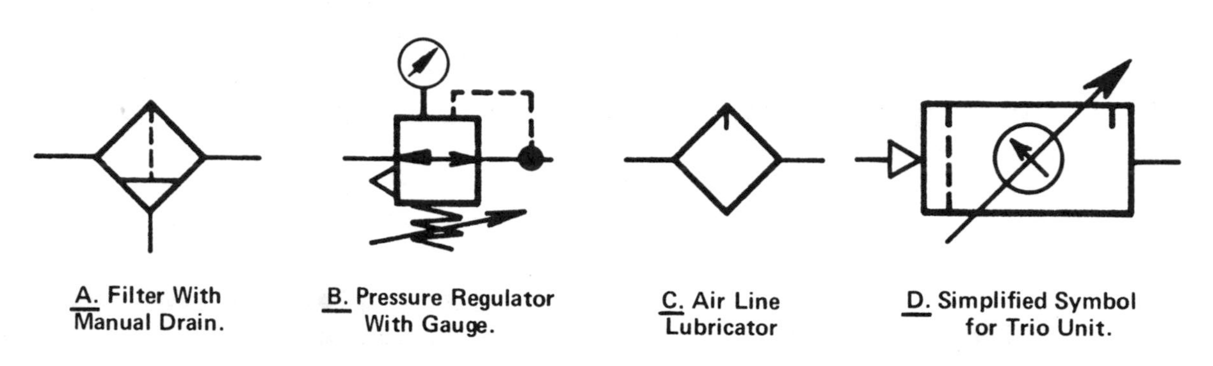

FIGURE 6-7. Graphic Symbols for Trio Units and Assemblies.

Figure 6-7. Filters, regulators, and lubricators used as individual components may be drawn on fluid diagrams as in Parts A, B, and C of the above illustration. When a complete trio assembly is used it may be more convenient to use the simplified symbol shown at D. The vertical dash line at the left is the filter. The large slash line is the adjustable regulator. The circle and small arrow indicate a pressure gauge (may be omitted if a gauge is not used). The short vertical line indicates a lubricator.

DRYERS FOR COMPRESSED AIR

If the air is lubricated a small amount of water condensation may not be harmful in cylinders and valves. But condensation cannot be tolerated on applications like air tools, shot peening, paint spraying, instrumentation, air gauging, in freezing environments, for blowing powdered materials, for medical and dental use, or where the air may come into contact with food. Water may condense in the air line when it passes from a warmer to a cooler area. Water is present in an air line in two forms:

Mist. Sometimes called fog. This is water in liquid form but in particles so small they will remain suspended in the flowing air. These particles can precipitate at any place in the air system. However, most of them can be removed with a mechanical filter of the type illustrated in Figure 6-2. A good mechanical filter can remove as much as 95 to 98% of these suspended water particles.

Vapor. This is water in gaseous form. It is invisible and can be compressed like air. It will transmit power just as efficiently as will air. If it would always stay in vapor form it would do no harm and would be a useful part of the fluid power energy. But when it condenses in the line or in certain components it is harmful. Water vapor cannot be filtered out; it must be removed with an air dryer.

Note in the following discussion that a certain quantity of air (1 SCF at 100 PSI for example) can hold less water vapor when cooled to a lower temperature. This principle is used in the refrigeration type air dryer. Also, this same air can hold more water vapor if allowed to expand to occupy a larger volume at a lower pressure. This principle is used in the regeneration type dessicant air dryer.

Cause of Water in An Air Line. Air drawn into a compressor intake is at a certain relative humidity. This is the percentage of water vapor actually in the air compared to the amount it could hold at that pressure and temperature before it becomes saturated (100% relative humidity). For example, at a relative humidity of 30%, the air contains only 30% of the water it could hold if it was saturated.

Suppose, for example, the intake air is at 30% relative humidity. When compressed to 100 PSI, this reduces its volume by 7.8 times. It is dry as it comes from the compressor because of its high temperature, but when it has cooled to its original temperature, its humidity will increase by 7.8 times 30% or to a theoretical 234%. While it is cooling, its relative humidity continues to increase. After the saturation point of 100% has been reached, water condenses as the cooling proceeds and it collects in the receiver tank or in the air line. For this reason the receiver tank should be regularly drained, and water drain legs should be provided at intervals along the distribution lines.

Now, air in the receiver tank is at 100% relative humidity at receiver tank temperature. But if this saturated air is put into the distribution system, additional water will condense at any place where the temperature may drop, even slightly, either because the air is piped into a colder area, or from expansion. For example, air discharged from a blowgun expands rapidly as it is released to atmosphere. This produces a cooling effect and may cause water or frost to condense around the nozzle.

Solution to Air Line Water Problem. To eliminate water condensation, the relative humidity of the air must be reduced, *after it has been compressed*, to less than 100%. First, a compressor aftercooler should be employed. This is usually installed between the compressor and the receiver tank. This should be followed by an air dryer installed downstream of the receiver tank. The aftercooler will remove a part of the water, reducing the load on the dryer, and will reduce air temperature to a level where the dryer becomes more efficient. But an aftercooler alone cannot reduce the humidity to less than saturation, and cannot completely solve the water problem in many systems.

Air Dryers. There are a number of types of air dryers which can be used, but we are describing the two types which appear to be more widely used and more suitable for industrial compressed air systems where a high volume of air must be handled. A dryer can be installed following the receiver tank to serve the entire plant, or can be installed in a branch to serve only one section, as for example the air supply to a spray painting room. On some air applications, completely dry air might not be needed in some branches, especially where oil mist lubrication is injected into the air stream.

Air dryers are more effective on cooler air, but in any case the inlet air temperature to a dryer should not exceed 120 degrees F, and on some types of dessicant dryers the inlet air must not exceed 100 degrees F.

For best system efficiency the air dryer need not reduce the vapor content of the air to a lower level than necessary, considering the worst ambient conditions of high humidity and low temperature to be encountered. There is no additional benefit to having the air any dryer than necessary to avoid condensation in the lines. Water removal to a very low degree of humidity can be costly, and system efficiency is reduced by removing water vapor which could be used to produce power output.

An air dryer is an excellent investment. In most plants it can pay for itself in less than a year, and its original cost and maintenance cost is far less than the damage caused by wet air.

Refrigeration-Type Air Dryers

Figure 6-8. This type is manufactured in sizes starting at 1/5th up to 10 HP or more. A 10 HP model will dry about 2000 SCFM flow, roughly equivalent to the output of a 400 HP air compressor. This will handle most systems or several dryers can be connected in parallel.

Dryers are self-contained, with all components housed in a sheet metal cabinet. Some models have pressure and temperature gauges.

FIGURE 6-8. Refrigeration-Type Compressed Air Dryer.

Water vapor is removed by chilling inlet air to force it to condense some of its water vapor. The humidity of the air, as it is warmed back up to room temperature, is reduced to less than 100%. This dryer is efficient and effective for producing dew points (the temperature at which water will start condensing again) as low as 39° F. But it cannot be used in locations where the dryer or the air will be exposed to lower temperatures.

Dryer Refrigeration System. Figure 6-9. The refrigeration system is a standard freon system very much like in a home refrigerator but with greater power. Starting with the freon compressor, the gas is compressed to a high pressure and becomes hot. It is then run through the condenser (hot coils) where it condenses to liquid freon. Next it is passed through an expansion valve into the evaporator (cold coils). As it evaporates it absorbs heat. The vaporized freon gas is returned to the compressor for re-cycling. The action of a dryer in removing humidity is as follows, referring to Figure 6-9:

The air to be dried enters the wet air inlet at the top of the diagram. It passes through coils which are in intimate contact with the cold coils of the freon system. As it cools, water begins to condense because cold air will not hold as much water vapor as warm air. Water

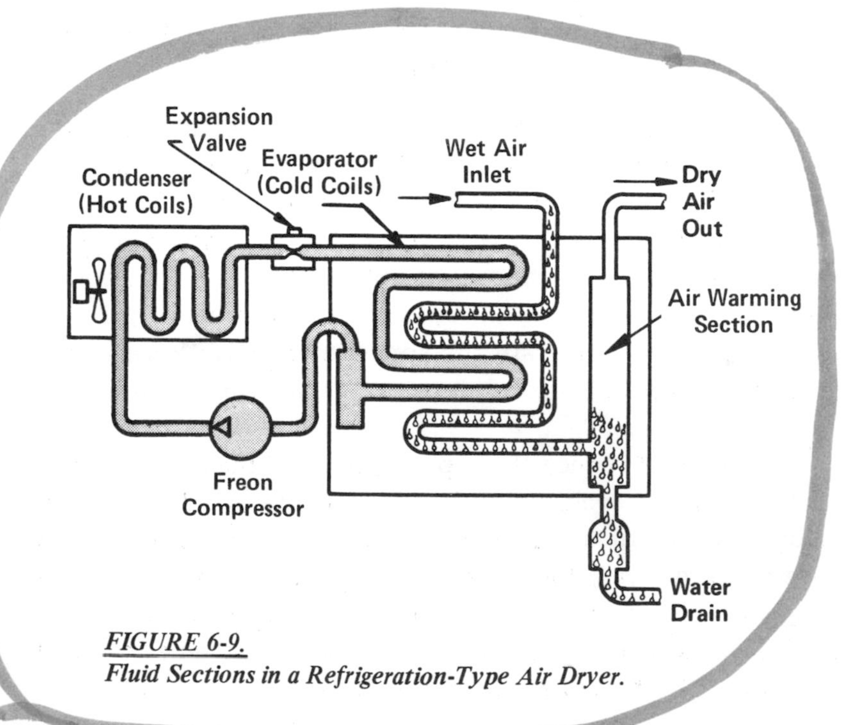

FIGURE 6-9.
Fluid Sections in a Refrigeration-Type Air Dryer.

condensate collects in the automatic water drain and is expelled. This drain can be seen in Figure 6-8. After being chilled the air is about 39° F and still at 100% relative humidity. It is then passed through a warming section which receives heat from the hot coils of the freon system. As it warms to room temperature its humidity decreases because when warm it will hold more water than when colder. As it is discharged and flows through the plant distribution system no more water will condense unless its temperature, either from the environment or through expansion, should drop below 39° F.

A refrigeration dryer should be allowed to run continuously day and night and over week ends. If shut down overnight, wet air may be discharged into the distribution system the next morning until the dryer has time to reach operating condition. Hermetically sealed refrigeration systems have been known to run continuously for 10 to 20 years with no maintenance.

Dessicant-Type Twin Tower Heatless Dryer

Figure 6-10. Two identical towers are filled with a dessicant similar to silica gel which will adsorb water without chemically combining with it. The water is simply trapped and held. It can be purged from the dessicant either by heat or by dry air. One tower is in service while the other tower is being dried by passing dry air through the dessicant. After a timed interval which may be 2½ to 5 minutes, the towers are switched so the first one can be dried while the second one is in service.

In the heatless dryer, water is removed from the dessicant during the regeneration cycle by passing a small flow of dry air, tapped from the dry air delivered from the opposite tower, through it. The dry air picks up moisture from the dessicant and carries it out to atmosphere. The towers are switched by a 4-way solenoid valve controlled by an electric timer. To understand how the drying process takes place, remember that every cubic foot of dry air tapped off for regeneration, when allowed to expand about 7 times (from 100 PSI) to atmospheric pressure through the regenerating tower, can carry out about 7 times as much water vapor as was carried in. Therefore, one cubic foot of dry air tapped off for regenerating the other tower, can carry off as much water vapor as 7 cubic feet of air can bring in.

Dessicant dryers of the heatless type are offered in small to very large sizes, with up to 2000 SCFM flow capacity. They are more wasteful of the air than are refrigerated dryers, losing about 10 to 15% of the air delivered from the compressor, this amount being used for drying the dessicant. They have several advantages: they require no regular maintenance; the dessicant is not

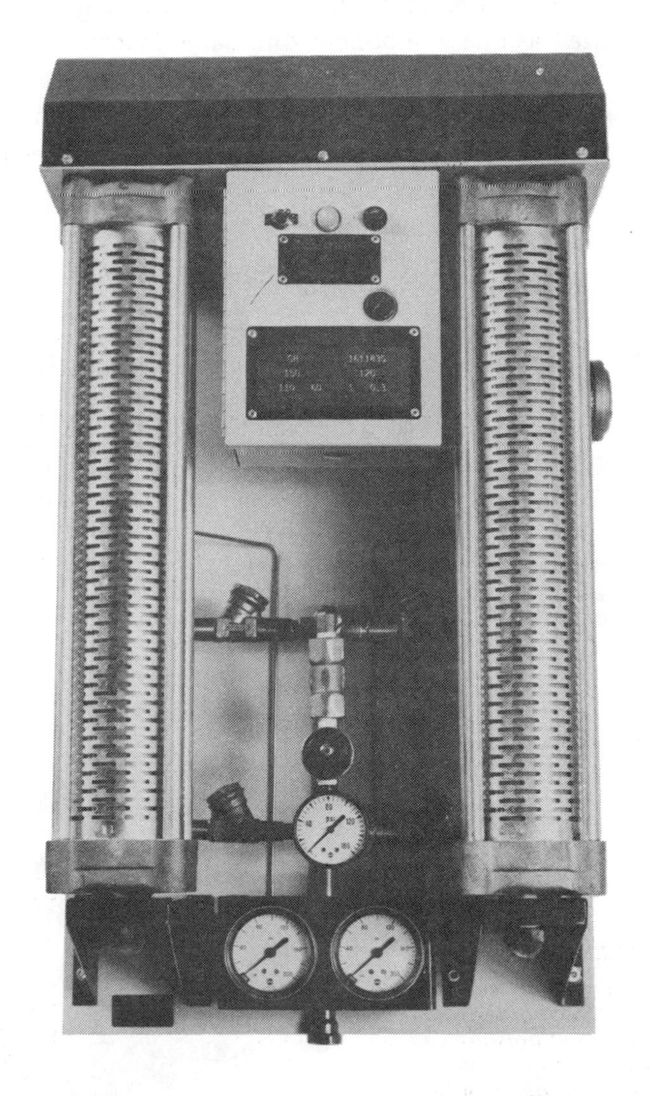

FIGURE 6-10. Twin Tower Dessicant Air Dryer.

consumed so it should not have to be replaced; no water drain is required because the purged water is carried out into the atmosphere; they consume very little electricity – no more than about 50 watts for operation of a solenoid air valve and timer; they can be shut down at night or on weekends and will immediately resume delivery of dry air when turned on; they can produce very dry air, down to -75° F pressure dewpoint depending on type of dessicant and volume of air tapped off for regeneration; they can operate in very cold environments, under arctic conditions if necessary. However, to save power and conserve air, they should be sized for or adjusted to a dewpoint no lower than necessary.

HYDRAULIC POWER UNITS

On most hydraulically operated machines the fluid power is generated by a unit assembly called a "hydraulic power unit" or "hydraulic pumping unit". On machines operating inside an industrial plant the power unit is designed to set alongside or overhead near the machine and connect to it through steel tubing or hose. Other power units are built to be portable so they can be moved around the plant and connected first to one machine, then another. For mobile applications the pump is often separated from the rest of the power unit to be belt driven from the power take-off shaft of the engine. Still other power units, usually small ones, are self-contained and built so they can be permanently mounted as one component on a large machine – a printing press for example. So, power units come in all shapes and sizes, but self-contained power units generally consist of hydraulic reservoir, electric motor, pump, and relief valve as basic items, sometimes supplemented by components such as magnetic motor starter, directional control valves, pressure control valves, pressure gauges, filters, accumulators, heat exchangers, etc. *(Continued on Page 220)*

FIGURE 6-11. This portable power unit can be moved around the plant and hooked up to various machines with quick connect couplers, for numerous jobs such as pressing bushings in gears, straightening shafts, crimping terminals, to name a few.

FIGURE 6-12. Vertical motor mounting. Pump operates in oil inside the reservoir. Solenoid valves and pressure control valves can be mounted on manifolds on the reservoir.

FIGURE 6-13. This flattop power unit features a 7½ HP motor externally coupled to a gear pump and using a 30-gallon reservoir as a mounting base. Reservoir construction closely follows the NFPA approved design which is described later.

FIGURE 6-14. Self-contained D-C power unit to operate tailgate lift cylinders. Includes electric motor, pump, control valve, reservoir, and pressure relief valve.

FIGURE 6-15. Self-contained A-C power unit for integral mounting on a large, hydraulically operated machine. Includes solenoid directional control valve.

FIGURE 6-16. Another example of flattop construction. A C-flange electric motor is used and the pump is mounted with a bracket on the motor end bell. Includes also a pressure relief valve and gauge.

FIGURE 6-17. Electric motors larger than 50 HP are usually too heavy to mount on top of the reservoir. Pump, motor and relief valve may be mounted alongside the reservoir while directional and pressure control valves, filters, and other components can be mounted on the side wall of the reservoir.

Some small power units such as those shown in Figures 6-12, 6-13, and 6-14, usually less than 5 HP, are standardized and are listed in manufacturers catalogs. But most larger power units are custom assembled for a specific machine or application. The examples shown on the preceding pages are but a few of the many variations to be seen in the field.

L-Shaped Hydraulic Power Units. Figure 6-17. Power units larger than 40 to 50 HP are commonly constructed with the electric motor and pump alongside the reservoir rather than mounted on top. Construction is more expensive but offers several advantages: a small pressure head is maintained on the pump inlet so cavitation is less likely; a hinged lid on the reservoir gives easy access for inspection, for cleaning or replacement of inlet filter, and for taking oil samples; the reservoir provides a mounting base for directional and pressure control valves, which reduces the amount of exposed plumbing.

An inverted T construction may be used which provides a shelf on each side of the reservoir for mounting an additional electric motor and pump drawing oil from the common reservoir.

Functions of an Oil Reservoir

Every hydraulic system must have an oil reservoir of some kind. Gallon capacity is usually selected in relation to rate of oil flow from the pump, or sometimes on the number of square feet of metal surface for radiation of a certain amount of heat. On most small to medium size systems the reservoir serves as a mounting base for other components and in addition, serves these important functions:

(1). Oil Storage. An extra supply of oil must be on hand to operate hydraulic rams (single-acting cylinders). The reservoir level falls while the ram is extending and rises again as it retracts. Double-acting cylinders which have a piston rod extending out one end have a different volume on opposite sides of the piston. So they, too, will cause the oil level to fall and rise as they extend and retract. The reservoir must supply oil to make up for that which is lost from the system due to leaks and spills.

(2). Oil Cooling. Heat is developed in a hydraulic system in several ways: from mechanical losses (friction) in pump and hydraulic motor elements and bearings; energy released as heat when oil discharges across relief, reducing, and flow control valves; and flow friction in plumbing lines and components. A large part of the heat so generated is carried by the flowing oil to the reservoir. On small systems all of the heat can be radiated from surfaces of the reservoir, surfaces of cylinder barrels, pump cases, and plumbing lines. On larger systems, surface areas may not be sufficient and a heat exchanger must be added to supplement the natural radiation cooling. See Pages 234 to 236 and Page 242 for further information on heat radiation.

(3). Dirt Precipitation. Oil circulating through the plumbing is traveling at 15 to 30 feet per second. Dirt generated in the system is carried to the reservoir. Oil velocity in the reservoir must be reduced so the dirt can settle. The larger the reservoir the lower the oil velocity in traveling through it. A guideline is given later for reservoir size.

(4). Air Purging. Entrained air and foam can escape from the oil if it is retained in the reservoir for a short time before being picked up and re-circulated. The larger the reservoir, the longer the oil will have to purge air and foam.

Flattop Reservoir Construction

Figure 6-18. The flattop hydraulic power units shown in Figures 6-13 and 6-16 represent by far the most common construction for power units, in the 7½ to 50 HP range, used in industrial plants for operation of presses of many kinds.

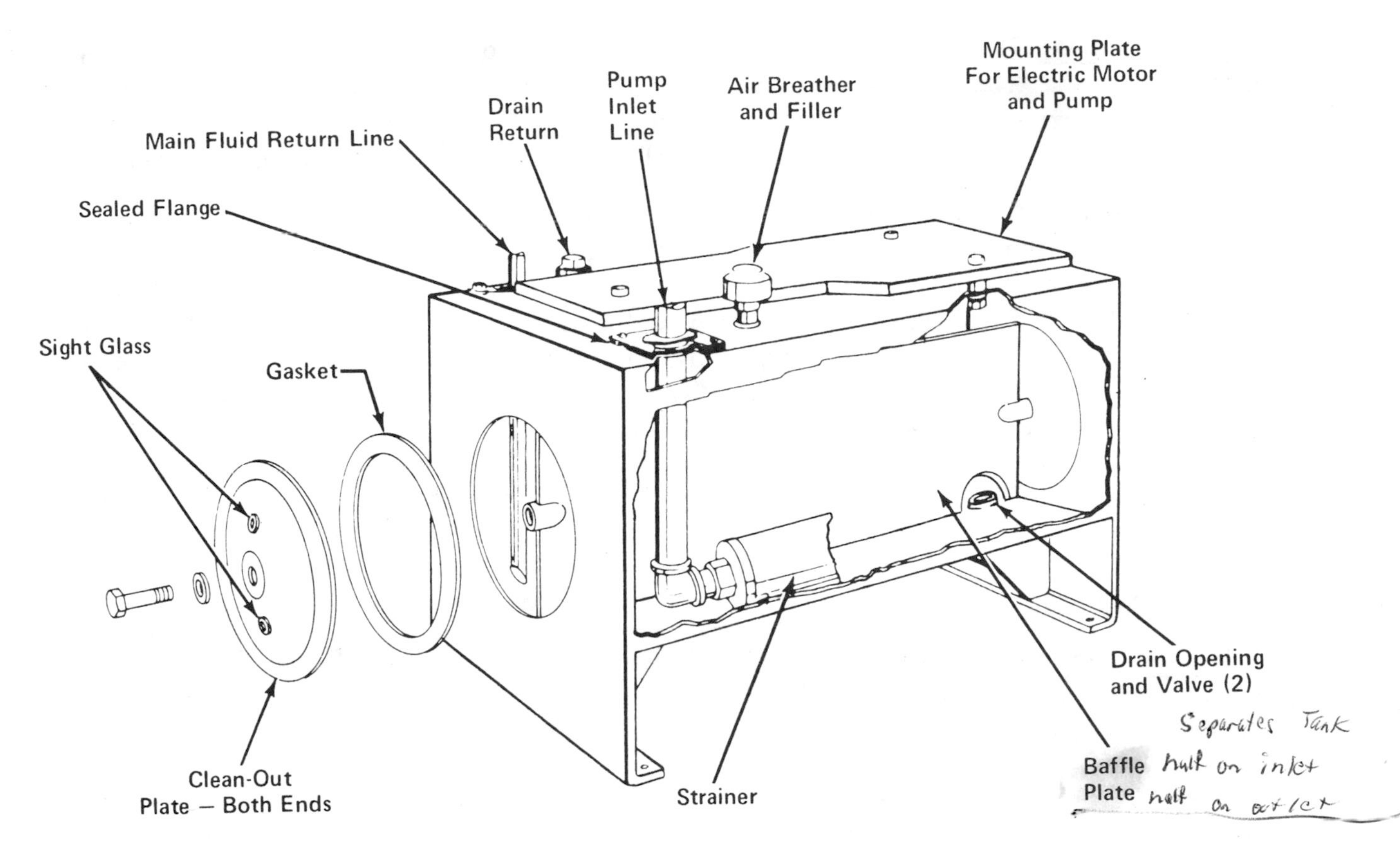

FIGURE 6-18. Construction Features of an NFPA Approved Flattop Hydraulic Reservoir.

NFPA Approved Flattop Hydraulic Oil Reservoir. Features shown in Figure 6-18 are taken from NFPA (National Fluid Power Association) Standards T3.16.69.2 dated December 1969. A copy of the complete standards may be purchased from the NFPA in Milwaukee, Wisconsin. These standards supercede the earlier JIC (Joint Industry Conference) standards which were last revised in 1959.

Specifications on reservoir construction are recommendations only, and show a good method of construction. They are not intended to prevent or discourage other configurations which could be equally reliable, nor are they intended to outlaw the economically built reservoirs which, although not as comprehensive would still satisfactorily serve the type of machine for which they were designed. However, to qualify as "NFPA approved", a reservoir would have to meet all of the qualifications set forth in the NFPA standard and which are briefly described here.

NFPA reservoirs as described here are intended for industrial applications. Specifications apply also to L-shaped units as well as to flattop units. Although some of the specifications might not apply to mobile hydraulic construction, nevertheless many features such as internal baffling, depth of oil above inlet strainer, sealing of the reservoir, apply to all hydraulic reservoirs.

Fluid Capacity. The gallon working capacity shall be a ***minimum*** of three times the gallon flow per minute from the pump or pumps unless other circumstances justify modification of this requirement, such as use of pressure compensated pumps, very intermittent use, forced oil cooling, or environmental conditions. Closed loop systems (see Volume 3) need only a reservoir capacity of three times the charge pump flow per minute. Gallon capacity shall be stamped on top of the reservoir.

In addition to working capacity, the reservoir shall have an additional 10% capacity for air space above the fluid for expansion and for separation of air from the fluid.

Reservoir Base. The bottom of the reservoir shall be supported on legs to a height of at least six inches above floor level to facilitate handling, draining, and to increase heat radiation from the bottom surface of the reservoir. The legs or support members shall have holes or slots a minimum of four places for foundation anchoring. Support area shall be sufficient for leveling by shims or wedges.

Interior Baffle. An interior baffle shall be provided to assist in precipitating large or heavy contaminants, separation of air, and improved oil cooling. The baffle shall separate the fluid return side from the pump intake side, and shall not restrict free air flow between compartments. It shall be so placed as to direct fluid flow along the side walls for better cooling. It shall have an oil flow passage near the bottom of the reservoir and sized sufficiently large so oil velocity through it does not exceed 2 feet per second. Square inch area of the flow passage can be calculated from the formula on Page 257. Baffle height shall be equal to or slightly above the maximum oil level.

Clean-Out Openings. Two clean-out openings shall be provided, one at each end, with minimum size of 6" diameter or 6" square. These openings shall provide complete access to all parts of the interior without disturbing any components or piping. Covers for these openings shall have re-usable gaskets, and shall be designed so they can be removed and replaced by one person.

Reservoir Bottom. The bottom shall be so shaped that all fluid will drain. Two drain openings shall be provided, one at each end and at the low point of the bottom. Drain openings shall be at least 3/4" pipe size and shall be equipped with shut-off valves of equal size.

Reservoir Top. If components are mounted on the top it shall be sufficiently rigid to maintain alignment of pump and motor and shall be suitable for the mounting of accessories. A heavy auxiliary mounting plate, shown in Figure 6-18, may be used but is not required. The reservoir top shall be attached to the body by perimeter welding, machine screws, or weld studs. All gaskets used to seal the reservoir shall be re-usable. Pipes extending through the top shall be sealed to a height of at least 3/4" above the top surface, either by a pipe coupling welded in place, through tubing bulkhead fittings, or by collars or plates with rubber compression seals.

Oil Lines. The pump intake (suction) line shall be sized for diameter as described on Page 255. See also the information on Page 268. The suction line shall be as free of joints and bends as possible, and shall extend to a minimum distance of 2" above the reservoir bottom. It shall be immersed below the ***minimum*** oil level at least 3" or 1½ pipe diameters, whichever is greater.

Return lines carrying main oil flow shall terminate below the fluid minimum level at a distance of at least 2" or 1½ pipe diameters, whichever is greater, from the tank bottom. The bottom end of the main return line shall be cut on a bias of 45° or a suitable diffuser shall be provided.

Drain lines from solenoid valves, flow control valves, counterbalance and sequence valves, and pressure reducing valves, may be externally combined with one another but shall not be combined with the main tank return line. The drain line must enter the reservoir separately, and may be discharged either above the oil level to eliminate siphoning if a component is removed, or in some cases should be run to below the oil level.

Finishes. The reservoir shall be painted or coated inside and out. Interior surfaces shall be thoroughly cleaned but should not be sand blasted. They should be covered with a paint which is compatible with the fluid to be handled. Or, a rust inhibitor shall be applied if an interior paint is not used. Check with fluid supplier for type of paint when fire resistant fluids are used.

Handling. The reservoir must be sufficiently sturdy so it can be hoisted while completely filled with fluid without suffering permanent distortion. Lifting lugs or holes shall be provided if required.

<u>Fluid Fillers. Figure 6-19.</u> Two filler openings shall be provided, located on opposite ends (or sides) of the reservoir. Each opening shall have a tight fitting cap which shall be permanently attached to the reservoir with a metal connector, like a chain. Top mounted and side mounted fillers are available. The cap shall either be non-vented or if vented, shall have a filtration rating of at least 40μm.

Each filler opening shall be protected with a metal strainer of maximum 30-mesh and of sufficient area to flow at least 5 GPM. The filler screen (not the top cap) shall require the use of hand tools for removal, and shall be protected with an internal metal guard. The filler openings, if on top of the reservoir, shall be sealed to a height of 3/4" above the surface. The entire filler assembly shall be sealed to the reservoir with a fluid-tight resilient gasket compatible with the system fluid.

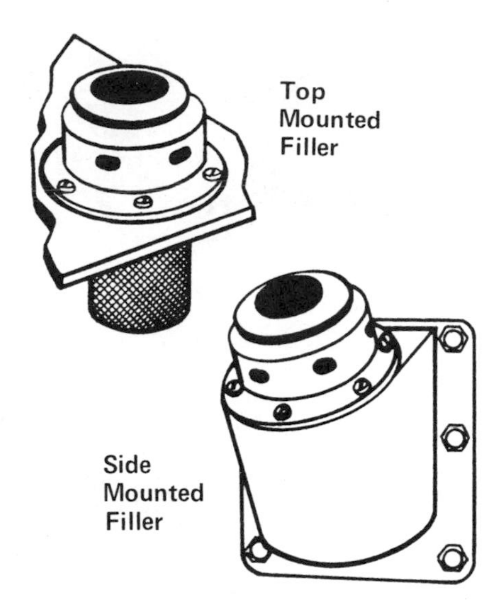

FIGURE 6-19. Filler Assemblies.

<u>Air Breather. Figure 6-20.</u> The reservoir fluid level rises and falls as cylinders retract and extend. On reservoirs exposed to atmospheric pressure the air space must be protected with an air breather to prevent intake of atmospheric dirt as the level falls. Small reservoirs may use a filler cap with built-in breather. Large reservoirs should use a separate breather with sufficient capacity to maintain approximate atmospheric pressure at all times. NFPA standards require the air breather to have maximum 40μm rating, and the element must be protected with a metal guard against physical damage.

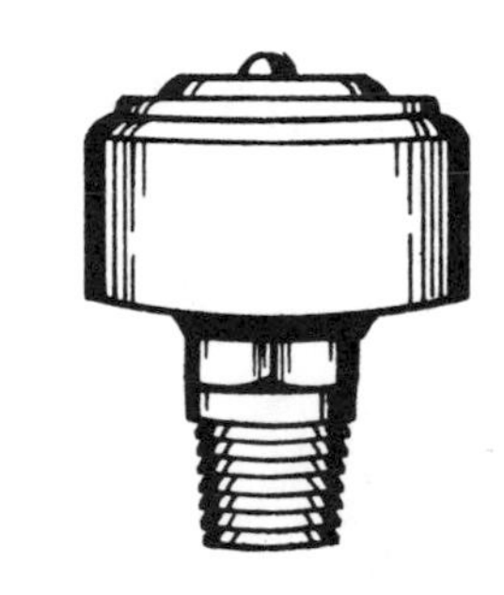

FIGURE 6-20. Reservoir Air Breather.

<u>Fluid Level Indicators. Figure 6-21.</u> Two indicators shall be provided, located near the two filler openings and visible while filling. A protected column indicator of the type shown is preferred, but two small round windows at each location are acceptable if they are correctly located to show high and low fluid levels.

The indicators shall be of sufficient length and so located that the LOW mark is at least 3" above the *top* of the intake strainer. The HIGH mark shall be located so that with pumps stopped and all cylinders retracted the air space above the fluid will be at least 10% of the total reservoir volume.

Dipsticks are not permitted because they may introduce contamination. At normal temperatures plastic windows are preferred over glass as they resist shattering. At high temperatures pyrex glass windows may be used. A thermometer may be included with the level indicator although not required by NFPA standards.

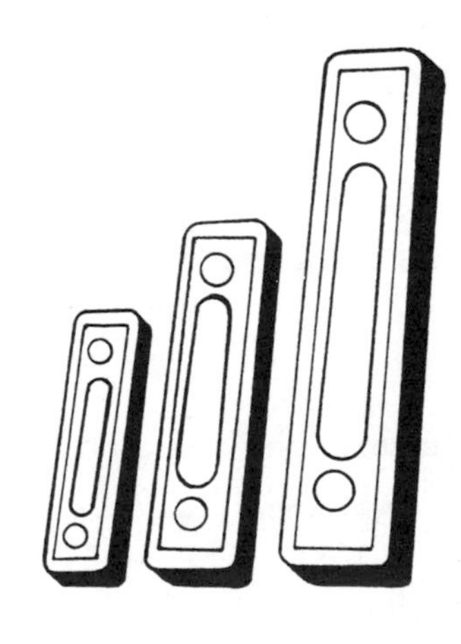

FIGURE 6-21. Oil Level Indicators.

FIGURE 6-22. Sump Type, Pump Intake Strainer.

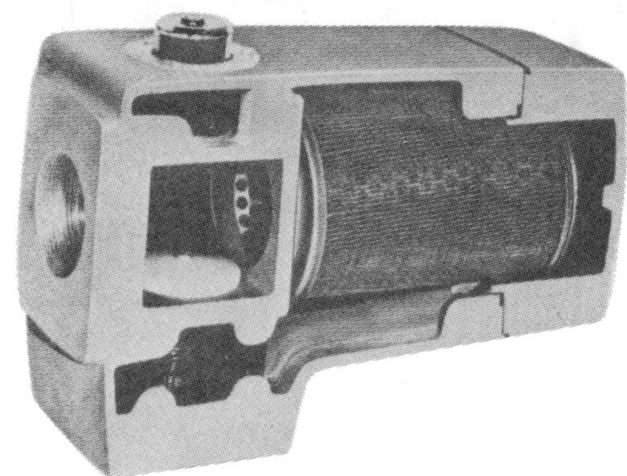

FIGURE 6-23.
In-Line Pump Strainer, With Condition Indicator.

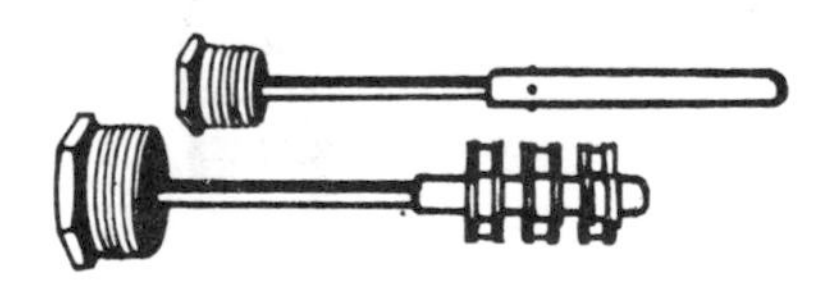

FIGURE 6-24. Magnet Assemblies.

Strainers and Filters. There is no clear distinction between a strainer and a filter, but a strainer is usually considered to be a coarse filter, like wire mesh. A filter is much finer, perhaps a composition of paper or fiber. A strainer can be cleaned and re-used but a filter element is replaced.

Pump Intake Strainer. Filtration in the pump intake line should be limited to course filtering because of the possibility of restricting the line to the point of cavitation which would damage the pump. Finer filtration, if used, is added elsewhere in the system.

The NFPA standards recommend a strainer in the intake line but do not specify the micron rating except that it shall not cause more pressure drop than is acceptable to the pump manufacturer.

For petroleum base hydraulic oil a stainless steel 100 mesh (100 wires per inch) strainer is recommended. It has a 150μm micron rating. About 10 square inches of filter surface should be provided for each 1 GPM of flow. All other fluids have a higher specific gravity than oil. These fluids include fire resistant synthetic, water base, and water/oil emulsions. A strainer with stainless steel 60 mesh element is recommended. It has a micron rating of 260μm. Of course, finer filtration can be used if sufficient area is provided to keep pressure drop to an acceptable value. However, finer strainers (or micronic filters) tend to clog up rapidly and require more frequent attention.

Sump Strainers. Figure 6-22. These protect the open end of the pump intake line and should be mounted in a horizontal position near the bottom of the reservoir, or in any case, with the top of the strainer at least 3'' below the LOW oil level.

In-Line Strainers. Figure 6-23. The strainer element is encased in a housing so it can be mounted outside the reservoir, in the pump intake line next to the pump. Many of these strainers, like the one pictured, have a vacuum indicator to tell the operator when the element needs attention. The indicator may be a pointer, a pop-up visual indicator, or an electrical switch which can be wired to a warning light.

Magnets. The use of magnets is not required by the NFPA reservoir standards but they do give beneficial results provided the magnets are taken out and cleaned once in a while. Magnets will remove iron particles too small to be picked up by any other means, but are not effective on non-magnetic material such as sand, aluminum, etc.

Pump inlet strainers can be purchased with built-in magnets, or magnet assemblies purchased separately consist of several magnets mounted on a long rod terminating in a pipe plug. They can be installed through holes in the reservoir top in which a pipe coupling has been welded. Recommended location is in the oil stream near the suction strainer.

FINE FILTRATION IN A HYDRAULIC SYSTEM

The Importance of Oil Cleanliness. In earlier days of industrial fluid power cleanliness in the hydraulic oil was not considered to be as important as it is today. Pumps were built with less precision and operated at much lower pressures. Cylinder barrels were not as finely finished and used "leathers" as piston and rod seals. A strainer on the pump intake was considered sufficient for most systems. Today, with pumps and cylinders built to closer tolerances and finer finishes, fluid systems operate at higher pressures and with greater efficiency. Cleanliness of the oil has become a very important consideration in the design and operation of hydraulic systems, at least on those which are expected to operate reliably for a long time with a minimum of maintenance and repair. If it were possible to keep hydraulic oil surgically clean, breakdowns in hydraulic systems would be greatly reduced.

The Problem of Cleanliness. All hydraulic systems accumulate dirt, some of it entering from external sources, some of it generated within the system. Therefore, every system should have a built-in means for removing this accumulated dirt. Cleanliness starts, of course, with the use of a properly constructed reservoir, the use of clean (or cleaned) pipe, fittings, and components, and careful construction practices such as thorough cleaning of weld slag, de-burring of pipe and tubing, etc. But even though a system is carefully constructed and is initially clean, dirt will start accumulating in a short time and should be removed on a continuing basis. Internal contamination is generated in the form of metal particles as a result of wear between metal-to-metal contact of moving parts in pumps, hydraulic motors, valve spools, and cylinders. Bits of packing material from cylinder and valve seals may circulate through the system jamming valve spools or lodging in small orifices which are present in many modern hydraulic components. Then, too, it is difficult and nearly impossible to completely exclude dirt from entering through cracks and crevices, through reservoir breather and filler openings, through worn or defective pump, valve, and cylinder rod seals, quick connect couplings, and through open pipes while they are disconnected for repair of the system.

How Much Filtering is Required? The best approach to good filtering is to use a suitable wire mesh strainer on the pump intake. This applies to all standard systems although there may be exceptional cases where an intake strainer is omitted. For some systems, this intake strainer may be sufficient filtering, depending on how dirt-sensitive the pump may be, whether there are components in the system having small internal orifices like pilot operated relief valves, 4-way valves, flow control valves, servo valves, etc. Also on expected duty cycle. For example, it might be a waste of money and time to install a sophisticated filtering system on a hydraulic system using a gear pump and a manual lever control valve and which operated only a few hours per month.

Then, in addition to an intake strainer, a micronic filter should be added to those systems which operate in a particularly dirty environment, on a long work schedule every day, and those using components like valve plate (swash plate) piston pumps which can suffer severe damage by small abrasive particles. Information in this section suggests how suitable filters can be designed into new systems or installed on existing systems.

Strainer and Filter Ratings. The "micro-metre" is the unit of measurement for defining the efficiency of a filter for removing small particles. It is one-millionth of a metre, or approximately 0.00003937 inch expressed in U.S. units. Note that 25 micro-metres = 0.001 inch. The micro-metre was formerly called a "micron". In this book it is represented by the symbol μm.

As a rule-of-thumb, any filter with rating coarser than $40 \mu m$ may be called a strainer. If rating is finer than $40 \mu m$ it may be called a filter. Filter elements are constructed with a porous medium which may consist of a formulated mixture of paper fiber, plastic fibers, vegetable fibers, etc. held together with a bonding material. Some elements are available in ratings down to 1μm, or finer.

Also as a rule-of-thumb, systems using gear pumps should be filtered to 40μm, those using vane pumps, to 25μm, those piston pumps which have shoes on each piston riding against a swash plate should be filtered to 10μm or finer.

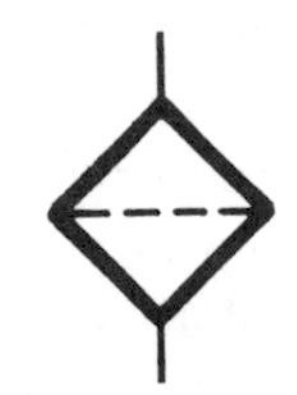

Graphic Symbol for Strainer or Filter. The symbol approved by the ANSI (American National Standards Institute) and the NFPA (National Fluid Power Association), as shown here, is used on fluid power diagrams to indicate all kinds of filters including both air and hydraulic, suction strainers, pressure line, and return line filters. Although not required, a notation of the micro-metre rating placed alongside the diagram may be useful.

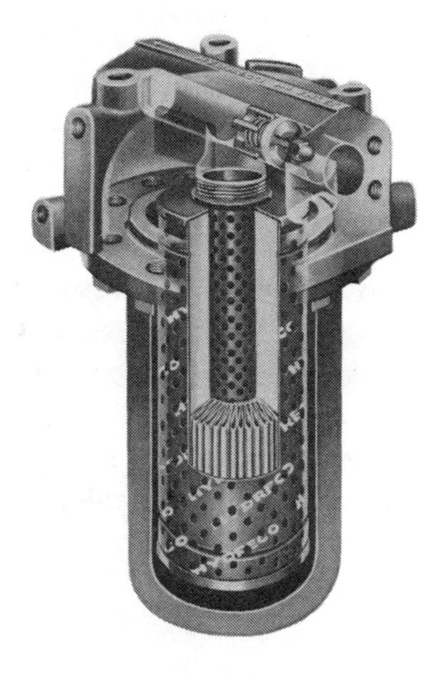

FIGURE 6-28. High Pressure Filter With Micronic Pleated Element.

FIGURE 6-26. Spin-On Filter Showing Internal By-Pass Valve.

FIGURE 6-27. Spin-On Filter With Vacuum Gauge Indicator.

Pressure and Return Line Filters

Spin-On Filters. Figures 6-26 and 6-27. The filtering element is permanently encased in a housing which screws on to the filter head. When replacing the element the head does not have to be removed from the line. The encased element is unscrewed and discarded; it cannot be cleaned and re-used.

These are low pressure filters for use in pump suction or tank return lines. Ratings vary from 100 to 200 and even up to 500 PSI. Nominal flows go to about 25 GPM so they can be used only for low to medium power systems.

Several kinds of elements are available including 10, 25, 33, or 40μm ratings. By-pass valves in the filter head are available with a 5 PSI spring when the filter is used in a pump intake line, and with a 15 PSI spring when it is used in a tank return line. A port is usually available for a vacuum gauge or other indicator to show when the filter should be replaced.

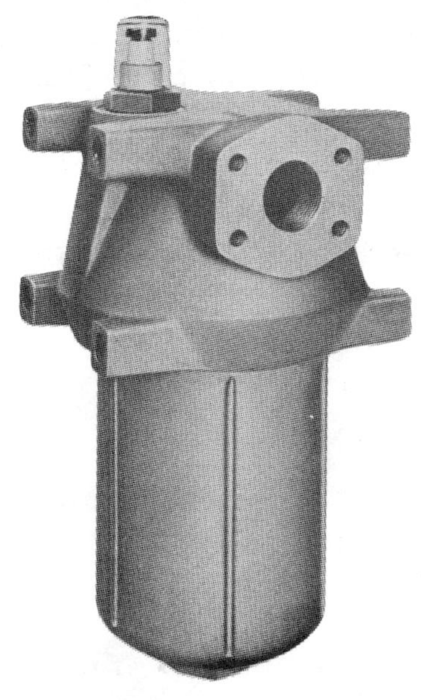

FIGURE 6-29. High Pressure Filter With Pop-Up Condition Indicator.

High Pressure Filters. Figures 6-28 and 6-29. Filters with removable elements are used for higher pressures and flows. The elements are a formulation of various fibers and available in many filtration ratings down to 1μm. Some of these can be cleaned and re-used; others must be discarded and replaced.

By-pass valves in the head have 15 PSI springs. A higher by-pass pressure is seldom used because it gives very little extra element life and introduces the risk of element collapse. Most of these filters have, or have available, some kind of condition indicator to show when the element should be changed. This is often a pop-up indicator as in Figure 6-29, or may be a switch for activating some kind of warning device or even shutting down the system.

Filter Locations in a Hydraulic Circuit

Figure 6-30. At Location 1 is the pump suction strainer which is recommended for almost every hydraulic system. This is the basic filter of the system but since it usually is not finer than 150μm, it may not be as effective as required for the type of components used and the life expectancy desired from the system. It can be supplemented with finer filters in other parts of the system.

Keep in mind that small particles, even within the rating of a fine filter, may not be picked up on their first pass through the filter but may circulate many times before being caught.

Location 2 is the pump line. This is a very good place to install a filter since a large part of internally generated contamination comes from the pump. Filter pressure rating must be equal to or greater than maximum system pressure. Any filter at Location 2 should have an internal by-pass valve with cracking pressure of 15 PSI, certainly no higher than 25 PSI. If the element becomes restricted, the fluid should be by-passed around it so it will not collapse and release all the trapped dirt plus pieces of element to the remainder of the system.

Location 3 is the tank return line. For most systems this is the best place to add a fine filter. Since this is a low pressure part of the circuit, a less expensive filter with a lower pressure rating can be used.

At this location a low pressure, ***non-adjustable*** relief valve should be placed across the filter to provide an additional path for excess flow, beyond the filter rating, which may appear while the system cylinders are retracting. Return flow from a cylinder (which has a single end rod) is greater than the flow from the hydraulic pump.

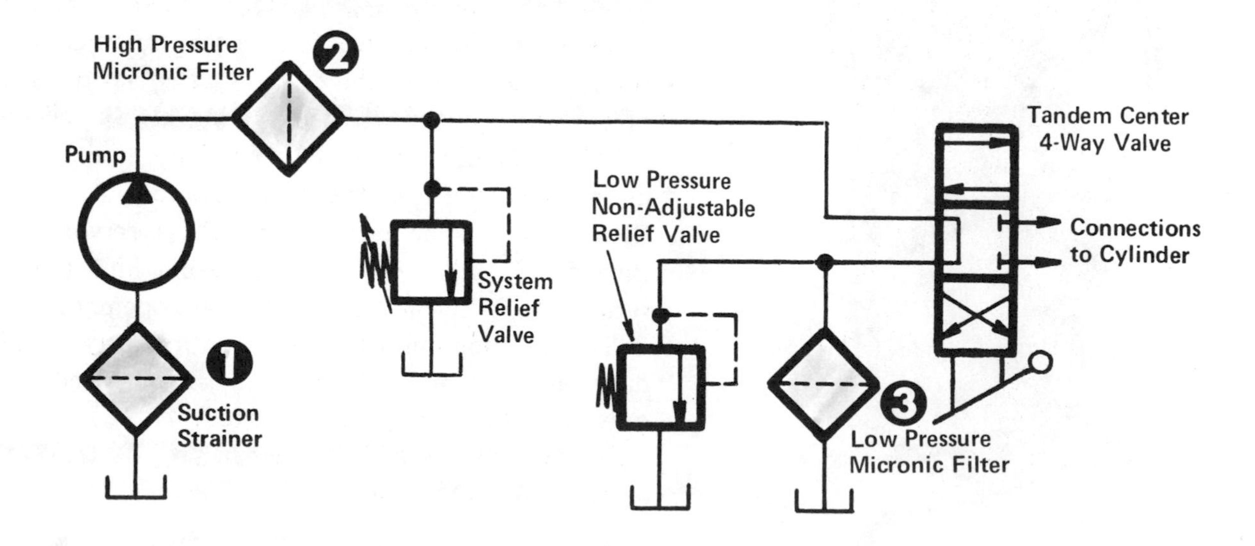

FIGURE 6-30. Hydraulic circuit showing three locations for installing oil filters.

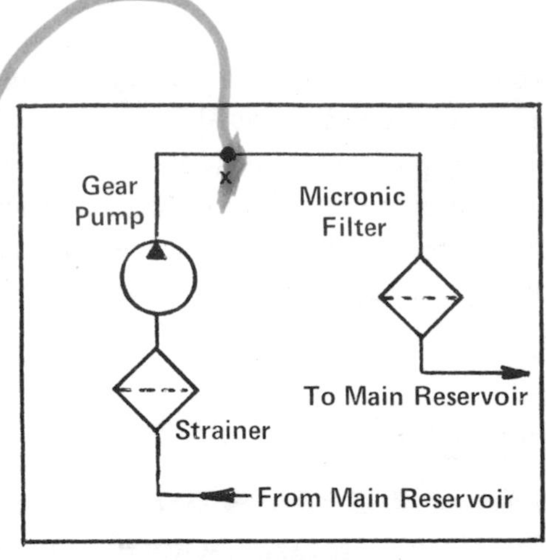

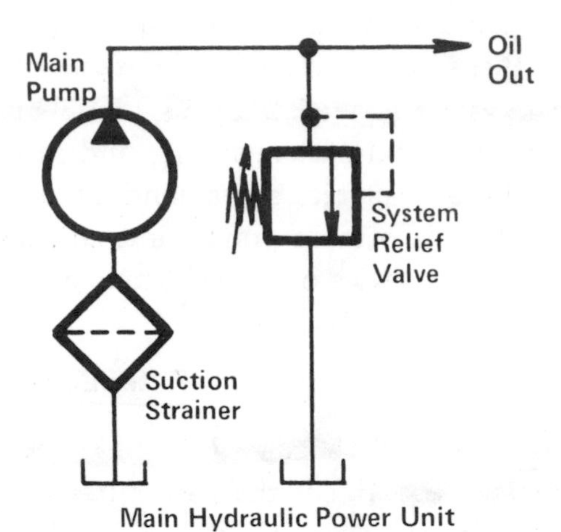

FIGURE 6-31. A small auxiliary filtration unit can be built to set alongside, and filter the oil in the reservoir of a hydraulic power unit. Oil is drawn from the main reservoir, passed through a filter, then returned to the reservoir. The auxiliary filtration unit has no reservoir of its own.

External Filtration for a Hydraulic System

Figure 6-31. On systems, especially large ones, where additional oil filtration must be added, this can be done more economically and with less disturbance to the system by adding an external filtration system. Filtration components can be mounted on a small picture-frame skid and set alongside the reservoir of the main system.

The external filtration system can include a 1/2 HP electric motor driving a 5 GPM gear pump. On very large systems a larger motor and pump can be used, large enough for a filtration flow of at least 5% of the pump flow on the main system. Other components include a suction strainer for the gear pump and a fine filter on the outlet of the pump. A relief valve may not be necessary if the fine filter has a built-in by-pass valve.

Oil to be filtered can be picked up at one of the drain connections on the main reservoir. After being filtered it can be returned to the reservoir either through the filler opening, by teeing into the tank return line, or through the other drain opening.

The filtration unit can be run continuously for a few days until the oil is clean, or it can be wired to run only when the main pump is started.

Incidentally, Point x in the auxiliary filtration unit is an excellent place to add a heat exchanger.

FIGURE 6-32. A commercially built scavenging-type portable filtration system.

Figure 6-32. In some plants the filtration skid, or a ready-built unit, is moved from machine to machine and connected to each one with quick connect couplers.

Hydraulic Filter Ratings

A test has been devised, called a "multi-pass" test to produce a more realistic rating for a hydraulic filter than the old "nominal" and "absolute" ratings previously used. The "beta ratio" (β) derived from the multi-pass test, is becoming an accepted method of expressing filter effectiveness. In Europe, other rating methods may be used such as CETOP or Code RP-70.

What is the Beta Ratio? The beta ratio is simply a way of stating the effectiveness of a filter for removing particles of a certain micro-metre size or larger. The effectiveness can also be expressed as efficiency. Beta ratio is expressed numerically while efficiency is expressed in per cent. Both methods use data derived from the multi-pass test and both give the same information.

Beta ratio is written with the Greek letter β. It is the ratio of the number of particles larger than a specified micro-metre size (μm) which enter the filter compared to the number of the same size particles which go through without being caught. For example, if 4000 particles per millilitre (ml) of fluid enter the filter and 750 of these pass through without being caught, the beta (filtration) ratio is $4000 \div 750 = 5.33$. The same identical information can be stated as filter efficiency. The percentage of particles getting through is $750 \div 4000 = .1875$ or 18.75%. Therefore, filter efficiency is 100% – 18.75% = 81.25% This is the same as a beta ratio of 5.33.

Beta ratio or percentage is always stated for a specified micro-metre size. For example a β10 ratio of 5.33 means that particles of 10μm and larger are removed with a filtration ratio of 5.33.

Some filter manufacturers give the β ratio at only one micro-metre size, usually 10μm, while others may give the β ratio at two or even three micro-metre sizes. A rating of [β15 & β30 = 2/20] means that at a micro-metre particle size of 15μm the β ratio is 2 (equivalent to 50% efficiency) and at a micro-metre particle size of 30μm the β ratio is 20 (equivalent to 95% efficiency).

For most of us it is easier to visualize the effectiveness of a filter in terms of per cent removal efficiency rather than in a β ratio. To convert a β ratio into percentage, take the reciprocal of the β ratio ($1 \div \beta$) and subtract it from 100%. In formula: $100\% - (1 \div \beta) = \%$ efficiency.

Filter Multi-Pass Test. This is a laboratory procedure on which standards have been prepared by the ANSI and NFPA. The purpose for developing this test procedure was to be able to test and to rate filter media under conditions which could be duplicated by every filter manufacturer, and which would be similar to operating conditions in the field. Since many particles are able to circulate several times before being caught in the filter, a single pass through the filter does not give a true indication of the filter's ability to catch particles, particularly small ones. It was deemed necessary to re-circulate the test fluid many times through the filter. Starting with clean oil the contaminate is added slowly as it would be under field conditions, until the filter becomes loaded with dirt up to a specified by-pass pressure, usually 25 PSI.

Although the multi-pass test does not completely define filter performance, it does deal with two of the more important areas of a filter's performance – its effectiveness for catching particles of various sizes, and its dirt holding capacity.

CROSS REFERENCE – BETA RATIO TO PER CENT EFFICIENCY

β	2.00	2.25	2.50	2.75	3.00	3.50	4.00	4.50	5.00	7.00	10.0	20.0	30.0	50.0	100
%	50	56	60	64	67	71	75	78	80	86	90	95	97	98	99

HYDRAULIC ACCUMULATORS

An accumulator is a storage vessel into which oil can be pumped under pressure and stored. Accumulators are used in hydraulic circuits in one of these ways:

(1). The stored oil can be used later to provide a high pressure flow either to perform an operation or to add to the flow from a hydraulic pump for a brief period. A hydraulic system with an accumulator acts much like a compressed air system – a relatively small pump, running continuously (regardless of whether the system cylinder is in motion or stopped) stores oil under pressure in the accumulator until it reaches a pre-set maximum, usually 3000 PSI. An unloading valve or a pressure switch then vents the pump to tank, letting it run idle until enough oil is used from the accumulator to cause its pressure to fall to a cut-in level. The unloading valve or pressure switch then causes the pump to bring the accumulator back up to full pressure before venting the pump again.

(2). A small accumulator is sometimes placed in the pressure line near a component to dampen the pressure shock produced by the component. Possible locations are at the pump pressure port, close to the pressure port on a 4-way valve, or close to a press cylinder.

(3). On a press holding application where the press platen must remain closed under pressure for a period of time, the hydraulic pump can be unloaded and an accumulator can feed whatever oil is needed into the press cylinder to make up for seal leakage in cylinder and 4-way valve.

(4). An accumulator can be used in any kind of application which requires a very high volume of oil flow for a very brief period.

An accumulator must be pre-charged with dry nitrogen from a pressure bottle through a charging hose. Pre-charging must be done with all oil discharged. The gas port is then closed and the charging hose and gas bottle removed. Air should not be used for precharging because oxygen in the air will deteriorate rubber seals. Also, under certain conditions, the oxygen could be a fire hazard.

Most circuits using accumulators should be designed to work at a maximum pressure of 3000 PSI, to take advantage of the maximum capacity of 3000 PSI accumulators. The gas pre-charge pressure is not critical for most applications, and a level of about 1500 PSI works out about right. At this pre-charge pressure, the accumulator can be pumped about half full of oil before the internal pressure reaches 3000 PSI. So, a 5-gallon size accumulator has a maximum oil capacity of about 2½ gallons from full 3000 PSI charge down to 1500 PSI pre-charge level. Pre-charging should be repeated when gas leakage has reduced it to about 1000 PSI.

Preferred mounting position for all types of accumulators is with the oil port down, to keep dirt purged out.

Accumulators with associated valving do add cost to a system, so they should not be used except on applications where they offer an advantage over more conventional circuits.

Types of Accumulators

Piston Accumulator. Figure 6-33. Built like a hydraulic cylinder except without a piston rod. The piston is sealed with a pair of O-rings, with the space between the rings vented to the oil side. The piston keeps the nitrogen gas pre-charge from dissolving in the oil and being carried away. On one end there is a large oil port. On the other end a gas valve. After the accumulator has been pre-charged with a high pressure bottle of nitrogen, the gas port is closed and the nitrogen bottle is removed.

Piston accumulators are available in standard sizes from a fraction of a gallon to about 20 gallons capacity. For greater capacity several accumulators can be connected in parallel.

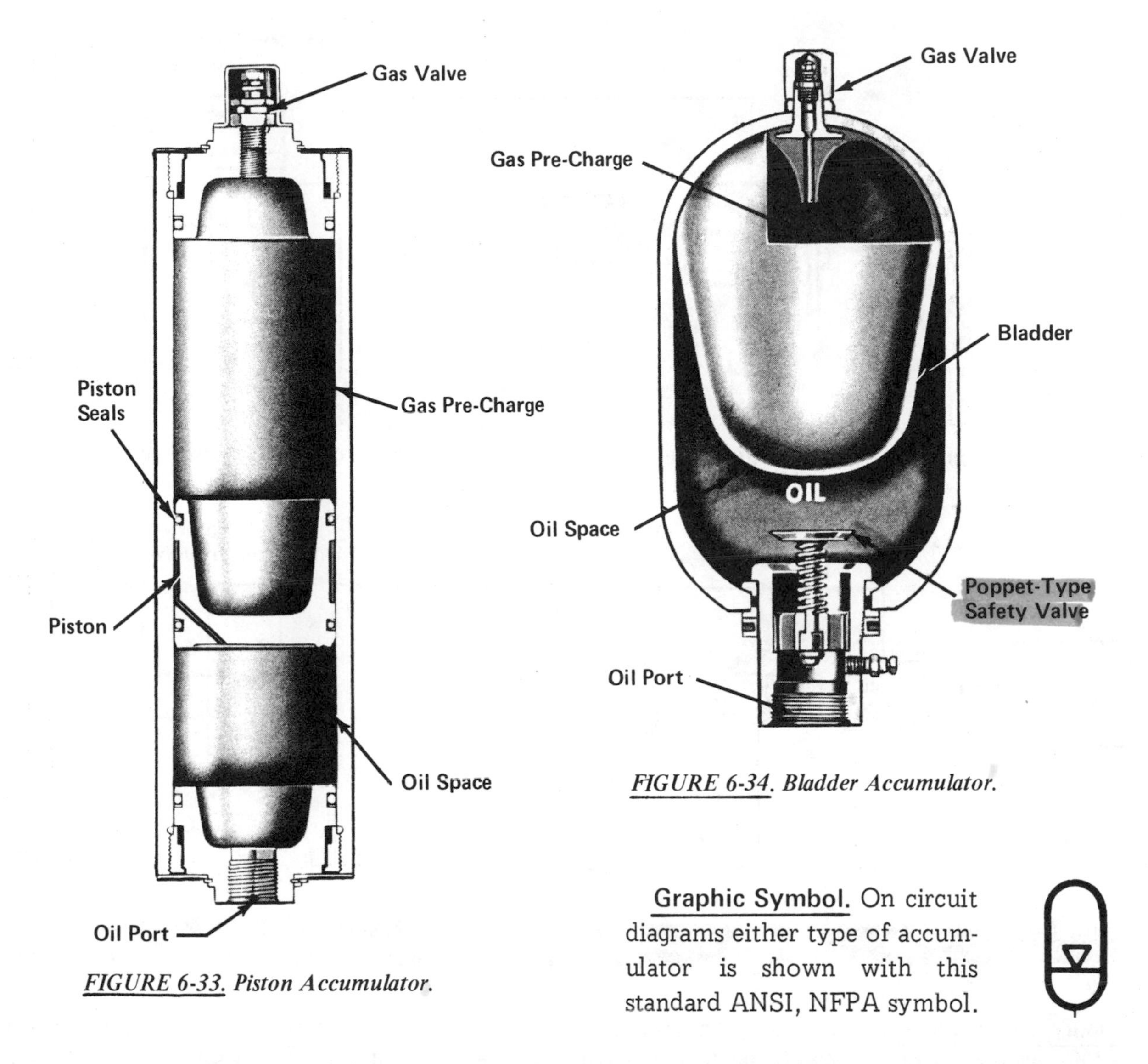

FIGURE 6-33. Piston Accumulator.

FIGURE 6-34. Bladder Accumulator.

Graphic Symbol. On circuit diagrams either type of accumulator is shown with this standard ANSI, NFPA symbol.

Bladder Accumulator. Figure 6-34. A synthetic rubber bag separates the nitrogen pre-charge from the hydraulic oil. The usual maximum pressure rating is 3000 PSI, and pre-charge at about 1500 PSI. Sizes start at a fraction of a gallon and go up to 10 gallons or more.

A poppet-type safety valve in the mouth of the oil port is pushed closed by the bag if all oil is discharged. This prevents extrusion of the bag through the oil port.

Accumulator Applications

Supplementing Pump Flow. This is probably the most important application in a hydraulic system. During resting periods of the pump, when oil is not needed to move a cylinder, the pump can store oil under pressure in an accumulator. Later, when the cylinder is ready to move, oil from the accumulator can add to the normal flow from the pump to produce a faster speed from the cylinder than with pump oil alone. Because of the cost of the accumulator and accessory equipment, this application is worthwhile only on applications where the pump remains idle for a large percentage of total working time – that is, while the cylinder is retracting and while parts are being loaded and unloaded from the machine. As a rule-of-thumb, an accumulator should be considered if the pump is working at full load less than 20% of the time and is unloaded more than 80% of the time.

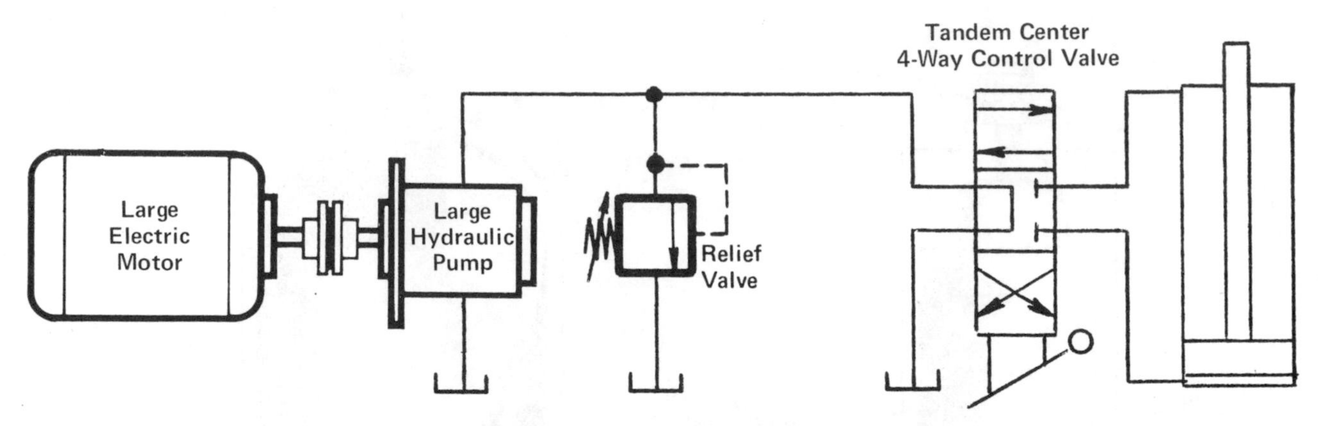

FIGURE 6-35. "Conventional" hydraulic circuit operating on a 20% duty cycle. Electric motor and pump must be large enough to supply full power continuously, so 80% of their potential capacity is unused. Compare with circuit below.

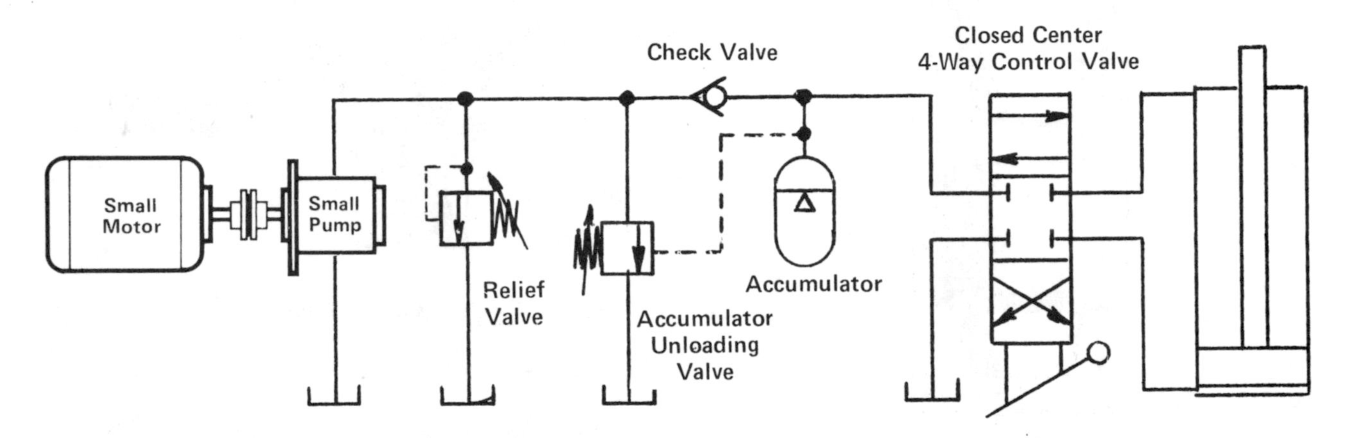

FIGURE 6-36. "Accumulator" hydraulic circuit operates on same 20% duty cycle as conventional circuit. Cylinder will produce same force and speed but the potential capacity of the smaller pump and motor are more efficiently used.

Figures 6-35 and 6-36. Compare these circuits. One of them (Figure 6-35) is a conventional hydraulic system, the other (Figure 6-36) is a circuit using an accumulator. They both produce the same cylinder force and speed. But on certain applications the accumulator circuit may be cheaper to build.

In the conventional circuit of Figure 6-35 the cylinder is in motion about 20% of the time. During the other 80% it is stopped, waiting for the operator to unload and re-load the machine. But the pump and electric motor must be sized large enough to produce full horsepower on a continuous basis even though it is being used only 20% of the time. This is a waste of 80% of the potential capacity of the pump and electric motor.

In the accumulator circuit of Figure 6-36 the cylinder will produce the same force and speed and operates on the same 20% : 80% cycle but can operate with an electric motor and pump 1/4th as large, supplemented with oil flow from an accumulator. The smaller pump is working most of the time at a steady rate, whether or not the cylinder is moving, storing oil under pressure in the accumulator. For brief periods after the pump has fully charged the accumulator it will be unloaded to tank through the unloading valve until the cylinder starts on its next cycle. Combined flow from the small pump plus the flow from the accumulator gives the same speed as in the conventional circuit. Even with the cost of the accumulator, unloading valve, check valve, and charging equipment, the system is cheaper because of the reduced cost of pump and electric motor. A further advantage is the relatively steady demand of power at a low level from the power line instead of high peak demand in short spurts.

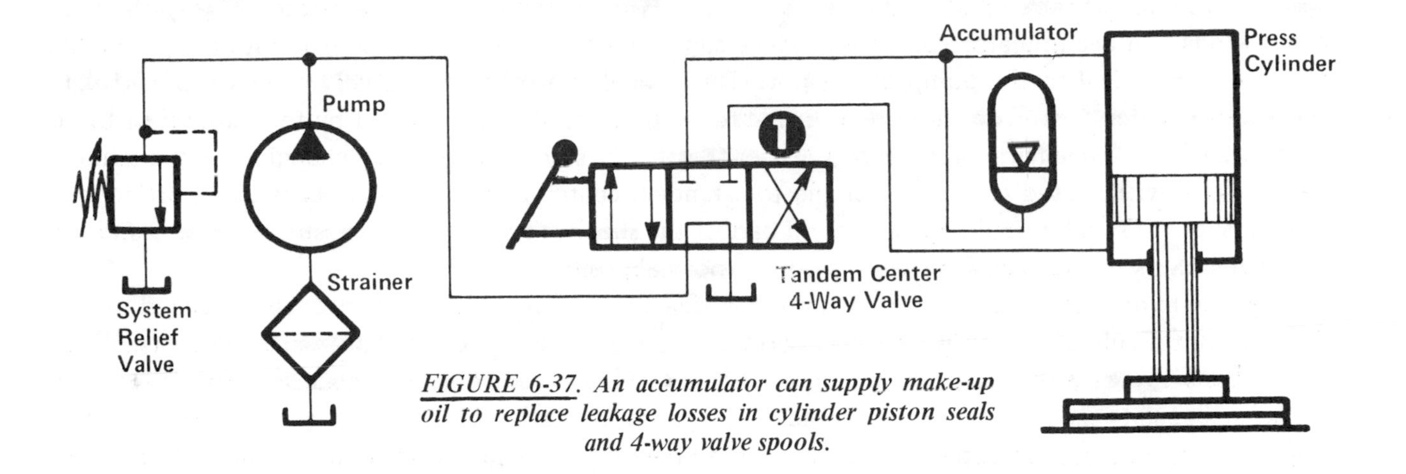

FIGURE 6-37. An accumulator can supply make-up oil to replace leakage losses in cylinder piston seals and 4-way valve spools.

Accumulator to Make Up Leakage. Figure 6-37. On hydraulic presses used for molding, laminating, bonding, curing, etc., it may be necessary to hold pressure on the work during the curing period. Active pressure must be maintained on the work during this time; otherwise, the piston force will relax due to small leakages across piston seals and valve spools. A pressure compensated pump can be used and it will reduce its pumping rate to just enough to make up leakage. Or if using a fixed displacement pump an accumulator across the cylinder port will supply leakage oil and permit the pump to be unloaded. The pump should not be allowed to discharge across a relief valve during this period.

Circuit operation is as follows: When the operator shifts the 4-way valve, the press closes and at stall, pressure builds up. When full pressure has been reached the accumulator is in a charged condition. The operator may then shift the valve to neutral to unload the pump. Leakage from the circuit will be made up from the accumulator. But remember that as oil discharges from the accumulator its pressure will fall according to the amount of oil fed out into the circuit.

Surge Dampening. Figure 6-38. Accumulators, properly placed, may sometimes be effective for reducing (not eliminating) shock waves or pressure spikes which pass through the system. To be most effective they should be placed *physically* as close as possible to the place or component where the shock is generated.

Most shock is generated at one of three locations: the pump, the 4-way control valve, or the press

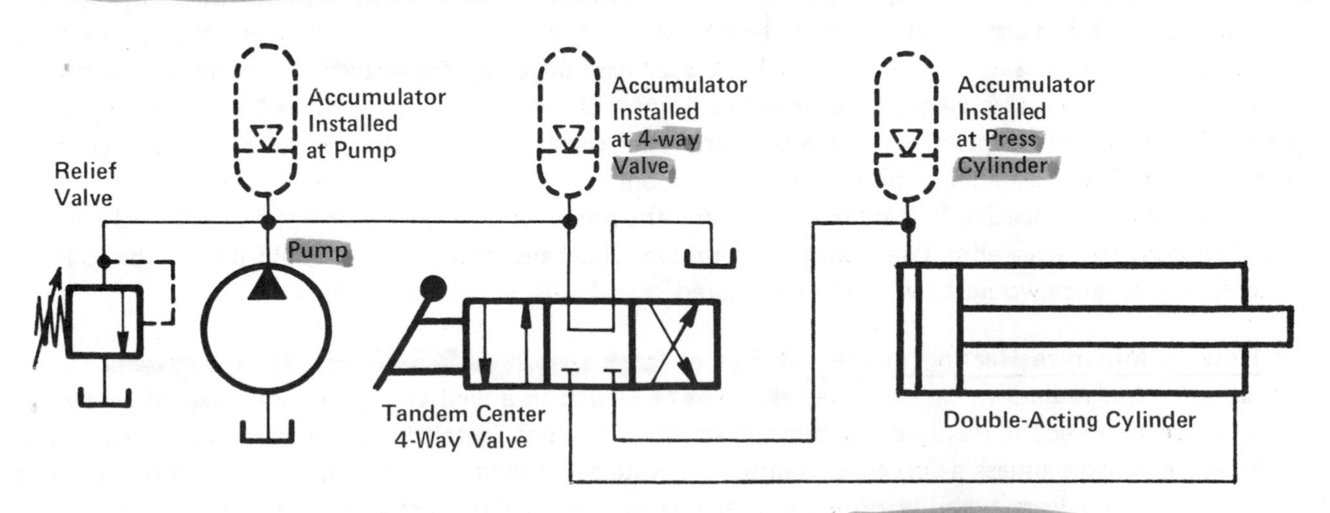

FIGURE 6-38. Possible locations for an accumulator to suppress shock.

cylinder as the pressure on the cylinder is relaxed. Piston pumps seem to generate more shock than other kinds. An accumulator can often reduce shock intensity, and should be teed into the pump line as close as practical to the pump outlet port. Directional control valves usually generate shock while the spool is being moved from a side position to center neutral, and is caused by the pump flow being momentarily deadheaded while the spool is in transit. On some applications, changing from a tandem center valve with closed crossover porting to a tandem center valve with open crossover porting may reduce shock. On a hydraulic press, a good part of the shock is produced as pressure is relaxed and the frame members spring back from a stressed to a normal position.

In most hydraulic systems, the accumulator size for reducing shock cannot be predicted or calculated. But a rule-of-thumb which may serve as a starting point is to provide an accumulator with an oil working capacity up to about 1/10th the GPM oil flow at the point where shock is generated. (Oil working capacity is about 1/2 the rated gas capacity). The level of pre-charge pressure may also influence accumulator effectiveness. Start with a pre-charge pressure (with all oil discharged) of about half the maximum system hydraulic pressure. With the system running, the pre-charge pressure can be bled down to see if shock suppression is more effective at a lower pre-charge pressure. But remember that an accumulator may destroy system rigidity by introducing compression in the oil circuit.

Bladder-type accumulators may work better than piston accumulators because they have a faster response to pressure surges.

HEAT EXCHANGERS FOR HYDRAULIC SYSTEMS

Even with good circuit design there are always power losses in converting mechanical energy into fluid power. Most of these losses generate heat and a great part of the heat may end up in the hydraulic oil. On smaller systems, say of less than 25 HP, the heat which finds its way into the oil can be radiated to the atmosphere from wall surfaces of reservoir, cylinder, pump, valves, and plumbing, and the oil temperature will not exceed an acceptable level. But on larger systems there may not be sufficient wall surface area to radiate the heat without the oil reaching an excessive temperature. On these systems a heat exchanger should be added.

Why Does Hydraulic Oil Get Hot? First of all, there is a mechanical power loss in the pump of 10 to 15%. A large part of the heat generated in the pump is carried by the flow of oil into the reservoir.

Then, as the pump produces a flow of oil under pressure, the energy level in the oil, at a given rate of flow, is directly proportional to its pressure. At any point in the system where any part of the flow is permitted to escape to a point of lower pressure without being converted into mechanical work (as it is by a cylinder), this part of the flowing energy must convert into heat which is carried into the reservoir. Heat can be produced by a pressure relief valve, a flow control valve, a pressure reducing valve, and in flow resistance in plumbing and components.

To eliminate all heating in a hydraulic system, the pump would have to operate at 100% efficiency, all flow control valves eliminated, and the plumbing lines and components would have to be so large that flow resistance would be virtually eliminated.

How to Minimize Heating in the Oil. Use as large a reservoir as practical for the system, to get maximum heat radiation. Be sure the reservoir is installed in a well ventilated area, away from walls, and where exposed, if possible, to moving air flow. Do not install it inside an enclosure such as a cabinet or console unless a forced air ventilation system is provided. Design the system so the pump is unloaded to the least possible pressure level during the time the actuators (cylinders or hydraulic motors) are not working. In some circuits, flow control and pressure reducing valves are necessary, but limit their use to places where they are vital for circuit operation. Bleed-off instead of series speed control methods may generate less heat if they are applicable. Use components and plumbing lines large enough to keep flow losses to a moderate level. Set the system relief valve to as low a level as will adequately do the job.

Types of Heat Exchangers

Air Cooled Heat Exchanger. Figure 6-39. Consists of a steel radiator core through which oil flows while a strong blast of air passes across the core. On industrial systems the air blast is produced by an electric-motor-driven fan. On mobile equipment only the core is used and it is usually mounted in front of the engine radiator to use the same air blast.

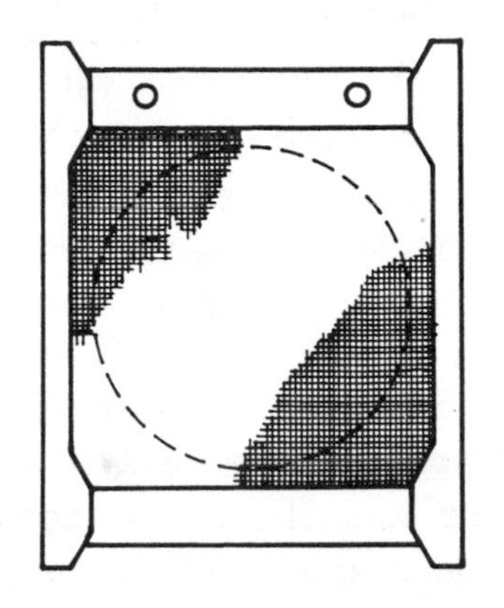

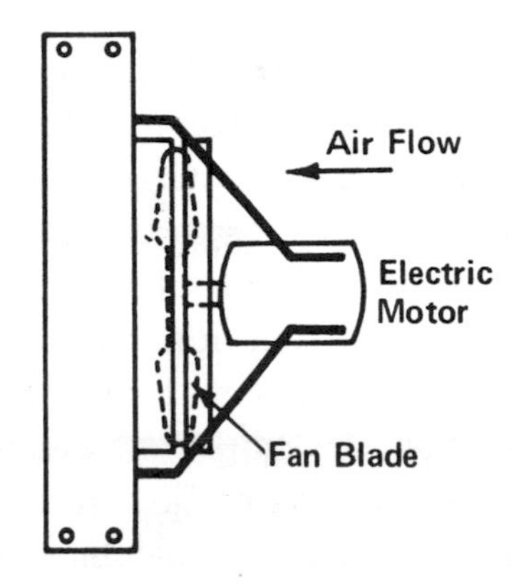

FIGURE 6-39. Air Cooled Heat Radiator With Electric Fan Motor.

Air cooled heat exchangers are more expensive than water cooled units of the same capacity but require no water supply. They work well if the atmospheric air is reasonably cool, but their capacity diminishes in proportion to the increase in surrounding air temperature. At temperatures over 100° F their effectiveness may be poor.

Water Cooled Heat Exchangers. Figure 6-40. If a source of cooling water is available, a shell and tube heat exchanger is usually preferred to an air cooled type because it is lower in first cost and cools efficiently no matter how high the ambient temperature, since its effectiveness is related to the temperature of the cooling water rather than the surrounding air temperature.

Cooling water is circulated through a bundle of bronze tubes from one end cap to the other. On "single-pass" models, water enters one end cap, flows through all the tubes and comes out through the other end cap. On "two-pass" models, water enters one end cap, flows through half of the tubes, is turned around and flows back through the other half of the tubes to the starting end cap. On "four-pass" models the water makes four passes, through one-quarter of the tubes on each pass. Whether to select a single-pass, two-pass, or four-pass model depends on the volume of water available. Single-pass models are more efficient but should have a water flow of about one-half the oil flow rate. A two-pass model of a slightly larger size may be used with a water rate of about one-fourth the oil flow rate, and a four-pass model of larger size should be used if less water than this is available.

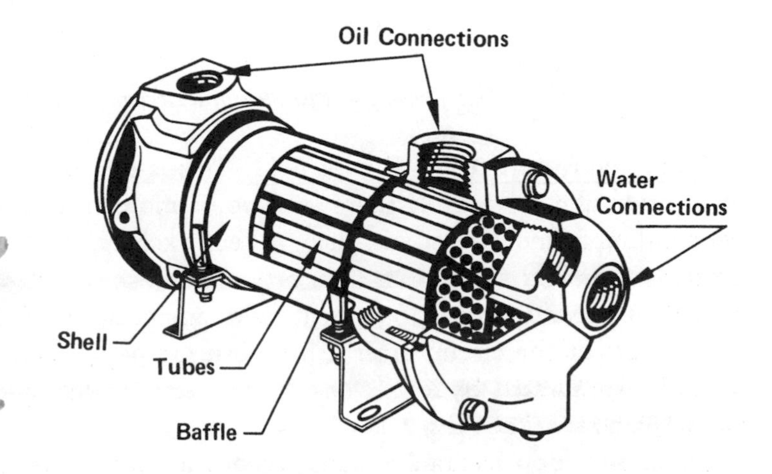

FIGURE 6-40. Water Cooled Shell and Tube Heat Exchanger.

Oil to be cooled is piped into the shell and flows around the outside of the tubes. Baffles are placed to direct the oil flow back and forth across the tubes as it flows from one end of the shell to the other.

To economize on water, a modulating type water flow valve should be used, with temperature sensing bulb installed under the oil in the reservoir. This valve cuts water flow to the minimum which will maintain a set maximum oil temperature. It cuts off water flow completely during intervals when all of the generated heat can be radiated from the reservoir walls.

How to Select Capacity of a Heat Exchanger

The *theoretical* maximum capacity of a heat exchanger will never have to be greater than the HP input to the system, and should be a great deal less. On an existing system the HP input to the electric motor can be determined with a loop ammeter and compared to the nameplate current rating. A rule-of-thumb is to provide a heat exchanger removal capacity of at least 25% of the HP input. Rarely would a capacity of more than 50% of the input power be required. Please remember this is a general rule and might not apply to your system.

Information to be furnished to your supplier is the rate of oil flow, in GPM, through the heat exchanger and BTU per hour or HP heat to be removed. On water cooled models, specify the maximum water flow rate which will be available, and the cooling water temperature. By allowing your supplier to calculate heat load on your application you may be able to use a smaller unit than indicated by the general rule given above. See also additional heat exchanger information on Page 242.

Heat Exchanger Installation

Heat exchangers will not stand high pressure. They are usually rated from 150 to 250 PSI maximum. Therefore, they must be installed at a location of relatively low pressure. In addition to the location shown in Figure 6-41, a good location is at Point x in the external filtering system of Figure 6-31.

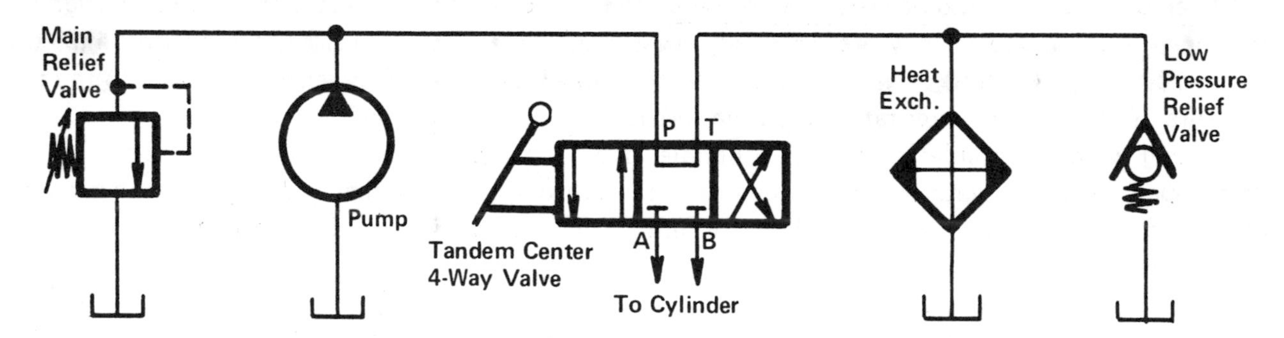

FIGURE 6-41. Heat Exchanger Installed in Reservoir Return Line. Low Pressure Relief Valve is Optional.

Reservoir Return Line. Figure 6-41. This is the most practical location for the heat exchanger in most hydraulic systems. If there are two or more control valves, tank return lines from all valves can be combined and routed through a heat exchanger. The low pressure relief valve is optional and protects the return line from high pressure surges which might damage the heat exchanger. If used, it should be ***non-adjustable*** to prevent field tampering. There is no way in the field to re-set a relief valve in this part of the circuit after it has been tampered with. It should have a cracking pressure of 20 to 50 PSI. A standard hydraulic check valve, or one with a heavier spring, makes a good low pressure, non-adjustable relief valve.

On closed loop hydraulic transmissions a heat exchanger may be installed in the combined flow from the low pressure relief valve plus the case drain oil from both pump and motor. See Volume 3.

REVIEW QUESTIONS – CHAPTER 6

1. What is a "trio" assembly? What individual units make up a trio? What units make up a duo assembly? *– See Page 209.*

2. If there is a filter on the air compressor intake, why is an additional filter necessary on each machine? *– See Page 208.*

3. What is the recommended fineness rating of an air filter in the trio? *– See Page 210.*

4. In a compressed air system what is the name of the valve which does the same thing as a pressure reducing valve in a hydraulic system? *– See Page 210.*

5. In what way is a self-relieving air regulator different from a non-relieving type? *– See Page 211.*

6. What are some of the hazards to be aware of when using clear plastic bowls, without bowl guards, on filters and lubricators? *– See Page 209.*

7. How is a pressure regulator different from a pressure relief valve? *– See Page 210.*

8. How is the automatic feed action created in a lubricator, so the oil feed stops when air flow stops? *– See Page 212.*

9. An atomizer type lubricator delivers smaller droplets of oil which carry farther down the line than the larger droplets from a standard lubricator. How is this extra fine particle size obtained? *– See Page 213.*

10. What is the difference between water "mist" and water "vapor"? *– See Page 214.*

11. How much of the water "vapor" can be removed with an air filter? *– See Pages 214 and 215.*

12. Filters and dryers both remove water from compressed air. What is the difference in their action? *– See Pages 210 and 215.*

13. On a hydraulic reservoir, what is the recommended gallon capacity in relation to pump flow? *– See Page 221.*

14. Describe four important functions of the reservoir in a hydraulic system. *– See Page 220.*

15. How much air space should be left above the maximum oil level in a hydraulic reservoir? *– See Page 221.*

16. What is the difference between a filter and a strainer? *– See Page 225.*

17. How large is a micro-metre (μm)? *– See Page 225.*

18. What is the recommended degree of filtration at the intake of a pump working on petroleum oil? On water base and other fire resistant fluids? *– See Page 224*

19. What are the three locations in a hydraulic system where strainers and filters are installed? *– See Page 227.*

20. What is the definition of "beta ratio" of a hydraulic filter? *– See Page 229.*

21. What is the most important use for an accumulator in a hydraulic system? *– See Page 231.*

22. What are the two main types of accumulators? *– See Page 230.*

23. What is the recommended operating pressure of a hydraulic system in which an accumulator is used? *– See Page 230.*

24. What is the recommended gas pre-charge pressure of an accumulator? *– See Page 230.*

25. What are the two most common types of heat exchangers? *– See Page 234.*

26. In a shell-and-tube heat exchanger which fluid is carried inside the tubes? *– See Page 235.*

27. At what location in a hydraulic system is a heat exchanger usually installed? *– See Page 236.*

28. What information must be furnished to your supplier to enable him to recommend a minimum size heat exchanger for your system? *– See Page 236.*

APPENDIX A
Design Calculations

SELECTING PLUMBING SIZE FOR HYDRAULIC SYSTEMS

The size of pipe, tubing, or hose for plumbing a hydraulic system is very important. If too small a size is used, oil velocity will be too high. This creates excessive power loss and heat generation in the oil. On the other hand, if piping size is larger than necessary, the losses will be low and the power transfer will be good but the cost of materials and time for installation will be greater than it needs to be.

Sometimes, especially on low power systems, one size and kind of pipe is used to plumb the entire system (except for the pump inlet line), and this is all right if the size chosen is large enough to keep flow losses down. The size of the pump outlet port should be the guide for the minimum pipe size. On larger systems, if the piping in various parts of the circuit is selected to meet the flow condition and with regard to the maximum operating pressure of the entire system, piping costs can be kept to a minimum and performance should be satisfactory. For additional information on plumbing materials and sizes, refer to the plumbing section, Pages 33 to 37.

Figure 7-1. This circuit is typical for many hydraulic systems. Various flow points have been marked with letters A, B, C, etc. The chart on the next page may be used as a guide to recommended pipe size for each of these lines. Flow areas are for a flow of 5 GPM. Multiply times the actual pump flow. For

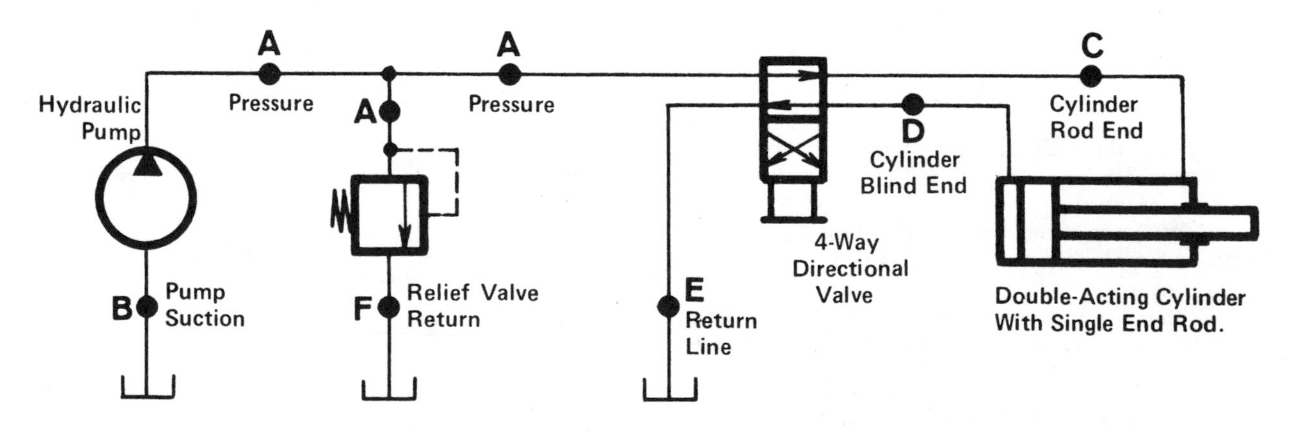

FIGURE 7-1. Flow points in a simple hydraulic system.

example, for pump flow of 35 GPM multiply chart values by 7. Tables of pipe, tubing, and hose flow areas are in Appendix C. Refer to chart on Page 268. Flow areas can also be calculated with the formula on Page 257.

Square Inch Flow Area for Each 5 GPM Pump Flow

System Pressure	Points A, C, & F	Point B	Points D & E*
0 - 1000 PSI	0.160 sq. in.	0.40 sq. in.	0.16 sq. in.
1000 - 2000 PSI	0.107 sq. in.	0.40 sq. in.	0.16 sq. in.
2000 - 3500 PSI	0.080 sq. in.	0.40 sq. in.	0.16 sq. in.
3500 - 5000 PSI	0.053 sq. in.	0.40 sq. in.	0.16 sq. in.

*Increase flow area at these points for cylinders with oversize rods.

Suggested flow areas for pressure lines A, C, and F are based on maximum design pressure of the system, with higher flows being permitted as acceptable on higher pressure systems. Flow areas are based on a flow velocity of 10 feet per second on high pressure lines in systems designed to operate in the 0 to 1000 PSI range, 15 feet per second for the 1000 to 2000 PSI range, 20 feet per second for the 2000 to 3500 PSI range, and 30 feet per second for the 3500 to 5000 PSI range. These flow areas can be increased to obtain less flow loss and higher efficiency or decreased for plumbing economy if a higher flow loss is acceptable.

Point A. (3 places). Flow volume in these lines will always be the same as the pump output flow. They can be sized according to the chart or sized equal to the flow area of the pump outlet port.

Point B. Pump Suction. This is a critical flow circuit. Oil velocity must be kept low to keep from damaging the pump by cavitation. Figures in the chart are bare minimum flow areas calculated at a velocity of 4 feet per second intended for applications where the suction line is no more than a few feet long and the suction head is less than 2 feet. For longer lines or greater suction head these areas should be increased up to twice the values shown.

Point C. Maximum flow in this line, while the cylinder is retracting, will be the same as pump flow and should be sized for this flow even though flow will be less than pump flow while the cylinder is extending. Plumbing material should have a pressure rating about 1½ to 2 times the system relief valve setting because pressure intensification can occur if cylinder flow should become restricted or blocked.

Point D. Maximum flow in this line will occur as the cylinder is retracting and will be greater than pump flow, up to twice the pump flow if the cylinder has a maximum size (2:1 ratio) piston rod. Figures in the chart are based on a flow velocity of 10 feet per second. Flow should be calculated and pipe size based on the calculated flow. Pressure rating of plumbing material should be equal to the pump relief valve pressure.

Point E. Oil return to tank. Maximum flow (while the cylinder is retracting) will be the same as at Point D, but low pressure plumbing materials may be used.

Point F. Maximum flow will be equal to pump flow. Use same pipe diameter as relief valve inlet but low pressure plumbing materials may be used.

Calculation Example

Using the diagram on the preceding page, calculate the maximum flow and the pressure rating of each line (or point) in a 3000 PSI system using a 4" bore cylinder with 2" rod. Pump flow is 50 GPM.

Solution: The first step is to calculate the maximum flow in each line, A, B, C, D, E, and F. Of course flow in Lines A, B, and F will always be 50 GPM. Flow in Line C will be 50 GPM while the cylinder is retracting, less while the cylinder is extending. Calculate plumbing size for 50 GPM flow. Flow in Lines D and E will be greater than pump flow while the cylinder is retracting. It will be greater by the same ratio that the cylinder full piston area is greater than its net area. For this particular cylinder, the full piston area is 12.57 square inches and the net area is 9.43. These values were taken from the chart on Page 260. The ratio is 1.33. Therefore, flow in Lines D and E will be 50 x 1.33 = approximately 65 GPM. Mark each line with its flow and pressure rating:

Point or Line	Maximum Flow, GPM	Necessary Flow* Area, Sq. Inches	Pressure Rating of Plumbing Materials
A	50 (10 times 5 GPM)	10 x 0.08 = 0.80	Same as relief valve setting
B	50 (10 times 5 GPM)	10 x 0.40 = 4.0	Vacuum or low pressure
C	50 (10 times 5 GPM)	10 x 0.08 = 0.80	1½ times relief valve setting
D	65 (13 times 5 GPM)	13 x 0.16 = 2.08	Same as relief valve setting
E	65 (13 times 5 GPM)	13 x 0.16 = 2.08	Low pressure
F	50 (10 times 5 GPM)	10 x 0.08 = 0.80	Low pressure

*Remember that the flow area chart is for each 5 GPM of flow. Therefore, 50 GPM will require 10 times the flow area shown in chart; 65 GPM will require 13 times the flow area shown in chart.

After flow areas and pressure ratings have been marked on your diagram, select size of plumbing materials from charts in Appendix C.

ESTIMATING INPUT POWER TO A HYDRAULIC SYSTEM

The input to a hydraulic system is nearly always mechanical power from an electric motor or an engine. First, this power must be converted into fluid power by a pump, and a certain amount of it will be lost to friction and flow resistance in the pump. Next, the fluid power must be transmitted through pipes and valves, and some additional part of it will be lost due to flow resistance. Finally, the power must be re-converted into mechanical power before it can be used. This conversion is by means of a cylinder, hydraulic motor, or rotary actuator, and some of it will be lost in mechanical friction. All of these power losses must be taken into consideration when estimating the amount of input power required to produce the mechanical output power required.

Output Force. The first step in designing any hydraulic system is to compute or estimate the amount of output *force* needed to just equal the resistance of the load to be moved. By choosing a suitable cylinder bore size, the fluid pressure to produce the output force can be calculated:

$$PSI = Output\ Force \div Piston\ Area$$

Add 5% additional pressure to make up friction loss in the cylinder. This is typical for most hydraulic cylinders. Charts on Pages 260, 262, 282, and 283 can be used to verify the above calculations.

In selecting a cylinder, there will be any number of combinations of bore diameter and pressure level which will produce the amount of force needed to balance the load resistance. The designer can choose the bore and pressure combination which seems most appropriate for the job.

Pump Pressure. Pressure produced by the pump will have to be higher than the pressure calculated to just balance the load resistance. Oil cannot move through the circuit from pump to cylinder unless there is enough additional pressure to push it through the flow resistance of piping and valves. On a new job it is seldom possible to accurately calculate the pressure needed to overcome flow resistance. A designer usually has to make an educated guess on the first model which may be a prototype for more machines to be built after the first one has proven out. After the first model has been tested for performance, the specifications on pump pressure and input horsepower can be adjusted to more accurate values. As a rule-of-thumb, to get the design started, an additional pressure of at least 15% or up to 20 to 25% above the load balance pressure should be allowed. On mobile equipment an allowance of about 30% would be more realistic because there are usually higher flow losses in a mobile system. If the oil must flow through components such as pressure reducing valves, flow dividers, flow combiners, or pressure compensated flow control valves, another 3 to 5% pressure should be added for each such component. Please remember these are only estimates based on experience with typical hydraulic systems. On the first model it is wise to make a generous allowance on pump pressure so the system will not end up under-powered.

Cylinder Speed. Using the cylinder bore previously selected, estimate the oil flow rate necessary to produce the desired speed from the cylinder. Calculate with this formula:

Oil Flow Rate (GPM) = Piston Area (Square Inches) x Travel Speed (Inches per Minute) ÷ 231.

Since there will be no loss of flow volume through the circuit, the calculated oil flow for the cylinder will also be the flow required from the pump. The above calculation can be checked against the charts on Pages 261 and 284.

Pump Selection and Input Power. From the above calculations a pump can now be selected which will deliver the required flow at the estimated pressure. From the combination of flow and pressure the output fluid HP of the pump can be calculated.

Fluid HP = PSI x GPM ÷ 1714

To find estimated mechanical input power to the pump, add 15% to make up mechanical and flow losses in the pump. (Divide output power by 85% efficiency). The results of this calculation can be verified with the chart on Page 269.

Calculation Example

Estimate the mechanical input power to a system in which the load is 50,000 lbs. on the cylinder extension stroke, and the extension speed must be 200 inches per minute.

Solution: We are picking at random a 6" bore cylinder, which has a piston area of 28.28 square inches. Theoretical pressure required = 50,000 ÷ 28.27 = 1857 PSI. Add 5% for cylinder loss: 1857 + (.05 x 1857) = 1950 PSI actually required across cylinder ports. Add 20% for estimated circuit losses between pump and cylinder: 1950 + (.20 x 1950) = 2340 PSI required at the pump.

Oil flow required for 200 inches per minute: 28.27 x 200 = 5654 cubic inches per minute. Divide by 231 (1 gallon = 231 cubic inches) to get GPM: 5654 ÷ 231 = 24.5 GPM.

Fluid HP output from pump: 2340 PSI x 24.5 GPM ÷ 1714 = 33.4 HP. Add 15% for pump conversion loss: 33.4 ÷ 0.85 = 39.3 mechanical HP input required on shaft of pump.

SELECTING SIZE OF A HYDRAULIC ACCUMULATOR

Information on this page can be used for selection of accumulator gallon capacity for the cylinder circuit of Figure 6-36 on Page 232 in which oil discharged from an accumulator adds to the steady flow from a pump to produce a cylinder speed faster than is possible with the pump alone.

Hydraulic systems using accumulators are usually designed for a working pressure, with accumulator fully charged, of 3000 PSI. At this pressure the maximum capability of the accumulator can be utilized and the system will produce the maximum capacity per dollar of cost.

Minimum Acceptable System PSI	Cubic Inch Discharge
2700	12
2600	17
2500	22
2400	27
2300	33
2200	40
2100	46
2000	55
1900	63
1800	73
1700	84
1600	96
1500	109

One problem with an accumulator is that, starting with full pressure charge, its pressure falls as oil is discharged. So, the starting pressure must be high enough so the necessary cylinder force can still be produced at the lowest pressure to which the accumulator will fall during its discharge.

The required accumulator gallon capacity is related to the bore and stroke of the cylinder, to the time allowed for the working stroke (assumed to be the extension stroke), and to the minimum pressure which must be available at the end of the cylinder stroke.

Figures in this chart show the number of cubic inches of oil which can be recovered from a ***1-gallon*** size accumulator as it discharges from a fully charged pressure of 3000 PSI to the lower pressure levels shown in the left column. A 2-gallon accumulator will deliver twice these amounts, a 5-gallon accumulator will deliver 5 times, etc.

How to Select Accumulator Size

For illustration we are assuming the accumulator discharge will be used only on the extension (or working) stroke and that the cylinder will be allowed to retract by pump oil alone. A fully charged pressure of 3000 PSI is also assumed.

Step 1. Select the bore of the cylinder for sufficient force not only at 3000 PSI but at some lower pressure to which it may fall during its discharge.

Step 2. Calculate the number of cubic inches of oil needed to fill the cylinder according to its bore and forward stroke.

Step 3. Using the time, in seconds, which is allowed for the forward stroke, calculate the number of cubic inches of oil which will be delivered from the pump during this time interval.

Step 4. Subtract calculated volume in Step 3 from calculated volume in Step 2. This is the volume of oil which must be supplied from the accumulator.

Step 5. Use the chart to find how much accumulator oil could be supplied from a 1-gallon size before its pressure dropped below the acceptable minimum pressure level. Divide this figure into the total cubic inches needed for the application, and this will be the minimum gallon capacity for a suitable accumulator. Choose the next larger standard size accumulator.

Formula. To solve for oil recovery from any size accumulator under any system pressure and any pre-charge level, use this formula: $D = [0.95 \times P_1 \times V_1 \div P_2] - [0.95 \times P_1 \times V_1 \div P_3]$, where D is cubic inch oil discharge, P_1 is pre-charge in PSI, P_2 is system pressure after Volume D has been discharged, P_3 is maximum system pressure, PSI, with accumulator fully charged, V_1 is the catalog-rated gas volume, cubic inches, and 0.95 is the assumed accumulator efficiency under temperature changes.

HEAT EXCHANGER SIZE SELECTION

Use this information along with the heat exchanger information on Pages 234 to 237. The chart on Page 243 has been prepared for the purpose of determining the HP or BTU per hour capacity of a heat exchanger which must be added to an existing system to reduce oil temperature from an undesirable higher level to a desired lower level.

Three measurements must be made before using the chart. (1), Measure oil temperature, degrees F, on the side wall of the reservoir below the oil level. If possible, measure with a thermocouple, and after the system has run under load for several hours. (2). Measure temperature of air surrounding the reservoir, degrees F. (3). Measure the area, in square feet, of all metal surfaces including top, bottom, and sides of reservoir, and cylinder walls which radiate heat. An estimate can be included for plumbing and valve surfaces.

Figures in the body of the chart show the amount of heat, in HP, which is now being radiated from the overheated system.

Instructions for Use of Chart. Take the temperature difference between atmospheric air and tank surface temperature and enter along the top of the chart in the appropriate column. Look down this column to the line for the number of square feet radiating surface of your system. The figure is the HP of heat now being radiated.

Next, decide what lower temperature you would like to maintain. Subtract this from surrounding air temperature to find the new temperature difference. Find the appropriate column and look down to the same line. This figure is the maximum radiating capacity of your present system for maintaining a 140° F tank surface temperature. The difference between these two HP figures is the help which must be provided with a heat exchanger.

Example: Presently a hydraulic system is running with a tank surface temperature of 190° F in a room where the air temperature is 80° F, so the temperature difference is 110° F. The reservoir and plumbing and component radiating surfaces have an estimated 90 square feet. What is the HP capacity of a heat exchanger which would have to be added to reduce the reservoir surface temperature to 140° F?

Solution. Enter the column in the chart marked "110° F. Look down the column to the line marked "90" square feet. The HP figure shown is 9.90. This is the radiation capacity of the tank to maintain a 190° F surface temperature.

Next, at the desired tank temperature of 140° F there will be a 60° F temperature difference between tank temperature and air temperature. Look down the 60° F column to the 90 square feet line. The figure shown is 8.10 HP. This is the maximum heat that can be radiated by the system itself without the aid of a heat exchanger if a 140° F tank temperature is to be maintained. Subtract 8.10 from 9.90 = 1.80 HP. This is the heat exchanger capacity which must be added. Convert to BTU per hour by multiplying HP times 2545.

Sq. Ft.	Temperature Difference, in Degrees F, Between Oil in Tank & Surrounding Air																	
Surface	30	35	40	45	50	55	60	65	70	75	80	85	90	95	100	110	120	130	140
10	0.30	0.35	0.40	0.45	0.50	0.55	0.60	0.65	0.70	0.75	0.80	0.85	0.90	0.95	1.00	1.10	1.20	1.30	1.40
15	0.45	0.53	0.60	0.68	0.75	0.83	0.90	0.98	1.05	1.13	1.20	1.28	1.35	1.43	1.50	1.65	1.80	1.95	2.10
20	0.60	0.70	0.80	0.90	1.00	1.10	1.20	1.30	1.40	1.50	1.60	1.70	1.80	1.90	2.00	2.20	2.40	2.60	2.80
25	0.75	0.88	1.00	1.13	1.25	1.38	1.50	1.63	1.75	1.88	2.00	2.13	2.25	2.38	2.50	2.75	3.00	3.25	3.50
30	0.90	1.05	1.20	1.35	1.50	1.65	1.80	1.95	2.10	2.25	2.40	2.55	2.70	2.85	3.00	3.30	3.60	3.90	4.20
35	1.05	1.23	1.40	1.58	1.75	1.93	2.10	2.28	2.45	2.63	2.80	2.98	3.15	3.33	3.50	3.85	4.20	4.55	4.90
40	1.20	1.40	1.60	1.80	2.00	2.20	2.40	2.60	2.80	3.00	3.20	3.40	3.60	3.80	4.00	4.40	4.80	5.20	5.60
45	1.35	1.58	1.80	2.03	2.25	2.48	2.70	2.93	3.15	3.38	3.60	3.83	4.05	4.28	4.50	4.95	5.40	5.85	6.30
50	1.50	1.75	2.00	2.25	2.50	2.75	3.00	3.25	3.50	3.75	4.00	4.25	4.50	4.75	5.00	5.50	6.00	6.50	7.00
60	1.80	2.10	2.40	2.70	3.00	3.30	3.60	3.90	4.20	4.50	4.80	5.10	5.40	5.70	6.00	6.60	7.20	7.80	8.40
70	2.10	2.45	2.80	3.15	3.50	3.85	4.20	4.55	4.90	5.25	5.60	5.95	6.30	6.65	7.00	7.70	8.40	9.10	9.80
80	2.40	2.80	3.20	3.60	4.00	4.40	4.80	5.20	5.60	6.00	6.40	6.80	7.20	7.60	8.00	8.80	9.60	10.4	11.2
90	2.70	3.15	3.60	4.05	4.50	4.95	5.4	5.85	6.30	6.75	7.20	7.65	8.10	8.55	9.00	9.90	10.8	11.7	12.6
100	3.00	3.50	4.00	4.50	5.00	5.50	6.00	6.50	7.00	7.50	8.00	8.50	9.00	9.50	10.0	11.0	12.0	13.0	14.0
110	3.30	3.85	4.40	4.95	5.50	6.05	6.60	7.15	7.70	8.25	8.80	9.35	9.90	10.5	11.0	12.1	13.2	14.3	15.4
120	3.60	4.20	4.80	5.40	6.00	6.60	7.20	7.80	8.40	9.00	9.60	10.2	10.8	11.4	12.0	13.2	14.4	15.6	16.8
130	3.90	4.55	5.20	5.85	6.50	7.15	7.80	8.45	9.10	9.75	10.4	11.1	11.7	12.4	13.0	14.3	15.6	16.9	18.2
140	4.20	4.90	5.60	6.30	7.00	7.70	8.40	9.10	9.80	10.5	11.6	11.9	12.6	13.3	14.0	15.4	16.8	18.2	19.6

APPENDIX B

Troubleshooting Procedures

Hydraulic System Check-Out

Figure 8-1. This section of the chapter describes a step-by-step check-out procedure for hydraulic systems which have previously been working satisfactorily but which have developed trouble, usually within the previous 24-hour working period, which has rendered them inoperative. This information is not intended as a diagnostic check for new systems which may have been incorrectly designed.

The diagram shows the basic circuit and basic major components typical of most hydraulic systems. System failures can usually be pinpointed to one of these components.

For check-out of a system at least one pressure gauge is necessary and should be installed in the pump pressure line as shown.

Symptoms of Trouble

Many of the failures in a hydraulic system show similar symptoms: a sudden or gradual loss of high pressure, resulting in a loss of cylinder speed and/or force. The cylinder(s) may not move at all, or if they do they may move too slowly or may stall under light loads. Often the loss of power is accompanied by an increase in pump noise, especially as it tries to build up pressure against a load.

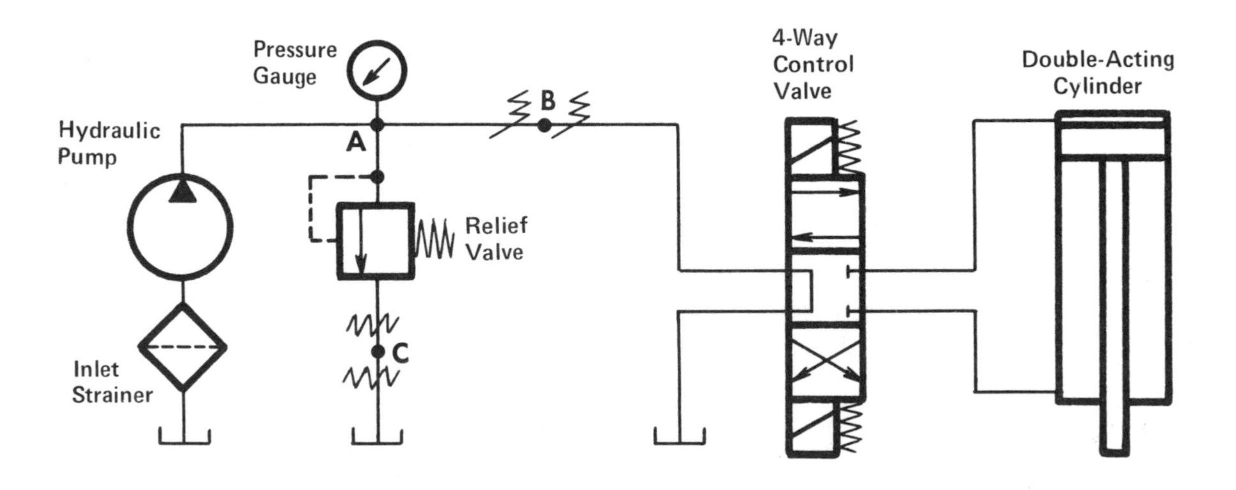

FIGURE 8-1. This circuit is typical of many industrial fluid power systems.

Of course, any major component – pump, relief valve, cylinder, 4-way valve, or filter could be at fault. And in a highly sophisticated system there are other minor components which could be at fault, but these possibilities are too numerous to be covered in this brief discussion.

By following an organized step-by-step procedure in the order given here, the problem can usually be traced to a general area, then if necessary, each component in that area can be tested or can be temporarily replaced with a similar component known to be good. It makes good sense to first check the areas which give the most frequent trouble on most systems, and that is the basis on which this procedure is based.

Step 1 – Pump Inlet Strainer

Probably the field trouble encountered most often is cavitation of the hydraulic pump caused by a build-up on the inlet strainer. Not only can this happen on a system which has been in service for a long time, it can happen on a new system after only a few hours of operation. It produces the symptoms described above: increased pump noise, loss of high pressure and/or cylinder speed.

If there is not a strainer located in the pump inlet line, it will usually be found immersed below the oil level in the reservoir. It can be removed for service by uncoupling the inlet line to the pump, removing the flanged cover where the line goes into the tank, then withdrawing the strainer.

Some operators are not aware of a strainer in the reservoir, or if they are, they do not clean it regularly. A dirty strainer restricts flow into the pump and may cause the pump to fail prematurely.

The inlet strainer should be removed and cleaned whether or not it appears dirty. Some clogging materials are hard to see. If there are holes in the mesh or other obvious physical damage, a new strainer should be installed. Wire mesh strainers can be cleaned effectively with an air hose, blowing from the inside out. They should be washed in a solvent, using a bristle brush. If possible, use a mineral spirits solvent. One brand is sold under the trade name of Stoddard Solvent. If no safe solvent is available, and if necessary to use a hydrocarbon solvent such as kerosene, work outdoors and far away from open flame or other heat source. Never use highly flammable solvents such as gasoline, lacquer thinner, naptha, etc. After cleaning with solvent use an air hose to blow out all remaining solvent. For water base or synthetic fluid systems, use some of the same fluid for cleaning the strainer.

When re-installing the strainer, inspect all joints in the inlet plumbing for air leaks, particularly at union joints. There must be no air leaks in the inlet line. Check the tank oil level to be sure it covers the top of the strainer by at least 3 inches at minimum oil level, which is with all cylinders extended. If it does not, there is danger of a vortex forming above the strainer which may allow air to enter the system when the pump is running.

Notice the condition of the inlet hose (if one is used). A partially collapsed hose or one with internal swelling has the same effect as a clogged inlet strainer.

Step 2 – Pump and Relief Valve

If cleaning the pump strainer does not correct the trouble, isolate the pump and relief valve from the rest of the system by disconnecting the plumbing at Point B and capping ***both*** ends of the disconnected lines. This deadheads the pump into the relief valve. First, back off the relief valve. Then, start the pump and watch the gauge for a pressure build-up as the relief valve adjustment is tightened. If full pressure can be developed, obviously the pump and relief valve are operating correctly and the trouble is further down the line. If full pressure cannot be developed, or if the pressure is erratic, continue with Step 3.

Step 3 – Pump or Relief Valve?

Further testing must be done to determine whether the pump is worn out or if the relief valve is malfunctioning.

Discharge from the relief valve tank port must be observed. If possible, disconnect the tank return line from the relief valve at Point C. Attach a short length of hose to the relief valve outlet. Hold the open end of the hose over the tank filler opening where the rate of flow can be observed. Start the pump and run the relief valve adjustment up and down while observing the relief valve discharge flow. If the pump is bad, a full stream of oil may possibly be observed when the relief valve is backed off but this stream will greatly diminish or stop as the relief valve setting is increased. If a flowmeter is available, the flow rate can be measured and compared with the catalog flow rating of the pump.

If a flowmeter is not available the flow can be estimated by discharging the stream into a clean container over a measured time interval. However, even without any measurement of the flow volume, a bad pump is indicated if discharge flow varies widely as the relief valve adjustment is run up and down. The discharge flow should be fairly constant at all pressure levels, dropping off slightly at higher pressures.

If the relief valve discharge line cannot be disconnected, the mechanic can place his hand near the discharge opening inside the tank and can detect a large change in the flow volume as the pressure is varied.

If the flow decreases as the relief valve setting is raised, and only a moderate, but not full pressure, can be developed, this may also indicate pump trouble. Proceed to Step 4.

During this test if gauge pressure does not rise above a low value, 100 to 200 PSI, and if the discharge flow remains constant as the relief valve adjustment is tightened, the relief valve may be at fault and should be cleaned or replaced as instructed in Step 5.

Step 4 – Pump

If a full stream of oil is not obtained in Step 3, or if the stream diminishes markedly as the relief setting is raised, the pump is probably worn out. Assuming that the inlet strainer has been cleaned and the inlet plumbing has been inspected for air leaks and collapsed hose, the pumped oil is slipping inside the pump from outlet back to inlet. The pump may be worn out or the oil may be too thin. High temperature in the oil will cause it to become thin and slip excessively. High slippage within the pump will cause it to run much hotter than the oil in the tank. In normal operation, with a good pump, the pump case may run 20 to 30°F higher than the temperature in the oil tank. If greater than this, excessive pump slippage may be the cause.

Check also for slipping belts, sheared shaft key, broken shaft, broken coupling, loosened set screw, and other possible mechanical causes.

Step 5 – Relief Valve

If Step 3 has indicated the relief valve may be at fault, the quickest proof is to temporarily replace it with one known to be good. The faulty valve may later be disassembled and cleaned. Pilot-operated relief valves have small internal orifices which may become blocked with dirt. Use an air hose to blow out all passages and pass a small wire through orifices. Check also for free movement of the spool or poppet. Pipe thread connections in the body may distort the body and cause the spool to bind. If possible, check for spool binding before unscrewing threaded connections, or, while testing on the bench, screw pipe fittings tightly into the port threads.

Step 6 – Cylinder

If the pump will develop full pressure while deadheaded into the relief valve in Step 2, both of these components can be assumed to be good. Test cylinder piston seals as described later in this chapter.

Step 7 – Directional (4-Way) Valve

If the cylinder has been tested for piston leakage and found to have reasonably tight piston seals, the 4-way control valve may be checked for excessive spool leakage. It is rare that a valve becomes so worn that the pump cannot build up full pressure, but it can happen. Symptoms of excessive leakage in the valve spool are a loss of cylinder speed together with difficulty in building up to full pressure even with the relief valve adjusted to a high setting. This condition would be more likely to happen when using a pump with small displacement operating at very high pressure, and might have developed gradually over a long time. Valve spool leakage can be checked by the method described later in this chapter.

Other Components

If the above procedure does not reveal the trouble, check other components individually. Usually the quickest and best troubleshooting procedure is to replace suspected components, one at a time, with similar ones known to be good. Pilot-operated solenoid valves which will not shift out of center position may have insufficient pilot pressure available. See more information later in this chapter.

TROUBLESHOOTING ON HYDRAULIC PUMPS

Some of the points summarized here were mentioned briefly in Chapter 5 and earlier in this chapter. The pump is the component most subject to wear and the one most likely to give trouble. On systems where the pump has to be replaced more often than seems necessary, one or more of the following problems may be the cause.

Pump Cavitation

Cavitation is the inability of a pump to draw a full charge of oil either because of air leaks or restrictions in the inlet line. When a pump starts to cavitate its noise level increases and it may become very hot around the shaft and front bearing. Other symptoms of cavitation are erratic movement of cylinders, difficulty in building up full pressure, and a milky appearance of the oil. If cavitation is suspected, check these points:

1. Check condition of the pump inlet strainer. Clean it even if it does not look dirty. Use a solvent and blow dry with an air hose. Varnish deposited in the wire mesh may be restricting the oil flow but may be almost invisible. If brown varnish deposits are found on internal surfaces of pumps or valves, this is a sure indication that the system has been operating at too high a temperature. A heat exchanger should be added.

2. Check for restricted or clogged pump inlet plumbing. If hoses are used, be sure they are not collapsed. Only hose designed for vacuum service should be used at the pump inlet. It has an internal wire braid to prevent collapse.

3. Be sure the air breather on the reservoir is not clogged with dirt or lint. On systems where the air space above the oil is relatively small, the pump could cavitate during its extension stroke if the breather became clogged.

4. Oil viscosity may be too high for the particular pump. Some pumps cannot pick up prime on heavy oil or will run in a cavitated condition.

Cold weather start-up is particularly damaging to a pump. Running a pump across a relief valve for several hours to warm up the oil can severely damage the pump if it is running in a cavitated condition during this time. On outdoor equipment use an oil with as high as possible ***viscosity index (V.I.).***

This minimizes the viscosity *change* from cold to hot oil operation and reduces cavitation on a cold start-up.

5. Check pump inlet strainer size. Be sure the original strainer has not been replaced with a smaller size. Increasing its size (number of square inches of filtering surface) may help on some systems where the original size selection was marginal.

6. The use of a higher quality oil may reduce formation of varnish and sludge.

7. Determine recommended speed of pump. Check pulley and gear ratios. Be sure the original electric motor has not been replaced with one which runs at a higher speed.

8. Be sure pump has not been replaced with one which delivers a higher flow which might overload the inlet strainer. Increase inlet size if possible.

Air Leaking Into the System

The air which is in a newly assembled system will purge itself after a short time. The system should be run for perhaps 15 to 30 minutes under very low pressure. Air will dissolve in the oil, a little at a time, and be carried to the reservoir from where it can escape. The air purging process can, of course, be accelerated by bleeding air from high points in the system by cracking fittings, especially on cylinders.

Air which comes into the system from continuous air leaks will cause the oil to assume a milky appearance a short time after the system is started, but the oil will usually become clear in about an hour after shut-down. To find where air is entering the system, check these points:

1. Be sure the oil reservoir is filled to its normal level, and that the pump intake is well below the minimum oil level. The NFPA reservoir specifications call for the highest point on the strainer to be at least 3 inches below minimum oil level.

Check oil level when all cylinders are extended to be sure it is not below the Low mark on the level gauge. However, do not overfill the reservoir when cylinders are extended; it may overflow when the cylinders retract.

2. Air may be entering around the pump shaft seal. Gear and vane pumps which are pulling oil by suction from a reservoir will have a slight vacuum behind the shaft seal. When this seal becomes badly worn, air may enter through the worn seal. Piston pumps usually have a small positive pressure, up to 15 PSI, inside the case and behind the shaft seal. Air is unlikely to enter through this route.

3. Check for air leaks in the pump inlet plumbing, especially at union joints. Check for leaks in hoses used in the inlet line. An easy way to check for leaks is to squirt oil over a suspected leak. If the pump noise diminishes, you have found your leak.

Check also around the inlet port. Screwing a tapered pipe fitting into a straight thread port will damage the thread, causing a permanent air leak which is next to impossible to repair.

4. Air may be entering through the rod seal of a cylinder. This could happen on cylinders mounted with the rod up, and which are not properly counterbalanced. On the downstroke, the gravity load may cause a partial vacuum in the rod end of the cylinder. Cylinder rod seals are not usually designed to seal air out, so even a good seal can leak under these conditions.

5. Be sure the main tank return line discharges well below the minimum oil level and not on top of the oil. On new designs it may be helpful to enlarge the diameter of the main return line a few feet before it enters the tank. This causes oil velocity to decrease which minimizes turbulence in the tank.

Water Leaking Into the System

Water leaking into the system will cause the oil to have a milky appearance while the system is running but the oil will usually clear up in a short time after the pump is shut down as water settles to the bottom of the reservoir. Water usually enters a system in these ways:

1. A leak in a water cooled (shell and tube) heat exchanger may leak water into the oil if water pressure is higher than pressure of the oil being cooled.

2. Condensation on the interior surfaces of the reservoir which are above oil level. This is almost unavoidable on systems operating in environments where ambient temperature changes from day to night. During periods when reservoir walls are cooler than surrounding air, condensation may take place if ambient humidity is fairly high. Since water settles to the bottom of the reservoir the practical solution is to daily or weekly tap off this water through the drain valve. This should be done after the pump has been shut down long enough for the water to settle.

3. Be sure that pipe or tubing which carries cooling water into the reservoir enters and leaves below the oil level. Cold pipe in the reservoir air space is a sure way to condense a large volume of water.

Oil Leakage Around the Pump

1. Leakage Around the Shaft. On piston pumps and on other pumps which take inlet oil from an overhead reservoir, there is usually a slight internal pressure behind the shaft seal. As the seal becomes well worn, external leakage may appear. This will usually be more pronounced while the pump is running, and may disappear while the pump is stopped.

Other pumps such as gear and vane types usually run with a slight vacuum behind the seal. A worn-out seal may allow air to leak into the oil while the pump is running and oil to leak out after the pump has been stopped.

Prematurely worn shaft seals may be caused by excessive oil temperature. At oil temperatures of 200° F and higher, a rubber shaft seal will have a very short life.

Abrasives in the oil may wear out shaft seals quickly, and may also produce circumferential scoring on the shaft. If abrasives are present they will settle out of an oil sample drawn from the reservoir after the sample has been allowed to stand for an hour. Check all crevices and cracks in the reservoir where dust could enter. The most common entry point is through the reservoir air breather. In extreme cases the reservoir may have to be sealed air tight and a slight air pressure of no more than 1 PSI maintained in the reservoir air space. This pressure can be taken from the shop air line through a low range pressure regulator.

2. Leakage Around a Pump Port. Sometimes leakage at these ports may be caused by damaged threads, as by screwing a taper pipe thread fitting into a straight thread port. Once the threads have been damaged it is very difficult ever to obtain a leaktight seal.

Check tightness of fittings in the ports. If dryseal (NPTF) pipe threads are used there should seldom be a need for any kind of thread sealant. If a sealant is used we recommend Teflon Sealant which comes in the form of a paste. We do not recommend Teflon tape. Beware of screwing taper pipe threads too tightly into a pump or valve body casting. In the past this has been the cause for many cracked pump housings.

3. If leakage is from a small crack in the body casting, this has most likely been caused by over-tightening a taper pipe fitting, or from operating the pump in a system where either the relief valve has, at some time, been adjusted too high, or where high pressure spikes have been generated as a result of shocks. While it is possible that the casting may have originally been defective, this has rarely turned out to be the cause.

Pump Delivering Too Little or No Flow

1. Shaft may be rotating in the wrong direction. Shut down immediately. Reversed leads on a 3-phase electric motor are a common cause for wrong rotation. Pumps must be run in the direction marked on their nameplate or case.
2. Pump inlet may be clogged or restricted. Check strainer for dirt and check for collapsed hose.
3. Oil may be low in the reservoir. Check level when all cylinders are extended.
4. Stuck vanes, valves, or pistons, either from varnish in the oil or from rust or corrosion. Varnish indicates the system is running much too hot. Rust or corrosion may indicate water in the oil.
5. Oil may be too thin either from wrong choice of oil or from thinning out at high temperature. A system with this problem may operate normally the first few hours after start-up, then gradually slow down as the oil gets overheated.
6. Mechanical trouble. Check for broken shaft or coupling, sheared key or pin, etc.
7. Pump running too slow. Most pumps will deliver a flow at all speeds proportional to RPM. But some vane pumps which depend solely on centrifugal force to extend the vanes into contact with the cam surface will deliver little or no flow at slow speeds such as engine idling speed.
8. If the driving electric motor has been replaced, make sure it is the correct speed for the pump.

Pump Noise Has Recently Increased

1. Cavitation of pump inlet. Refer to corrective measures previously described.
2. Air leaking into the system from low oil or other causes previously described.
3. Mechanical noise caused by loose or worn coupling, loose set screw, badly worn internal parts, etc.
4. System may be running with oil temperature too high.
5. Pump may be running at higher than rated speed.

Short Pump Life

1. Pump may be operating above catalog maximum pressure rating, especially if pump must maintain this pressure for a high percentage of total running time.
2. Oil of incorrect viscosity or of poor quality.
3. Running the system at excessively high temperatures.
4. Inadequate filtering. Check oil for contamination and add micronic filters if necessary.
5. Improper maintenance, particularly failure to regularly clean pump inlet strainer.
6. Mis-alignment of pump shaft with motor or engine shaft. Note: When replacing a bracket mounted pump, leave the bracket and replace only the pump. Chances are that if an exact replacement pump is installed it will not have to be re-aligned with motor.
7. Air or water may be leaking into the system.
8. The pump may be running too fast (or too slow).
9. Inlet cavitation from causes other than listed above.

LEAKAGE TESTING OF CYLINDERS AND VALVES

Air Cylinders. Figure 8-2. Piston seals and rod seals eventually become worn to the point that cylinder performance or efficiency is affected. Leakage around the rod seal can easily be detected and the seal can be replaced. At that time the rod bearing which is usually brass, bronze, or Teflon, should also be replaced. On some cylinders the seal and bearing are in a cartridge which can be replaced as a single unit. Many cylinders are now being constructed so the rod cartridge can be replaced without removing the cylinder from its mounting. After the piston rod has been uncoupled from the load, the cartridge can be unbolted or unscrewed and pulled out without further dis-assembly. Continuous air leaks in whatever part of the system should be repaired without undue delay because even a small continuous leak can waste power at the rate of 1 HP or more.

Leakage of the piston seals may not be so easy to detect but will usually show up as a continuous escape of air from the 4-way valve exhaust ports while the cylinder is stopped. Leakage from a valve exhaust port can also indicate a badly worn spool in the control valve, and procedure for determining whether the leak is in the cylinder piston or the control valve will be described later.

Air cylinders and many air valves are designed with soft seals for leaktight operation, and there should be no appreciable leakage from valve exhausts. However, a small amount of air leakage is normal for those air valves with metal-to-metal spool fit.

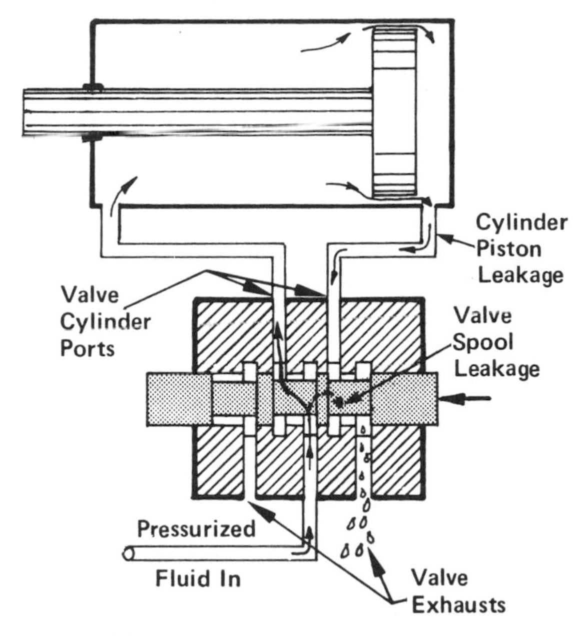

FIGURE 8-2.
Leakage paths through cylinder and valve.

In the majority of cases of valve exhaust seepage, cylinder piston seals are the cause and they should be tested first.

Leak Testing of an Air Cylinder. Figure 8-3. In the preceding figure the leak paths through the cylinder and valve are shown. To determine if the leakage is at the cylinder piston, proceed as follows:

With the air pressure shut off, remove one of the connecting lines between valve and cylinder. Plug the port on the valve but leave the cylinder line open. Re-connect the air pressure and with the pressure set at a low level, shift the 4-way valve in the direction to move the cylinder to the end of its stroke. Raise the pressure to its normal level. Air which comes out of the open line is air which has leaked past the piston. A very small leakage may be acceptable, but leakage greater than a whisper should be repaired. Re-connect the cylinder line and repeat the test with the opposite line. A cylinder may leak in one direction and not the other because a separate selaing lip is active in each direction. Occasionally a cylinder may be leaktight at both ends of its stroke but may leak at some intermediate point, possibly due to a scratch or dent in the barrel. If this is suspected, block the piston rod in this position and make the above test. Once in a while, usually on a large bore cylinder, a piston seal may leak intermittently. This may be caused by a soft seal or O-ring moving slightly or rolling or twisting as the piston strokes.

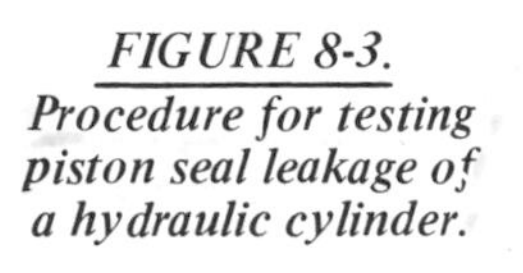

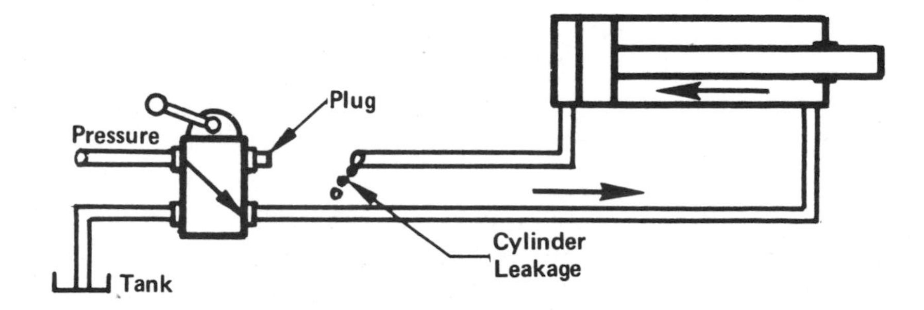

FIGURE 8-3.
Procedure for testing piston seal leakage of a hydraulic cylinder.

Hydraulic Cylinders. The leak test described for air cylinders can also be used for hydraulic cylinders. A small amount of by-pass flow (leakage) is normal for most hydraulic cylinders, and a substantial by-pass flow may be observed on cylinders with metal piston ring seals.

Air Valve Testing. Refer to Figure 8-2 for flow path of air which leaks across the spool. To test an air valve, shut off the air supply and disconnect both lines to the cylinder and plug the ports on the valve. Turn on the pressure and shift the valve first to one side position and then the other. Air coming out either exhaust is spool leakage.

Hydraulic Valves. If the tank return line can be disconnected from the valve so leakage flow can be observed, a hydraulic valve can be tested as described above. If the tank return line cannot be disconnected to observe leakage flow, the valve should be removed and checked on a test bench.

SOLENOID VALVE PROBLEMS

Checklist for Recurrent Solenoid Burn-Out of A-C Valves

Coil burn-out is more common on A-C than on D-C valves because of the high inrush current. Until the armature can pull in and close the air gap in the magnetic loop, the inrush current is often 5 times as high as the steady state, or holding, current after the armature has seated. Inrush is approximately the same as holding current on a D-C valve.

Coil Does Not Match Operating Voltage. Improper match between electrical source and the coil is sometimes a cause for coil burn-out. Check these possible causes:

1. Voltage Too High. The operating voltage should not be more than 10% higher than the coil voltage rating. Excessive voltage causes excessive coil current which may overheat the coil.

2. Voltage Too Low. Voltage should be no less than 10% below coil rating. Low voltage reduces mechanical force of the solenoid. It may continue to draw inrush current, being unable to pull in.

The low voltage test should be made by measuring the voltage directly on the coil wires while the solenoid is energized and with its armature blocked open so it is drawing inrush current. Energize the solenoid just long enough to take a voltage reading. Also take a no-load reading with the solenoid coil disconnected from the feed wires. A difference of more than 5% between these two readings indicates excessive resistance in the machine wiring circuit or insufficient volt-ampere capacity in the control transformer, if one is used.

3. Frequency. Operation of a 60 Hz coil on 50 Hz causes the coil to draw above normal current. Operation of a 50 Hz coil on 60 Hz causes the coil to draw less than its rated current and it may burn out from inability to pull the armature in.

Overlap in Energization. On some double solenoid valves, if both solenoids are energized at the same time and held in this state for a short time, the last coil to be energized will burn out from the excessive inrush current, which may be 5 times the normal holding current.

This burn-out condition will occur only on double solenoid valves where the two solenoids are yoked to opposite ends of a common spool. If each solenoid is free to immediately close its air gap, neither will burn out if both are simultaneously energized.

Careful attention must be given to electrical circuit design to make certain that the machine operator, through accident, cannot energize both solenoids at the same time.

Even with correct circuit design and interlock circuits a relay with sticking contacts or slow release could be responsible for a momentary overlap of energization on each cycle and eventual coil burn-out.

Too Rapid Cycling. Since inrush current may be up to 5 times the holding current, a standard A-C coil on an air gap solenoid may overheat and burn out if required to cycle too frequently. The extra heat generated during inrush periods cannot dissipate fast enough. The gradual build-up of heat inside the coil winding may, in time, damage the coil insulation.

High cycling applications can be roughly defined as those requiring the solenoid to be energized more than 5 to 10 times per minute. On those applications, oil immersed solenoid structures should be used. The coil operates cooler because heat is conducted more rapidly from the winding through the oil.

Dirt in Oil or Atmosphere. A small solid particle lodging under the solenoid armature may keep it from fully seating against the core, causing coil current to remain higher than normal during the holding period. Be sure that solenoid dust covers remain tightly in place to protect against dust deposited from the air.

Small dirt particles in the oil may lodge on the surface of the spool, glued there by "varnish" circulating in the oil, or the varnish itself may cause excessive spool drag and excessive coil current. "Varnish" forms in systems where the oil is allowed to run too hot.

Environmental Conditions. Abnormally high or low ambient temperatures to which a solenoid is exposed for an extended time may cause a coil to burn out.

1. High Temperature. Coil insulation may be damaged and one layer of wire may short to the next layer. A heat shield between valve and heat source may give some protection against radiated heat. Oil immersed solenoids are the best protection against heat conducted either through metal surfaces or from surrounding high temperature air.

2. Low Temperature. Cold ambient temperatures cause oil to become more viscous, possibly overloading solenoid valve capacity. Mechanical parts of the valve or solenoid structure may distort, causing the valve spool to stick and burn out the coil. Use an oil more suitable for the low temperature or use an oil immersed solenoid.

Dead End Service. Fluid circulating through a solenoid valve helps to carry away electrical heat. Some valves depend on fluid flow to keep excessive heat from accumulating, and if used on dead end service, where the solenoid may remain energized for a long time without fluid flow, the coil may burn out from this effect, possibly in combination with other problems.

Atmospheric Moisture. High humidity, along with frequently changing ambient temperature, may form corrosion on metal parts of the solenoid structure, causing the armature to drag or the spool to stick. Humidity also tends to deteriorate standard solenoid coils, causing shorts in windings.

Change to molded coils or oil immersed solenoids. Keep solenoid protective covers tightly in place, and perhaps seal the electrical conduit openings after the wiring is installed.

Pilot-Operated Solenoid Hydraulic Valves

Valve description starts on Page 151. A problem sometimes encountered with these valves, especially on newly-built systems, is their failure to shift out of center position when the "Start" button is pressed. Use the manual override buttons on the solenoid end caps to determine whether the problem is lack of electrical current or lack of sufficient hydraulic pilot pressure. With the pump running, press one of the override buttons. If the valve spool shifts and the cylinder starts forward, chances are that the problem is in the electrical circuit. If the valve spool does not shift in response to the override button, the problem may be low hydraulic pilot pressure which should be 50 to 100 PSI minimum.

APPENDIX C

Fluid Power Design Data

GRAPHIC SYMBOLS FOR FLUID POWER DIAGRAMS

Common symbols are shown here. A complete list of approved ANSI (American National Standards Institute) graphic symbols can be obtained from the National Fluid Power Association. Ask for Publication ANSI Y32.10-1967.

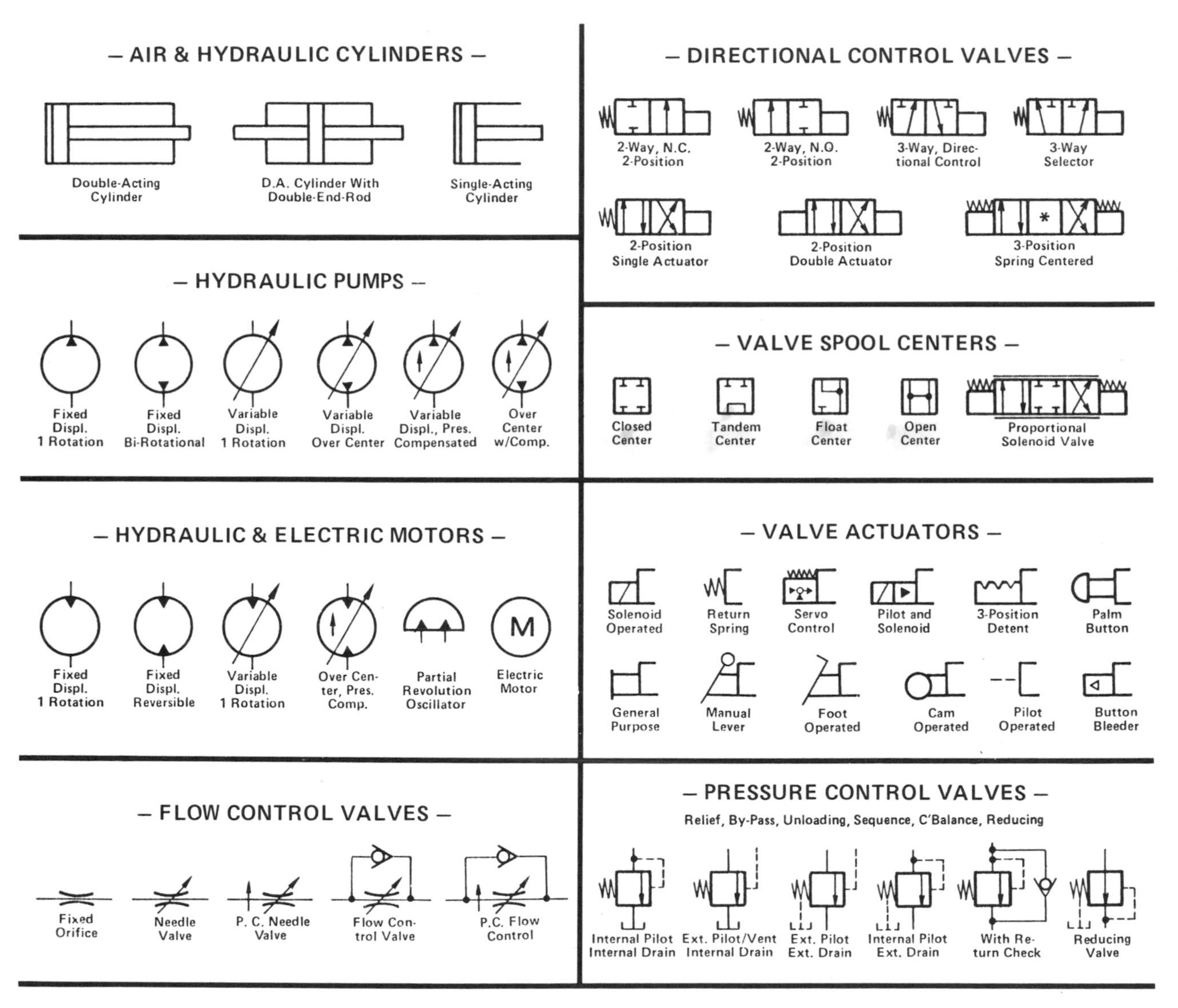

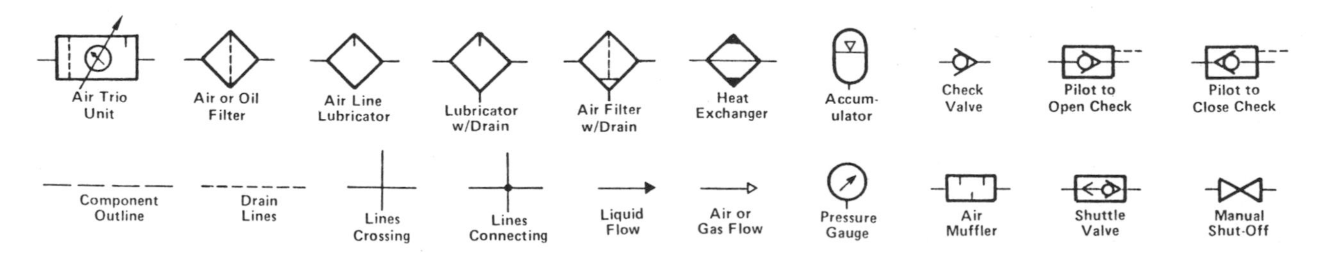

FLUID POWER DEFINITIONS

For a more complete list of definitions for fluid power terms, the student should obtain a copy of American Standards Association Glossary of Terms from the National Fluid Power Association. Ask for Publication ASA B93.2-1965.

ACCUMULATOR – A container in which liquid is stored under pressure as a source of fluid power.

AIR, COMPRESSED – Air at any pressure greater than atmospheric.

AIR, FREE – Air under the pressure, only, due to atmospheric conditions at any specific location.

AIR, STANDARD – Air at a temperature of 68°F, a pressure of 14.7 pounds per square inch absolute, and a relative humidity of 36% (0.075 lbs/cu. ft.).

ANNULUS – A ring-shaped or doughnut-shaped section, most commonly used to denote the cross section of the rod end of a fluid power cylinder, with the cylinder bore as the outside diameter and the rod as the inside diameter of the annulus.

BOYLE'S LAW – The absolute pressure of a confined body of gas varies inversely as its volume provided its temperature remains constant. (Formula Page 257).

CHARLES' LAW – The volume of a fixed mass of gas varies directly with absolute temperature provided its pressure remains constant. Or, the pressure of a fixed mass of gas varies directly with absolute temperature provided its volume remains constant. See formula on Page 257.

CAVITATION – A local gaseous condition in a liquid stream such as the inlet of a hydraulic pump.

COMPRESSOR – A device which causes a gas to flow against a pressure. It converts mechanical energy into fluid energy or fluid power.

CYCLE – A single complete operation consisting of progressive phases starting and ending at the neutral position. For example a cylinder cycle consists of a forward and a return stroke.

CYLINDER – A device which converts fluid force or power into linear mechanical force or power. It usually consists of a movable element such as a piston and piston rod, plunger or ram, operating within a cylindrical bore, enclosed with end caps.

CYLINDER, DOUBLE-ACTING – A cylinder in which fluid force can be applied to the movable element in either direction.

CYLINDER, SINGLE-ACTING – A cylinder in which the fluid force can be applied to the movable element in only one direction.

CYLINDER, RAM – A cylinder in which the movable element has the same cross sectional area as the piston rod.

CYLINDER, DOUBLE-ROD – A cylinder with a single piston, and a piston rod extending from each end.

CYLINDER, SINGLE-ROD – A cylinder with a piston rod extending from only one end.

FLOW, LAMINAR – A flow situation in which motion occurs as a movement of one layer of fluid on another.

FLOW, TURBULENT – A flow situation in which the fluid particles move in a random manner.

FLUID, FIRE-RESISTANT – A fluid difficult to ignite and which shows little tendency to propagate flame.

FLUID POWER – Energy transmitted and controlled through use of a pressurized fluid.

FLUID POWER SYSTEM – A system that transmits and controls power through use of a pressurized fluid within an enclosed circuit.

FILTER – A device through which fluid is passed to separate matter held in suspension.

HEAD – The height of a column or body of fluid above a given point expressed in linear units. It is often used to indicate gauge pressure. Pressure is equal to the height times density or specific gravity of fluid.

HEAD, SUCTION – The distance from the surface of the supply source to the free discharge surface.

HEAT EXCHANGER – A device which transfers heat through a conducting wall from one fluid to another.

HYDRAULIC POWER UNIT – A combination of componentry to facilitate fluid storage and conditioning, and delivery of the fluid under conditions of controlled pressure and flow to the discharge port of the pump, including maximum pressure controls and sensing devices when applicable. Circuitry components, although sometimes mounted on the reservoir, are not considered a part of the power unit.

HYDRAULICS – The engineering science pertaining to liquid pressure and flow.

MOTOR – A device which converts fluid power into mechanical power. It usually delivers torque and rotary motion to a shaft.

MOTOR, FIXED DISPLACEMENT – A motor in which the displacement per cycle cannot be varied.

MOTOR, VARIABLE DISPLACEMENT – A motor in which the displacement per cycle can be varied.

MOTOR, ROTARY – A motor having a shaft capable of continuous rotation.

MOTOR, LINEAR – Some cylinder manufacturers use this term for a cylinder.

ORIFICE – A restricted passageway in a fluid power system, usually a small hole drilled for the purpose of metering a flow.

PASCAL'S LAW – A pressure applied to a confined fluid at rest is transmitted with equal intensity throughout the fluid. See definition, Page 15.

PIPE THREADS, DRYSEAL – Taper pipe threads in which sealing is a function of root and crest interference.

PORT – An internal or external terminus of a passage in a component.

PRESSURE – Force per unit area, usually expressed in pounds per square inch in the U.S. system of units.

PRESSURE, ABSOLUTE – The sum of atmospheric and gauge pressures. (Varies with altitude).

PRESSURE, GAUGE – Pressure differential above or below atmospheric pressure.

PRESSURE, ATMOSPHERIC – Pressure exerted by the atmosphere at any specific location. (Sea level atmospheric pressure is approximately 14.7 lbs/sq. inch).

PRESSURE, BACK – The pressure encountered on the return side of a fluid system.

PRESSURE, DIFFERENTIAL – The difference in pressure between any two points of a system or component.

PRESSURE, SYSTEM – The pressure which overcomes the total resistance in a system. It includes all losses as well as useful work.

PUMP – A device which causes a liquid to flow against a pressure. It converts mechanical energy into fluid energy.

PUMP, FIXED DISPLACEMENT – A pump in which the displacement per cycle cannot be varied.

PUMP, VARIABLE DISPLACEMENT – A pump in which the displacement per cycle can be varied.

PUMP, AXIAL PISTON – A pump having multiple pistons disposed with their axes parallel to shaft axis.

PUMP, RADIAL PISTON – A pump having multiple pistons disposed radially to the shaft axis, actuated by an eccentric element.

PUMP, GEAR – A pump having two or more intermeshing gears or lobed members enclosed in a housing.

PUMP, VANE – A pump having multiple radial vanes within a supporting rotor.

PUMP, HAND – A pump operated one stroke at a time by manual effort of the operator.

STRAINER – A device through which a fluid is passed to separate solids in suspension. A strainer is usually 40 micro-metres or courser while a filter is usually finer than 40 micro-metres.

VISCOSITY – A measure of the internal friction or the resistance to fluid flow.

VISCOSITY, SSU or SUS – Viscosity expressed in Saybolt Universal Seconds, which is time, in seconds, for 60 cubic centimetres of oil to flow through a standard orifice at a given temperature.

VISCOSITY INDEX – A measure of the viscosity change as the temperature changes as referred to other fluids.

EQUIVALENT MEASUREMENTS

1 U. S. Gallon = 231 Cubic Inches
= 4 Quarts; or 8 Pints
= 8.3356 Pounds
= 3.785 Litres

1 Imperial Gallon = 1.2 U. S. Gallons

1 Litre = 0.2642 U. S. Gallons

1 Cubic Foot = 7.48 Gallons
= 1728 Cubic Inches

1 Cubic Foot of Water Weighs 62.4 Pounds

1 Bar (at Sea Level) = 14.5 PSI (Approx. 1 Atmos.)
= 33.8 Foot Water Column
= 42 Foot Oil Column
= 29.92 "Hg.

Approx. ½ PSI Decrease Each 1000 Ft. Altitude

1 Horsepower = 33,000 Ft. Lbs. per Minute
= 550 Ft. Lbs. per Second
= 42.4 BTU per Minute
= 2545 BTU per Hour
= 746 Watts, or 0.746 Kilowatts

1 PSIG = 2.0416 "Hg
= 27.71 "Water

1 Foot Column of Water = 0.433 PSI

1 Foot Column of Oil = 0.390 PSI

1 "Hg = 0.491 PSI
= 1.132 Ft. Water

1 Barrel of Oil = 42 Gallons

1 Micro-metre = 1 Millionth of a Metre (Micron)
= 1 Thousandth of a Centimetre
= 0.00004 Inch

25 Micro-metres = 0.001 Inch

APPROXIMATE MEASUREMENTS

1 Pint = 2 cups = 16 fluid ounces = 1 pound
1 Pint = 96 teaspoons = 32 tablespoons = 16 fluid ounces
1 Quart = 4 cups = 2 pints = 32 fluid ounces = 2 pounds
1 Quart = 192 teaspoons = 64 tablespoons = 32 fluid ounces
1 Gallon = 16 cups = 4 quarts = 8 pints
1 Gallon = 768 teaspoons = 256 tablespoons = 128 fl. oz.

1 Gallon = 231 cubic inches = 76,800 drops
1 Cup = 16 tablespoons = 48 teaspoons
1 Tablespoon = 3 teaspoons
2 Tablespoons = 1 fluid ounce
1 Fluid ounce (volume) = 660 drops (hydraulic oil)
1 Cubic inch = 330 drops

ABBREVIATIONS – Commonly Used in Fluid Power Work

abs	absolute (as in psia)
A-C	alternating current
Bhn	Brinell hardness number
Btu	British thermal unit
C	degrees Centigrade
cc	closed center
ccw	counter clockwise
cfm	cubic feet per minute
cfs	cubic feet per second
cir.	cubic inches/revolution
cim	cubic inches per minute
com	common
cpm	cycles per minute
cps	cycles per second
cu in/rev	cubic inches/revolution
cw	clockwise
cyl	cylinder
dia	diameter
ext	external
F	degrees Fahrenheit
fl	fluid
fpm	feet per minute
ft	foot
ft-lb	foot pound
gal	gallon
gpm	gallons per minute
Hg	Mercury
HP	horsepower
Hz	Hertz
ID	inside diameter
in	inch
in-lb	inch pound
int	internal
ipm	inches per minute
ips	inches per second
lb	pound
max	maximum
min	minimum
mtd	mounted
NC	normally closed
NO	normally open
NPT	national pipe thread
NPTF	dryseal pipe threads
oc	open center
oz	ounce
P.O.	pilot operated
pres	pressure
psi	pounds/square inch
psia	psi absolute
psig	psi gauge
pt	pint
qt	quart
r	radius
rms	root mean square
rpm	revolutions per minute
rps	revolutions per second
scfm	standard cu.ft./minute
Smls	seamless
sol	solenoid
SSU	Saybolt seconds universal
SUS	Saybolt universal seconds
u	micro-meters or microns
T	torque or thrust
vac	vacuum
VI	viscosity index
visc	viscosity

FLUID POWER FORMULAE (Using Abbreviations Above)

Torque and horsepower relations:

$T = HP \times 5252 \div RPM$

$HP = T \times RPM \div 5252$, where:

Torque values must be in foot pounds.

Hydraulic (fluid power) horsepower:

$HP = PSI \times GPM \div 1714$, where:

PSI is gauge pressure in the system in pounds per square inch, GPM is flow in gallons per minute.

Velocity of oil flow in pipe or through an opening:

$V = GPM \times 0.3208 \div A$, where:

V is oil velocity in feet per second, GPM is flow in gallons per minute, A is inside area of pipe in square inches.

Charles' Law for behavior of gases:

$T_1V_2 = T_2V_1$, or, $T_1P_2 = T_2P_1$, where:

T_1, P_1, and V_1 are initial temperature, pressure, and volume, and T_2, P_2, and V_2 are final values.

Boyle's Law for behavior of gases:

$P_1V_1 = P_2V_2$, where:

P_1 and V_1 are initial pressure and volume, and P_2 and V_2 are final values of pressure and volume.

Circle formulae:

Area $= \pi r^2$ or $\pi d^2 \div 4$

Circumference $= 2\pi r$ or πd, where:

r is circle radius; d is circle diameter; π is 3.14

Heat equivalent of fluid power:

BTU per hour $= PSI \times GPM \times 1\frac{1}{2}$

Hydraulic cylinder travel speed of piston:

$S = CIM \div A$, where:

S is piston travel speed, inches per minute, CIM is oil flow into cylinder in cubic inches per minute, A is piston area in square inches.

Thrust or force of an air or hydraulic cylinder:

$T = A \times PSI$, where:

T is thrust in pounds, A is piston or "net" area in square inches, PSI is gauge pressure in fluid circuit.

Force for piercing or shearing sheet metal:

$F = P \times T \times PSI$, where:

F is shearing force in pounds, P is hole perimeter in inches, T is metal thickness in inches, PSI is the shear strength of the material in pounds per square inch.

Side load on pump or motor shaft:

$F = (HP \times 63024) \div (RPM \times R)$, where:

F is side load, in lbs., against the shaft; R is pitch radius, in inches, of sheave on pump shaft; HP is driving power applied to the shaft.

Effective force of a cylinder working at an angle to the direction of the load travel:

$F = T \times \sin A$, where:

T is the total cylinder thrust, in pounds, F is the part of the thrust which is effective, in pounds, A is the least angle, in degrees, between cylinder axis and load direction. See Page 59.

PORT MARKINGS FOR VALVES

To avoid mistakes in plumbing, port identification markings on air and hydraulic valves should be understood. These markings will usually be found stamped on the valve body near the port being identified. Port markings have been standardized to a certain extent for many years, but some markings are inconclusive or ambiguous, like the letters X and Y which can be used for any one of several kinds of ports. These letters have been used over many years by various manufacturers for specific designations. Therefore, it may be necessary at times to refer to manufacturers literature on markings which may be inconclusive.

Simple valves have one inlet and one outlet. On some valves either port can be used as inlet, but usually there is a preferred or mandatory inlet, and unless the valve is correctly connected it may not function properly. More complex valves have many ports, and on these valves, correct port identification is very important.

Regardless of port markings, it is always wise to refer to the manufacturers literature if it is available. Some valves have internal passages which must either be plugged or unplugged according to how the valve is to be used. This operation must sometimes be done in the field, and if not done correctly, the circuit will not perform properly.

Information on this page was extracted from Standards ANSI B93.9-1969 published by the American National Standards Institute, and sponsored by the National Fluid Power Association in their Standard NFPA T3.5.6.68.2.

PORT MARKINGS . . .

Port markings are applicable for air, oil, or water valves except where otherwise indicated.

A & B Outlet or cylinder ports on valves which have no more than two such ports.

K1, K2, K3, etc. for outlet ports on valves having three or more such ports.

CA, CB, C1, etc. identifies pilot ports on an air valve which are used for shifting the spool. The last letter or number indicates the outlet port which becomes pressurized when pilot pressure is applied.

D . . . . Drain (not tank) port on a hydraulic valve. Also a vent port on an air valve. This port must not be subjected to restriction or back pressure.

E . . . . Exhaust to atmosphere on an air valve. May be marked EA, EB, E1, etc. to show which outlet or cylinder port exhausts through this port.

F . . . . Controlled flow port of a flow control valve.

P . . . . Pressure inlet on all valves. May be marked P1, P2, etc. if there is more than one inlet. Also indicates the free flow port of a flow control valve.

T . . . Tank return on a hydraulic valve. May be marked TA, TB, T1, etc. to indicate the working port it serves.

VA, VB, V1, etc. identifies a vent port for shifting the spoól as on a button bleeder valve. Last letter indicates the outlet port which becomes pressurized when the port is vented.

XA, XB, X1, YA, YB, Y1, etc. The letters X and Y are not standardized. Manufacturers catalog must be consulted. May indicate external drain on a hydraulic valve, or a pilot supply or control port on an air or hydraulic valve.

1, 2, 3, etc. Numerals may be used to identify ports not otherwise described. Manufacturers literature will specify use.

ELECTRICAL LEADS . . .

All solenoid valve coil leads should be identified when the solenoid from which they originate cannot be determined visually.

It is recommended, however, that wires be identified in all cases. They should be marked with the same symbol as the cylinder port which becomes pressurized when the coil is energized.

Example:

FLOW ARROWS . . .

A single arrow (⟶) is sufficient to indicate the direction of preferred flow.

A single arrow (⟶) is sufficient to indicate the direction of free flow.

A double headed arrow (⟷) is sufficient to indicate that flow may be in either direction.

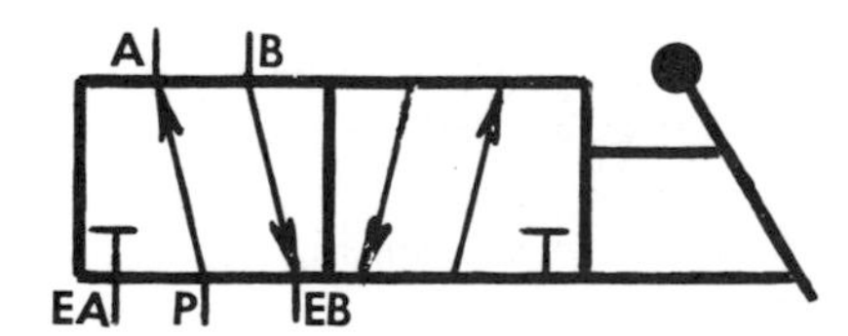

Port Markings for Dual Exhaust Air Valve

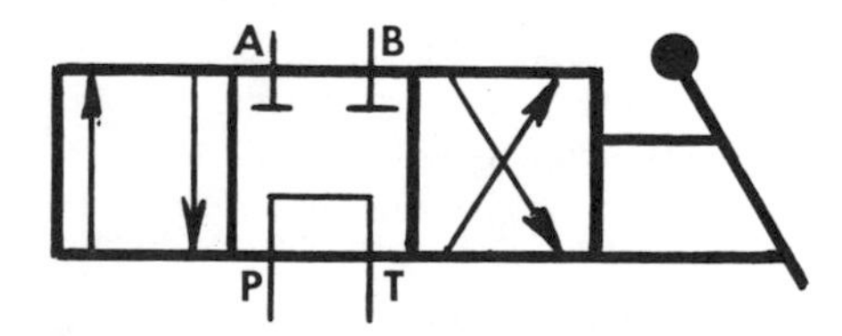

Port Markings for Hydraulic Valve

INDUSTRIAL SQUARE HEAD CYLINDERS – BORE AND ROD COMBINATIONS

Tables below show cylinder bore and rod combinations adopted as standard by the major manufacturers who supply most of the square head industrial cylinders made in the U.S.A. Data in these tables were taken from ANSI Standard B93.8.

In both these tables, standard bore diameters are shown in the left column. Then, the standard rod diameter for each bore is shown in the next column. This rod is normally furnished unless a larger rod is needed for strength or for a regenerative circuit. The far right column shows the largest rod available for each bore. Its piston rod has approximately one-half the full piston area. These rods should be used for regenerative circuits. Intermediate rod sizes are available as shown in the center columns.

High Pressure Models – 3000 to 5000 PSI

All Dimensions Are In Inches

Bore Diameter	Standard Rod Dia.	Oversize Rod Diameters Available			Largest Rod (2:1 Ratio)
1.50	*0.625*				*1.000*
2.00	*1.000*				*1.375*
2.50	*1.000*	*1.375*			*1.750*
3.25	*1.375*	*1.750*			*2.000*
4.00	*1.750*	*2.000*			*2.500*
5.00	*2.000*	*2.500*	*3.000*		*3.500*
6.00	*2.500*	*3.000*	*3.500*		*4.000*
7.00	*3.000*	*3.500*	*4.000*	*4.500*	*5.000*
8.00	*3.500*	*4.000*	*4.500*	*5.000*	*5.500*
10.00	*4.500*	*5.000*	*5.500*		*7.000*
12.00	*5.500*	*7.000*			*8.500*
14.00	*7.000*	*8.500*			*10.000*

Pneumatic and Low Pressure Models

All Dimensions Are In Inches

Bore Diameter	Standard Rod Dia.	Oversize Rod Diameters Available			Largest Rod (2:1 Ratio)
1.50	*0.625*				*1.000*
2.00	*1.000*				*1.375*
2.50	*1.000*	*1.375*			*1.750*
3.25	*1.375*	*1.750*			*2.000*
4.00	*1.750*	*2.000*			*2.500*
5.00	*2.000*	*2.500*	*3.000*		*3.500*
6.00	*2.500*	*3.000*	*3.500*		*4.000*
7.00	*3.000*	*3.500*	*4.000*	*4.500*	*5.000*
8.00	*3.500*	*4.000*	*4.500*	*5.000*	*5.500*
10.00	*4.500*	*5.000*	*5.500*		*7.000*
12.00	*5.500*	*7.000*			*8.500*
14.00	*7.000*	*8.500*			*10.000*

TONNAGE REQUIRED FOR PUNCHING HOLES

Cylinders can be used for numerous jobs of hole punching in sheet metal, riveting, staking, marking, notching, and others. An approximation can be made of the cylinder force needed for shearing sheet metal, using this method:

(1). Find the length of cut to be made, in inches. This is the perimeter or circumference of the hole or length of notch. Multiply this times the thickness of the material to be sheared, also in inches. This is the number of square inches to be sheared.

(2). Find the shearing force required by multiplying the number of square inches to be sheared times the mechanical shear strength of the material. For approximation, values in this table may be used. But for more accurate results, the material supplier should be asked to furnish an exact value for its shear strength, in PSI.

Material	Shear Strength, PSI
Aluminum, half-hard sheet	*20,000*
Brass, half-hard sheet	*35,000*
Dural	*40,000*
Copper, rolled	*28,000*
Steel, mild	*50,000*
Steel, 50 carbon	*70,000*
Steel, cold drawn	*60,000*
Stainless 18-8	*70,000*
Lead	*4,000*

Example: Calculate force to punch a 1/4" x 3/8" rectangular hole in .093" material having a shear strength of 25,000 PSI.
Solution: Perimeter of hole is 1.25". Area to be sheared = 1.25 x .093 = .1163. Force = .1163 x 25,000 = 2906 lbs.

HYDRAULIC CYLINDER FORCE – EXTENSION AND RETRACTION

Lines in bold type show extension force, in pounds; italic lines show retraction force, in pounds with various rod diameters. Force values are theoretical; allow at least 5% for cylinder friction. Pressures along top of chart are pressure differentials across cylinder ports, not pressure in the pump line. Allow sufficient extra pump pressure to take care of flow losses through lines and valves as well as back pressure on cylinder exhaust line to tank.

For pressures not shown, use effective area of piston and multiply times pressure differential across ports.

Bore Dia., Ins.	Rod Dia., Ins.	Effec. Area, Sq.Ins.	Pressure Differential Across Cylinder Ports 500 PSI	750 PSI	1000 PSI	1250 PSI	1500 PSI	2000 PSI	2500 PSI	3000 PSI	5000 PSI
1-1/2	None*	1.7672	884	1325	1767	2209	2651	3534	4418	5302	8836
	5/8	1.4604	730	1095	1460	1826	2191	2921	3651	4381	7302
	1	0.9818	491	736	982	1227	1473	1964	2455	2945	4909
2	None*	3.1416	1571	2356	3142	3927	4712	6283	7854	9425	15,708
	1	2.3562	1178	1767	2356	2945	3534	4712	5891	7069	11,781
	1-3/8	1.6567	828	1243	1657	2071	2485	3313	4142	4970	8284
2-1/2	None*	4.9087	2454	3682	4909	6136	7363	9817	12,272	14,726	24,544
	1	4.1233	2062	3092	4123	5154	6185	8247	10,308	12,370	20,617
	1-3/8	3.4238	1712	2568	3424	4280	5136	6848	8560	10,271	17,119
3-1/4	None*	8.2958	4148	6222	8296	10,370	12,444	16,592	20,740	24,887	41,479
	1-3/8	6.8109	3405	5108	6811	8514	10,216	13,622	17,027	20,433	34,055
	1-3/4	5.8905	2945	4418	5891	7363	8836	11,781	14,726	17,672	29,453
	2	5.1542	2577	3866	5154	6443	7731	10,308	12,886	15,463	25,771
4	None*	12.567	6284	9425	12,567	15,709	18,851	25,134	31,418	37,701	62,835
	1-3/4	10.162	5081	7622	10,162	12,703	15,243	20,324	25,405	30,846	50,810
	2	9.4254	4713	7069	9425	11,782	14,138	18,851	23,564	28,266	47,127
	2-1/2	7.6583	3829	5744	7658	9573	11,487	15,317	19,146	22,975	38,292
5	None*	19.635	9818	14,726	19,635	24,544	29,453	39,270	49,088	58,905	98,175
	2	16.493	8247	12,370	16,493	20,616	24,740	32,986	41,233	49,479	82,465
	2-1/2	14.726	7363	11,045	14,726	18,408	22,089	29,450	36,815	44,178	73,630
	3	12.566	6283	9425	12,566	15,708	18,849	25,132	31,415	37,698	62,830
	3-1/2	10.014	5007	7511	10,014	12,518	15,021	20,028	25,035	30,042	50,070
6	None*	28.274	14,137	21,206	28,274	35,343	42,411	56,548	70,685	84,822	141,370
	2-1/2	23.365	11,683	17,524	23,365	29,206	35,048	46,730	58,413	70,095	116,825
	3	21.205	10,603	15,904	21,205	26,506	31,808	42,410	53,013	63,615	106,025
	3-1/2	18.653	9327	13,990	18,653	23,316	27,980	37,306	46,633	55,959	93,265
	4	15.707	7854	11,780	15,707	19,634	23,561	31,414	39,268	47,121	78,535
7	None*	38.485	19,243	28,864	38,485	48,106	57,728	76,970	96,213	115,455	192,425
	3	31.416	15,708	23,562	31,416	39,270	47,124	62,832	78,540	94,248	157,080
	3-1/2	28.864	14,432	21,648	28,864	36,080	43,296	57,728	72,160	86,592	144,320
	4	25.918	12;959	19,439	25,918	32,398	38,877	51,836	64,795	77,754	129,590
	4-1/2	22.581	11,291	16,936	22,581	28,226	33,872	45,162	56,453	67,743	112,905
	5	18.850	9425	14,138	18,850	23,563	28,275	37,700	47,125	56,550	94,250
8	None*	50.266	25,133	37,700	50,266	62,833	75,399	100,532	125,665	150,798	251,330
	3-1/2	40.645	20,323	30,484	40,645	50,806	60,968	81,290	101,613	121,935	203,225
	4	37.699	18,850	28,274	37,699	47,124	56,549	75,398	94,248	113,097	188,495
	4-1/2	34.362	17,181	25,772	34,362	42,953	51,543	68,724	85,905	103,086	171,810
	5	30.631	15,316	22,973	30,631	38,289	45,947	61,262	76,578	91,893	153,155
	5-1/2	26.508	13,254	19,881	26,508	33,135	39,762	53,016	66,270	79,524	132,540
10	None*	78.540	39,270	58,905	78,540	98,175	117,810	157,080	196,350	235,620	392,700
	4-1/2	62.636	31,318	46,977	62,636	78,295	93,954	125,272	156,590	187,908	313,180
	5	58.905	29,453	44,179	58,905	73,631	88,358	117,810	147,263	176,715	294,525
	5-1/2	54.782	27,391	41,087	54,782	68,478	82,173	109,564	136,955	164,346	273,910
	7	40.055	20,028	30,041	40,055	50,069	60,082	80,110	100,138	120,165	200,275
12	None*	113.10	56,550	84,825	113,100	141,375	169,650	226,200	282,750	339,300	565,500
	5-1/2	89.339	44,670	67,004	89,339	111,374	134,009	178,678	223,348	268,017	446,695
14	None*	153.94	76,970	115,455	153,940	192,425	230,910	307,880	384,850	461,820	769,700
	7	115.46	57,730	86,595	115,460	144,325	173,190	230,920	288,650	346,380	577,300

***These lines are for extension force. No piston rod diameter is involved.**

HYDRAULIC CYLINDER SPEEDS – EXTENSION AND RETRACTION

Lines with bold type show cylinder piston extension speeds in "Inches per Minute". Lines with italic type show piston retraction speeds in "Inches per Minute" with various size piston rods. The largest diameter rod in each bore size is a "2:1" rod in which piston area is approximately twice rod area and net area. Since speed is directly proportional to GPM, for flows not shown, multiply speed in the "1 GPM" column times the desired GPM.

Piston Diam	Rod Diam.	1 GPM	3 GPM	5 GPM	8 GPM	12 GPM	15 GPM	20 GPM	25 GPM	30 GPM	40 GPM	50 GPM	60 GPM	75 GPM	100 GPM
1½"	None*	131	392	654	----	----	----	----	----	----	----	----	----	----	----
	5/8"	158	475	791	----	----	----	----	----	----	----	----	----	----	----
	1	235	706	----	----	----	----	----	--						----
2	None*	74	221	368	588	882	----	----	--						----
	3/4"	86	257	428	684	----	----	----	----	----	----	----	----	----	----
	1	98	294	490	784	----	----	----	----	----	----	----	----	----	----
	1-3/8	139	418	697	----	----	----	----	----	----	----	----	----	----	----
2½	None*	47	141	235	376	565	706	941	----	----	----	----	----	----	----
	1"	56	168	280	448	672	840	----	----	----	----	----	----	----	----
	1-3/8	67	202	337	540	810	----	----	----	----	----	----	----	----	----
	1-3/4	92	277	461	738	----	----	----	----	----	----	----	----	----	----
3¼	None*	28	84	139	223	334	418	557	696	835	----	----	----	----	----
	1-3/8	34	102	170	271	407	509	678	848	----	----	----	----	----	----
	1-3/4	39	118	196	314	471	588	784	980	----	----	----	----	----	----
	2	45	134	224	359	538	672	896	----	----	----	----	----	----	----
	2-1/4	53	160	267	428	642	802	----	----	----	----	----	----	----	----
4	None*	18	55	92	147	221	276	368	460	551	735	919	----	----	----
	1-1/4	20	61	102	163	244	306	407	509	611	815	----	----	----	----
	1-3/4	23	68	114	182	273	341	455	568	682	909	----	----	----	----
	2	25	74	123	196	294	368	490	613	735	980	----	----	----	----
	2-1/2	30	90	151	241	362	452	603	754	905	----	----	----	----	----
	2-3/4	35	105	174	279	418	523	697	871	----	----	----	----	----	----
5	None*	12	35	59	94	141	176	235	294	353	471	588	706	882	----
	1-1/2	13	39	65	103	155	194	259	323	388	517	646	776	970	----
	2	14	42	70	112	168	210	280	350	420	560	700	840	----	----
	2-1/2	16	47	78	125	188	235	314	392	471	627	784	941	----	----
	3	18	55	92	147	221	276	368	460	551	735	919	----	----	----
	3-1/2	23	69	115	185	277	346	461	577	692	923	----	----	----	----
6	None*	8.2	25	41	65	98	123	163	204	245	327	409	490	613	817
	1-3/4	8.9	27	45	71	107	134	179	223	268	357	446	536	670	893
	2-1/2	10	30	49	79	119	148	198	247	297	395	494	593	741	989
	3	11	33	54	87	131	163	218	272	327	436	545	654	817	----
	3-1/2	12	37	62	99	149	186	248	310	372	495	619	743	929	----
	4	15	44	74	118	176	221	294	368	441	588	735	882	----	----
7	None*	6.0	18	30	48	72	90	120	150	180	240	300	360	450	600
	3	7.4	22	37	59	88	110	147	184	221	294	368	441	551	735
	3-1/2	8.0	24	40	64	96	120	160	200	240	320	400	480	600	800
	4	8.9	27	45	71	107	134	178	223	267	357	446	535	668	891
	4-1/2	10	31	51	82	123	153	205	256	307	409	511	614	767	----
	5	12	37	61	98	147	184	245	306	368	490	613	735	919	----
8	None*	4.6	14	23	37	55	69	92	115	138	184	230	270	345	460
	3-1/2	5.7	17	28	45	68	85	114	142	170	227	284	341	420	568
	4	6.1	18	31	49	74	92	123	153	184	245	306	368	460	613
	4-1/2	6.7	20	34	54	81	101	134	168	202	269	336	403	504	672
	5	7.5	23	38	60	90	113	151	189	226	302	377	452	566	754
	5-1/2	8.7	26	44	70	105	131	174	218	261	349	436	523	654	871
10	None*	2.9	8.8	15	24	35	44	59	74	88	118	147	176	221	294
	4-1/2	3.7	11	18	30	44	55	74	92	111	148	184	221	277	369
	5	3.9	12	20	31	47	59	78	98	118	157	196	235	294	392
	5-1/2	4.2	13	21	34	51	63	84	105	127	169	211	253	316	422
	7	5.8	17	29	46	69	87	115	144	173	231	288	346	433	577

Figures in the body of this chart are piston speeds in "Inches per Minute".

***These lines are for extension speed. No piston rod diameter is involved.**

AIR CYLINDER FORCE – EXTENSION AND RETRACTION

Lines in bold type show extension force, in pounds; italic lines show retraction force, in pounds, with various rod diameters. Force values are theoretical; allow about 5% for cylinder friction. Pressures across top of chart are differentials across cylinder ports and not necessarily air line gauge pressures, because of back pressure in the exhaust line to atmosphere. When designing an air circuit, remember to allow a feed line air pressure at least 25% higher than required by the load to make up for back pressure and flow losses while the cylinder is moving.

For pressures not shown, use effective area of piston and multiply times pressure differential across ports.

Piston Dia., Ins.	Rod Dia., Ins.	Travel Direction	Effec. Area, Sq.Ins.	Pressure Differential Across Cylinder Ports								
				60 PSI	70 PSI	80 PSI	90 PSI	100 PSI	110 PSI	120 PSI	130 PSI	150 PSI
1-1/2	**None**	**Extend**	**1.77**	**106**	**124**	**142**	**159**	**177**	**195**	**212**	**230**	**266**
	5/8	*Retract*	*1.46*	*88*	*102*	*117*	*132*	*146*	*161*	*176*	*190*	*205*
	1	*Retract*	*.985*	*59*	*69*	*79*	*89*	*98*	*108*	*118*	*128*	*148*
1-3/4	**None**	**Extend**	**2.41**	**144**	**168**	**192**	**216**	**241**	**265**	**289**	**313**	**361**
	5/8	*Retract*	*2.10*	*126*	*147*	*168*	*189*	*210*	*231*	*252*	*273*	*315*
	1-1/4	*Retract*	*1.18*	*71*	*83*	*95*	*106*	*118*	*130*	*142*	*154*	*177*
2	**None**	**Extend**	**3.14**	**188**	**220**	**251**	**283**	**314**	**345**	**377**	**408**	**471**
	5/8	*Retract*	*2.83*	*170*	*198*	*227*	*255*	*283*	*312*	*340*	*368*	*425*
	1	*Retract*	*2.35*	*141*	*165*	*188*	*212*	*235*	*259*	*283*	*306*	*353*
2-1/2	**None**	**Extend**	**4.91**	**295**	**344**	**393**	**442**	**491**	**540**	**589**	**638**	**737**
	5/8	*Retract*	*4.60*	*276*	*322*	*368*	*414*	*460*	*506*	*552*	*598*	*690*
	1	*Retract*	*4.12*	*247*	*289*	*330*	*371*	*412*	*454*	*495*	*536*	*619*
	1-3/8	*Retract*	*3.43*	*206*	*240*	*274*	*308*	*343*	*377*	*411*	*445*	*514*
3	**None**	**Extend**	**7.07**	**424**	**495**	**565**	**636**	**707**	**778**	**848**	**919**	**1060**
	1	*Retract*	*6.28*	*377*	*440*	*503*	*565*	*628*	*691*	*754*	*817*	*942*
	1-3/4	*Retract*	*4.66*	*280*	*326*	*373*	*420*	*466*	*513*	*560*	*606*	*699*
3-1/4	**None**	**Extend**	**8.30**	**498**	**581**	**664**	**747**	**830**	**913**	**996**	**1079**	**1245**
	1	*Retract*	*7.51*	*451*	*526*	*601*	*676*	*751*	*827*	*902*	*977*	*1127*
	1-3/8	*Retract*	*6.82*	*409*	*477*	*545*	*613*	*681*	*750*	*818*	*886*	*1022*
	1-3/4	*Retract*	*5.89*	*354*	*413*	*472*	*531*	*589*	*648*	*707*	*766*	*884*
3-1/2	**None**	**Extend**	**9.62**	**577**	**674**	**770**	**866**	**962**	**1058**	**1155**	**1251**	**1443**
	1	*Retract*	*8.84*	*530*	*618*	*707*	*795*	*884*	*972*	*1060*	*1149*	*1325*
4	**None**	**Extend**	**12.57**	**754**	**880**	**1006**	**1131**	**1257**	**1283**	**1508**	**1634**	**1886**
	1	*Retract*	*11.78*	*707*	*825*	*943*	*1061*	*1178*	*1296*	*1414*	*1532*	*1768*
	1-3/8	*Retract*	*11.09*	*665*	*776*	*887*	*998*	*1109*	*1219*	*1330*	*1441*	*1663*
	1-3/4	*Retract*	*10.16*	*610*	*712*	*813*	*915*	*1016*	*1118*	*1220*	*1321*	*1525*
5	**None**	**Extend**	**19.64**	**1178**	**1375**	**1571**	**1768**	**1964**	**2160**	**2357**	**2553**	**2946**
	1	*Retract*	*18.85*	*1131*	*1320*	*1508*	*1697*	*1885*	*2074*	*2263*	*2451*	*2828*
	1-3/8	*Retract*	*18.16*	*1089*	*1271*	*1452*	*1634*	*1816*	*1997*	*2179*	*2360*	*2723*
6	**None**	**Extend**	**28.27**	**1696**	**1979**	**2262**	**2544**	**2827**	**3110**	**3392**	**3675**	**4241**
	1-3/8	*Retract*	*26.79*	*1607*	*1875*	*2143*	*2411*	*2679*	*2946*	*3214*	*3482*	*4018*
	1-3/4	*Retract*	*25.9*	*1552*	*1811*	*2069*	*2328*	*2586*	*2845*	*3104*	*3362*	*3880*
7	**None**	**Extend**	**38.49**	**2309**	**2694**	**3079**	**3464**	**3849**	**4234**	**4619**	**5004**	**5774**
	1-3/8	*Retract*	*37.01*	*2220*	*2590*	*2960*	*3331*	*3701*	*4071*	*4441*	*4811*	*5551*
8	**None**	**Extend**	**50.27**	**3016**	**3519**	**4022**	**4524**	**5027**	**5530**	**6032**	**6535**	**7541**
	1-3/8	*Retract*	*48.79*	*2927*	*3415*	*3903*	*4391*	*4879*	*5366*	*5854*	*6342*	*7318*
	1-3/4	*Retract*	*47.90*	*2872*	*3351*	*3829*	*4308*	*4786*	*5265*	*5744*	*6222*	*7180*
10	**None**	**Extend**	**78.54**	**4712**	**5498**	**6283**	**7069**	**7854**	**8639**	**9425**	**10,210**	**11,781**
	1-3/4	*Retract*	*76.14*	*4568*	*5329*	*6091*	*6852*	*7614*	*8375*	*9136*	*9898*	*11,420*
	2	*Retract*	*75.40*	*4524*	*5278*	*6032*	*6786*	*7540*	*8294*	*9048*	*9802*	*11,310*
12	**None**	**Extend**	**113.1**	**6786**	**7917**	**9048**	**10,179**	**11,310**	**12,441**	**13,572**	**14,703**	**16,965**
	2	*Retract*	*110.0*	*6598*	*7697*	*8797*	*9896*	*10,996*	*12,095*	*13,195*	*14,295*	*16,494*
	2-1/2	*Retract*	*108.2*	*6491*	*7573*	*8655*	*9737*	*10,819*	*11,901*	*12,983*	*14,065*	*16,229*
14	**None**	**Extend**	**153.9**	**9234**	**10,773**	**12,312**	**13,851**	**15,390**	**16,929**	**18,468**	**20,007**	**23,085**
	2-1/2	*Retract*	*149.0*	*8939*	*10,429*	*11,919*	*13,409*	*14,899*	*16,389*	*17,879*	*19,369*	*22,349*
	3	*Retract*	*146.8*	*8810*	*10,278*	*11,747*	*13,215*	*14,683*	*16,151*	*17,620*	*19,088*	*22,025*

AIR CONSUMPTION OF CYLINDERS

When designing a machine using air cylinders which operate on a fast, continuously reciprocating program, it may be necessary to estimate the compressor HP capacity to keep the cylinders in continuous operation.

To estimate compressor load, first calculate air consumption of the cylinder(s), then convert this into compressor HP.

Air consumption should be calculated at total pressure required to balance the load plus additional pressure to make up for circuit losses in the piping, directional valve, flow control valves, mufflers, etc. In the chart the pressure values across the top are those set on the pressure regulator serving the circuit. This will automatically include circuit losses in the calculation. The system pressure regulator should, of course, be set no higher than necessary to produce the travel speed required. The total pressure required for load plus losses is often about 25% more than load balance pressure.

How to Use the Chart. The chart at the foot of this page is set up for standard bore sizes with standard (smallest) rod. The use of a piston rod larger than standard size for a given bore will cause the air consumption to be slightly less than shown, but the difference will be so slight that it can be disregarded.

To use the chart, find cylinder bore in the left column. Follow this line to the column with your expected pressure regulator setting. Figures in the body of the chart are SCF (standard cubic feet) of free air to operate a 1-inch stroke cylinder through one cycle, forward and return. Multiply this figure times your actual cylinder stroke, then times the number of complete cycles per minute. The final result will be the SCFM (standard cubic feet per minute) which must be supplied by the air compressor on a continuous basis.

This chart was calculated using compression ratios from the chart at the right. Consumption was calculated assuming the cylinder piston will be allowed to stall, at least momentarily, at each end of its stroke, giving it time to fill with air to full pressure regulator setting. If reversed at either end before reaching full stall, air consumption will be slightly less than shown.

Calculation Example. Estimate the SCFM consumption of a 4-inch bore air cylinder having a 28-inch stroke, to cycle 11 times a minute. Assume the load balance pressure for this cylinder is 65 PSI.

Solution: The pressure regulator will probably have to be set for a pressure of 80 PSI to take care of circuit flow losses in addition to load balance.

Use the 80 PSI column and the 4-inch bore line. The chart shows .091 SCF for a 1-inch stroke. Then SCFM air flow required will be:

SCFM = .091 x 28 x 11 = 17.9

Converting SCFM Into Compressor HP. Use the chart on Page 276 to convert SCFM into HP according to the kind of compressor in use.

PRESSURE CONVERSIONS

Pres., PSIG	Pres., PSIA	No. of Atmos.*	No. of Bars
50	64.7	4.40	4.46
55	69.7	4.74	4.81
60	74.7	5.08	5.15
65	79.7	5.42	5.50
70	84.7	5.76	5.84
75	89.7	6.10	6.19
80	94.7	6.44	6.53
85	99.7	6.78	6.88
90	104.7	7.12	7.22
95	109.7	7.46	7.57
100	114.7	7.80	7.91
105	119.7	8.14	8.26
110	124.7	8.48	8.60
115	129.7	8.82	8.94
120	134.7	9.16	9.29
125	139.7	9.50	9.63
130	144.7	9.84	9.98
135	149.7	10.2	10.3
140	154.7	10.5	10.7
145	159.7	10.9	11.0
150	164.7	11.2	11.4
155	169.7	11.5	11.7
160	174.7	11.9	12.0
165	179.7	12.2	12.4
170	184.7	12.6	12.7

**Atmospheres are used to determine the quantity of free air contained in a given SCF volume when under compression, and were used to calculate the chart below.*

Cylinder Air Consumption per 1-Inch Stroke, Forward and Return

Figures in body of chart are SCF air consumption of cylinders with 1-inch stroke, at various air pressures.

Cylin. Bore, Ins.	Rod Dia., Ins.	PSI Pressure Setting on Regulator Outlet											
		50 PSI	60 PSI	70 PSI	80 PSI	90 PSI	100 PSI	110 PSI	120 PSI	130 PSI	140 PSI	150 PSI	160 PSI
1.50	.625	.008	.010	.011	.012	.013	.015	.016	.017	.018	.020	.021	.022
2.00	.625	.015	.018	.020	.022	.025	.027	.029	.032	.034	.036	.039	.041
2.50	.625	.024	.028	.032	.035	.039	.043	.047	.050	.054	.058	.062	.065
3.00	1.00	.035	.040	.046	.051	.056	.062	.067	.071	.076	.081	.087	.092
3.25	1.00	.040	.047	.053	.059	.065	.071	.077	.084	.090	.096	.102	.109
4.00	1.00	.062	.072	.081	.091	.100	.110	.120	.129	.139	.148	.158	.168
5.00	1.00	.098	.113	.128	.143	.159	.174	.189	.204	.219	.233	.249	.265
6.00	1.38	.140	.162	.184	.205	.227	.249	.270	.292	.313	.334	.357	.379
8.00	1.38	.252	.291	.330	.369	.408	.447	.486	.525	.564	.602	.642	.682
10.00	1.75	.394	.455	.516	.576	.637	.698	.759	.820	.881	.940	1.00	1.07
12.00	2.00	.568	.656	.744	.831	.919	1.01	1.09	1.18	1.27	1.36	1.45	1.54
14.00	2.50	.771	.891	1.01	1.13	1.25	1.37	1.49	1.61	1.74	1.85	1.98	2.10

STEEL TUBING – PRESSURE AND FLOW RATINGS FOR HYDRAULIC PLUMBING

Pressure ratings are for annealed steel tubing "hydraulic grade" having a tensile strength of 55,000 PSI (the kind most often used for plumbing hydraulic systems). See notes at foot of next page for adjustment to other steels.

Pressure ratings are shown for S.F. (safety factors) from 2 to 6. A factor of 4 is recommended for general service. On shockless systems a smaller factor is sometimes used.

GPM ratings are shown for flow velocities from 10 through 30 f/s (feet per second). A general guideline is to use 10 f/s on systems operating at a maximum pressure of 1000 PSI; 15 f/s on systems having maximum pressures from 1000 to 2000 PSI; 20 f/s if system pressure is in the range of 2000 to 3500 PSI; and a velocity of 30 f/s on systems designed for pressures higher than 3500 PSI.

Tubing	Wall, Inches	.025	.032	.035	.042	.049	.058	.065	.072	.083	.095	.109
5/16" O.D. TUBING *55,000 PSI Tensile*	PSI @ S.F. = 6	1470	1880	2050	2460	2880	3400	3800	----	----	----	----
	PSI @ S.F. = 5	1760	2250	2460	2960	3450	4080	4580	----	----	----	----
	PSI @ S.F. = 4	2200	2820	3080	3700	4310	5100	5720	----	----	----	----
	PSI @ S.F. = 3	2930	3750	4100	4930	5750	6800	7630	----	----	----	----
	PSI @ S.F. = 2	4400	5630	6160	7390	8630	10,200	11,440	----	----	----	----
	GPM @ 10 f/s	*1.69*	*1.51*	*1.44*	*1.28*	*1.13*	*0.95*	*0.82*	----	----	----	----
	GPM @ 15 f/s	*2.53*	*2.27*	*2.16*	*1.92*	*1.69*	*1.42*	*1.23*	----	----	----	----
	GPM @ 20 f/s	*3.37*	*3.02*	*2.88*	*2.56*	*2.25*	*2.25*	*1.89*	----	----	----	----
	GPM @ 30 f/s	*5.06*	*4.54*	*4.32*	*3.83*	*3.38*	*2.83*	*2.45*	----	----	----	----
3/8" O. D. TUBING *55,000 PSI Tensile*	PSI @ S.F. = 6	1220	1560	1710	2050	2400	2840	3180	3520	----	----	----
	PSI @ S.F. = 5	1470	1880	2050	2460	2880	3400	3800	4220	----	----	----
	PSI @ S.F. = 4	1830	2350	2570	3080	3590	4250	4770	5280	----	----	----
	PSI @ S.F. = 3	2440	3130	3420	4100	4790	5670	6350	7040	----	----	----
	PSI @ S.F. = 2	3670	4690	5130	6160	7190	8500	9530	10,550	----	----	----
	GPM @ 10 f/s	*2.59*	*2.37*	*2.28*	*2.07*	*1.88*	*1.64*	*1.47*	*1.31*	----	----	----
	GPM @ 15 f/s	*3.88*	*3.55*	*3.42*	*3.11*	*2.82*	*2.46*	*2.20*	*1.96*	----	----	----
	GPM @ 20 f/s	*5.17*	*4.74*	*4.56*	*4.15*	*3.76*	*3.29*	*2.94*	*2.61*	----	----	----
	GPM @ 30 f/s	*7.76*	*7.11*	*6.84*	*6.22*	*5.64*	*4.93*	*4.40*	*3.92*	----	----	----
1/2" O. D. TUBING *55,000 PSI Tensile*	PSI @ S.F. = 6	920	1170	1280	1540	1800	2130	2380	2640	3040	----	----
	PSI @ S.F. = 5	1100	1400	1540	1850	2150	2550	2860	3170	3650	----	----
	PSI @ S.F. = 4	1380	1760	1920	2310	2700	3190	3580	3960	4560	----	----
	PSI @ S.F. = 3	1830	2350	2570	3080	3590	4250	4770	5280	6090	----	----
	PSI @ S.F. = 2	2750	3520	3850	4620	5390	6380	7150	7920	9130	----	----
	GPM @ 10 f/s	*4.96*	*4.65*	*4.53*	*4.24*	*3.96*	*3.61*	*3.35*	*3.10*	*2.73*	----	----
	GPM @ 15 f/s	*7.43*	*6.98*	*6.79*	*6.35*	*5.93*	*5.41*	*5.03*	*4.65*	*4.10*	----	----
	GPM @ 20 f/s	*9.91*	*9.31*	*9.05*	*8.47*	*7.91*	*7.22*	*6.70*	*6.20*	*5.46*	----	----
	GPM @ 30 f/s	*14.9*	*14.0*	*13.6*	*12.7*	*11.9*	*10.8*	*10.1*	*9.30*	*8.19*	----	----
5/8" O. D. TUBING *55,000 PSI Tensile*	PSI @ S.F. = 6	730	940	1030	1230	1440	1700	1910	2110	2435	2790	3200
	PSI @ S.F. = 5	880	1130	1230	1480	1720	2040	2290	2530	2920	3340	3840
	PSI @ S.F. = 4	1100	1400	1540	1850	2160	2550	2860	3170	3650	4180	4800
	PSI @ S.F. = 3	1470	1880	2050	2460	2870	3400	3810	4220	4870	5570	6390
	PSI @ S.F. = 2	2200	2800	3080	3700	4320	5100	5720	6340	7300	4360	9600
	GPM @ 10 f/s	*8.10*	*7.71*	*7.54*	*7.17*	*6.80*	*6.34*	*6.00*	*5.66*	*5.16*	*4.63*	*4.06*
	GPM @ 15 f/s	*12.1*	*11.6*	*11.3*	*10.7*	*10.2*	*9.52*	*9.00*	*8.50*	*7.38*	*6.95*	*6.08*
	GPM @ 20 f/s	*16.2*	*15.4*	*15.1*	*14.3*	*13.6*	*12.7*	*12.0*	*11.3*	*10.3*	*9.26*	*8.11*
	GPM @ 30 f/s	*24.3*	*23.1*	*22.6*	*21.5*	*20.4*	*19.0*	*18.0*	*17.0*	*15.5*	*13.9*	*12.2*
3/4" O. D. TUBING *55,000 PSI Tensile*	PSI @ S.F. = 6	610	780	860	1030	1200	1420	1590	1760	2030	2320	2660
	PSI @ S.F. = 5	730	940	1030	1230	1440	1700	1900	2110	2430	2790	3200
	PSI @ S.F. = 4	920	1170	1280	1540	1800	2130	2380	2640	3040	3480	4000
	PSI @ S.F. = 3	1220	1560	1710	2050	2400	2840	3180	3520	4060	4640	5330
	PSI @ S.F. = 2	1840	2340	2560	3080	3600	4260	4760	5280	6080	6960	8000
	GPM @ 10 f/s	*12.0*	*11.5*	*11.3*	*10.9*	*10.4*	*9.84*	*9.41*	*8.99*	*8.35*	*7.68*	*6.93*
	GPM @ 15 f/s	*18.0*	*17.3*	*17.0*	*16.3*	*15.6*	*14.8*	*14.1*	*13.5*	*12.5*	*11.5*	*10.4*
	GPM @ 20 f/s	*24.0*	*23.0*	*22.6*	*21.7*	*20.8*	*19.7*	*18.8*	*18.0*	*16.7*	*15.4*	*13.9*
	GPM @ 30 f/s	*36.0*	*34.6*	*34.0*	*32.6*	*31.2*	*29.5*	*28.2*	*27.0*	*25.1*	*23.0*	*20.8*

(This Table is Continued on the Next Page)

(This Table is Continued From the Preceding Page)

Wall, Inches	.025	.032	.035	.042	.049	.058	.065	.072	.083	.095	.109
7/8″ O.D. TUBING 55,000 PSI Tensile											
Inside Area	.5346	.5166	.5090	.4914	.4742	.4525	.4359	.4197	.3948	.3685	.3390
PSI @ S.F. = 6	**520**	**670**	**730**	**880**	**1030**	**1210**	**1360**	**1500**	**1740**	**1990**	**2280**
PSI @ S.F. = 5	**630**	**800**	**880**	**1060**	**1230**	**1460**	**1630**	**1810**	**2090**	**2390**	**2740**
PSI @ S.F. = 4	**790**	**1000**	**1100**	**1320**	**1540**	**1820**	**2040**	**2260**	**2600**	**2990**	**3420**
PSI @ S.F. = 3	**1050**	**1340**	**1470**	**1760**	**2050**	**2430**	**2720**	**3020**	**3480**	**3980**	**4570**
PSI @ S.F. = 2	**1580**	**2000**	**2200**	**2640**	**3080**	**3640**	**4080**	**4520**	**5200**	**5980**	**6840**
GPM @ 10 f/s	*16.7*	*16.1*	*15.9*	*15.3*	*14.8*	*14.1*	*13.6*	*13.1*	*12.3*	*11.5*	*10.6*
GPM @ 15 f/s	*25.0*	*24.2*	*23.8*	*23.0*	*22.2*	*21.2*	*20.4*	*19.6*	*18.5*	*17.2*	*15.9*
GPM @ 20 f/s	*33.3*	*32.2*	*31.7*	*30.6*	*29.6*	*28.2*	*27.2*	*26.2*	*24.6*	*23.0*	*21.1*
GPM @ 30 f/s	*50.0*	*48.3*	*47.6*	*46.0*	*44.3*	*42.3*	*40.8*	*39.2*	*36.9*	*34.5*	*31.7*
1″ O.D. TUBING 55,000 PSI Tensile											
Inside Area	.7088	.6881	.6793	.6590	.6390	.6138	.5945	.5755	.5463	.5153	.4803
PSI @ S.F. = 6	**460**	**590**	**640**	**770**	**900**	**1060**	**1190**	**1320**	**1520**	**1740**	**2000**
PSI @ S.F. = 5	**550**	**700**	**770**	**920**	**1080**	**1280**	**1430**	**1580**	**1830**	**2090**	**2400**
PSI @ S.F. = 4	**690**	**880**	**960**	**1150**	**1350**	**1590**	**1790**	**1980**	**2280**	**2610**	**3000**
PSI @ S.F. = 3	**920**	**1170**	**1280**	**1540**	**1800**	**2130**	**2380**	**2640**	**3040**	**3480**	**4000**
PSI @ S.F. = 2	**1380**	**1760**	**1920**	**2300**	**2700**	**3180**	**3580**	**3960**	**4560**	**5220**	**6000**
GPM @ 10 f/s	*22.1*	*21.4*	*21.2*	*20.5*	*19.9*	*19.1*	*18.5*	*17.9*	*17.0*	*16.1*	*15.0*
GPM @ 15 f/s	*33.1*	*32.2*	*31.8*	*30.8*	*29.9*	*28.7*	*27.8*	*26.9*	*25.5*	*24.1*	*22.5*
GPM @ 20 f/s	*44.2*	*42.9*	*42.4*	*41.1*	*39.8*	*38.3*	*37.1*	*35.9*	*34.1*	*32.1*	*29.9*
GPM @ 30 f/s	*66.3*	*64.3*	*63.5*	*61.6*	*59.8*	*57.4*	*55.6*	*53.8*	*51.1*	*48.2*	*44.9*
1¼″ O.D. TUBING 55,000 PSI Tensile											
Inside Area	1.131	1.105	1.094	1.068	1.042	1.010	.9852	.9607	.9229	.8825	.8365
PSI @ S.F. = 6	**365**	**470**	**515**	**615**	**720**	**850**	**950**	**1060**	**1220**	**1390**	**1600**
PSI @ S.F. = 5	**440**	**560**	**610**	**740**	**860**	**1020**	**1140**	**1270**	**1460**	**1670**	**1920**
PSI @ S.F. = 4	**550**	**700**	**770**	**920**	**1080**	**1280**	**1430**	**1580**	**1820**	**2090**	**2400**
PSI @ S.F. = 3	**730**	**940**	**1030**	**1230**	**1440**	**1700**	**1900**	**2110**	**2430**	**2790**	**3200**
PSI @ S.F. = 2	**1100**	**1400**	**1540**	**1840**	**2160**	**2560**	**2860**	**3160**	**3640**	**4180**	**4800**
GPM @ 10 f/s	*35.3*	*34.4*	*34.1*	*33.3*	*32.5*	*31.5*	*30.7*	*29.9*	*28.8*	*27.5*	*26.1*
GPM @ 15 f/s	*52.9*	*51.7*	*51.2*	*49.9*	*48.7*	*47.2*	*46.1*	*44.9*	*43.2*	*41.3*	*39.1*
GPM @ 20 f/s	*70.5*	*68.9*	*68.2*	*66.6*	*65.0*	*63.0*	*61.4*	*59.9*	*57.5*	*55.0*	*52.2*
GPM @ 30 f/s	*106*	*103*	*102*	*100*	*97.4*	*94.5*	*92.1*	*89.8*	*86.3*	*85.5*	*78.2*
1½″ O.D. TUBING 55,000 PSI Tensile											
Inside Area	1.651	1.612	1.606	1.575	1.544	1.504	1.474	1.444	1.398	1.348	1.291
PSI @ S.F. = 6	**305**	**390**	**430**	**515**	**600**	**710**	**795**	**880**	**1014**	**1161**	**1332**
PSI @ S.F. = 5	**370**	**470**	**510**	**610**	**720**	**850**	**950**	**1050**	**1210**	**1390**	**1600**
PSI @ S.F. = 4	**460**	**590**	**640**	**770**	**900**	**1060**	**1190**	**1320**	**1520**	**1740**	**2000**
PSI @ S.F. = 3	**610**	**780**	**850**	**1030**	**1200**	**1420**	**1590**	**1760**	**2030**	**2320**	**2660**
PSI @ S.F. = 2	**920**	**1180**	**1280**	**1540**	**1800**	**2120**	**2380**	**2640**	**3040**	**3480**	**4000**
GPM @ 10 f/s	*51.5*	*50.2*	*50.1*	*49.1*	*48.1*	*46.9*	*45.9*	*45.0*	*43.6*	*42.0*	*40.2*
GPM @ 15 f/s	*77.2*	*75.4*	*75.1*	*73.6*	*72.2*	*70.3*	*68.9*	*67.5*	*65.4*	*63.0*	*60.4*
GPM @ 20 f/s	*103*	*101*	*100*	*98.2*	*96.3*	*93.8*	*91.9*	*90.0*	*87.2*	*84.0*	*80.5*
GPM @ 30 f/s	*154*	*151*	*150*	*147*	*144*	*141*	*138*	*135*	*131*	*126*	*121*
2″ O.D. TUBING 55,000 PSI Tensile											
Inside Area	2.986	2.944	2.926	2.883	2.841	2.788	2.746	2.705	2.642	2.573	2.490
PSI @ S.F. = 6	**230**	**295**	**320**	**385**	**450**	**530**	**595**	**660**	**760**	**870**	**1000**
PSI @ S.F. = 5	**270**	**350**	**390**	**460**	**540**	**640**	**710**	**790**	**910**	**1040**	**1200**
PSI @ S.F. = 4	**340**	**440**	**480**	**580**	**670**	**800**	**890**	**990**	**1140**	**1300**	**1500**
PSI @ S.F. = 3	**460**	**590**	**640**	**770**	**900**	**1060**	**1190**	**1320**	**1520**	**1740**	**2000**
PSI @ S.F. = 2	**680**	**880**	**960**	**1160**	**1340**	**1600**	**1780**	**1980**	**2280**	**2600**	**3000**
GPM @ 10 f/s	*93.1*	*91.8*	*91.2*	*89.4*	*88.6*	*86.9*	*85.6*	*84.3*	*82.4*	*80.2*	*77.6*
GPM @ 15 f/s	*139*	*138*	*137*	*135*	*133*	*130*	*128*	*126*	*123*	*120*	*116*
GPM @ 20 f/s	*186*	*184*	*182*	*180*	*177*	*174*	*171*	*169*	*165*	*160*	*233*
GPM @ 30 f/s	*279*	*275*	*274*	*270*	*266*	*261*	*257*	*253*	*247*	*241*	*233*

For steels with tensile strength other than 55,000 PSI, pressure rating is proportional to tensile strength.

Barlow's Formula *(P = 2t x S ÷ O)* was used for pressure calculations, in which P = burst pressure, PSI; t = wall thickness, inches; S = tensile strength of material, in PSI; and O = outside diameter of tube, in inches. Working pressure = burst pressure ÷ safety factor.

The formula: *GPM = V x A ÷ 0.3208* was used for calculating flow velocity, in which V = velocity in feet per second, and A = inside area, in square inches.

PRESSURE AND FLOW RATINGS OF IRON PIPE

Pressure ratings are for wrought steel, butt welded pipe, the kind used most often in hydraulic plumbing. It is a low carbon steel with 40,000 PSI tensile strength. High carbon steel pipe, with tensile ratings up to 60,000 PSI is also available. Its pressure rating is higher in proportion to the increase in its tensile rating.

Schedule 40 is "standard weight" pipe. Schedules 80 and 160 have the same O. D. for a given size, but have heavier walls and a smaller internal flow area. Double extra strength pipe is also available in sizes of 1/2" and larger where higher pressure ratings are necessary.

Taper pipe threads, if used, should be NPTF (dryseal type) to minimize leakage around the crest of the threads.

Abbreviations used in the table are: S.F. is safety factor on the pressure ratings; f/s is flow velocity in feet per second.

	Pipe Schedule →	40	80	160
1/4" NPT PIPE 40,000 PSI Tensile	PSI @ S.F. = 6	2170	2940	- - - -
	PSI @ S.F. = 5	2610	3530	- - - -
	PSI @ S.F. = 4	3260	4410	- - - -
	PSI @ S.F. = 3	4340	5880	- - - -
	PSI @ S.F. = 2	6520	8820	- - - -
	GPM @ 10 f/s	*3.00*	*2.40*	- - - -
	GPM @ 15 f/s	*4.50*	*3.60*	- - - -
	GPM @ 20 f/s	*6.00*	*4.80*	- - - -
	GPM @ 30 f/s	*9.00*	*7.20*	- - - -
3/8" NPT PIPE 40,000 PSI Tensile	PSI @ S.F. = 6	1800	2490	- - - -
	PSI @ S.F. = 5	2160	2990	- - - -
	PSI @ S.F. = 4	2700	3730	- - - -
	PSI @ S.F. = 3	3590	4980	- - - -
	PSI @ S.F. = 2	5390	7470	- - - -
	GPM @ 10 f/s	*6.00*	*4.20*	- - - -
	GPM @ 15 f/s	*9.00*	*5.30*	- - - -
	GPM @ 20 f/s	*12.0*	*8.40*	- - - -
	GPM @ 30 f/s	*18.0*	*11.0*	- - - -
1/2" NPT PIPE 40,000 PSI Tensile	PSI @ S.F. = 6	1730	2330	2980
	PSI @ S.F. = 5	2080	2800	3580
	PSI @ S.F. = 4	2600	3500	4480
	PSI @ S.F. = 3	3460	4670	5970
	PSI @ S.F. = 2	5190	7000	8950
	GPM @ 10 f/s	*9.50*	*7.20*	*5.33*
	GPM @ 15 f/s	*14.0*	*11.0*	*8.00*
	GPM @ 20 f/s	*19.0*	*14.0*	*10.6*
	GPM @ 30 f/s	*29.0*	*22.0*	*16.0*
3/4" NPT PIPE 40,000 PSI Tensile	PSI @ S.F. = 6	1430	1950	2780
	PSI @ S.F. = 5	1720	2350	3340
	PSI @ S.F. = 4	2150	2930	4170
	PSI @ S.F. = 3	2870	3910	5560
	PSI @ S.F. = 2	4300	5870	8340
	GPM @ 10 f/s	*16.0*	*15.0*	*9.33*
	GPM @ 15 f/s	*25.0*	*22.0*	*14.0*
	GPM @ 20 f/s	*33.0*	*30.0*	*18.6*
	GPM @ 30 f/s	*50.0*	*44.0*	*28.0*
1" NPT PIPE 40,000 PSI Tensile	PSI @ S.F. = 6	1350	1810	2530
	PSI @ S.F. = 5	1620	2180	3040
	PSI @ S.F. = 4	2020	2720	3800
	PSI @ S.F. = 3	2700	3630	5070
	PSI @ S.F. = 2	4040	5440	7600
	GPM @ 10 f/s	*27.0*	*22.2*	*16.0*
	GPM @ 15 f/s	*41.0*	*33.3*	*24.0*
	GPM @ 20 f/s	*55.0*	*44.0*	*32.0*
	GPM @ 30 f/s	*83.0*	*66.0*	*48.0*

	Pipe Schedule →	40	80	160
1¼" NPT PIPE 40,000 PSI Tensile	PSI @ S.F. = 6	1120	1530	2000
	PSI @ S.F. = 5	1350	1840	2410
	PSI @ S.F. = 4	1690	2300	3010
	PSI @ S.F. = 3	2250	3070	4020
	PSI @ S.F. = 2	3370	4600	6020
	GPM @ 10 f/s	*47.0*	*40.2*	*32.6*
	GPM @ 15 f/s	*70.0*	*60.0*	*49.0*
	GPM @ 20 f/s	*94.0*	*68.0*	*65.0*
	GPM @ 30 f/s	*140*	*120*	*98.0*
1½" NPT PIPE 40,000 PSI Tensile	PSI @ S.F. = 6	1020	1400	1970
	PSI @ S.F. = 5	1220	1680	2370
	PSI @ S.F. = 4	1530	2100	2960
	PSI @ S.F. = 3	2030	2810	3940
	PSI @ S.F. = 2	3050	4210	5920
	GPM @ 10 f/s	*65.0*	*55.2*	*43.3*
	GPM @ 15 f/s	*95.0*	*83.0*	*65.0*
	GPM @ 20 f/s	*130*	*110*	*87.0*
	GPM @ 30 f/s	*190*	*166*	*130*
2" NPT PIPE 40,000 PSI Tensile	PSI @ S.F. = 6	865	1220	1930
	PSI @ S.F. = 5	1040	1470	2320
	PSI @ S.F. = 4	1300	1840	2900
	PSI @ S.F. = 3	1730	2450	3860
	PSI @ S.F. = 2	2590	3670	5790
	GPM @ 10 f/s	*105*	*91.2*	*68.6*
	GPM @ 15 f/s	*156*	*138*	*103*
	GPM @ 20 f/s	*210*	*185*	*137*
	GPM @ 30 f/s	*312*	*276*	*206*
2½" NPT PIPE 40,000 PSI Tensile	PSI @ S.F. = 6	940	1280	1740
	PSI @ S.F. = 5	1130	1540	2090
	PSI @ S.F. = 4	1410	1920	2610
	PSI @ S.F. = 3	1880	2560	3480
	PSI @ S.F. = 2	2820	3840	5220
	GPM @ 10 f/s	*120*	*132*	*111*
	GPM @ 15 f/s	*222*	*198*	*166*
	GPM @ 20 f/s	*300*	*265*	*220*
	GPM @ 30 f/s	*444*	*396*	*332*
3" NPT PIPE 40,000 PSI Tensile	PSI @ S.F. = 6	820	1140	1670
	PSI @ S.F. = 5	985	1370	2000
	PSI @ S.F. = 4	1230	1710	2500
	PSI @ S.F. = 3	1650	2290	3340
	PSI @ S.F. = 2	2470	3430	5000
	GPM @ 10 f/s	*225*	*206*	*167*
	GPM @ 15 f/s	*345*	*310*	*250*
	GPM @ 20 f/s	*450*	*412*	*334*
	GPM @ 30 f/s	*690*	*620*	*500*

DIMENSIONS AND FLOW AREAS OF PIPES

Nom. Pipe Size	Outside Diam., Inches	Circumference, Inches	Schedule 40 Inside Diam., Inches	Schedule 40 Inside Area, Sq. Ins.	Schedule 80 Inside Diam., Inches	Schedule 80 Inside Area, Sq. In.	Schedule 160 Inside Diam., Inches	Schedule 160 Inside Area, Sq. Ins.	Schedule XXS Inside Diam., Inches	Schedule XXS Inside Area, Sq. Ins.
1/8 NPTF	.405	1.27	.269	.057	.215	.036	- - - -	- - - -	- - - -	- - - -
1/4 NPTF	.540	1.70	.364	.104	.302	.072	- - - -	- - - -	- - - -	- - - -
3/8 NPTF	.675	2.12	.493	.191	.423	.141	- - - -	- - - -	- - - -	- - - -
1/2 NPTF	.840	2.64	.622	.304	.546	.234	.464	.169	.252	.050
3/4 NPTF	1.05	3.30	.824	.533	.742	.432	.612	.294	.434	.148
1" NPTF	1.32	4.13	1.05	.866	.957	.719	.815	.522	.599	.282
1¼ NPTF	1.66	5.22	1.38	1.50	1.28	1.29	1.16	1.06	.896	.631
1½ NPTF	1.90	5.97	1.61	2.04	1.50	1.77	1.34	1.41	1.10	.950
2" NPTF	2.38	7.46	2.07	3.37	1.94	2.96	1.69	2.24	1.50	1.77
2½ NPTF	2.88	9.03	2.47	4.79	2.32	4.23	2.13	3.56	1.77	2.46
3" NPTF	3.50	11.0	3.07	7.40	2.90	6.61	2.62	5.39	2.30	4.15

FLOW AREAS OF STEEL TUBING

Figures in the body of this chart are internal flow areas, in square inches, of steel tubing. When connecting from pipe into steel tubing, use this chart to find equivalent internal flow area to match areas of pipe in the chart above.

Tube O.D.	Wall Thickness, Inches *.025*	*.032*	*.035*	*.042*	*.049*	*.058*	*.065*	*.072*	*.083*	*.095*	*.109*
5/16	.0541	.0485	.0462	.0410	.0361	.0303	.0262	- - - -	- - - -	- - - -	- - - -
3/8	.0830	.0760	.0731	.0665	.0603	.0527	.0471	.0419	- - - -	- - - -	- - - -
1/2	.1590	.1493	.1452	.1359	.1269	.1158	.1075	.0995	.0876	- - - -	- - - -
5/8	.2597	.2472	.2419	.2299	.2181	.2035	.1924	.1817	.1655	.1486	.1301
3/4	.3848	.3696	.3632	.3484	.3339	.3157	.3019	.2884	.2679	.2463	.2223
7/8	.5346	.5166	.5090	.4914	.4742	.4525	.4359	.4197	.3948	.3685	.3390
1	.7088	.6881	.6793	.6590	.6390	.6138	.5945	.5755	.5463	.5153	.4803
1¼	1.131	1.105	1.094	1.068	1.042	1.010	.9852	.9607	.9229	.8825	.8365
1½	1.651	1.612	1.606	1.575	1.544	1.504	1.474	1.444	1.398	1.348	1.291
2	2.986	2.944	2.926	2.883	2.841	2.788	2.746	2.705	2.642	2.573	2.490

COPPER TUBING TO SCHEDULE 40 PIPE – EQUIVALENT FLOW CAPACITY

Copper tubing is a good plumbing medium for compressed air but is not recommended for hydraulic oil plumbing. This chart shows the size tubing which should be used to connect into components which have NPTF pipe thread portholes. For example, if the porthole size is 1/4" NPTF, 3/8" O.D. tubing must be used if full flow capacity is to be maintained. Usually brass fittings or braze-type fittings are recommended for permanent installations. The SAE flare angle of 45° is most often used rather than the JIC angle of 37° as used on steel tubing for hydraulics. Ferrule type or plastic fittings are not recommended for permanent plumbing.

If Pipe Size is:	**1/8" NPT**	**1/4" NPT**	**3/8" NPT**	**1/2" NPT**	**3/4" NPT**	**1" NPT**
Nearest Equivalent Tubing Size is:	1/4" or 5/16"	3/8" O.D.	1/2" O.D.	5/8" O.D.	3/4" O.D.	1" O.D.

HOSE – EQUIVALENT PIPE AND TUBING SIZES

Hose is not standarized to the same extent as pipe and tubing. There is a wide variation in dimensions, pressure ratings, and availability between manufacturers.

Hose is specified by its inside diameter, overall length fitting to fitting, type and size of end fittings, pressure rating, composition, and temperature range. This table is limited to showing characteristics of one popular brand of low pressure, small diameter hose. Space does not permit listing the wide variety of sizes and types which are available. Consult catalogs of hose manufacturers.

Hose I.D.	**Hose O.D.**	**Minimum Bend Radius**	**Working Pressure**	**Min. Burst Pressure**	**Equivalent Pipe Size**	**Nearest Equiv. Tubing Size**
1/4"	.50"	4"	400 PSI	1250 PSI	1/8" NPT	5/16" O.D.
3/8"	.63"	4"	300 PSI	1000 PSI	1/4" NPT	3/8" or 1/2"
1/2"	.78"	6"	150 PSI	750 PSI	3/8" NPT	1/2" or 5/8"
5/8"	.91"	6"	140 PSI	700 PSI	1/2" NPT	5/8" or 3/4"

CHOOSING THE BEST PIPE SIZE FOR SYSTEM OIL FLOW

The piping size in a hydraulic system should be carefully considered. If it is too small, the oil velocity will be excessive and this will produce abnormally high power losses due to flow friction. But if the piping size is larger than it needs to be, although the flow losses will be low, the cost of the plumbing material and the labor cost for installing it will be high. The key to a good balance between system power loss and cost of plumbing is the flow velocity permitted through each section of pipe.

The flow velocity method of selecting optimum pipe size is explained on Pages 238 to 240. After deciding on flow velocities, various charts in this appendix can be used to determine pipe or tubing size for desired flow. Refer to charts on Pages 264 to 267. Also, the charts on Page 270 can be used for estimating actual pressure loss, if desired.

The tables on this page are flow capacities of iron pipe, Schedules 40, 80, and 160 at selected velocities. Tables were calculated from the formula:

$$V = GPM \times 0.3208 \div A, \text{ where:}$$

V is flow velocity in feet per second; GPM is the flow volume in gallons per minute; and A is inside area of pipe in square inches.

Figures in the body of these charts are GPM flow capacities. For example, a 1¼" Schedule 40 pipe will carry 70 GPM at 15 feet per second.

Pressure rating of the plumbing material must also be considered, and this is indicated for various parts of the circuit in the chart on Page 240. Pressure in some parts of the circuit could be higher than the system relief valve setting.

Flow Capacity of Schedule 40 Pipe

Pipe Size	2 Ft/Sec.	4 Ft/Sec.	10 Ft/Sec.	15 Ft/Sec.	20 Ft/Sec.	30 Ft/Sec.
1/8" NPT	0.36 GPM	0.72 GPM	1.80 GPM	2.70 GPM	3.60 GPM	5.40
1/4	0.60	1.20	3.00	4.50	6.00	9.00
3/8	1.20	2.40	6.00	9.00	12.0	18.0
1/2	1.92	3.80	9.50	14.0	19.0	29.0
3/4	3.36	6.60	16.0	25.0	33.0	50.0
1	5.50	11.0	27.0	41.0	55.0	83.0
1¼	9.40	19.0	47.0	70.0	94.0	140
1½	13.0	26.0	65.0	95.0	130	190
2	21.0	42.0	105	156	210	312
2½	30.0	60.0	150	222	300	444
3	45.0	92.0	225	345	450	690

Flow Capacity of Schedule 80 Pipe

Pipe Size	2 Ft/Sec	4 Ft/Sec	10 Ft/Sec	15 Ft/Sec	20 Ft/Sec	30 Ft/Sec
1/8" NPT	0.24 GPM	0.48 GPM	1.20 GPM	1.80 GPM	2.40 GPM	3.60
1/4	0.48	0.96	2.40	3.60	4.80	7.20
3/8	0.84	1.68	4.20	5.30	8.40	11.0
1/2	1.44	2.88	7.20	11.0	14.0	22.0
3/4	3.00	6.00	15.0	22.0	30.0	44.0
1	4.44	8.88	22.2	33.0	44.0	66.0
1¼	6.80	13.6	40.2	60.0	68.0	120
1½	11.0	22.0	55.2	83.0	110	166
2	18.0	37.0	91.2	138	185	276
2½	26.0	53.0	132	198	265	396
3	41.0	82.0	206	310	412	620

Flow Capacity of Schedule 160 Pipe

Pipe Size	2 Ft/Sec	4 Ft/Sec	10 Ft/Sec	15 Ft/Sec	20 Ft/Sec	30 Ft/Sec
1/2" NPT	1.06 GPM	2.13 GPM	5.33 GPM	8.00 GPM	10.6 GPM	16.0
3/4	1.86	3.73	9.33	14.0	18.6	28.0
1	3.20	6.40	16.0	24.0	32.0	48.0
1¼	6.50	13.0	32.6	49.0	65.0	98.0
1½	8.70	17.3	43.3	65.0	87.0	130
2	13.7	27.5	68.6	103	137	206
2½	22.0	44.0	111	166	220	332
3	33.4	67.0	167	250	334	500

NEMA FRAME ASSIGNMENTS – 3-PHASE INDUCTION-TYPE ELECTRIC MOTORS

Open Dripproof Frame Electric Motors

HP	Speed, RPM	NEMA Frame	Shaft Diam.	Shaft Length	Shaft Height
1	1200	145T	7/8	2-1/4	3-1/2
1	1800	143T	7/8	2-1/4	3-1/2
1½	1200	182T	1-1/8	2-3/4	4-1/2
1½	1800	145T	7/8	2-1/4	3-1/2
1½	3600	143T	7/8	2-1/4	3-1/2
2	1200	184T	1-1/8	2-3/4	4-1/2
2	1800	145T	7/8	2-1/4	3-1/2
2	3600	145T	7/8	2-1/4	3-1/2
3	1200	213T	1-3/8	3-3/8	5-1/4
3	1800	182T	1-1/8	2-3/4	4-1/2
3	3600	145T	7/8	2-1/4	3-1/2
5	1200	215T	1-3/8	3-3/8	5-1/4
5	1800	184T	1-1/8	2-3/4	4-1/2
5	3600	182T	1-1/8	2-3/4	4-1/2
7½	1200	254T	1-5/8	4	6-1/4
7½	1800	213T	1-3/8	3-3/8	5-1/4
7½	3600	184T	1-1/8	2-3/4	4-1/2
10	1200	256T	1-5/8	4	6-1/4
10	1800	215T	1-3/8	3-3/8	5-1/4
10	3600	213T	1-3/8	3-3/8	5-1/4
15	1200	284T	1-7/8	4-5/8	7
15	1800	254T	1-5/8	4	6-1/4
15	3600	215T	1-3/8	3-3/8	5-1/4
20	1200	286T	1-7/8	4-5/8	7
20	1800	256T	1-5/8	4	6-1/4
20	3600	254T	1-5/8	4	6-1/4
25	1200	324T	2-1/8	5-1/4	8
25	1800	284T	1-7/8	4-5/8	7
25	3600	256T	1-5/8	4	6-1/4
30	1200	326T	2-1/8	5-1/4	8
30	1800	286T	1-7/8	4-5/8	7
30	3600	284TS	1-5/8	3-1/4	7
40	1200	364T	2-3/8	5-7/8	9
40	1800	324T	2-1/8	5-1/4	8
40	3600	286TS	1-5/8	3-1/4	7
50	1200	365T	2-3/8	5-7/8	9
50	1800	326T	2-1/8	5-1/4	8
50	3600	324TS	1-7/8	3-3/4	8
60	1200	404T	2-7/8	7-1/4	10
60	1800	364TS	1-7/8	3-3/4	9
60	3600	326TS	1-7/8	3-3/4	8
75	1200	405T	2-7/8	7-1/4	10
75	1800	365TS	1-7/8	3-3/4	9
75	3600	364TS	1-7/8	3-3/4	9
100	1200	444T	3-3/8	8-1/2	11
100	1800	404TS	2-1/8	4-1/4	10
100	3600	365TS	1-7/8	3-3/4	9
125	1200	445T	3-3/8	8-1/2	11
125	1800	405TS	2-1/8	4-1/4	10
125	3600	404TS	2-1/8	4-1/4	10
150	1800	444TS	2-3/8	4-3/4	11
150	3600	405TS	2-1/8	4-1/4	10
200	1800	445TS	2-3/8	4-3/4	11
200	3600	444TS	2-3/8	4-3/4	11
250	3600	445TS	2-3/8	4-3/4	11

Totally Enclosed Motors

HP	Speed, RPM	NEMA Frame
1	1200	145T
1	1800	143T
1½	1200	182T
1½	1800	145T
1½	3600	143T
2	1200	184T
2	1800	145T
2	3600	145T
3	1200	213T
3	1800	182T
3	3600	182T
5	1200	215T
5	1800	184T
5	3600	184T
7½	1200	254T
7½	1800	213T
7½	3600	213T
10	1200	256T
10	1800	215T
10	3600	215T
15	1200	284T
15	1800	254T
15	3600	254T
20	1200	286T
20	1800	256T
20	3600	256T
25	1200	324T
25	1800	284T
25	3600	284TS
30	1200	326T
30	1800	286T
30	3600	286TS
40	1200	364T
40	1800	324T
40	3600	324TS
50	1200	365T
50	1800	326T
50	3600	326TS
60	1200	404T
60	1800	364TS
60	3600	364TS
75	1200	405T
75	1800	365TS
75	3600	365TS
100	1200	444T
100	1800	405TS
100	3600	405TS
125	1200	445T
125	1800	444TS
125	3600	444TS
150	1800	445TS
150	3600	445TS

Note: TS indicates standard short shaft. For belt or gear drive, the standard long shaft, Symbol T, may be preferred.

PRESSURE LOSS DUE TO OIL FLOW THROUGH PIPES

This table has been calculated from a formula published by the Crane Company on Page 3-12 of Technical Paper 410. It shows the ***approximate*** pressure loss per 100 feet of Schedule 40 pipe with hydraulic oil of known specific gravity and known viscosity flowing through it.

The formula used is: $\Delta P = 0.0668\ \mu\nu \div D^2$, in which: ΔP is pressure loss per 100 feet of pipe; μ is viscosity in centipoises (not SSU); ν is the flow velocity in feet per second; D is inside diameter of pipe, in inches.

Note: Absolute viscosity in centipoises must be used in the formula. For any fluid this is kinematic viscosity in centistokes times the specific gravity. An absolute viscosity of 40 centipoises was used for calculating the table. This corresponds approximately to a hydraulic oil with 0.9 specific gravity and a viscosity of 220 SSU (or 44.4 centistokes). See next page for other fluids.

TABLE 1. Pressure Loss per 100 Feet of Schedule 40 Pipe With Oil of 220 SSU and 0.9 Specific Gravity

See Next Page for Adjustment to Other Fluids and Viscosities

GPM	Pipe Size*	Pres. Drop**	Flow Veloc†
3	1/8	624	17
	1/4	187	9.3
	3/8	55	5.0
	1/2	22	3.2
	3/4	7.1	1.8
6	1/4	373	19
	3/8	111	10
	1/2	44	6.3
	3/4	14	3.6
	1	5.4	2.2
10	3/8	185	17
	1/2	73	11
	3/4	24	6.0
	1	9.0	3.7
	1¼	3.0	2.2
15	1/2	109	16
	3/4	36	9.0
	1	14	5.6
	1¼	4.5	3.2
	1½	2.4	2.4
20	1/2	146	21
	3/4	47	12
	1	18	7.4
	1¼	6.0	4.3
	1½	3.2	3.2
25	1/2	180	26
	3/4	59	15
	1	23	9.3
	1¼	7.6	5.4
	1½	4.0	3.9
30	1/2	214	31
	3/4	71	18
	1	27	11
	1¼	9.0	6.4
	1½	4.8	4.7
35	1/2	249	36
	3/4	83	21
	1	32	13
	1¼	11	7.5
	1½	5.7	5.5
40	3/4	95	24
	1	36	15
	1¼	12	8.6
	1½	6.5	6.3
	2	2.4	3.8
45	3/4	106	27
	1	41	17
	1¼	14	9.7
	1½	4.4	7.1
	2	2.7	4.3
50	3/4	122	31
	1	46	19
	1¼	15	11
	1½	8.1	7.9
	2	3.0	4.8
55	3/4	130	33
	1	50	20
	1¼	17	12
	1½	8.9	8.7
	2	3.3	5.3
60	3/4	142	36
	1	53	22
	1¼	18	13
	1½	9.8	9.5
	2	3.6	5.7
65	3/4	154	39
	1	59	24
	1¼	20	14
	1½	11	10
	2	3.9	6.2
70	3/4	205	42
	1	63	26
	1¼	21	15
	1½	11	11
	2	4.2	6.7
75	1	68	28
	1¼	23	16
	1½	12	12
	2	4.5	7.2
	2½	2.2	5.0
80	1	75	31
	1¼	24	17
	1½	13	13
	2	4.8	7.7
	2½	2.3	5.4
90	1	80	33
	1¼	27	19
	1½	15	14
	2	5.4	8.6
	2½	2.6	6.0
100	1	92	38
	1¼	30	22
	1½	16	16
	2	6.0	9.6
	2½	2.9	6.7
125	1	114	47
	1¼	38	27
	1½	20	20
	2	7.5	12
	2½	9.8	8.4
150	1¼	44	31
	1½	24	24
	2	8.9	14
	2½	4.4	10
	3	1.8	6.4
175	1¼	53	38
	1½	29	28
	2	10	17
	2½	5.1	12
	3	2.2	7.6
200	1¼	60	43
	1½	32	31
	2	12	19
	2½	5.9	13
	3	2.5	8.7
225	1¼	69	49
	1½	37	36
	2	13	22
	2½	6.6	15
	3	2.8	9.8

*Standard Schedule 40 pipe. **Pressure loss per 100 feet of pipe. †Oil flow velocity in feet per second.

TABLE 2. Factors for Converting Table 1 for Use With Steel Tubing

For pressure loss per 100 feet of steel tubing, use the nearest NPT size shown in this table. Find pressure loss from Table 1 on preceding page. Then multiply this loss times the factor shown in the last column of this table.

Example: For flow of 50 GPM through a 1½" O.D. tube with .095 wall, use the 1¼" pipe size under 50 GPM in Table 1. This shows a 265 PSI loss per 100 feet. Multiply this times 1.11 from Table 2 = 295 PSI per 100 feet loss.

Tube O.D.	Wall Thick.	Tube I.D.	Use NPT	Mult. by
3/16	.032	.124	1/4	8.69
1/4	.035	.180	1/4	4.09
	.042	.166	1/4	4.81
	.049	.152	1/4	5.73
	.058	.134	1/4	7.38
	.065	.120	1/4	9.20
3/8	.035	.305	1/4	1.42
	.042	.291	1/4	1.56
	.049	.277	1/4	1.73
	.058	.259	1/4	1.97
	.065	.245	1/4	2.21
1/2	.035	.430	3/8	1.31
	.042	.416	3/8	1.40
	.049	.402	3/8	1.50
	.058	.384	3/8	1.65
	.065	.370	3/8	1.78
	.072	.356	3/8	2.01
	.083	.334	3/8	2.18

Tube O.D.	Wall Thick.	Tube I.D.	Use NPT	Mult. by
3/4	.049	.652	1/2	.910
	.058	.634	1/2	.962
	.065	.620	1/2	1.01
	.072	.606	1/2	1.08
	.083	.584	1/2	1.13
	.095	.560	1/2	1.23
	.109	.532	1/2	1.37
1	.049	.902	3/4	.835
	.058	.884	3/4	.869
	.065	.870	3/4	.897
	.072	.856	3/4	.927
	.083	.834	3/4	.976
	.095	.810	3/4	1.03
	.109	.782	3/4	1.11
	.120	.760	3/4	1.18
1¼	.049	1.152	1	.830
	.058	1.134	1	.857
	.065	1.120	1	.878

Tube O.D.	Wall Thick.	Tube I.D.	Use NPT	Mult. by
1¼	.072	1.106	1	.901
	.083	1.084	1	.938
	.095	1.060	1	.981
	.109	1.032	1	1.03
	.120	1.010	1	1.08
1½	.065	1.370	1¼	1.01
	.072	1.356	1¼	1.04
	.083	1.334	1¼	1.07
	.095	1.310	1¼	1.11
	.109	1.282	1¼	1.16
	.120	1.260	1¼	1.20
2	.065	1.870	2	1.22
	.072	1.856	2	1.24
	.083	1.834	2	1.27
	.095	1.810	2	1.30
	.109	1.782	2	1.35
	.120	1.760	2	1.38
	.134	1.732	2	1.42

How to Adjust for Other Fluids and Conditions

First use Table 2, if necessary, to find factor to convert a tube size to equivalent pipe size. Next, use Table 1 to find pressure loss per 100 feet. Next, use Table 3 to adjust for viscosities other than 220 SSU. If using a fluid other than oil, adjust for its gravity as explained elsewhere on this page.

Generally, as shown by the formula at the top of the preceding page, pressure loss increases in direct proportion to a velocity increase. This can also be seen in the velocity columns of Table 1.

Always remember that ***centistoke*** viscosity defines only the flow resistance to shear in the fluid. ***Centipoise*** viscosity defines the combined flow resistance including both shear in the fluid and specific gravity. Centipoises = centistokes x specific gravity.

Adjusting for Other Gravities

Pressure loss through a pipe is directly proportional to specific gravity of the fluid. Other hydraulic fluids have a higher specific gravity than petroleum oil and (at the same viscosity) will have a higher pressure loss. Water/oil emulsion will have a 7% higher, water/glycol a 14%, and phosphate ester will have a 22% higher pressure loss than petroleum oil.

TABLE 4. Pressure Loss Through Fittings

Pressure loss through common fittings is shown in terms of the equivalent length of straight pipe of the same size. *Example:* The flow from the side outlet of a 1½" tee suffers approximately the same pressure loss as if it were flowing through a 9-foot straight length of the same size pipe. For pipe sizes less than 1/2", pressure loss through fittings is little more than for a straight section of the same length (The Crane Company).

	NPT Pipe Size							
	3/4"	1"	1¼"	1½"	2"	2½"	3	3½"
Tee, Side →	4½	5½	7½	9	11½	14	16½	20
45° El	1	1¼	1¾	2	2½	3	3¾	4½
90° El	2	2¾	3¼	4¼	5	6	8	9½

Adjusting for Other Viscosities

Pressure loss through a pipe is directly proportional to viscosity in centistokes (for a given specific gravity). This table may be used with Table 1 to adjust pressure loss per 100 feet to oil with viscosity other than 220 SSU (44.4 centistokes).

Example: A hydraulic oil of 500 SSU will have a higher pressure loss than shown in the table by a factor of 2.48 for the same size pipe and the same flow.

In using Table 3, multiply factor in 3rd column times the pressure loss taken from Table 1.

TABLE 3

SSU Vis.	Centistokes	Factor
80	15.8	.356
100	20.8	.468
150	33	.743
300	65	1.46
400	87	1.96
500	110	2.48
750	163	3.67
1000	220	4.95
2000	420	9.46
3000	630	14.2
4000	850	19.1

Water is a special case. For plain water, pressure loss will be approximately half the values shown in Table 1.

SIZE AND NUMBERING SYSTEM OF TUBE FITTINGS AND HOSE ENDS

Contributed by Mr. Ron Watson, Fluid Power Instructor and Metric Coordinator
Pritchard Engineering Co., Ltd. – Winnipeg, Manitoba

When automobiles were first built in the United States, some of the material needed was various tube assemblies to carry fuel and lubrication. Originally, the easiest material from which to manufacture tubing was copper. It is soft enough to be rolled or extruded. It was usually made into sizes identified by the O.D. measurement, and the most common sizes were from 1/8" to 1" O.D. in 1/16" increments. However, 11/16", 13/16", and 15/16" sizes were seldom used and for all practical purposes have been dropped. The wall thickness was .030, so the inside diameter was approximately 1/16 less than the O.D.

The Society of Automotive Engineers

The SAE (Society of Automotive Engineers), formed in 1910, made itself responsible for the manufacturing standards of all items used in automobile manufacture, and copper tubing was one item to be so controlled. A set of dash numbers was introduced so that the makers part number followed by a dash and a number would identify it as being made to SAE standards. Dash numbers such as -5 for 5/16" O.D., were assigned with similar dash numbers for other sizes.

Any inlet or outlet thread to which copper tubing was to be connected required a brass fitting which would make a leakproof connection. First, a hand flaring tool was developed so mechanics could make an outward flare on the end of a piece of copper tubing, and the inside flare surface always finished off with an exact 45° angle. This became the angle for an SAE flare. A long nut, made with a tapered neck, was machined to fit exactly over the outside of the flare, and it was then screwed on to an adaptor. The adaptor had one end with a male taper pipe thread and the other end with a fine SAE thread, plus a male 45° chamfer face angle which would seat with the inside of the copper tube flare. The nuts and adaptors were also identified with the SAE dash number to identify the size. The thread, identifying dash numbers, and adaptor to be used with each size tube was regulated by SAE. This was the beginning of all the dash sizes we now have in use in fluid power connectors.

Development of Steel Tubing

When hydraulic brakes were developed for trucks and cars, the copper tubing was no longer strong enough to stand the brake line pressure. Steel tubing was made, using the same outside and inside diameters as copper tubing. It carried the same identifying dash numbers used for copper tubing. The same brass nuts and adaptors were used on flared ends formed to the same flare angle, so these still had the SAE numbering system.

An examination of the rather odd I.D. sizes of tubing made with 1/32" wall will show that it does not match with the standard I.D. sizes of iron pipe. Therefore, the flow area is different.

Development of Hose

When the first hydraulic hose was made to be used in fluid power for the higher hydraulic pressures, the I.D. was made to match the I.D. of SAE copper and steel tubing so the flow velocity between hose and tube would match. Thus, the dash system was carried over into this rapidly developing field of medium pressure hoses, and manufacturers could identify a hose product with a dash number for ready reference to SAE sizing. Likewise, when hose end fittings were developed for attachment of hose to hose or hose to pipe thread, the same dash numbering system was used on hose end fitting numbers, and all had the SAE 45° chamfer or seat face.

High Pressure Hose

However, when hoses were improved by adding a layer of steel braid wire, the SAE inside sizes were dropped in favor of even increments of inch fraction sizes including 1/4", 3/8", 1/2", 3/4", 1" and larger. This could not match up with SAE so the industry became confused from one maker to another, each with his own idea for hose ends. The confusion was especially felt in the oil well drilling business as the drilling rigs were moved from one location to another. The American made hose ends and adaptor fittings did not match with those made in other countries, especially in Europe and Asia.

Joint Industry Conference

With American support, arrangements were made with manufacturers in many other countries for an all-world seminar to arrange common standards. This seminar was called the Joint Industry Conference and the standards which came out of it were called JIC standards. The first thing to be done was to make an easily noticed difference between an SAE and a JIC fitting. The face chamfer angle on the female hose ends was changed to 37°, enough difference to be noticed by the eye. Also, many thread sizes were made to different diameters so JIC fittings would not mate with SAE fittings. In some cases the number of threads per inch was made different. Each mechanic could tell that a JIC fitting could not be used with an SAE adaptor. The very fact that JIC was adopted in most countries using hydraulics was enough to encourage the use of hydraulics in all industrial nations. The total result was a rapid growth in the industry. However, a variation of the dash numbering system was agreed upon for the hose manufacturing trade, so that the I.D. sizes were in sixteenths of an inch. Thus, 1/2" I.D. became 8/16" or -8 size. The JIC fittings or hose ends which have female JIC swivel ends are identified with the same dash numbers, but of course the JIC fittings have a 37° chamfer.

Tapered Pipe Threads

The tapered pipe thread, either U.S. or British, has always had a sealing problem. The U.S. thread is NPT (National Pipe Taper) and the British thread is BSP. A modified thread form the NPTF (National Pipe Taper Fuel) has been developed to solve part of the problem. See Page 34. But the wedging action of the pipe taper will always be a problem. Any poorly trained mechanic can take a large wrench to a pipe fitting and by putting too much torque on the wrench he can split the casting of an expensive component because of the wedging action of the taper.

To solve these problems a straight thread has been developed which seals with an O-ring. The thread which is machined into a housing is an SAE fine thread, but at the top, a chamfer is machined which will accommodate an O-ring. When the fitting is screwed into a porthole in a housing it can be tightened until the fitting makes a metal-to-metal seat on the housing and the O-ring provides the fluid seal. Since there is no taper on the threads, there is no wedging action to split the housing. A full range of hose ends and tubing connectors are now made with this straight thread. In the last few years, many component manufacturers are providing straight thread ports as "standard". The tubing size to use for connecting to a port thread size is shown in this chart:

Dash No.	Tubing O.D.	Th'd Size	O-Ring I.D.	O-Ring Thick.
-4	1/4"	7/16-20	.351	.072
-5	5/16"	1/2-20	.414	.072
-6	3/8"	9/16-18	.468	.078
-8	1/2"	3/4-16	.644	.087
-10	5/8"	7/8-14	.755	.097
-12	3/4"	1-1/16-12	.924	.116
-14	7/8"	1-3/16-12	1.048	.116
-16	1"	1-5/16-12	1.171	.116
-20	1¼"	1-5/8-12	1.475	.118
-24	1½"	1-7/8-12	1.720	.118
-32	2"	2-1/2-12	2.337	.118

Split Flange Hose Connectors

The SAE flange hose end fitting has come into wide usage in the last several years. It is primarily intended for high pressure connection of a hose to a component port pad. It is only for hose and is not used on rigid plumbing. The flange end fitting is crimped on to the end of a hose. See illustration below. These fittings come either as a straight-through connection or with various angles from 22½° to 90°. The flange is clamped to a machined port pad on the component, usually a valve, pump or motor, by two half flanges using hex head cap screws. An O-ring fits into a circular groove in the hose flange and makes a seal against the port pad on the component.

These flange fittings are made in several pressure ratings, for pressures from 2500 to 6000 PSI.They are available in sizes as small as 1/2" and up to 3 inches. Although they were originally built for mobile hydraulic systems, they are being used on industrial systems as well. Size identification is with dash numbers similar to those used on other fittings, expressed in sixteenths of an inch.

Tubing to Pipe Connectors

Single dash numbers connect tubing and pipe sizes which have approximately the same flow area and are preferred sizes.

Dash Numbers	Tubing O.D.	Pipe NPTF	Dash Numbers	Tubing O.D.	Pipe NPTF
-2	1/8"	1/8"	-12	3/4"	3/4"
-2-4	1/8"	1/4"	-12-6	3/4"	3/8"
-3	3/16"	1/8"	-12-8	3/4"	1/2"
-3-4	3/16"	1/4"	-12-16	7/8"	1"
-4	1/4"	1/8"	-14	7/8"	3/4"
-4-4	1/4"	1/4"	-14-8	7/8"	1/2"
-5	5/16"	1/8"	-14-16	7/8"	1"
-5-4	5/16"	1/4"	-16	1"	1"
-6	3/8"	1/4"	-16-12	1"	3/4"
-6-2	3/8"	1/8"	-16-20	1"	1¼"
-6-6	3/8"	3/8"	-20	1¼"	1¼"
-6-8	3/8"	1/2"	-20-16	1¼"	1"
-8	1/2"	3/8"	-20-24	1¼"	1½"
-8-4	1/2"	1/4"	-24	1½"	1½"
-8-8	1/2"	1/2"	-24-20	1½"	1¼"
-8-12	1/2"	3/4"	-24-32	1½"	2"
-10	5/8"	1/2"	-32	2"	2"
-10-6	5/8"	3/8"	-32-20	2"	1¼"
-10-12	5/8"	3/4"	-32-24	2"	1½"

Tube to Tube Connectors

Single dash numbers indicate transition into a tube of the same diameter. Double dash numbers are for transition into a tube of a different diameter.

Dash Numbers	Tubing O.D.	Tubing O.D.	Dash Numbers	Tubing O.D.	Tubing O.D.
-2	1/8"	Same	-10-8	5/8"	1/2"
-3	3/16"	Same	-12	3/4"	Same
-4	1/4"	Same	-12-6	3/4"	3/8"
-4-2	1/4"	1/8"	-12-8	3/4"	1/2"
-4-3	1/4"	3/16"	-12-10	3/4"	5/8"
-5-4	5/16"	1/4"	-14	7/8"	Same
-6	3/8"	Same	-16	1"	Same
-6-4	3/8"	1/4"	-16-10	1"	5/8"
-6-5	3/8"	5/16"	-16-12	1"	3/4"
-8	1/2"	Same	-20	1¼"	Same
-8-4	1/2"	1/4"	-20-12	1¼"	3/4"
-8-6	1/2"	3/8"	-20-16	1¼"	1"
-10	5/8"	Same	-24	1½"	Same
-10-6	5/8"	3/8"	-32	2"	Same

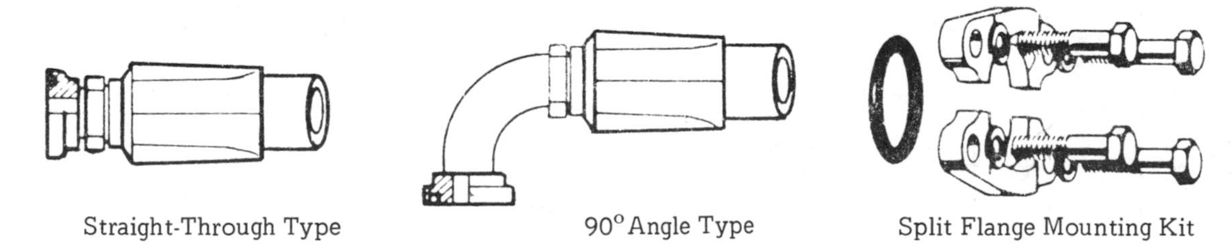

Straight-Through Type — 90° Angle Type — Split Flange Mounting Kit

SAE hose flange ends for bolting to a port pad with split flanges.

NFPA STANDARD DIMENSIONS ON MOUNTING FLANGES AND SHAFTS

For Positive Displacement Hydraulic Pumps and Motors

For many years the only standardization on pump and motor shafts and mounting flanges has been the SAE (Society of Automotive Engineers) Standard J744a. It was widely accepted by other industries as well as the automotive and is still a valid standard. The NFPA (National Fluid Power Association) started work in 1962 on a set of standards which would more directly apply to industrial fluid power pumps and motors, and would be more complete. This set of standards was unanimously approved by the NFPA Board of Directors and issued in 1965 as NFPA Recommended Standard STD T3.9.65.1. In 1966 it was also adopted by the ANSI (American National Standards Institute) and issued as ANSI Standard B93.6-1966. It has since been revised and now carries the numbers ANSI B93.6-1972, and NFPA T3.9.2 R1. Copies may be purchased from the NFPA.

The purpose of the standard is to encourage manufacturers to use interchangeable dimensions on shafts and mounting flanges as far as possible, to simplify dimensional interchangeability for the user. No standards exist at this time on foot mounting dimensions. No performance specifications exist although recommended methods of testing pumps and motors, and the manner of presenting test data are given in NFPA Recommended Standard T3.9.17-1971.

The new NFPA and ANSI standards are similar to the SAE standards but differ in these respects: Additional shaft diameters and an alternate long length shaft have been added to the straight-shaft-without-thread listings. Additional mounting flange sizes have been added to provide a wider selection for the designer. NFPA has not assigned any horsepower ratings, making it the responsibility of the pump designer. However, the SAE horsepower ratings are shown for mounting flanges.

Drawings and charts are only for identification in the field. Complete decimal dimensions and tolerances are contained in the NFPA standards, and copies of the standards mentioned may be purchased from the NFPA.

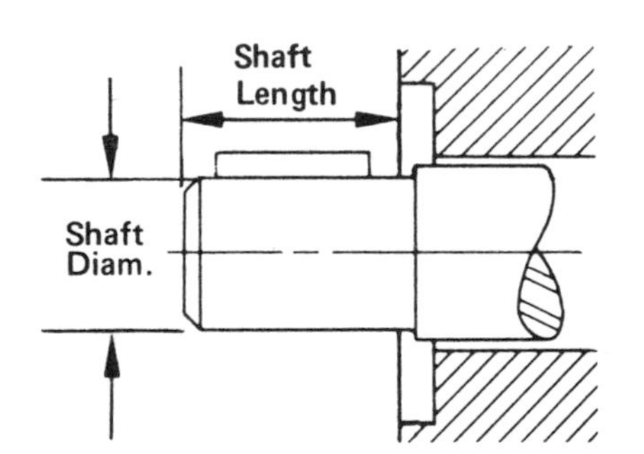

NFPA STANDARD
STRAIGHT SHAFTS WITHOUT THREAD

Shaft Ident. Code	Shaft Diam.	Short Shaft Lgth.	Long Shaft Lgth.	Key Width Inches	SAE Reference*
13-1	1/2"	3/4"	- - - -	1/8"	- - - -
16-1	5/8	15/16	2"	5/32	A
22-1	7/8	1-5/16	2-1/2	1/4	B
25-1	1	1-1/2	2-3/4	1/4	- - - -
32-1	1-1/4	1-7/8	3	5/16	C
38-1	1-1/2	2-1/8	3-1/4	3/8	- - - -
44-1	1-3/4	2-5/8	3-5/8	7/16	D,E

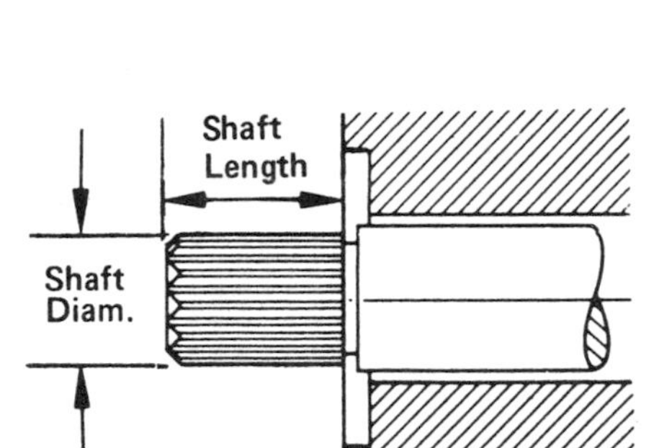

NFPA STANDARD
30-DEGREE INVOLUTE SPLINE SHAFTS

Shaft Code	Shaft Diam.	Shaft Lgth.	Spline Specifications	SAE Ref.*
13-4	1/2"	3/4"	9T, 20/40 DP	- - - -
16-4	5/8	15/16	9T, 16/32 DP	A
22-4	7/8	1-5/16	13T, 16/32 DP	B
25-4	1	1-1/2	15T, 16/32 DP	- - - -
32-4	1-1/4	1-7/8	14T, 12/24 DP	C
38-4	1-1/2	2-1/8	17T, 12/24 DP	- - - -
44-4	1-3/4	2-5/8	13T, 8/16 DP	D,E
50-4	2	3-1/8	15T, 8/16 DP	- - - -

*Indicates matching SAE front flange for each shaft diameter

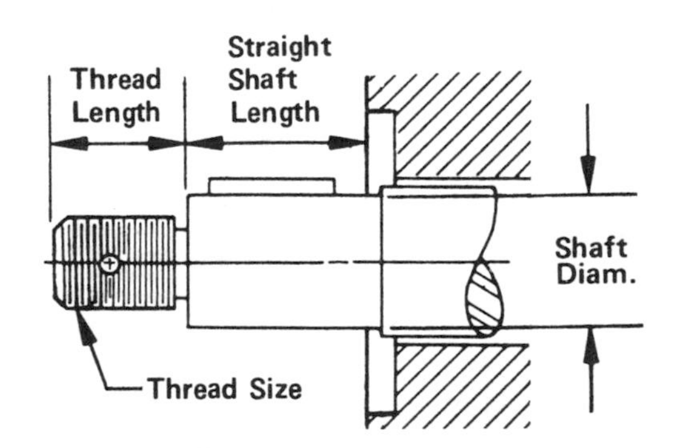

NFPA STANDARD
STRAIGHT SHAFTS WITH THREAD

Shaft Code	Shaft Diam.	Str. Shaft	Th'd Size	Th'd Lgth.	Key Width
13-2	1/2"	3/4"	3/8 - 24	9/16"	1/8"
16-2	5/8	15/16	1/2 - 20	23/32	5/32
22-2	7/8	1-5/16	5/8 - 18	29/32	1/4
25-2	1	1-1/2	3/4 - 16	1-1/16	1/4
32-2	1-1/4	1-7/8	1 - 12	1-7/32	5/16
38-2	1-1/2	2-1/8	1-1/8-12	1-3/8	3/8
44-2	1-3/4	2-5/8	1-1/4-12	1-9/16	7/16

NFPA STANDARD
TAPERED SHAFTS WITH THREAD

Shaft Ident. Code	Shaft Diam.	Tprd. Shaft Lgth.	Thd'd Shaft Lgth.	Th'd Size	Key Width
13-3	1/2"	11/16"	1/2"	5/16-32	1/8"
16-3	5/8	11/16	23/32	1/2-20	5/32
22-3	7/8	1-1/8	29/32	5/8-18	1/4
25-3	1	1-3/8	1-1/16	3/4-16	1/4
32-3	1-1/4	1-3/8	1-7/32	1-12	5/16
38-3	1-1/2	1-7/8	1-3/8	1-1/8-12	3/8
44-3	1-3/4	2-1/8	1-9/16	1-1/4-12	7/16
50-3	2	2-7/8	1-9/16	1-1/4-12	1/2

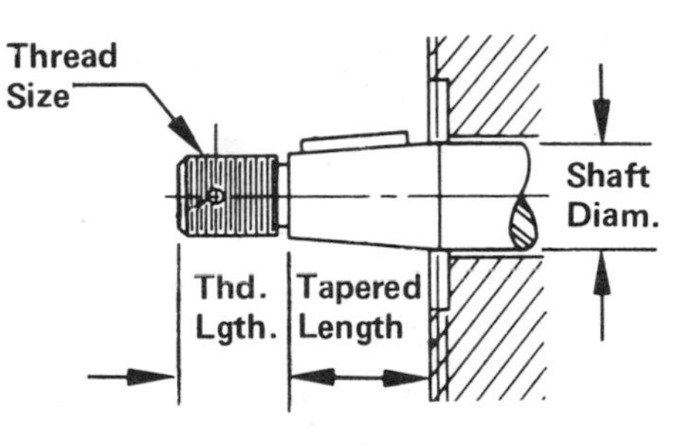

NFPA STANDARD
TWO-BOLT MOUNTING FLANGES

Flange Code	SAE No.	SAE HP Rating	M'tg. Bolt Circle	M'tg. Hole Diam.	Pilot Diam.	Pilot Hgt.
50-2	- - - -	- - - -	3-1/4"	13/32"	2"	1/4"
82-2	A	10	4-3/16	7/16	3-1/4	1/4
101-2	B	25	5-3/4	9/16	4	3/8
127-2	C	50	7-1/8	11/16	5	1/2
152-2	D	100	9	13/16	6	1/2
165-2	E	200	12-1/2	1-1/16	6-1/2	5/8
177-2	F	300	13-25/32	1-1/16	7	5/8

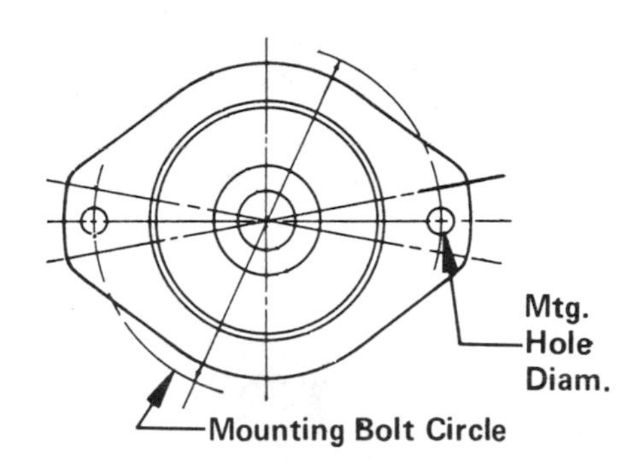

NFPA STANDARD
FOUR-BOLT MOUNTING FLANGES

Flange Code	SAE No.	SAE HP Rating	M'tg. Bolt Circle	M'tg. Hole Diam.	Pilot Diam.	Pilot Hgt.
101-4	B	25	5"	9/16"	4	3/8"
127-4	C	50	6-3/8	9/16	5	1/2
152-4	D	100	9	13/16	6	1/2
165-4	E	200	12-1/2	13/16	6-1/2	5/8
177-4	F	300	13-25/32	1-1/16	7	5/8

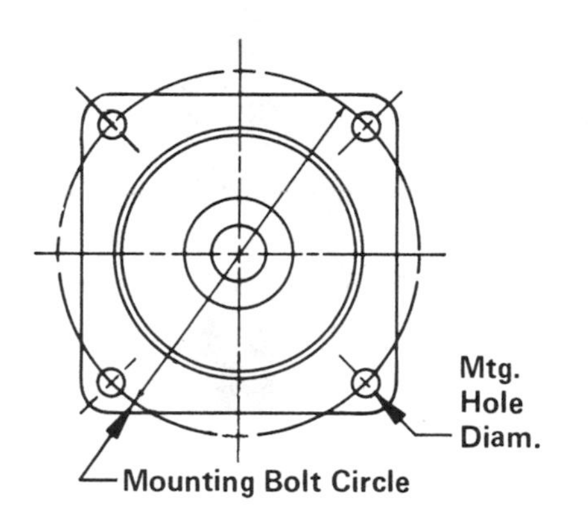

TEMPERATURE CONVERSION – FAHRENHEIT & CELSIUS

Enter the table in the column marked "Temp" with the temperature either Fahrenheit or Celsius that you wish to convert. If converting to Celsius read the equivalent value in the column to the left. If converting to Fahrenheit, read the value in the column to the right. Table calculated from formulae: °F = [°C x 9/5] + 32, or °C = 5/9 x [°F - 32].

°C	Temp	°F	°C	Temp	°F	°C	Temp	°F	°C	Temp	°F
-17.7	0	32.0	26.8	80	176.0	100	212	413	188	370	698
-15.0	5	41.0	29.3	85	185.0	104	220	428	193	380	716
-12.2	10	50.0	32.1	90	194.0	110	230	446	199	390	734
-9.4	15	59.0	34.9	95	203.0	115	240	464	204	400	752
-6.6	20	68.0	38.0	100	212.0	121	250	482	210	410	770
-3.9	25	77.0	43.0	110	230.0	127	260	500	215	420	788
-1.1	30	86.0	49.0	120	248.0	132	270	518	221	430	806
1.6	35	95.0	54.0	130	266.0	138	280	536	226	440	824
4.4	40	104.0	60.0	140	284.0	143	290	554	232	450	842
7.1	45	113.0	65.0	150	302.0	149	300	572	238	460	860
9.9	50	122.0	71.0	160	320.0	154	310	590	243	470	878
12.6	55	131.0	76.0	170	338.0	160	320	608	249	480	896
15.6	60	140.0	83.0	180	356.0	165	330	626	254	490	914
18.2	65	149.0	88.0	190	374.0	171	340	644	260	500	932
21.0	70	158.0	93.0	200	392.0	177	350	662	265	510	950
23.8	75	167.0	99.0	210	410.0	182	360	680	271	520	968

HORSEPOWER REQUIRED FOR COMPRESSING AIR

Values in this 3-part table are for single-stage, 2-stage, and 3-stage piston-type air compressors operating at 85% efficiency, and working on intake air under approximately standard conditions. Compression conditions intermediate between adiabatic and isothermal have been assumed, which we believe to be more nearly representative of conditions existing in the shop air supply of most plants. (See explanation of these terms below). The table was prepared from information in Machinery's Handbook. Please refer to your copy of the Handbook for additional information on air compression and for formulae from which the tables were calculated.

If your compressor operates with an efficiency other than 85%, an allowance can be made in table values.

Explanation of the Table

The table is useful either in determining the HP for a new application or for checking the capacity of an existing system for addition of more air operated equipment, especially large air cylinders reciprocating continuously for long periods. Figures in the tables are HP to compress 1 SCFM from 0 PSIG to the gauge pressures shown. Calculate cylinder air consumption from Page 263 and multiply times the figure in the table.

Adiabatic Compression. This is defined as compression taking place without allowing the escape of heat of compression. This is a theoretical condition because cooling starts immediately after compression.

Isothermal Compression. This is compression which takes place while allowing dissipation of all heat of compression. This is also a theoretical condition.

Single-Stage Compressor, 85% Eff.		*Two-Stage Compressor, 85% Eff.*		*Three-Stage Compressor, 85% Eff.*	
PSIG	HP*	PSIG	HP*	PSIG	HP*
5	.021	50	.116	100	.159
10	.033	60	.128	150	.190
15	.056	70	.138	200	.212
20	.067	80	.148	250	.230
25	.079	90	.155	300	.240
30	.089	100	.164	350	.258
35	.099	110	.171	400	.269
40	.108	120	.178	450	.279
45	.116	130	.185	500	.289
50	.123	140	.190	550	.297
55	.130	150	.196	600	.305
60	.136	160	.201	650	.311
65	.143	170	.206	700	.317
70	.148	180	.211	750	.323
75	.155	190	.216	800	.329
80	.160	200	.220	850	.335
85	.165	210	.224	900	.340
90	.170	220	.228	950	.345
95	.175	230	.232	1000	.290
100	.179	240	.236	1050	.354
110	.191	250	.239	1100	.358
120	.196	260	.243	1150	.362
130	.204	270	.246	1200	.366
140	.211	280	.250	1250	.370
150	.218	290	.253	1300	.374
160	.225	300	.255	1350	.378
170	.232	350	.269	1400	.380
180	.239	400	.282	1450	.383
190	.244	450	.289	1500	.386
200	.250	500	.303	1550	.390

**HP to compress 1 SCFM from 0 PSIG to the values shown.*

PUMP-DOWN TIME FOR EVACUATING TANKS

Figures in the body of this chart show the running time, in minutes, of a 1 SCFM vacuum pump to evacuate a tank with cubic foot or gallon capacity shown at the top of the chart, to the degree of vacuum in the left column. Divide chart values by the free running displacement of your vacuum pump. If your pump has 5 SCFM displacement, running time will be 1/5th the values in the chart.

Formula for Any Vacuum Pump. This formula, published by Gast, can be used for any vacuum pump, any size tank, and to any degree of vacuum:

$$T = \frac{V}{D} \times Log_e \left(\frac{A}{A - B}\right)$$

T is pumping time, in minutes. V is volume of tank, in cubic feet. D is free running displacement of pump. A is deadhead rating of vacuum pump (with inlet blocked). B is desired vacuum level in tank, in "Hg.

	Tank Volume, in Cubic Feet and Gallons											
Cu. Feet	**1½**	**2¼**	**3½**	**5**	**7½**	**10**	**17½**	**25**	**40**	**55**	**85**	**130**
Gallons	*11.2*	*16.8*	*26.2*	*37.4*	*56.1*	*74.8*	*131*	*187*	*300*	*411*	*636*	*972*
Vac. "Hg	Time, in Minutes, to Evacuate Tank											
10	.66	.99	1.5	2.2	3.3	4.4	7.7	11.1	17.7	24.3	37.6	57.5
12	.84	1.3	2.0	2.8	4.2	5.6	9.8	14.0	22.4	30.8	47.6	72.7
14	1.0	1.6	2.4	3.5	5.2	6.9	12.1	17.3	27.7	38.1	58.9	90.1
16	1.3	1.9	3.0	4.2	6.4	8.5	14.8	21.2	33.9	46.6	72.0	110
18	1.5	2.3	3.6	5.1	7.1	10.3	18.0	25.7	41.2	56.6	87.5	134
20	1.9	2.8	4.4	6.3	9.4	12.5	21.9	31.3	50.1	68.9	106	163
22	2.3	3.5	5.4	7.7	11.6	15.4	27.0	38.5	61.6	84.7	131	200
24	2.9	4.4	6.8	9.7	14.6	19.4	34.1	48.6	77.8	107	165	253
26	4.0	5.9	9.2	13.2	19.8	26.4	46.2	66.0	106	145	224	343

DRIVE HORSEPOWER FOR A HYDRAULIC PUMP

Figures in the body of this table show the horsepower needed to drive a hydraulic pump having an efficiency of 85%. Most positive displacement pumps (gear, vane, piston) fall in the range of 80% to 90% efficiency so this chart should be accurate to within 5% for almost any pump. The table was calculated from the formula:

$$HP = PSI \times GPM \div [1714 \times 0.85]$$

For pumps with other than 85% efficiency, the formula can be used, substituting actual efficiency, in decimals, in place of 0.85.

Using the Table. The range of 500 to 5000 PSI covers most hydraulic systems, but power requirements can be determined for conditions outside the range of the table, or for intermediate values. For example, power at 4000 PSI will be exactly 2 times the figure shown for 2000 PSI. At 77 GPM, power will be the sum of the figures shown in the 75 and 2 GPM lines, etc. For systems operating below 500 PSI, horsepower calculations tend to become inaccurate because mechanical friction losses reduce pump efficiency.

Rules-of-Thumb. Approximate HP requirement can be estimated by our "rule of 1500" which states that 1 HP is required for each multiple of 1500 when multiplying PSI x GPM. For example, a 5 GPM pump at 1500 PSI would require 5 HP, or at 3000 PSI would require 10 HP. A 10 GPM pump at 1000 PSI would require 6-2/3 HP or the same pump at 1500 PSI would require 10 HP, etc.

Another rule-of-thumb states that about 5% of the pump maximum rated HP is required to idle the pump when it is "unloaded" and the full flow is circulating at near 0 PSI. This amount of power is consumed in flow losses plus mechanical friction losses in bearings and pumping elements.

Figures in Body of Table are HP's Required to Drive a Hydraulic Pump (A Pump Efficiency of 85% is Assumed)

GPM	500 PSI	750 PSI	1000 PSI	1250 PSI	1500 PSI	1750 PSI	2000 PSI	2500 PSI	3000 PSI	3500 PSI	4000 PSI	5000 PSI
3	1.03	1.54	2.06	2.57	3.09	3.60	4.12	5.15	6.18	7.21	8.24	10.3
5	1.72	2.57	3.43	4.29	5.15	6.00	6.86	8.58	10.3	12.0	13.7	17.2
7½	2.57	3.86	5.15	6.43	7.72	9.01	10.3	12.9	15.4	18.0	20.6	25.7
10	3.43	5.15	6.86	8.58	10.3	12.0	13.7	17.2	20.6	24.0	27.5	34.3
12½	4.29	6.43	8.58	10.7	12.9	15.0	17.2	21.4	25.7	30.0	34.3	42.9
15	5.15	7.72	10.3	12.9	15.4	18.0	20.6	25.7	30.9	36.0	41.2	51.5
17½	6.01	9.01	12.0	15.0	18.0	21.0	24.0	30.0	36.0	42.0	48.0	60.1
20	6.86	10.3	13.7	17.2	20.6	24.0	27.5	34.3	41.2	48.0	54.9	68.6
22½	7.72	11.6	15.4	19.3	23.2	27.0	30.9	38.6	46.3	54.1	61.8	77.2
25	8.58	12.9	17.2	21.4	25.7	30.0	34.3	42.9	51.5	60.1	68.6	85.8
30	10.3	15.4	20.6	25.7	30.9	36.0	41.2	51.5	61.8	72.1	82.4	103
35	12.0	18.0	24.0	30.0	36.0	42.0	48.0	60.1	72.1	84.1	96.1	120
40	13.7	20.6	27.5	34.3	41.2	48.0	54.9	68.6	82.4	96.1	110	137
45	15.4	23.2	30.9	38.6	46.3	54.1	61.8	77.2	92.7	108	124	154
50	17.2	25.7	34.3	42.9	51.5	60.1	68.6	85.8	103	120	137	172
55	18.9	28.3	37.8	47.2	56.6	66.1	75.5	94.4	113	132	151	189
60	20.6	30.9	41.2	51.5	61.8	72.1	82.4	103	124	144	165	206
65	22.3	33.5	44.6	55.8	66.9	78.1	89.2	112	134	156	178	223
70	24.0	36.0	48.0	60.1	72.1	84.1	96.1	120	144	168	192	240
75	25.7	38.6	51.5	64.3	77.2	90.1	103	129	154	180	206	257
80	27.5	41.2	54.9	68.7	82.4	96.1	110	137	165	192	220	275
85	29.2	43.8	58.3	72.9	87.5	102	117	146	175	204	233	292
90	30.9	46.3	61.8	77.2	92.7	108	124	154	185	216	247	309
100	34.3	51.5	68.6	85.8	103	120	137	172	206	240	275	343

Oversizing or Undersizing of an Induction-Type Electric Motor

Optimum results are obtained if HP rating of an electric motor is neither too far oversize or undersize for the job. Some effects of motor power mismatch are:

Oversize Motor. Using a 20 HP motor to do a 10 HP job, for example, will give good results as far as running the fluid power system is concerned, but it will consume a little more current for the same 10 HP output. It will also cause the power factor of the plant electric system to be poorer, with higher power cost. Idling current will be higher so more power will be wasted during periods in the cycle when the motor is running at idle condition.

Undersize Motor. A 3-phase induction motor can usually be safely overloaded during peak parts of the cycle as explained on Page 195, but during these peak periods, motor current will be all out of proportion to the excess power being produced with a considerable amount of overheating. If the motor is too far undersize it will, of course, burn out in a short time.

APPENDIX D
Metric Design Data

INTRODUCTION TO METRIC SYSTEM INTERNATIONAL

Contributed by Mr. Ron Watson, Fluid Power Instructor and Metric Coordinator
Pritchard Engineering Co., Ltd. – Winnipeg, Manitoba

In North America, the governments of both U.S.A. and Canada have passed acts which authorize the change of measuring units in each country to metric S.I. This is part of the international attempt to bring the world into one common system of measuring. At present, almost 75% of other countries already use the metric system, either the older system before 1960, or the newer S.I. system. This will mean that exports from North America can fit and match applications to both industrial and mobile machinery.

Representative committees from all countries have prepared the greater part of standard tables that will be known as System International, which we call S.I. for short. The ANSI (American National Standards Institute) has published a Metric Practice Guide known as ANSI Z210-1-1973, and this will be followed by later revisions as further S.I. agreements occur within working committees. In Canada, the CSA (Canadian Standards Association) has published in 1976 the latest printing of CSA No. CAN-3 Z234.1-76 and will also reprint as required. Countries in Central and South America have their own metric standards.

In the following paragraphs we attempt to give a brief illustration of the measurements that will affect us in the use of Fluid Power. First, we look at the method of using a handy prefix onto a measurement unit, that characterizes all metric units, and how the prefixes help everyone.

Essentially, metric measurement is built around the use of the "ten" unit and its multiples; such as 100, 1000, etc., when working with whole numbers. When dealing with fractions, primarily it is the decimal fraction that is used. So each decimal place will represent a "ten" unit, for 1/10, 1/100, 1/1000, etc. This does away with the former awkwardness of changing yards to feet, feet to inches, or inches to sixteenths. It also equips all computer and calculator programs with the ability to use their fastest speed for the figuring of large sums in multiplication and division.

Notice in the following table how whole numbers that have become very large can be reduced to a size easy to write and remember, and this reduction is done by removing one or more zeros. At the same time this is done, we change the unit name by adding a prefix as shown:

Reduce By:	Add Prefix	As in These Examples	Prefix Abbrev
1 zero	deka	Dekagrams, dekametres	da
2 zeros	hecto	hectograms. hectometres	h

The above two examples may not be commonly used, as most abbreviation is done when removing 3 or more zeros, as follows:

3 zeros	kilo	kilograms, kilometres	k
6 zeros	mega	megagrams, megametres	M
9 zeros	giga	gigagrams, gigametres	G
12 zeros	tera	teragrams, terametres	T

Similarly, to change a measurement to a smaller unit (in the same way we change a yard to feet or feet to inches) a decimal fraction of the unit is made by adding a prefix to the unit name:

Fraction	Decimal Fraction	Prefix	Example	Prefix Abbrev.
1/10	.1	deci	decigram	d
1/100	.01	centi	centigram	c
1/1000	.001	milli	milligram	m
1/1000000	.000001	micro	microgram	μ
1/1000000000	.000000001	nano	nanogram	n

Any mechanic or technician will be able to tell within minutes of his first study that this system when used for multiplying or dividing, means he only has to add or delete zeros, or add or delete decimal places.

Next, for an examination of the standard figure units, we study the definitions from which we obtain all other units. Note that certain of these units are named after the scientist who did the earliest research in each field.

Temperature will be changed to Celsius readings. There is also a Kelvin scale, but we will seldom see Kelvin readings in fluid power work. The Fahrenheit scale, as we use it, will be converted to Celsius. On the Celsius scale there is exactly 100° between the new freezing point of 0°C and the new boiling point of water 100°C., so once more we are dealing in groups of tens.

Length will be taken as the basic unit of a metre (note that this is the correct S.I. spelling). A world standard of a metre length is kept in Paris and also in Washington and other world capitols. It consists of a flat bar made of platinum-iridium on which one metre is marked out between two fine hairlines etched into the bar. The bar is kept in a climate controlled cabinet at a temperature of 20°C. (68°F). Scientists have also specified the metre as equal to a stated number of light wavelengths emanating from a mercury or krypton arc lamp. Conversions from foot measurements are given in the conversion tables, Page 280.

Mass will be taken as the basic unit of a gram, or multiples. The standard of a kilogram is also kept as a platinum-iridium cylinder, one kept in Paris and another in Washington, under bar-bells in a climate controlled cabinet kept at 4°C (39.2°F). The gram is specified as the 1/1000th part of this cylinder. Scientists also have a reference under the older C.G.S. equivalents, that the gram is equal to the gravity weight of 1.001 cubic centimetres of water, at sea level. However, where possible we should attempt to stop any conversions of ounces or pounds into kilograms because these are basically not mass but force units. Instead, we should convert U. S. force and weight units into Newtons.

Time will be retained as our present system where 60 seconds equals one minute and 60 minutes equals one hour, and 24 hours equals one solar day (mean average).

Force as created by any means, fluid power, mechanical, or gravity weight will be measured in units of Newtons. A Newton is defined as the force which, when applied to any body with a mass of one kilogram, is able to give the mass body an acceleration equal to one metre per second per second. Formula: $1\ N = 1\ kg{\cdot}m{\cdot}s^2$.

Pressure in liquids or gases will be measured in units of a Pascal (abbreviated Pa) which will be equal to the pressure created by the force on one Newton over the area of one square metre. The Pa will be one of the units used on metric pressure gauges. The bar is another pressure unit not recommended by S.I., but being used by many engineers. Note that 1 bar = 100 kPa (kiloPascals) This same pressure measurement will also apply to mechanical stress such as tensile strength, shear strength, compression strength, and adhesion strength. Formula: $1\ Pa = 1\ N{\cdot}m^2$. (1 Newton per square metre).

Work will be measured by a unit called a Joule. This is equal to the work done at the actual point of applying a force of 1 Newton, to cause that point to move a distance of 1 metre in the same direction as the force. The scientist Joule was one who originally did research into a means of measuring how much mechanical energy was required to raise the temperature of water, by means of stirring paddles inside an insulated calorimeter pot. Therefore, he made the original specification of work done in mechanical machines. Formula: 1 Joule = 1 N·m.

Power will be measured as an electrical power equivalent which will be the Watt (from here on spelled with capital W). A Watt is defined as the power producing energy at the rate of one Joule per second. Formula: 1 W = 1 J·s. (1 Joule per second).

Note: All horsepower measurements, whether for fluid, mechanical, or electrical power, steam or internal combustion engines, rotating shafts or flywheels, or rubber tired wheels will be converted and rated in Watts.

Heat. Any measured amount of heat will be calculated as the equivalent of Joules, for every degree of temperature rise. The base temperature for this measurement was originally set at 15.5°C for a rise of one degree. Heat energy was originally stated in metric calories, but there is also a conversion from calories to Joules. For practical purposes both horsepower and heat will be expressed in Watts. 1 BTU per hour = 0.293 071 Watts.

METRIC S. I. CONVERSIONS

It is accepted practice in metric S.I. to print either whole figures or decimal place figures in groups of three numerals separated by spaces, as in this example:

12 345 678.234 765

It is also accepted that abbreviations for unit names such as: kilograms per metre per second, be shown as: kg·m·s with the period (or decimal point) halfway above the line. However, in typewritten material the decimal point or period may be printed on the line because the keys cannot be changed.

1. Temperature. Fahrenheit readings may be converted to Celsius as shown on Page 275.

2. Length.

1 inch = 25.4 mm (millimetres) = 0.0254 m (metres) = 0.254 dm (decimetres). To convert inches to mm, multiply inches times 25.4.

1 foot = 304.8 mm = 0.304 8 m. To convert feet to mm, multiply feet times 304.8; To convert feet to m, multiply feet times 0.3048.

1 mile = 1.609 344 km (kilometres). To convert miles to km, multiply miles times 1.609 344.

To convert fluid velocity of a liquid or gas: 1 foot per second = 0.3048 m·s (metres per second).

3. Area.

1 square inch = 0.000 645 16 square metres (m^2). To convert square inches to square metres, multiply square inches times 0.000 645 16.

1 square foot = 0.092 030 4 square metres. To convert square feet to square metres, multiply square feet times 0.092 030 4.

4. Force. (Including Gravity Weight)

1 pound force or weight = 4.448 222 Newtons (N). To convert pounds to Newtons, multiply pounds times 4.448 222.

5. Liquid Volume.

In handwritten documents, litre should be abbreviated with a script "ℓ" to avoid confusion with the Figure 1. This is difficult, sometimes impossible, in typing or type setting. In food and medical measurements the ml (millilitre) is used. It is one-thousandth of a litre.

1 litre (l) has been accepted by the S.I. as a volume of 1 cubic decimetre (dm^3).

Therefore, a tank or reservoir with a volume measured in dm^3 holds the same number of litres.

1 U.S. gallon = 3.785 412 dm^3 = 3.785 412 litres. To convert U.S. gallons to dm^3, multiply gallons times 3.785 412.

You may notice certain car makers are now stating engine displacement in cubic metres instead of litres.

6. Pressure.

1 PSIG = 6.894 757 kPaG (kilopascals gauge)
14.5 PSIG = 1 bar gauge
100 kPa gauge = 1 bar gauge

1 atmosphere (14.7 PSIG) = 101.325 kPa, which is just slightly larger than 1 bar.

7. Heat.

1 BTU = 1 059.67 Joule.

1 BTU per hour = 0.293 071 W (Watts). To convert BTU/hr to W, multiply times 0.293 071.

All power of fluid systems, mechanical systems, electrical systems, engines, machinery, shafts, flywheels, etc. will be rated in Watts. 1 HP = 0.746 kiloWatts.

8. Date.

Parts of our country will gradually convert to the metric dating on all letters, documents, drawings, etc. The Manitoba (Canada) government is now using it on letters and checks as follows:

Example: 79.04.25.13.45 (first the year, then month, then day, then time, if needed, on a 24-hr clock.

INTERCHANGE BETWEEN SI, METRIC, AND U. S. CUSTOMARY UNITS

International Standard (SI) units are shown in the first column of each chart. Values with exponents can be handled directly on a pocket calculator which has an exponent key.

For manual calculations, remember that the + or - sign in front of an exponent tells whether to move the decimal point to the right (for a + sign), or to the left (for a - sign), and the exponent tells how far to move it. Examples: $2.540 \times 10^{-5} = .0000254$, and $3.048 \times 10^{2} = 304.8$, etc.

Equivalent values of all units are shown on the same horizontal line. Perhaps the easiest way to use the charts is to look down the column of the unit to be converted and find the line on which the "1" appears. Then move to the left or right to the column of the desired new unit. That figure is a multiplier.

Example: Look down the "Inch" column to the "1" line. The chart shows 1 inch = 1.578×10^{-5} mile. Therefore, 627 inches would be: $627 \times [1.578 \times 10^{-5}]$ miles, etc.

TORQUE

Newton-Metres	*Kilopond-Mtrs.*	*Foot-Pounds*	*Inch-Lbs.*
1	1.020×10^{-1}	7.376×10^{-1}	8.851
9.807	1	7.233	86.80
1.356	1.382×10^{-1}	1	12
1.130×10^{-1}	1.152×10^{-2}	8.333×10^{-2}	1

LENGTH (Linear Measurement)

Metre	*Centimetre*	*Millimetre*	*Kilometre*	*Mile*	*Inch*	*Foot*
1	100	1000	1×10^{-3}	6.214×10^{-4}	39.370	3.281
0.01	1	10	1×10^{-5}	6.214×10^{-6}	3.937×10^{-1}	3.281×10^{-2}
1×10^{-3}	0.10	1	1×10^{-6}	6.214×10^{-7}	3.937×10^{-2}	3.281×10^{-3}
1×10^{3}	1×10^{5}	1×10^{6}	1	6.214×10^{-1}	3.937×10^{4}	3.281×10^{3}
1.609×10^{3}	1.609×10^{5}	1.609×10^{6}	1.609	1	6.336×10^{4}	5280
2.540×10^{-2}	2.540	25.40	2.540×10^{-5}	1.578×10^{-5}	1	8.333×10^{-2}
3.048×10^{-1}	30.479	3.048×10^{2}	3.048×10^{-4}	1.894×10^{-4}	12	1

VOLUME – (Cubic)

Cubic Metre	*Cu. Decimetre (Litre)*	*Cu. Centimetre*	*Imperial Gallon*	*U.S. Gallon*	*Cubic Inch*	*Cubic Foot*
1	1×10^{3}	1×10^{6}	2.20×10^{2}	2.642×10^{2}	6.102×10^{4}	35.314
1×10^{-3}	1	1×10^{3}	2.20×10^{-1}	2.642×10^{-1}	61.024	3.531×10^{-2}
1×10^{-6}	1×10^{-3}	1	2.20×10^{-4}	2.642×10^{-4}	6.102×10^{-2}	3.531×10^{-5}
4.546×10^{-3}	4.546	4.546×10^{3}	1	1.200	2.774×10^{2}	1.605×10^{-1}
3.785×10^{-3}	3.785	3.785×10^{3}	8.327×10^{-1}	1	2.310×10^{2}	1.337×10^{-1}
1.639×10^{-5}	1.639×10^{-2}	16.387	3.605×10^{-3}	4.329×10^{-3}	1	5.787×10^{-4}
2.832×10^{-2}	28.317	2.832×10^{4}	6.229	7.481	1.728×10^{3}	1

AREA – (Square measurement)

Square Metre	*Sq. Centimetre*	*Sq. Millimetre*	*Sq. Kilometre*	*Square Inch*	*Square Foot*	*Square Mile*
1	1×10^{4}	1×10^{6}	1×10^{-6}	1.550×10^{3}	10.764	3.861×10^{-7}
1×10^{-4}	1	100	1×10^{-10}	1.550×10^{-1}	1.076×10^{-3}	3.861×10^{-11}
1×10^{-6}	1×10^{-2}	1	1×10^{-12}	1.550×10^{-3}	1.076×10^{-5}	3.861×10^{-13}
1×10^{6}	1×10^{-10}	1×10^{12}	1	1.550×10^{9}	1.076×10^{7}	3.861×10^{-1}
6.452×10^{-4}	6.452	6.452×10^{2}	6.452×10^{-10}	1	6.944×10^{-3}	2.491×10^{-10}
9.290×10^{-2}	9.290×10^{2}	9.290×10^{4}	9.290×10^{-8}	144	1	3.587×10^{-8}
2.590×10^{6}	2.590×10^{10}	2.590×10^{12}	2.590	4.014×10^{9}	2.788×10^{7}	1

FORCE – (Including force due to weight)

Newton	*Dyne*	*Kilopond*	*Metric Ton (Tonne)*	*Long Ton*	*U.S. Ton*	*Pound*
1	1×10^{5}	1.020×10^{-1}	1.020×10^{-4}	1.004×10^{-4}	1.124×10^{-4}	2.248×10^{-1}
1×10^{-5}	1	1.020×10^{-6}	1.020×10^{-9}	1.004×10^{-9}	1.124×10^{-9}	2.248×10^{-6}
9.807	9.807×10^{5}	1	1×10^{-3}	9.842×10^{-4}	1.102×10^{-3}	2.205
9.807×10^{3}	9.807×10^{8}	1000	1	9.842×10^{-1}	1.102	2.205×10^{3}
9.964×10^{3}	9.964×10^{8}	1.016×10^{3}	1.016	1	1.120	2.240×10^{3}
8.896×10^{3}	8.896×10^{8}	9.072×10^{2}	9.072×10^{-1}	8.929×10^{-1}	1	2000
4.448	4.448×10^{5}	4.536×10^{-1}	4.536×10^{-4}	4.464×10^{-4}	5×10^{-4}	1

MASS –(Not Weight)

Kilogram	*Gram*	*Metric Ton (Tonne)*	*Newton*	*Pound*	*Slug*	*U.S. Ton*
1	1000	1×10^{-3}	9.807	2.205	6.853×10^{-2}	1.102×10^{-3}
1×10^{-3}	1	1×10^{-6}	9.807×10^{-3}	2.205×10^{-3}	6.853×10^{-5}	1.102×10^{-6}
1×10^{3}	1×10^{6}	1	9.807×10^{3}	2.205×10^{3}	68.530	1.102
1.020×10^{-1}	1.020×10^{2}	1.020×10^{-4}	1	2.248×10^{-1}	6.988×10^{-3}	1.124×10^{-4}
4.536×10^{-1}	4.536×10^{2}	4.536×10^{-4}	4.448	1	3.108×10^{-2}	5×10^{-4}
14.594	1.459×10^{4}	1.459×10^{-2}	1.431×10^{2}	32.170	1	1.609×10^{-2}
9.072×10^{2}	9.072×10^{5}	9.072×10^{-1}	8.896×10^{3}	2000	62.170	1

VELOCITY

Metres/Sec.	*Decimetres/Sec.*	*Kilometres/Hr.*	*Miles/Hr.*	*Feet/Min.*	*Feet/Sec*	*Inches/Min.*
1	10	3.6	2.237	1.968×10^{2}	3.281	2.362×10^{3}
1×10^{-1}	1	1×10^{-4}	6.214×10^{-5}	5.468×10^{-3}	9.113×10^{-5}	6.562×10^{-2}
2.778×10^{-1}	2.278	1	6.214×10^{-1}	5.468×10^{1}	9.113×10^{-1}	6.562×10^{2}
4.470×10^{-1}	4.470	1.609	1	88	1.467	1.056×10^{3}
5.080×10^{-3}	5.080×10^{-2}	1.829×10^{-2}	1.136×10^{-2}	1	1.667×10^{-2}	12
3.048×10^{-1}	3.048	1.097	6.818×10^{-1}	60	1	7.2×10^{2}
4.233×10^{-4}	4.233×10^{-3}	1.524×10^{-3}	9.470×10^{-4}	8.333×10^{-2}	1.389×10^{-3}	1

UNIT PRESSURE (Either fluid or mechanical)

Bar	*Newtons/m²(Pascal)*	*Kilopond/m²*	*Kilopond/cm²*	*Atmosphere*	*Pounds/Ft²*	*Pounds/Inch²*
1×10^{-5}	1	1.020×10^{-1}	1.020×10^{-5}	9.869×10^{-6}	2.088×10^{-2}	1.45×10^{-4}
1	1×10^{5}	1.020×10^{4}	1.020	9.869×10^{-1}	2.088×10^{3}	14.5
9.807×10^{-5}	9.807	1	1×10^{-4}	9.678×10^{-5}	2.048×10^{-1}	1.422×10^{-3}
9.807×10^{-1}	9.807×10^{4}	1×10^{4}	1	9.678×10^{-1}	2.048×10^{3}	14.220
1.013	1.013×10^{5}	1.033×10^{4}	1.033	1	2.116×10^{3}	14.693
4.789×10^{-4}	47.893	4.884	4.884×10^{-4}	4.726×10^{-4}	1	6.944×10^{-3}
6.897×10^{-2}	6.897×10^{3}	7.033×10^{2}	7.033×10^{-2}	6.806×10^{-2}	1.440×10^{2}	1

POWER – (Fluid, Electrical, or Mechanical)

Kilowatt	*Watt, Joules/s and N-m/s*	*U.S. & U.K Horsepower*	*Foot-Pounds per Minute*	*Foot-Pounds per Second*	*BTU per Hour*	*BTU per Min.*
1	1000	1.340	4.425×10^{4}	7.376×10^{2}	3.412×10^{3}	56.862
1×10^{-3}	1	1.340×10^{-3}	44.254	7.376×10^{-1}	3.412	5.686×10^{-2}
7.461×10^{-1}	746	1	3.300×10^{4}	5.500×10^{2}	2.545×10^{3}	42.44
2.260×10^{-5}	2.260×10^{-2}	3.029×10^{-5}	1	1.667×10^{-2}	7.710×10^{-2}	1.285×10^{-3}
1.356×10^{-3}	1.356	1.817×10^{-3}	60	1	4.626	7.710×10^{-2}
2.931×10^{-4}	2.931×10^{-1}	3.928×10^{-4}	12.971	2.162×10^{-1}	1	1.667×10^{-2}
1.759×10^{-2}	17.586	2.357×10^{-2}	7.783×10^{2}	12.971	60	1

ENERGY OR WORK

Kilowatt-Hour	*Watt-second Joule, or N-m*	*Dyne-Cm. or Erg*	*Horsepower-Hr.*	*Foot-Pound*	*Inch-Pound*	*BTU*
1	3.6×10^{6}	3.6×10^{13}	1.341	2.655×10^{6}	3.187×10^{7}	3.412×10^{3}
2.778×10^{-7}	1	1×10^{7}	3.725×10^{-7}	7.376×10^{-1}	8.851	9.477×10^{-4}
2.778×10^{-14}	1×10^{-7}	1	3.725×10^{-14}	7.376×10^{-8}	8.851×10^{-7}	9.477×10^{-11}
7.457×10^{-1}	2.685×10^{6}	2.685×10^{13}	1	1.980×10^{6}	2.376×10^{7}	2.544×10^{3}
3.766×10^{-7}	1.356	1.356×10^{7}	5.051×10^{-7}	1	12	1.285×10^{-3}
3.138×10^{-8}	1.130×10^{-1}	1.130×10^{6}	4.209×10^{-8}	8.333×10^{-2}	1	1.071×10^{-4}
2.931×10^{-4}	1.055×10^{3}	1.055×10^{10}	3.931×10^{-4}	7.783×10^{2}	9.339×10^{3}	1

WORKING WITH METRIC CYLINDERS

Sometime in the future, cylinders as well as other fluid power components will be built to ISO (International Standards Organization) dimensions in which metric measurements are used. Conversion to international standards has been slow in the United States, and at this writing the availability of metric dimension cylinders is quite limited, but complete conversion will come in due time.

Cylinder charts in this section cover standardized bore and rod combinations from 25mm through 200 mm bore and with minimum and maximum size piston rods. Intermediate size piston rods will no doubt be offered by most manufacturers. ISO standard sizes also include bore sizes of 8, 10, 12, 16, 20, 250, 320, and 400 mm.

Calculations of cylinder force and velocity are not quite as straightforward as in the U. S. system because of extra conversions between units which become necessary. The ISO units which will be used in cylinder calculations are these:

FORCE. Force values are in Newtons (N). One Newton is equal to about ¼ pound (.2248 lb. to be exact), or 1 pound is equal to about 4½ N (4.448 N to be exact). This unit should serve for most cylinder calculations except where very large forces are involved in which the kilo Newton (kN) equal to 1000 N may be used.

PISTON AREA. Piston bore is cataloged in units of millimetres (mm) as shown in the charts. For area, the mm^2 is too small for convenient calculations, so the unit for piston surface area will be the square centimetre (cm^2). To calculate piston area change bore diameter to cm. by dividing by 10. Then find cm^2 piston area with the formula: $A = \pi r^2$ in the usual manner.

PRESSURE. Fluid pressure will usually be expressed in kilo Pascals (kPa) because the Pascal which is defined as one Newton of force per square metre, is such a small unit that it is hard to work with in making calculations.

METRIC AIR CYLINDERS – FORCE CHART – 3 TO 6 BARS PRESSURE

		Bars →	3	3½	4	4½	5	5½	6
		Kilo Pascals →	300	350	400	450	500	550	600
		PSI →	43.5	50.8	58.0	65.3	72.5	79.8	87.0
Bore mm.	*Bore cm.*	*Area sq. cm.*	Theoretical Cylinder Force in Newtons						
25	2.5	4.91	147	172	196	221	245	270	294
32	3.2	8.04	241	281	322	362	402	442	483
40	4.0	12.57	377	440	503	565	628	691	754
50	5.0	19.63	589	687	785	884	982	1 080	1 178
63	6.3	31.17	935	1 091	1 247	1 403	1 559	1 714	1 870
80	8.0	50.27	1 508	1 759	2 011	2 262	2 513	2 765	3 016
100	10.0	78.54	2 356	2 749	3 142	3 534	3 927	4 320	4 712
125	12.5	122.72	3 681	4 295	4 909	5 522	6 136	6 749	7 363
160	16.0	201.06	6 032	7 037	8 042	9 048	10 053	11 058	12 064
200	20.0	314.16	9 225	10 996	12 566	14 137	15 708	17 279	18 850

METRIC AIR CYLINDERS – FORCE CHART – 6½ TO 11 BARS PRESSURE

		Bars	6½	7	7½	8	9	10	11
		Kilo Pascals	650	700	750	800	900	1000	1100
		PSI	94.3	102	109	116	131	145	160
Bore mm.	*Bore cm.*	*Area sq. cm.*	Theoretical Cylinder Force in Newtons						
25	2.5	4.91	319	343	368	393	442	491	540
32	3.2	8.04	523	563	603	643	724	804	885
40	4.0	12.57	817	880	942	1 005	1 131	1 257	1 382
50	5.0	19.63	1 276	1 374	1 473	1 571	1 767	1 963	2 160
63	6.3	31.17	2 026	2 182	2 338	2 494	2 805	3 117	3 429
80	8.0	50.27	3 267	3 519	3 770	4 021	4 524	5 027	5 529
100	10.0	78.54	5 105	5 498	5 890	6 283	7 069	7 854	8 639
125	12.5	122.72	7 977	8 590	9 204	9 817	11 044	12 272	13 499
160	16.0	201.06	13 069	14 072	15 080	16 085	18 095	20 106	22 117
200	20.0	314.16	20 420	21 991	23 562	25 133	28 274	31 416	34 557

One kPa is equal to 1000 Pa.

The bar is a more convenient unit for fluid pressure and will be allowed, at least for a limited time. The bar is related to the Pascal. One bar = 100,000 Pascals or 100 kPa. It is also equal to 14.5 PSI which is very close to one atmosphere. Pressure values in the following charts are given in three pressure units, bars, kPa, and PSI, to help the student get a "feel" for the way metric pressure units compare with PSI units he has been using.

PUMP FLOW. Oil flow from a hydraulic pump is expressed in litres per minute (l/min). A litre is defined as one cubic decimetre (dm^3), and is roughly ¼ gallon (.2642 gal. to be exact). Or, 1 gal. = 3.785 litres. On very large flows units of litres per second (l/s) can be used.

Cylinder Calculations

FORCE CALCULATION. Cylinder force is calculated by multiplying piston surface area times fluid pressure.

$$F = A \times P \div 10, \text{ in which:}$$

F = force, in Newtons (N).
A = piston area in square centimetres (cm^2).
P = differential pressure across ports in kPa.
10 is a necessary metric conversion constant.

When working with pressure in bars, the formula becomes:

$$F = A \times P \times 10, \text{ in which,}$$

P is pressure differential in bars.

VELOCITY CALCULATIONS. The travel speed of a cylinder piston is calculated with this formula:

$$S = V \times 10 \div A, \text{ in which:}$$

S = travel speed expressed in metres per minute.
V = Pump oil flow in l/min. (dm^3/min).
A = Piston or net area, in cm^2.
10 is a necessary metric conversion between cm and dm.

See next page for cylinder piston velocity charts

METRIC HYDRAULIC CYLINDERS – FORCE CHART – 25 TO 175 BARS PRESSURE

		Bars →	25	50	75	100	125	150	175
		Kilo Pascals →	2500	5000	7500	10 000	12 500	15 000	17 500
		PSI →	363	725	1 088	1 450	1 813	2 175	2 538
Bore mm.	*Bore cm.*	*Area sq. cm.*	Theoretical Cylinder Force in Newtons						
25	2.5	4.91	1 227	2 454	3 680	4 907	6 138	7 361	8 587
32	3.2	8.04	2 011	4 021	6 032	8 042	10 053	10 053	12 063
40	4.0	12.57	3 142	6 283	9 425	12 566	15 708	18 849	21 991
50	5.0	19.63	4 909	9 817	14 726	19 634	24 543	29 451	34 360
63	6.3	31.17	7 793	15 585	23 378	31 170	38 963	46 755	54 548
80	8.0	50.27	12 566	25 133	37 699	50 265	62 831	75 398	87 964
100	10.0	78.54	19 635	39 270	58 904	78 539	98 174	117 809	137 443
125	12.5	122.72	30 679	61 358	92 037	122 716	153 395	184 074	214 753
160	16.0	201.06	50 265	100 531	150 796	201 061	251 326	301 592	351 857
200	20.0	314.16	78 540	157 080	235 619	314 159	392 699	471 239	549 778

METRIC HYDRAULIC CYLINDERS – FORCE CHART – 200 TO 350 BARS PRESSURE

		Bars →	200	225	250	275	300	325	350
		Kilo Pascals	20 000	22 500	25 000	27 500	30 000	32 500	35 000
		PSI →	2 900	3 263	3 625	3 988	4 350	4 713	5 075
Bore mm.	*Bore cm.*	*Area sq. cm.*	Theoretical Cylinder Force in Newtons						
25	2.5	4.91	9 814	11 041	12 268	13 494	14 721	15 948	17 175
32	3.2	8.04	16 084	18 095	20 105	22 116	24 126	26 137	28 147
40	4.0	12.57	25 132	28 274	31 415	34 557	37 698	40 840	43 981
50	5.0	19.63	39 268	44 177	49 085	53 994	58 902	63 811	68 719
63	6.3	31.17	62 340	70 133	77 925	85 718	93 510	101 303	109 095
80	8.0	50.27	100 530	113 096	125 663	138 229	150 795	163 361	175 928
100	10.0	78.54	157 078	176 713	196 348	215 982	235 617	255 252	274 887
125	12.5	122.72	245 432	276 111	306 790	337 469	368 148	398 827	429 506
160	16.0	201.06	402 122	452 387	502 653	552 918	603 183	653 448	703 714
200	20.0	314.16	628 318	706 858	785 398	863 937	942 477	1 021 017	1 099 557

METRIC HYDRAULIC CYLINDERS – PISTON VELOCITY – 3 TO 75 LITRES/MINUTE FLOW

Piston Diam., mm.	Rod Diam., mm.	3 l/min.	5 l/min.	10 l/min.	15 l/min.	20 l/min.	25 l/min.	30 l/min.	40 l/min.	50 l/min.	75 l/min.
25	None*	6.111	10.19	20.37	----	----	----	----	----	----	----
	12	7.938	13.23	26.46	----	----	----	----	----	----	----
	18	12.68	21.14	42.27	----						----
32	None*	3.732	6.220	12.44	18.66						----
	14	4.614	7.690	15.38	23.07	----	----	----	----	----	----
	22	7.077	11.80	23.59	35.39	----	----	----	----	----	----
40	None*	2.387	3.978	7.955	11.93	15.91	19.89	----	----	----	----
	18	2.992	4.987	9.974	14.96	19.95	24.94	----	----	----	----
	28	4.677	7.795	15.59	23.39	31.18	38.98	----	----	----	----
50	None*	1.528	2.547	5.094	7.641	10.19	12.74	15.28	20.39	----	----
	22	1.895	3.159	6.316	9.476	12.63	15.79	18.95	25.28	----	----
	36	3.174	5.290	10.58	15.87	21.16	26.45	31.74	42.34	----	----
63	None*	.9624	1.604	3.208	4.812	6.415	8.020	9.624	12.84	16.03	----
	28	1.199	1.999	3.998	5.997	7.995	9.995	11.99	15.00	19.82	----
	45	1.965	3.275	6.550	9.824	13.10	16.37	19.65	26.21	32.74	----
80	None*	.5968	.9947	1.989	2.984	3.978	4.973	5.968	7.961	9.943	14.92
	36	.7483	1.247	2.494	3.742	4.988	6.236	7.483	9.982	12.47	18.71
	56	1.170	1.950	3.900	5.850	7.800	9.751	11.70	15.61	19.49	29.25
100	None*	.3820	.6366	1.273	1.910	2.546	3.183	3.820	5.095	6.364	9.549
	45	.4790	.7983	1.597	2.395	3.193	3.991	4.790	6.389	7.979	11.97
	70	.7490	1.248	2.497	3.745	4.993	6.241	7.490	9.991	12.49	18.72
125	None*	.2445	.4074	.8149	1.222	1.630	2.037	2.445	3.261	4.073	6.112
	56	.3060	.5100	1.020	1.529	2.039	2.549	3.059	4.080	5.096	7.646
	90	.5076	.8460	1.692	2.538	3.384	4.230	5.076	6.771	8.457	12.69
160	None*	.1492	.2487	.4974	.7460	.9946	1.243	1.492	1.990	2.486	3.730
	70	.1845	.3080	.6151	.9227	1.230	1.538	1.845	2.462	3.074	4.613
	110	.2830	.4716	.9432	1.415	1.887	1.358	2.830	3.776	4.714	7.074
200	None*	.0955	.1592	.3183	.4775	.6366	.7958	.9549	1.284	1.591	2.387
	90	.1197	.1996	.3991	.5987	.7982	.9978	1.197	1.597	1.995	2.994
	140	.1872	.3121	.6241	.9362	1.248	1.560	1.872	2.498	3.119	4.681

Figures in the body of this chart are piston speeds in metres per minute (m/min).

METRIC HYDRAULIC CYLINDERS – PISTON VELOCITY – 100 TO 450 LITRES PER MINUTE

Piston Diam., mm.	Rod Diam., mm.	100 l/min.	125 l/min.	150 l/min.	175 l/min.	200 l/min.	250 l/min.	300 l/min.	350 l/min.	400 l/min.	450 l/min.
80	None*	19.89	24.87	----	----	----	----	----	----	----	----
	36	24.94	31.18	----	----	----					----
	56	39.00	48.75	----	----	----					----
100	None*	12.73	15.92	19.10	22.28	25.46	----	----	----	----	----
	45	15.97	19.96	23.95	27.94	31.93	----	----	----	----	----
	70	24.97	31.21	37.45	43.69	49.93	----	----	----	----	----
125	None*	8.149	10.19	12.22	14.26	16.30	20.37	24.44	28.52	----	----
	56	10.20	12.74	15.29	17.84	20.39	25.49	30.59	35.68	----	----
	90	16.92	21.15	25.38	29.61	33.84	42.30	50.76	59.22	----	----
160	None*	4.974	6.217	7.460	8.704	9.947	12.43	14.92	17.41	19.89	22.38
	70	6.151	7.689	9.227	10.76	12.30	15.38	18.45	21.53	24.60	27.68
	110	9.432	11.79	14.15	16.51	18.86	23.58	28.29	33.01	37.73	42.44
200	None*	3.183	3.979	4.775	5.570	6.366	7.958	9.549	11.14	12.73	14.32
	90	3.991	4.989	5.987	6.985	7.983	9.978	11.97	13.97	15.97	17.96
	140	6.241	7.802	9.362	10.92	12.48	15.60	18.72	21.84	24.97	28.09

Figures in the body of this chart are piston speeds in metres per minute (m/min).

*Lines with bold typs show extension speeds, metres per minute – Lines with italic type show retraction speeds.

ISO METRIC PIPE THREAD AND PORT THREAD SIZES

Metric Pipe Threads. These pipe thread sizes have been adopted by the ISO (International Standards Organization), and are primarily intended for pipes or tubes which screw together or into ISO pipe ports on valves, cylinders, pumps, and other components. An appropriate thread sealant can be used, if necessary, to ensure pressure-tight joints.

However, except on very small ports, straight threads with O-ring seal may be preferred to pipe threads as component ports. ISO metric machine threads should be used instead of pipe threads for mechanical fastenings such as nuts and bolts.

BSP British Standard Pipe threads, using the Whitworth thread form, have been adopted as the ISO standard for pipe threads, and will hereafter be called out as ISO and not as BSP. ISO taper pipe threads are similar to the American NPT thread, having the same nominal sizes based on the pipe O.D. in inches, but with a slightly different thread pitch so the two systems are not interchangeable. Pitch (the distance between threads) is expressed as the number of threads per 25.4 mm (1 inch). As with NPT threads, the nominal size does not represent the true outside pipe diameter but in the larger sizes it comes very close. Sizes start at 1/16" and go up through 6 inches. Note: there is no ISO thread modification such as the NPTF (National Pipe Thread Fuel) for improved sealing between threads. Usually a thread sealant is required.

There are two variations of ISO pipe threads. Parallel threads (also called straight threads) and taper threads. Taper threads have a 1:16 taper and the same reamers can be used as for NPT threads. Reamers are not strictly necessary but in the larger sizes they enable a better thread to be produced. ISO pipe threads, when specified, are assumed to be parallel unless a taper thread is called out. Taper threads seal on the threads themselves. Parallel threads, if a leaktight fluid seal is required, seal with a gasket, washer, or O-ring against the female seat. Since there is only one thread pitch, the pitch is not usually specified and threads are called out only by nominal size as noted below.

ISO Specification 7/1, 1982 issue, contains information on threads in which the fluid seal is made on the threads and covers both taper and parallel threads. ISO Specification 228/1, 1982 issue, contains information on threads which do NOT seal on the threads. The specifications give information on thread form, machining dimensions, and tolerances. Information on inspection and gauging is in Part 2 of the same specifications. Copies of ISO specifications can be purchased from the ANSI (American National Standards Institute).

Taper external threads (Specification 7/1 can be mated to either taper or parallel internal threads. However, parallel external threads do not make a tight fluid seal on the threads and should not be used for jointing. They should not be mated with internal pipe threads except where a leaktight joint is not required.

Internal parallel threads (Specification 228/1) can be mated to external parallel threads. If a fluid tight seal is required, a sealing element (washer or gasket) must be used.

ISO Specification 7/1 applies to pipe threads which make a fluid tight seal on the threads. The correct way of calling them out is:

Internal parallel threads: "Pipe thread ISO 7/1-Rp 1½" (for example).

Internal taper threads: "Pipe thread ISO 7/1-Rc 1½" (for example).

External threads (always tapered for ISO 7/1): "Pipe thread ISO 7/1-R 1½" (for example). Under Spec. 7/1 there is no external parallel thread. See Specification 228/1).

ISO Specification 228/1 applies to pipe threads which do NOT make a fluid tight seal on the threads. The correct way of calling them out is:

Internal thread (one tolerance only): "Pipe thread ISO 228/1-G1½" (for example).

External thread (closer tolerance): "Pipe thread ISO 228/1-G1½A" (for example).

External thread (with wider tolerance): "Pipe thread ISO 228/1-G1½B" (for example).

ISO Standard Pipe Sizes (for both Parallel and Taper Threads)

ISO Specification 228/1		ISO Specification 7/1		Pipe Diameters	
Parallel Thread Only	Suggested Tap Drill	Tapered or Parallel	Suggested Tap Drill	Pipe O.D. Inches	Pipe O.D. mm*
1/8-28	Q	1/8-28	21/64	.3830	9,73
1/4-19	29/64	1/4-19	7/16	.5180	13,16
3/8-19	37/64	3/8-19	9/16	.6560	16,66
1/2-14	47/64	1/2-14	23/32	.8250	20,96
5/8-14	54/64	----	----	.9020	22,91
3/4-14	61/64	3/4-14	59/64	1.041	26,44
7/8-14	1-1/16	----	----	1.189	30,20
1-11	1-3/16	1-11	1-11/64	1.309	33,25
1-1/8-11	1-13/64	----	----	1.492	37,90
1-1/4-11	1-17/32	1-1/4-11	1-1/2	1.650	41,91
1-3/8-11	1-41/64	----	----	1.745	44,32
1-1/2-11	1-49/64	1-1/2-11	1-47/64	1.882	47,80
1-3/4-11	2	----	----	2.116	53,75
2-11	2-15/64	2-11	2-3/16	2.347	59,61

*In the ISO metric system a decimal point is written with a comma instead of a period.

Index to Volume 1

Other Womack Books

on Industrial Fluid Power

If you now have any of these books, please check to be sure you have the latest edition shown in listings on the opposite page. These books are frequently revised. Each edition is a complete re-write, updated, and enlarged with a substantial amount of new material. Prices shown are current but are subject to increase when a new and enlarged edition is published. Please contact the publisher for the latest price list on books and other educational material.

VOLUME 2 – INDUSTRIAL FLUID POWER. This is one of our advanced textbooks. From the foundation laid in Volume 1, this book goes deeper into greater depth on certain valves and circuits.

Some of the topics covered include: how to calculate many kinds of cylinder loads, vertical, horizontal, and at an angle. Direction, force, and speed control of cylinders using some of the more sophisticated 4-way directional, pressure compensated flow control, and pressure control valves such as sequence, by-pass, unloading, reducing, and counterbalance valves. Pressure intensification with piston-type and rotary type intensifiers. Sample circuit diagrams, design charts, and formulae are shown for each application. This book is limited to applications using air or hydraulic cylinders to deliver power output. Rotary mechanical output from motors and rotary actuators is covered in Volume 3.

Other material includes a chapter on air-over-oil circuits in which compressed shop air is used for power to produce a hydraulic oil flow. The last chapter illustrates many ideas, some of them novel, for the use of cylinders.

VOLUME 3 – INDUSTRIAL FLUID POWER. Although covering many advanced applications which require rotary mechanical output, the presentation is simple as it is in all the Womack books, and can easily be understood by anyone with mechanical or electrical aptitude. Mathematics are limited to very simple formulae, and in most cases the charts and tables can be used instead of the formulae.

The format includes first an explanation of each component and how it works, using photos and sectional views. Then circuit diagrams plus design charts show how to design the component into circuits.

Among the components covered in this particular book are common types of hydraulic motors – gear, vane, and piston types. Shows how to match a hydraulic motor to its load; circuits for direction and speed control; closed loop hydrostatic transmissions; motor starting and running torque, efficiency, HP, life expectancy, and installation; air motor operation and circuit design; power steering design; rotary and spool-type flow dividers and how to use them; bootstrapping principles and circuits for saving power; and many other topics.

FLUID POWER IN PLANT AND FIELD. The new Second Edition is a revision with 56 new pages added to include more information on installation and start-up of hydrostatic and open loop systems; more information on troubleshooting of both hydrostatic and open loop hydraulic systems; and more reference data in the Appendix on pipe sizes, pump and motor shafts and flanges, electric motor frame sizes, and much more of interest to installers.

Although this is not one of the regular textbook series, it has much practical information that could not be included in a textbook. Some companies use it as a training manual for mechanics, field service personnel, and installers. Contains many practical ideas on the best way to install cylinders, hydraulic pumps and motors, air and hydraulic valves, hydraulic reservoirs, heat exchangers, accumulators, air trio units, air dryers, hydrostatic systems, and vacuum systems. Unique ideas on increasing speed and efficiency of air and hydraulic cylinders, causes of recurring problems, etc. Material in this book is of a practical nature which is usually not included in an ordinary fluid power design handbook. We recommend this book as a companion reference to students of the 3-volume industrial fluid power textbooks.

ELECTRICAL CONTROL OF FLUID POWER. We are offering the new Second Edition in which the previous section on fluidics has been replaced with a section on electronic control with programmable controllers.

Some of the topics include . . . A description of popular fluid and electrical components . . . How to draw and read JIC ladder diagrams . . . Directional control circuits to cause a cylinder to reciprocate . . . Directional control circuits for programming two or more cylinders to operate in sequence . . . Pressure control in air and hydraulic circuits with solenoid valves . . . Solving design problems in electrical circuitry, for special functions . . . Miscellaneous applications for the use of timers, counters, peck drilling circuits, index table operation, etc. . . . Safety circuits to protect the operator and the machine . . . Electric motors and motor starters . . . NEMA motor designs, motor duty cycle, etc. Electronic control with programmable controllers . . . Electrical ladder diagrams for controllers . . . Switching action in a programmable controller . . . Conversion of JIC diagrams to controller ladder diagrams, etc.

HOW TO ORDER . . . Send personal or company check to publisher. Prices include postage. Texas residents add 6% sales tax. Foreign shipments $1.00 per book extra. Prices are in U.S.A. funds. Company purchase orders for more than $40.00 accepted for 30-day invoicing, and may be placed by phone. Call (214) 357-3871 for information or to place book orders. Prices may change without notice. Request latest price list.